About Island Press

Island Press is the only nonprofit organization in the United States whose principal purpose is the publication of books on environmental issues and natural resource management. We provide solutions-oriented information to professionals, public officials, business and community leaders, and concerned citizens who are shaping responses to environmental problems.

In 2005, Island Press celebrates its twenty-first anniversary as the leading provider of timely and practical books that take a multidisciplinary approach to critical environmental concerns. Our growing list of titles reflects our commitment to bringing the best of an expanding body of literature to the environmental community throughout North America and the world.

Support for Island Press is provided by the Agua Fund, Brainerd Foundation, Geraldine R. Dodge Foundation, Doris Duke Charitable Foundation, Educational Foundation of America, The Ford Foundation, The George Gund Foundation, The William and Flora Hewlett Foundation, Henry Luce Foundation, The John D. and Catherine T. MacArthur Foundation, The Andrew W. Mellon Foundation, The Curtis and Edith Munson Foundation, National Environmental Trust, The New-Land Foundation, Oak Foundation, The Overbrook Foundation, The David and Lucile Packard Foundation, The Pew Charitable Trusts, The Rockefeller Foundation, The Winslow Foundation, and other generous donors.

The opinions expressed in this book are those of the author(s) and do not necessarily reflect the views of these foundations.

D1212064

Large Carnivores
and the Conservation
of Biodiversity

Large Carnivores and the Conservation of Biodiversity

EDITED BY

Justina C. Ray, Kent H. Redford,
Robert S. Steneck, and Joel Berger

ISLANDPRESS

Washington • Covelo • London

Copyright © 2005 Island Press

All rights reserved under International and Pan-American
Copyright Conventions. No part of this book may be
reproduced in any form or by any means without permission
in writing from the publisher: Island Press, 1718 Connecticut
Ave., NW, Suite 300, Washington, DC 20009.

ISLAND PRESS is a trademark of The Center for Resource
Economics.

No copyright claim is made in the work of James A. Estes,
William J. McShea, Douglas W. Smith, and S. Joseph Wright,
employees of the federal government.

Library of Congress Cataloging-in-Publication data.

Large carnivores and the conservation of biodiversity /
edited by Justina C. Ray . . . [et al.].
    p.   cm.
Includes bibliographical references and index.
ISBN 1-55963-079-5 (cloth : alk. paper) — ISBN 1-55963-080-9
(pbk. : alk. paper)
   1. Carnivora—Conservation.   2. Carnivora—Ecology.
3. Predation (Biology)   4. Biological diversity conservation.
I. Ray, Justina C.
  QL737.C2C794 2005
  591.716—dc22   2004021349

British Cataloguing-in-Publication data available.

Printed on recycled, acid-free paper ♼

Design by Paul Hotvedt

Manufactured in the United States of America
10 9 8 7 6 5 4 3 2 1

Contents

Acknowledgments   *xi*

CHAPTER 1

Introduction: How to Value Large Carnivorous Animals   1

Kent H. Redford

PART I

Setting the Stage   7

CHAPTER 2

An Ecological Context for the Role of Large Carnivores in
     Conserving Biodiversity   9

Robert S. Steneck

"The World Is Green" Revolution   10

Paradigm Gained—Predation as a Structuring Process   11

Top-Down Forces in Food Webs: Keystones to Trophic Cascades   15

Variability of Trophic Cascades   17

Do Marine Systems Have the Strongest Trophic Cascades?   23

Why Big Fierce Animals Are Rare: Top-Down Vulnerability and Chronically
     Sliding Baselines   26

Conserving Biodiversity   30

Summary   32

CHAPTER 3

Large Carnivorous Animals as Tools for Conserving Biodiversity: Assumptions
     and Uncertainties   34

Justina C. Ray

Large Carnivorous Animals as Conservation Tools   37

Assumptions Underlying the Use of Large Carnivores as Conservation Tools   45

Testing the Assumptions behind Using Large Carnivores as Conservation Tools   48

Conservation Implications: Large Carnivorous Animals in Conservation Practice   53

Summary   55

PART II

The Scientific Context for Understanding the Role of Predation   57

CHAPTER 4

Carnivory and Trophic Connectivity in Kelp Forests   61

James A. Estes

The Sea Otter–Kelp Forest Ecosystem   63

Implications for Other Species and Ecosystems   74

Implications for Conservation and Management   78

Summary   80

CHAPTER 5

The Green World Hypothesis Revisited   82

John Terborgh

Tests of the Plant Self-Defense Hypothesis   84

Predator Exclusion Experiments   88

Nature Reduced to an Artifact   94

Conservation Recommendations   97

Summary   97

CHAPTER 6

Restoring Functionality in Yellowstone with Recovering Carnivores: Gains and Uncertainties   100

Joel Berger and Douglas W. Smith

Ecology in Yellowstone National Park with and without Wolves   101

Ecology beyond Yellowstone National Park with and without Wolves   107

Summary   108

CHAPTER 7

Large Marine Carnivores: Trophic Cascades and Top-Down Controls in Coastal
    Ecosystems Past and Present   110

Robert S. Steneck and Enric Sala

Predation Theory and Evidence of Effects   111

We Eat Large Marine Carnivores: Fisheries-Induced Declines in
    Predator Abundance   112

Evidence for Past and Present Top-Down Predator Effects Altering Trophic
    Cascades in Major Benthic Marine Ecosystems   116

Evidence from Other Marine Ecosystems   132

General Consequences of the Loss of Large Carnivores, and Implications for
    Conservation   134

Summary   136

CHAPTER 8

Forest Ecosystems without Carnivores: When Ungulates Rule
    the World   138

William J. McShea

Extent of the Problem of High-Density Ungulate Populations   138

Ungulate Effects on Biodiversity   142

Is There Evidence That Reducing Ungulate Density Restores Biodiversity?   147

Hunters versus Large Carnivores   148

Conservation Recommendations   150

Summary   152

CHAPTER 9

King of the Beasts? Evidence for Guild Redundancy among Large
    Mammalian Carnivores   154

Rosie Woodroffe and Joshua R. Ginsberg

Large Carnivore Assemblages and the Forces That Structure Them   159

When Guilds Collapse: Rules for the Disassembly of Large Carnivore
    Assemblages   160

Guild Redundancy and Compensation   162

Conserving Complete Large Carnivore Guilds in Fragmented Landscapes: Some
  Complex and Surprising Predictions   169
Conclusions and Conservation Recommendations   172
Summary   174

PART III
From Largely Intact to Human-Dominated Systems: Insight on the Role of
  Predation Derived from Long-Term Studies   177

CHAPTER 10
Tigers and Wolves in the Russian Far East: Competitive Exclusion, Functional
  Redundancy, and Conservation Implications   179
*Dale G. Miquelle, Philip A. Stephens, Evgeny N. Smirnov, John M. Goodrich, Olga J.
  Zaumyslova, and Alexander E. Myslenkov*

Study Areas   180
Data Analysis and Modeling Methods   182
Research Findings   186
Competitive Exclusion and Functional Redundancy in Tigers and Wolves   200
Conservation Implications   203
Summary   205

CHAPTER 11
Large Carnivores and Biodiversity in African Savanna Ecosystems   208
M. G. L. Mills

Study Areas   209
Predator–Prey Relationships in Various Systems   212
Intraguild Relationships   223
Biodiversity Implications of Large Carnivore Ecology   223
Conserving Carnivores and Carnivory in Ecosystems   225
Ecosystems without Carnivores and Other Conservation Implications   227
Summary   228

CHAPTER 12

Large Carnivores and Ungulates in European Temperate Forest Ecosystems:
    Bottom-Up and Top-Down Control   230
Bogumiła Jędrzejewska and Włodzimierz Jędrzejewski

Białowieża Primeval Forest: Study Area and Methods   231
Wolf and Lynx Predation on Ungulates   233
Long-Term Data on Large Carnivores and Ungulates   236
Top-Down and Bottom-Up Forces in Diverse Guilds of Predators and Prey   240
Trophic Cascades: Indirect Effects of Large Carnivores on Forest Ecosystems   242
Conservation Implications   244
Summary   245

CHAPTER 13

Recovery of Carnivores, Trophic Cascades, and Diversity in Coral Reef
    Marine Parks   247
Tim R. McClanahan

Statement of the Problem   249
Study Sites and History of the Coral Reef Parks   249
Field Sampling and Data Analysis   251
Research Findings   254
Discussion   261
Conservation Recommendations and Concluding Thoughts   264
Summary   266

CHAPTER 14

Human-Induced Changes in the Effect of Top Carnivores on Biodiversity in
    the Patagonian Steppe   268
Andrés J. Novaro and R. Susan Walker

Native Carnivore and Prey Communities of the Patagonian Steppe   269
Human Impact on Patagonian Wildlife   271
Effects on Wildlife of Reduction of Sheep and Hunting   273
Possible Top-Down Control of Native Herbivores by Pumas and Culpeos   276
Predation as a Potential Threat to Wildlife Conservation in Patagonia   282

Conservation Recommendations   284

Concluding Thoughts: How Unique Is the Patagonian Example?   287

Summary   287

PART IV

Achieving Conservation and Management Goals through Focus on Large
Carnivorous Animals   289

CHAPTER 15

Large Carnivores, Herbivores, and Omnivores in South Florida:
An Evolutionary Approach to Conserving Landscapes and
Biodiversity   293

David S. Maehr, Michael A. Orlando, and John J. Cox

Study Area   295

The Bear and the Weevil   297

Felid Predators and Deer   301

Wolves and Big Cypress Deer   306

The Challenge to Managers   309

Summary   313

CHAPTER 16

Hunting by Carnivores and Humans: Does Functional Redundancy Occur and
Does It Matter?   315

Joel Berger

Limitations of Approach   316

Current Overlap between Hunting Humans and Carnivores   317

Predictions: Concordance in Effects of Human and Carnivore Hunting   319

Conservation Recommendations: Functionality in Systems with Carnivores
and Humans   337

Summary   340

CHAPTER 17

Detecting Top-Down versus Bottom-Up Regulation of Ungulates by Large
    Carnivores: Implications for Conservation of Biodiversity   342
R. Terry Bowyer, David K. Person, and Becky M. Pierce

Conceptual Models of Predator–Prey Dynamics   344
Failure to Consider Effects of *K*   346
Prey to Predator Ratios   349
Kill Rates   351
A Prey-Based Approach for Understanding Top-Down and Bottom-Up Processes   354
Future Directions for Predator–Prey Modeling   358
Linking Predator–Prey Dynamics to Ecosystem Processes and Biodiversity   359
Summary   360

CHAPTER 18

Top Carnivores and Biodiversity Conservation in the Boreal Forest   362
Stan Boutin

The Boreal Forest Context   364
How Do Carnivores Affect the Boreal Forest?   366
Top Carnivores as Umbrellas for Biodiversity   377
A Biodiversity Conservation Approach Focused on Maintaining the Range of
    Natural Variability   378
Summary   379

CHAPTER 19

The Linkage between Conservation Strategies for Large Carnivores and
    Biodiversity: The View from the "Half-Full" Forests of Europe   381
*John D. C. Linnell, Christoph Promberger, Luigi Boitani, Jon E. Swenson, Urs
    Breitenmoser, and Reidar Andersen*

The Nature of Biodiversity   382
Europe: A Continent Shaped by Humans   382
European Large Carnivore Populations   386
Goals for Large Carnivore Conservation in Europe   389
How Does Conserving Carnivores Conserve Biodiversity in Europe?   394

Is Europe Unique?   396

Conservation Recommendations   397

Summary   398

CHAPTER 20

Conclusion: Is Large Carnivore Conservation Equivalent to Biodiversity
    Conservation and How Can We Achieve Both?   400
Justina C. Ray, Kent H. Redford, Joel Berger, and Robert Steneck

Framing the Question   401

The Link between Large Carnivorous Animals and Biodiversity   401

Where There Is Strong Evidence for Biodiversity Impacts through Predation by Large
    Carnivorous Animals   403

Where the Evidence Is Less Compelling or Absent   405

Unknowns and Unknowable   409

Is Carnivore Conservation Compatible with Biodiversity Conservation?   411

Large Carnivores as Conservation Tools   415

Conservation Recommendations: Achieving Carnivore and
    Biodiversity Conservation   419

Concluding Thoughts   425

Summary   426

References   *429*

List of Contributors   *509*

Index   *512*

Acknowledgments

We are grateful to the attendants of the workshop held at the White Oak Plantation in May 2003, of which this volume is a direct result. They generously contributed their time and thoughts to addressing the complicated set of questions we have posed here, and then spent considerable energy formulating these beautifully written essays in a relatively rapid timeframe. Thanks also to nonattending coauthors who enthusiastically shared their expertise, often with very short notice, to fill critical gaps in our coverage. It was our sincere pleasure and honor to work with such high-caliber scientists.

All chapters of this volume benefited enormously from rigorous reviews provided by the following individuals, who generously donated their time and creative energy to improve clarity and content of each contribution: Liz Bennett, Richard Bodmer, Luigi Boitani, Stan Boutin, Terry Bowyer, Rodrigo Bustamante, Carlos Carroll, Emmett Duffy, Sarah Durant, Jim Estes, Graham Forbes, Todd Fuller, Josh Ginsberg, Jodi Hilty, Luke Hunter, John Linnell, David Maehr, Tim McClanahan, Bill McShea, Brian Miller, Reed Noss, François Messier, Andrés Novaro, Tim O'Brien, John Robinson, Mel Sunquist, Rick Sweitzer, and Adrian Treves. To all we are grateful.

Barbara Dean at Island Press was a bottomless source of enthusiasm from the moment she heard the first seeds of our proposal. She and Barbara Youngblood responded to each and every query, guiding us with steady hands through the small details and the complexities alike. The preparation of this volume in the relatively short timeframe from conception to printing would simply not have been possible without the cheerful and able assistance of Joanna Zigouris, whose careful attention to detail and understanding of the process helped at every turn of the way.

The workshop was made possible through the generous support of the White Oak Conservation Center of Gilman International Conservation. In particular we thank John Lukas. We gratefully acknowledge the Wildlife Conservation Society for its support through the development of the volume. Finally, we extend our deep appreciation for the patience and encouragement of our families.

CHAPTER 1

Introduction: How to Value Large Carnivorous Animals

Kent H. Redford

According to a quote attributed to Marjory Stoneham Douglas, "The Everglades is a test, if we pass, we get to keep the planet." This evocative challenge can be applied equally to the conservation of large carnivorous animals. Over the entire surface of the globe, these animals, wolves (*Canis* spp.), bears (*Ursus* spp.), large cats, sharks, and orcas (*Orcinus orca*) are fighting a rearguard action for survival. As the world increasingly becomes a handmaiden to the human race, saving these species has become one of the most difficult tests we face in biological conservation. The urgent need to develop and implement strategies to conserve such creatures has led to two approaches, one based on the ecological roles or services played by these species in maintaining biodiversity, and the other on their intrinsic value as a component of biodiversity.

In this volume we probe the relationship between these two approaches and the science underlying them, seeking to understand the relationship between the presence of large carnivorous animals and the conservation of all attributes and components of biodiversity. We specifically address the conservation challenges of these species, whose diets consist mainly or exclusively of large animal prey. As such, they are often in direct and indirect conflict with humans (c.f. Treves and Karanth 2003). We describe the species of interest as large and carnivorous because we are interested in the conservation problems and opportunities posed by species with these characteristics, and not in their taxonomic position per se. We have worked to include studies from the marine as well as the terrestrial realm. We have been only partially successful in this effort; although the loss of large carnivores in the marine realm has been well documented (e.g., Myers and Worm 2003), the effects of such loss have only recently begun to be examined. We were unable to find cases that include reptiles and avian species and welcome further analysis that extends the scope of this volume and its conclusions. The book is intended as a representative review of what is currently known about the relationship between carnivores and biodiversity and how it relates to the conservation of both. It is aimed at practitioners and academics, with a hope that the work of the former can more effectively inform the work of the latter.

This volume is based on a conference convened by the Wildlife Conservation Society (WCS) at the White Oak Conservation Center of the Gilman Foundation in 2003. We invited professionals who had worked on the relationship between large carnivorous animals and biodiversity or had long-term data sets that might be used to examine this issue. WCS is a conservation organization with a mission of conserving wildlife and wildlands using science-based, field-grounded work. As well as operating the world's largest set of urban zoos in New York City, WCS is engaged in field conservation at over 300 sites in over 50 countries. In many places WCS works with large carnivorous animals and everywhere finds this work complicated.

Conservation organizations and individual conservation biologists have been very effective in drawing attention to the plight of these animals by arguing that they play critical roles in the conservation of their ecosystems. As such, conservation of these species is said to be a prerequisite for achieving larger-scale conservation.

This is a very important claim. As practitioners of science-based conservation, and strong supporters of the value of large carnivorous animals, we thought it a critical time to bring together experts from the scientific community to assess this link. Therefore, we convened the White Oak meeting to address the question: Is the conservation of large carnivorous animals equivalent to the conservation of biological diversity? Aware of the complicated history of ecological thought that addresses versions of this question, we have worked to place our volume in the perspective of larger ecological theory. Also aware of the efforts of others addressing similar questions, we have placed our work in the context of the work of others. We have also found that as we worked on this book the question we had posed to the workshop has proven more complicated than expected to answer, and our efforts to answer it have led us into unexpected quarters (see the concluding chapter). We have found that our search for a simple answer has been frustrated by, but also informed by, the paucity of scientific investigation addressing the role of large carnivores; the conclusion that even when there is sufficient science, the answer will depend on context; and the rich, complicated mix of ethics, values, and science that envelops and obscures virtually everything having to do with the interactions between humans and large carnivores. But despite this ambiguity, we have worked throughout this book to bring to the surface the management implications of and actions connected with conserving both large carnivorous animals and the biodiversity that enrobes them. This constant eye on conservation action makes this book different from many others. We hope this book will be of use to those charged with the conservation and management of both wildlife and wildlands.

The book is organized into four parts. The first part, Setting the Stage, lays out the theoretical and practical issues underlying the question of whether conservation of carnivores is equivalent to the conservation of biodiversity. The two chapters review both the ecological foundations that are at the core of this question and the assumptions and uncertainties underlying the ways in which large carnivorous animals have been used as tools for conserving biodiversity. Part Two, The Scientific Context for Understanding the Role of Predation, consists of six chapters. The first set presents several of the best-known research projects that have examined the ecosystem-structuring role of large carnivorous animals including sea otter (*Enhydra lutris*)–kelp systems, Lago Guri, Venezuela, and wolves (*Canis lupus*) of the Greater Yellowstone Ecosystem. The remaining chapters contribute through examining research results from a set of systems less frequently appreciated as central to the topic of this book. These include examining the general phenomenon of trophic cascades and top-down controls in large marine carnivores, the forests of the northeastern United States where large carnivores are gone and ungulates "rule the world," and what is known about redundant roles in groups of large carnivores, focusing particularly on the African guilds.

In Part Three, From Largely Intact to Human-Dominated Systems: Insight on the Role of Predation Derived from Long-Term Studies, five case studies are presented by ecologists who have worked on a long-term basis in various systems and provide information essential for determining whether the functional importance of carnivores necessarily means that focusing conservation efforts on them will achieve conservation of biodiversity. Their contributions are arranged from those that examine relatively intact ecosystems to those heavily influenced by humans, in the Russian Far East, African savannas, European temperate forests, tropical coral reefs, and Patagonian Steppe. In Part Four, Achieving Conservation and Management Goals through Focus on Large Carnivorous Animals, five chapters address the practical applications that may be derived from the science of understanding carnivory. These include discussions of how long-term studies on carnivores designed to address management issues can play a role in conserving landscapes and biodiversity, an analysis of whether hunting by humans and hunting by other large carnivorous animals are functionally redundant, and a conceptual framework for assessing whether populations of large herbivores are regulated by top-down or bottom-up processes. The final two chapters in Part Four offer contrasting perspectives on how top carnivores are related to biodiversity conservation in boreal forest ecosystems and the "half-full" forests of Europe.

This book is not meant to be another book about carnivores. It is intended to be a book about the relationship between carnivores and conservation. All authors were asked specifically to address the conservation implications of their work. The book concludes with Chapter 20, an overall synthesis that draws the conservation implications from the rich mix of chapters, making the point that despite the lack of a simple answer to a complicated question, there are ways to improve our thinking and action to conserve both large carnivorous animals and biodiversity.

There has been a good deal of ecological work done on the impact of biodiversity loss on ecosystem structure and function (Scheffer and Carpenter 2003), with trophic interactions appearing to play important roles in these processes (Worm and Duffy 2003). But there continues to be debate about the relative role of consumer-driven (top-down) versus resource-driven (bottom-up) control, with both appearing to operate at some times, in some systems (Meserve et al. 2003; Sinclair et al. 2003). Yet little of this work has provided tools that would help conservation practitioners in their efforts to conserve biodiversity.

To us, the question, Is the conservation of large carnivorous animals equivalent to the conservation of biological diversity? is a vital one for the conservation community to address head-on. It is fashionable to argue in some quarters that large carnivorous animals are a "tool" whose presence is required in order to achieve conservation of all components and attributes of biodiversity. And further, this argument states that restoration of this full spectrum of biodiversity is not possible without reintroduction of large carnivorous animals. If this utilitarian approach to large carnivore conservation is correct, then we must be able to prove the vital role played by these species. If it is not correct, then we must proceed with caution (c.f. Warren et al. 1990), for these species may not be necessary (in this utilitarian sense) and, given the negative costs of their presence and the conservative nature of scientific proof, a limited version of conservation success might be easier to achieve in their absence. Difficult though this question is, it exists as a reality in the world of the "Designer Ark" (Weber in press).

A different, though perhaps complementary, argument for the conservation of large carnivorous animals is value based and draws on the long-intertwined history of humans and these species and the roles they played and play in the human psyche. As Quammen (2003: 13) has written: "For as long as *Homo sapiens* has been sapient . . . alpha predators have kept us acutely aware of our membership within the natural world. They've done it by reminding us that to them we're just another flavor of meat." The power that large carnivorous animals had over humans is bred in the bone and has resulted in complex accounting of the relationship between the two (Redford and Robinson 2002). Origin myths place humans descended from jaguars (*Panthera onca*) (Benson 1997) or sharing the same mother as tigers (*Panthera tigris*) (Wessing 1986). And a common theme is the blurred boundary between the two with lycanthropy, or humans turning into wolves, found in Europe (Otten 1986), echoing beliefs from throughout the world that humans transform into jaguars, pumas (*Puma concolor*), leopards (*Panthera pardus*), lions (*Panthera leo*), tigers, and bears (Boomgaard 2001).

The power of the relationship between humans and large carnivorous animals lies in its ambiguity and blurring of boundaries (Wessing 1986; Benson 1997; Boomgaard 2001). For example, in some of the early European illustrations of the New World—such as a Dutch woodcut published in 1695—there is a conflating of human and jaguar, with the jaguar pictured standing in a human position (Saunders 1990). Large carnivorous animals are symbols of the nonhuman world both within and outside of the human body, as illustrated by the human–lion hybrid, the sphinx. The nature of this relationship between such animals and the nonhuman world is well illustrated in a Javanese tale related by Crawford (1967) (in Wessing 1986):

Make choice of an equal friend, and do not like the tiger and the forest. A tiger and a forest had united in close friendship, and they afforded each other mutual protection. When men wanted to take wood or leaves from the forest, they were dissuaded by their fear of the tiger, and when they would take the tiger, he was concealed by the forest. After a long time, the forest was rendered foul by the residence of the tiger and it began to be estranged from him. The tiger thereupon quitted the forest, and men having found out that it was no longer guarded came in numbers and cut down the wood, and robbed the leaves, so that, in a short time, the forest was destroyed, and became a bare place. The tiger, leaving the forest, was seen and although he attempted to hide himself in clefts and valleys, men attacked him and killed him, and thus, by their disagreement the forest was terminated, and the tiger lost his life.

Undoubtedly, there is no single unifying theory to tell us when the tiger and the forest are locked into this symbiotic relationship. We do not know enough to predict the role of large carnivorous animals in the ecology of every place, time, and circumstance. We must therefore be careful not to assume that we know when and where such species must be conserved in order to conserve other components and attributes of biodiversity. Their existence is worth more than just the role they play in ecosystems.

In an evolutionarily abrupt turning of the tables, humans are now responsible for the survival of large carnivorous animals. Will Quammen (2003) be correct in his prediction of the year 2150 as a probable end point to the special relationship between humans and alpha predators? We certainly hope not. Boomgaard (2001) recounts early Dutch reports from Indonesia documenting the existence of a kind of tiger called the *volgtiger*, literally a following or attendant tiger or a "familiar." The concept of a familiar (meaning a spirit, usually taking the form of an animal but also a close friend or companion) that helps someone (often a witch) is apt in this context. Large carnivorous animals are a part of humans and of our past. But they are also a test of our humanity and of our ability to save the earth. Perhaps it is true, as told in a Colombian indigenous myth, that "the jaguar was sent to the world as a test of the will and integrity of the first humans" (Davis 1996). If we are to save ourselves, we must save all the parts of our humanity. As go these wild animals, so goes the human soul.

PART I

Setting the Stage

There are both theoretical and practical aspects to the question regarding the relationship between large carnivorous animals and biodiversity posed by this book. Although the link between the two is often acknowledged, conservation scientists and practitioners have generally remained in two separate camps. Scientific inquiries examining the role of large carnivorous animals in structuring biological communities do not generally delve into how the science translates into practical terms. By the same token, practitioners utilizing large carnivores as tools to increase the efficiency of attaining conservation goals do not often probe deeper into the labyrinth of exceptions and uncertainties that form the scientific basis of the work.

This introductory section is composed of two chapters that lay out the theoretical and practical foundations of this topic. The first, by Robert Steneck (Chapter 2), provides a theoretical framework for exploring the ecosystem role of large carnivorous animals. Although most research on this topic has focused on small-bodied predatory animals in relatively closed systems, there is a strong theoretical basis for extending many conclusions from this work to large-bodied predators. Large carnivores can affect local and regional biodiversity, but it is important to consider the conditions that might be necessary for their influence on ecosystem properties to be strong. Such questions are central to the scientific basis for conserving biodiversity. Justina Ray (Chapter 3) takes a first step in considering the conservation applications stemming from the growing body of research on the relationship between large carnivorous animals and biodiversity. Although a substantial shift in attitude toward top predators from obstacles to instruments for achieving conservation goals has enabled their increasing use as centerpieces of conservation strategy, the assumptions behind this use have undergone little scrutiny. Dr. Ray examines the rationale and underlying assumptions that characterize conservation tools that have been developed using large carnivores in both terrestrial and marine settings.

Together, these two chapters set the stage for the remainder of the volume in which the scientific context and practical implications of the role of large carnivores in conserving biodiversity are explored in finer detail.

CHAPTER 2

An Ecological Context for the Role of Large Carnivores in Conserving Biodiversity

Robert S. Steneck

How important are large carnivorous animals for conserving biodiversity? Today they are rare or absent from most terrestrial, aquatic, and marine ecosystems. Should we invest heavily in political and real capital to restore them? These questions require that we understand their ecological roles in ecosystems. However, most studies are too limited in scope to provide answers to such broad questions. One way around this lack of data that would allow a more holistic perspective is to apply ecological theory to help sort out which concepts are most appropriate, most compelling, and most robust.

Most general ecological concepts begin with observations made in nature. Fortunately, people have always been keen observers of predators. Cave paintings in France made 35,000 years ago depict large carnivores stalking prey; some of which were humans. Obviously, our preoccupation with carnivores is primal, and it has resulted in a wealth of knowledge about them and their effects. However, over time perceptions of their roles in ecosystems have changed and, accordingly, observations could have been colored by prevailing dogma and existing social and scientific paradigms. Therefore, we can better understand contemporary concepts by knowing their conceptual history.

Considerable empirical and theoretical ecological research supports the thesis that large predators *can* affect community structure and biodiversity. It is less well known under which conditions predators *do* exert, or could exert, major influences on the structure and functioning of ecosystems. Today, relatively few ecosystems have large predators that play important roles, often due to extirpations induced by hunting and habitat change. But it is also possible that some habitats and ecosystems never had ecologically significant large predators. Under what

9

conditions should we expect predators to be important regulators of biodiversity? Obviously there is no point in trying to restore large carnivores if they were never major players in the system.

The present is often not the key to the past because baselines for most structuring processes have slid so far that they may now be unrecognizable from their former "selves" and, worse, they may be unrestorable. This is not wholly a question for science because it depends on how we weigh human values relative to other conservation values. To help sort out what we can do from what we should do, this chapter considers the effects of large carnivores on ecosystem biodiversity within the context of contemporary ecological theory. Specifically I will review the origin and evolution of ecological theories that led to our current understanding of effects of large carnivores on biodiversity. I will consider under what conditions and in which ecosystems predator impacts are greatest, and where and when those impacts translate to lower trophic levels. That is, where predator impacts affect the biodiversity of the entire ecosystem. Finally, I will discuss the seductive nature of predator baselines that have been sliding for centuries and causing generations of people to redefine what we see as "natural."

"The World Is Green" Revolution

In 1960 Hairston, Smith, and Slobodkin (HSS) wrote a deceptively simple paper entitled "Community structure, population control and competition" (Hairston et al. 1960). This may be the ecologist's only parallel to Albert Einstein's "thought experiments" in physics, since both were based entirely on a logical interpretation of the world as understood by the authors. HSS argued that, since the world looks green, it is not overgrazed by herbivores. They pointed to coal deposits as an indication of accumulating plant matter over geological time. If herbivores are seldom food-limited, they are most likely to be predator-limited. Thus they concluded that density-dependent processes regulate carnivores at the top of food webs and producers at the base, but density-independent processes (i.e., carnivory) regulate the herbivores in the middle.

The context and the consequences of the HSS paper are underappreciated today. At the time the paper was written, density-dependent processes (specifically competition) were thought to be the primary processes structuring natural populations. This idea was championed by the luminaries of the day, including Andrewartha, Birch, Hutchinson, and MacArthur. HSS elegantly sensitized ecologists to interactions among trophic levels as well as to the fact that different processes may act at different trophic levels. They also pointed out that predators at upper trophic levels might control the distribution and abundance of consumers at lower trophic levels.

The HSS paper evoked numerous responses and stimulated several avenues of research that are still actively pursued (e.g., Terborgh, this volume). Arguably, their paper defined how we now consider biodiversity (Box 2.1). It also spawned contemporary concepts such as the intermediate disturbance hypothesis, trophic cascades, top-down forces in food web structure and facilitation, and positive and indirect interactions. To understand contemporary concepts on the role of carnivory in preserving biodiversity, it is useful to appreciate where these ideas originated and how they have been shaped over the past several decades.

Paradigm Gained—Predation as a Structuring Process

A year after HSS's publication, Connell (1961b) wrote a paper entitled: "The influence of interspecific competition and other factors on the distribution of the barnacle *Chthamalus stellatus.*" This classic paper demonstrated the importance of competition in determining dominance among intertidal barnacles. There was no reference to HSS, although Nelson Hairston was thanked for critiquing it. Nevertheless, one of the "other factors" in the title turned out to be predation by snails. Connell showed that only in the absence of their predator did carpets of barnacles (i.e., prey) cover the rocks and intensely compete with each other for space. From this, he and others generalized that the intensity of competition varies inversely with the intensity of predation. That conclusion may have initiated a paradigm shift that focused on the role of consumers in structuring natural communities.

A few years later Paine (1966) (Fig. 2.1) observed that intertidal mussels and a few other herbivorous, suspension-feeding animals form "monopolies" that will outcompete other organisms unless they get eaten by predators. He advanced the hypothesis that: "Local species diversity is directly related to the efficiency with

Box 2.1

Defining Biodiversity (a brief history)

A simple question posed by G. Evenly Hutchinson (1959), "[W]hy are there so many kinds of animals?" is difficult to answer because it can be interpreted in so many ways. How do we count the "many kinds of animals"? Today, "biodiversity" has ecological and evolutionary connotations. The term itself has evolved from Hutchinson's focus on simply the number of species found at a site (now called species richness) to what is now almost synonymous with some definitions of "ecology."

Today's term "biodiversity" has its roots in the phrase "species diversity." Beginning in the 1940s, information indices were developed to go beyond simple lists of the number of species (i.e., species richness or species density for a fixed area) to integrate taxa abundance. These indices were useful to demonstrate that a biota was dominated by few species (e.g., Simpson's dominance index, Simpson 1949), or if they were more evenly distributed (e.g., the Shannon diversity index or index of evenness, Shannon and Weaver 1949). Information indices, however, had their critics too. They were thought by some to represent hyperreductionism that was generating index numbers to complex natural communities. Statistics could not be performed on them; they could not be arranged in linear order along a diversity scale (see Hurlbert 1971). What did it mean if a peat bog in New England had the same diversity index as a mudflat in New Zealand?

Ecologists focusing on the structure of communities and ecosystems (e.g., Whittaker 1975) saw "species diversity" (defined as the abundance-weighted distribution of species) as being useful if it was grouped by habitat type. Whittaker identified within-habitat diversity ("alpha diversity"), as different from between-habitat diversity ("beta diversity") and different still from the entire pool of species within a region ("gamma diversity"). Although the latter results from evolutionary processes over geological time, the former results from habitat and process-driven ecological differences (Huston 1994). Thus patterns in species diversity may reflect regional or local pools of species but not necessarily processes driving those patterns. Some ecologists saw this as a serious shortcoming. They were particularly frustrated that interspecific interactions were not part of the species diversity indices (Hurlbert 1971).

The term "biodiversity" is a contraction of the phrase "biological diversity" (Wilson and Peter 1988). It was intended to encompass all scales of diversity from genomic to species, populations, communities, ecosystems, and landscapes. Significantly, it also includes ecological interactions among the species (Huston 1994). Thus, "Biodiversity refers to the natural variety and variability among living organisms, the ecological complexes in which they naturally occur, and the ways in which they interact with each other and with the physical environment" (Redford and Richter 1999: 1247). Although this is close to many definitions of "ecology," it is commonly used by conservation biologists and it allows us to consider holistically the relative role of large carnivorous animals as they interact with species at lower trophic levels all in the context of the ecosystem's other physical and biological processes.

which predators prevent the monopolization of the major environmental requisites by one species (Paine 1966: 65)."

Monopoly busting was only one part of the species diversity story. As predation pressure or other forms of disturbance increase to high levels, fewer species can persist. Therefore, at very low or very high levels of disturbance, diversity is low; so it follows that the highest diversity will be between those two extremes. This became known as the Intermediate Disturbance Hypothesis (its origins have been attributed to several papers; Paine and Vadas 1969; Connell 1978; Lubchenco 1978; Huston 1979), and it established a strong link between the process of predation and the local species diversity (see Fig. 2.1).

By the 1970s, community ecologists were becoming increasingly convinced that predators could control community structure. Some viewed this paradigm shift from competition-based to predator-based control of community structure as revolutionary. After all, niche theory was founded on the notion that "animal communities appear qualitatively to be constructed as if competition were regulating their structure" (Hutchinson 1957: 419). However, Hutchinson himself in his "Homage to Santa Rosalia" (Hutchinson 1959) suggested diversity relates at least partially to the trophic organization of the community. He cited Odum's (1953) textbook treatment of trophic structure (called a "predator chain") and remarked "the lowest link is a green plant, the next a herbivorous animal, the next

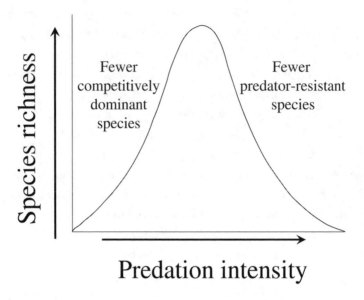

Figure 2.1
The intermediate disturbance hypothesis (after Lubchenco 1978). Maximum species diversity falls between the extremes of no predation where one or a few competitively dominant species thrive and high predation pressure where only a few predator-resistant species persist.

a primary carnivore and the next a secondary carnivore, etc." Hutchinson (1959: 147). Hutchinson further pointed out the assertion of both Wallace (1858) and Elton (1927) that food webs were constructed such that "the predator at each level is larger and rarer than its prey" Hutchinson (1959: 147). This pattern became known as the "Eltonian food pyramid" and it developed into the field of "trophic-dynamics" (*sensu* Lindeman 1942). In this view, each trophic level is "successively dependent upon the preceding level as a source of energy" (Lindeman 1942: 415). In other words, the primary interactions resulted from lower tropic levels fueling those at the top. Today, this is called bottom-up control of community structure (*sensu* Power 1992). What Hairston et al. (1960) proposed to the world was decidedly different. Rather than resources at lower trophic levels fueling higher trophic levels (bottom up), consumers at higher trophic levels limit the abundance of lower trophic levels (top-down) (*sensu* Power 1992). This paradigm shift is much more than changing terminology. While predators had long been considered part of natural communities, they had been thought of as "passengers" carried by the resources available in the ecosystem. What had been underappreciated was that predators could be "drivers" of the system by limiting the abundance of their prey. This new way of thinking opened new avenues of ecological theory focusing on the communitywide impacts of higher-order predators on organisms at lower trophic levels.

Top-Down Forces in Food Webs: Keystones to Trophic Cascades

The top-down manner by which predators drive the structure of ecosystems was illustrated in several compelling studies published in rapid succession, beginning in the mid-1960s. These early studies from widely divergent ecosystems all found that a single predator can control the distribution, abundance, body size, and species diversity of all other species in the system. One classic example came from Robert Paine who had been a student of Frederick Smith (the first "S" of HSS). Paine observed that the intertidal sea star, *Pisaster ochraceus*, controlled the abundance of the competitively dominant large mussel, *Mytilus californianus* in the Pacific Northwest (Paine 1966). Without the carpet of mussels, a variety of algae and other organisms flourished. About the same time Brooks and Dodson (1965) observed in freshwater lakes that a planktivorous predatory fish, the alewife (*Alosa pseudoharengus*), consumed most large herbivorous zooplanktons, thereby allowing small, nonpreferred, competitively inferior species to thrive. Finally, in the Aleutian Islands of Alaska, the sea otter (*Enhydra lutris*) was shown to control the distribution and abundance of herbivorous sea urchins, which in turn control kelp forest development (Estes and Palmisano 1974; reviewed in Estes, this volume). In all of these examples, a single predator affected the entire community by removing either a dominant spatial competitor or a dominant herbivore. Thus the larger impacts resulted from a release of ecological control by those competitors or herbivores.

Single species that greatly affect communities but constitute only a low proportion of the community biomass are called "keystone" species (*sensu* Paine 1966; Power et al. 1996; Fig. 2.2). Most widely recognized keystone species are apex predators, such as the sea stars and sea otters (already described), large predator snails (*Concholepas concholepas*) (Castilla and Paine 1987), and freshwater bass (Power et al. 1996). Curiously, these and most examples of keystone predators are from either marine or freshwater aquatic ecosystems. It could be that terrestrial predator–prey interactions are more difficult to observe because they play out over much larger areas and over a much longer period of time. Nevertheless, the effects of keystone species can become evident when they are reintroduced to isolated terrestrial ecosystems such National Parks (Ripple and Larsen 2000; Berger et al. 2001a; Berger and Smith, this volume) or islands. For example, fluctuating population densities of wolves (*Canis lupus*) on an island in Lake Superior control the abundance of moose that in turn control the abundance of the island's balsam fir (*Abes balsamea*) trees (McLaren and Peterson 1994).

Keystone species need not be carnivores, but most are, because of the stipulation that they have a great impact at low abundance. Other species that have a large impact but are abundant in the system are called "foundation" (Dayton 1975; Soulé et al. 2003) or just "dominant" species (Power et al. 1996; see Fig. 2.2). Often, herbivores rather than carnivores were the dominant species in ecosystems. Although large herds of wildebeest and other ungulates in Africa (Sinclair and Norton-Griffths 1979) or sea urchins in numerous shallow marine habitats (Steneck et al. 2002) control the structure of lower trophic levels (i.e., plant communities), they do so by the brute force of numbers and as such they do not qualify as keystone species (see Fig. 2.2).

Usually there are relatively few carnivorous species at the highest trophic levels. These "apex" predators are so named because no predator controls their abundance (i.e., they are resource limited according to HSS). Thus all keystone predators are apex predators, but the reverse is not true. There are relatively few keystone predators in the world. They are rare or absent from most highly diverse ecosystems. Arguably, it is immaterial whether a single or several predators are controlling prey densities. What matters most is that carnivores at or near the top level control consumers at lower trophic levels, thus creating ripple effects throughout the food web.

Robert Paine pointed out that strong interactions by consumers "cascade through the community, transmitted by a chain of strongly interacting links" (Paine 1980: 674). Such "trophic cascades"(Paine 1980: 676) result from the top-down control of consumers on their prey. If prey are themselves strong interactors, then their prey, at yet lower trophic levels, are also affected. In this way top-down

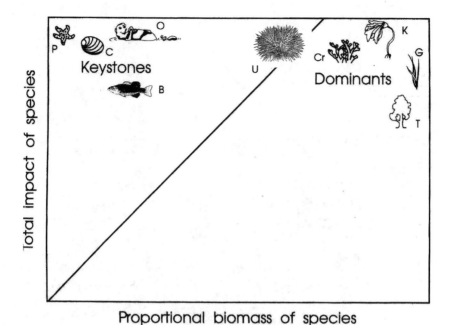

Figure 2.2

Keystone and dominant species. Their functional importance of species relative to their abundance (after Power et al. 1996, with permission—Copyright, American Institute of Biological Sciences). Important species include *Pisaster* (P), sea otter (O), the Chilean predatory whelk *Concholepas* (C), sea urchins (U), trees (T), kelp (*K*), grass (G), and reef-building corals (Cr). Note keystone species are only those with a high impact relative to their abundance.

impacts cascade from apex predators to primary producers. Typically, trophic cascades must show inverse patterns of abundance between a consumer and its prey across more than one trophic level in a food web.

Variability of Trophic Cascades

HSS described a hypothetical trophic cascade that became the classical standard: predators regulate herbivores allowing edible plants to be limited only by resources available to them. This was a food web of three trophic levels (i.e., an odd number).

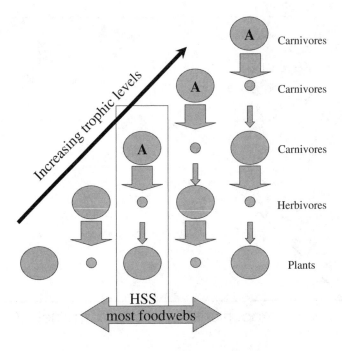

Figure 2.3
Number of trophic levels and effects. Large arrows indicate large effect from strong interactors. Small arrows indicate small effect. Relative abundance of organisms in any given trophic level is indicated by the circle diameter. "A" indicates apex predators.

However, food webs can have fewer or more than three levels (Morin and Lawler 1995). If top-down forces dominate the system, then a higher-order apex predator representing a fourth trophic level will effectively control the predators of herbivores (Fig. 2.3). In even-numbered trophic-level food webs with four or more levels, herbivore populations can expand and overgraze plant communities (Fretwell 1977, 1987; Oksanen et al. 1981). An excellent example is the tri-trophic sea otter/sea urchin/kelp system described by Estes (this volume; Estes and Palmisano 1974; Estes and Duggins 1995). The sea otter in Alaska is a reintroduced apex predator that controlled sea urchin population densities until otter-eating killer whales (*Orcinus orca*) entered this coastal ecosystem in the 1990s. The addition of this fourth trophic level eliminated sea otters, causing herbivorous sea urchin populations to explode and denude kelp forests over vast areas (Estes et al. 1998). This also illustrates the context-dependent nature of top-down controls (Pace et al. 1999). With the unprecedented attacks on sea otters by killer whales, beginning in the 1990s, sea otters lost their status as the system's apex predator.

Not all predation from upper trophic levels cascades to lower levels. There are several reasons for this "attenuation" (*sensu* Schmitz et al. 2000). Edibility (that something can be consumed) and palatability (that something edible is chosen to be consumed) control what is eaten. For example, many woody mature plants are inedible, so changes in herbivores will have little immediate impact on them. In contrast, saplings are usually edible and thus are more likely to show the effects of herbivory.

Megaherbivores can grow to a size at which they are relatively inedible and thus immune to apex predators (Owen-Smith 1988; Sinclair et al. 2003). Herbivory can weed out the most edible and palatable plants from a community, leaving plants that are avoided or impossible to eat. Thus the world can be green and herbivores could be trophically limited (reviewed in Terborgh, this volume).

Even among undefended prey in highly diverse ecosystems, the effect of the predator and herbivore guilds can become so diffuse that their per capita impacts become very low (Duffy 2002). In that case, loss of a species may have modest or undetectable communitywide implications. Attenuation of top-down effects in highly diverse ecosystems caused some to question whether trophic cascades are important or even possible there (Strong 1992; Polis and Strong 1996). Most examples of cascading effects are at the community level (e.g., those already described here), but there are also cases where top predators strongly affect one or two species but because those species are either not strong interactors or constitute a small fraction of the community, there is little or no communitywide impact (Polis 1999). It is possible that greater prey diversity can reduce the penetration of trophic cascades beyond one trophic level (Duffy 2002). However, explicit tests showing biodiversity per se can attenuate trophic cascades are generally lacking (Duffy 2002). Classic marine examples of trophic cascades are primarily confined to ecosystems with naturally low biodiversity (e.g., kelp forests of Maine and Alaska; Steneck et al. 2002) or more diverse systems that have lost functional diversity (e.g., Caribbean coral reefs due to overfishing and disease; Hughes 1994). Nevertheless, even some highly diverse ecosystems have been shown to have trophic cascades (e.g., Pace et al. 1999; Terborgh et al. 2001).

Perhaps the more important question about variability of trophic cascades is why they are so evident in some ecosystems but not in others. There are at least four factors that can diminish the expression of strong trophic cascades in which predator effects conspicuously translate to change among plants or other basal trophic level organisms within the community. They include, in increasing order of importance: (1) compensatory community changes initiated by top-down forces that result in an environment hostile to herbivores, (2) poorly defined trophic structure resulting from widespread omnivory that blurs trophic-level distinctions and functions, (3) reduced interaction strengths of predators or herbivores due to reduced consumer body size from biogeographic or anthropogenic effects, and (4) environmental regulation of interaction strengths via physiological stress. Following here, I will describe each of these factors.

Compensatory Community Changes

Several compensatory mechanisms can dampen or eliminate trophic cascades (Pace et al. 1999). These include changes to ecosystems that reduce the effectiveness of consumers. Some good examples come from shallow marine ecosystems where the vegetation responds quickly to changes in herbivory. Predator-induced declines in herbivores can result in rapid increases of macroalgae that change habitat architecture, creating a predator-free refuge for small herbivores and other mesopredators in which to hide from visual predators (Hacker and Steneck 1990; McNaught 1999). Similarly, some distasteful, toxic or heavily armored plants that are avoided by consumers create an effective defense for organisms closely associated with them (Hay 1986; Pfister and Hay 1988; Bruno et al. 2003). On coral reefs, reductions in herbivory resulting in marked increases in vegetation reduce the susceptibility of the plant community to subsequent herbivory (McClanahan et al. 2001b). Thus consumer-driven changes to ecosystems can reduce the effectiveness and impact of other consumers that drive local trophic cascades.

Poorly Defined Trophic Structures

Consumers' structuring effects on food webs can be difficult to characterize because trophic levels can be hard to define. Omnivores and detritivores are ubiquitous and can switch facultatively among trophic levels, resulting in trophic interactions that are more "reticulate" than those classically structured into distinct trophic levels (Polis 1999). This is well known and was specifically addressed by Paine (1980) when he first described trophic cascades. He was clear to dispel the notion that the study of food webs and community structure possesses the crisp determinism of physics. He considered food webs as idealized local abstractions or "nontrivial determinism" (*sensu* Pascual and Levin 1999) of dynamic, complex trophic interactions. Thus these and many other food web studies do show communitywide effects on lower trophic levels from functionally distinct higher trophic levels, even if their exact placement in the food web remains unclear. Further, Menge and Sutherland (1976) suggested that it matters less in the abstraction of food webs if consumers eat meat or plants because, in most systems, larger consumers eat smaller species. Because apex predators are often the largest consumers in the community preying on smaller carnivores, omnivores, and herbivores, the cascading effects resulting from them will vary primarily as a function of their interaction strength.

Reduced Consumer Body Size

The functional role of consumers often scales with their body size. Apex predators that are "strong interactors" (*sensu* MacArthur 1972) often initiate the top-down control leading to trophic cascades. Such predators, by definition, have high per capita interaction strength (Sala and Graham 2002). The strength of carnivore effects often relates to their body size (Sinclair et al. 2003). Body size scaling dictates both predatory and competitive dominance (e.g., Connell 1983). Larger predators can eat larger prey. For example, there is a strong linear relationship between terrestrial predators ranging from 10^{-4} to 10^3 kg and their prey ranging from 10^{-6} and 10^3 kg (Peters 1983). Thus large predatory mammals and birds scale to the mass of their prey in the same manner as small predatory lizards, amphibians, and birds. Large predators also consume the widest range of prey sizes (Peters 1983), which magnifies their per capita impact to the structure of the food web. However, body sizes and predation capacity change ontogenetically, resulting in ecological niche shifts (Werner and Gilliam 1984). Slow-growing predators that are hunted or fished may not attain the size necessary to be strong interactors in the community (to be discussed further).

Cascades also vary geographically. Although it is the strong interactors that often define the web's structure and function (called "interaction webs" by Menge and Sutherland 1987), those same species may be weak interactors in other parts of their geographic range (Paine 1980; Menge and Sutherland 1987). For example, on the coast of Washington State the sea star (*Pisaster* sp.) is the keystone predator (e.g., see Fig. 2.2) that limits the abundance of the dominant space competitor, the mussel (*Mytilus californianus*). In Alaska, those same species are present, but their interaction strength is low and thus neither species is a major player in that system (Paine 1980). In this example, the physical environment regulates the interaction strength of a keystone predator and thus regulates its structuring role in the community.

Interaction Strengths via Physiological Stress

Regional differences in predator effects can vary due to differences in productivity in the system. As was the case in coastal Alaska versus Washington State, predation potential can be controlled by environmental stress (Menge et al. 1994; Fig. 2.4a,b). Such consumer control of prey falls along a continuum of primary productivity of the system (Oksanen 1990). Under environmentally harsh physical conditions, predation becomes less important (Lubchenco 1978). This is obvious in deserts where evapotransporation is more important than top-down trophic cascades in controlling community structure. Where resources are more limited, they become more limiting to species. Under such conditions, competition drives the ecosystem, overriding predation effects (Menge et al. 1994) (see Fig. 2.4). In this view, physical factors and competition drive interactions at the highest trophic levels, but predation becomes increasingly important at lower levels. In systems with high environmental stress, the role of predation in community regulation is low (see Fig. 2.4b). However, some sessile organisms, such as terrestrial and intertidal vegetation or intertidal barnacles and mussels, create positive feedbacks in which abiotic stress is reduced, which increases competition strength (Bruno et al. 2003). Other organisms can reduce predation potential by providing associational defenses or refugia (Fig. 2.4c). Thus many aspects of trophic cascades are context dependent. Although this should give pause to some sweeping generalizations, some of the variability just described is more the exception than the rule. Other factors, such as low environmental stress (i.e., high productivity potential) driving consumption (e.g., see Fig. 2.4), may be much more important to the ubiquity and strength of trophic cascades in benthic marine and aquatic ecosystems.

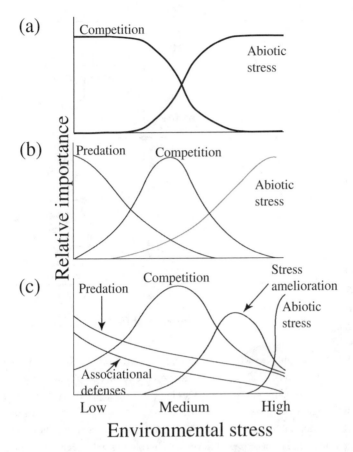

Figure 2.4
Ecological processes structuring apex carnivores (a), mesopredators, herbivores, and producers in isolation (modified from Menge and Sutherland 1987) (b), and with positive feedback ("facilitations") (c) such as associational refugia (defenses) and stress amelioration (from Bruno and Bertness 2001, with permission).

Do Marine Systems Have the Strongest Trophic Cascades?

Trophic cascades described from benthic aquatic ecosystems (both marine and freshwater), often have a higher impact on lower trophic levels (i.e., lower "attenuation") than those from terrestrial ecosystems (Fig. 2.5) (Shurin et al. 2002). In fact, the only examples of predator change cascading to the complete denuding of all canopy-forming vegetation are from marine ecosystems (Polis

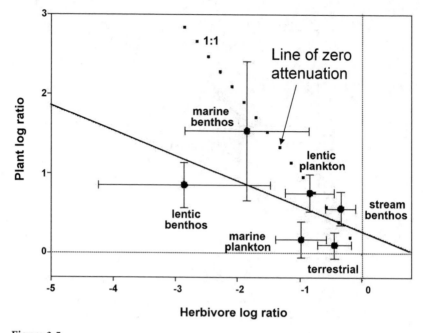

Figure 2.5

Attenuation of trophic cascade effects among different biomes (from Shurin et al. 2002 with permission). Changes in predators that result in changes in herbivore abundance are given in the x-axis. Changes in plant abundance that result from subsequent changes in herbivore abundance are given in the y-axis. If predators have relatively little impact in the system, they will be plotted near the origin (bottom right). If there is no attenuation of predator impacts, herbivores will decline and plants will increase in proportion with changes in carnivore abundance. This "line of zero attenuation" is represented by the dotted line. The solid line represents the average line of attenuation.

1999). The sea otter/sea urchin/kelp example of predator control of vegetation-denuding herbivores has many other marine parallels in tropical (McClanahan and Muthiga 1989; Sala et al. 1998) and temperate to arctic (Steneck et al. 2002) systems, whereas the best tri-trophic terrestrial cases have relatively modest vegetational impacts.

Predator effects on herbivores that cascade to plants exist in both marine and terrestrial systems, but the changes in higher-order terrestrial predators translate to relatively modest or undetectable cascading changes to plants (Fig. 2.5). In a review of 60 terrestrial studies, Schmitz et al. (2000) found evidence for trophic cascades in 45 of them. However, the evidence was strongest when measured as injuries to the plants rather than as changes in biomass. They suggested attenuation could have resulted from induced antiherbivore deterrents in grazed plants, which could slow biomass loss.

It is also possible that trophic cascades simply take longer to show themselves in terrestrial ecosystems. Could complete deforestation be occurring in areas where large predators have been extirpated, but it will take centuries to observe it? Several studies suggested historical declines in wolf populations in the Rocky Mountains of North America resulted in increased moose (*Alces alces*) and elk (*Cervus elaphus*) populations and decreased aspen (*Populus tremuloides*) and willow (*Salix* spp.) tree abundance over the past century (Berger et al. 2001a; Ripple et al. 2001). On an island in Lake Superior, cyclic population flushes of wolves limit herbivory by moose and thus allow balsam fir trees on the island to prosper in concordant cycles (McLaren and Peterson 1994). In all of these examples, the presence of predators changed the landscape's vegetation but without evidence that the system was heading toward complete deforestation as is commonly seen in the sea.

Perhaps the best example of strong top-down control in a diverse tropical terrestrial ecosystem comes from a study on islands in an artificial lake in Venezuela where only some of the islands had predators (Terborgh et al. 2001; Terborgh, this volume). Islands without predators had 10 to 100 times greater population densities of herbivores, which limited the density of seedlings and saplings of canopy trees. The system may well have been moving toward a plant-denuded state until the lake level dropped, thereby ending the experiment. Nevertheless, this is a rare example compared to the many aquatic ones. Unquestionably, the vegetation has changed in concert with changes in predator populations, but the magnitude of change between marine and terrestrial systems appears to vary markedly.

There are four major hypotheses for this apparent marine/terrestrial difference (reviewed by Polis 1999; Shurin et al. 2002): (1) biomaterials of canopy-forming marine vegetation are more edible and nutritious than their terrestrial counterparts; (2) body size ratios between herbivores and plants are greater in aquatic systems; the largest marine herbivores swim and can graze canopies from the top down (Steneck and Dethier 1995); (3) herbivores consume three times more primary production in lentic than in terrestrial food webs (Cyr and Pace 1993); and (4) higher inherent per area and per mass production potential of benthic marine versus terrestrial ecosystems (Mann 1973; Wiebe et al. 1987; Shurin et al. 2002) reduces environmental stress and pushes those systems toward predator control (e.g., see left side of Fig. 2.4b). None of these hypotheses are mutually exclusive. Also trophic cascades may be more apparent in benthic marine ecosystems because they scale well for ecological studies. Small-predator exclusion cages yield results in months or years and thus find their way into Ph.D. dissertations and to the published literature. However the undeniable strength of some marine and freshwater trophic cascades carries with it an additional conservation consequence because, in those systems, top carnivores are actively hunted for food, which is relatively rare in modern terrestrial ecosystems.

Why Big Fierce Animals Are Rare: Top-Down Vulnerability and Chronically Sliding Baselines

Paul Colinvaux, in his book *Why Big Fierce Animals Are Rare* (1979), explains that this rarity is the natural result of Eltonian pyramids. After all, food webs can support only so many apex predators due to inefficiencies of converting energy initially produced by plants and consumed up the food web. However, it is not the energetics of food webs that limit the abundance of large predators today; it is direct and indirect negative interactions with humans. Nevertheless both of these are only proximate factors nested within a larger evolutionary context. In this section I'll consider why large predators have always been rare, what their top-down effects may have been prior to humans, how and why human interactions on land and in the sea have led to their decline, and what all of this might mean for conserving biodiversity.

Extinction Rates

There are also relatively few species of "big fierce animals" because both being big and being fierce (i.e., obligate carnivores) increase rates of extinction. Extinction rates are commonly greatest among large taxa, as demonstrated for myriad organisms, including Cretaceous bivalves (Jablonski and Raup 1995) and Cenozoic mammals (MacFadden 2000). In fact, when mostly herbivorous mammals from

North and South America commingled for the first time 2.5 million years ago, following the formation of the Isthmus of Panama, extinction rates of primarily large mammals increased (Lessa et al. 1997). In fact, body mass was the only factor strongly associated with the probability of extinction following this great American biotic interchange (Lessa et al. 1997). This conforms with evolutionary theory that predicts extinction rates will be higher due to morphological specializations necessary to attain large size (Stanley 1973).

Large carnivores have even higher extinction rates than do other large consumers (Duffy 2002). For example, large obligate carnivores, called hypercarnivores, had the highest extinction rates among all mammals of the Middle and Late Miocene (Viranta 2003). Reviewing the 46-million-year evolutionary history of carnivorous mammals of North America and Eurasia, Van Valkenburgh (1999) concluded that large carnivorous mammals produced more specialized species, which were inherently more vulnerable to extinctions.

Top-Down Controls Prior to Humans

Before humans began to proliferate, large carnivores and large herbivores were much more abundant and diverse. A "snapshot" of this is evident in the megafauna of North America that has been preserved in the Late Pleistocene tar pits of La Brea, California (Van Valkenburgh and Hertel 1993). The large predators included the extinct American lion (*Panthera atrox*), saber-toothed tiger (*Smiliodon fatalis*), dire wolf (*Canis dirus*), and giant running bears together with the extant coyote (*Canis latrans*), puma (*Puma concolor*), and bobcat (*Lynx rufus*). This large predator guild coexisted with megaherbivores much larger and more diverse than those extant today: there were 56 herbivore species exceeding 30 kg compared to only 11 today; there were 29 species of megaherbivores exceeding 300 kg in size (i.e., moose size), whereas today there are only three. There were also at least seven species of elephant-size mastodons and mammoths, which have no modern counterpart in North America. The Late Pleistocene North American megafauna exceeded not only that found in North America today but also the megafauna found anywhere today (including Africa) (Van Valkenburgh and Hertel 1993).

Could large Pleistocene carnivores have been sufficiently abundant to limit herbivores and thus control community structure in ways predicted by HSS?

Evidence from tooth breakage patterns (Van Valkenburgh and Hertel 1993) suggests that large Pleistocene carnivores may have been food-limited and probably had to "feed more rapidly, guard their kills more aggressively, and completely consume their prey, often ingesting bone in the process" (Van Valkenburgh and Hertel 1993: 456). This resource-limited competition among top predators (Van Valkenburgh 2001), although rare or absent today, was predicted by HSS in their landmark paper (Hairston et al. 1960). Specifically, they suggested that top predators and plants would be resource-limited if herbivores were predator-limited. Intense competition among carnivores indicated by tooth breakage patterns is consistent with this HSS prediction.

However, this megapredator–prey relationship may also have been relatively fragile because of the disproportionate susceptibility of megafauna to extinction. In other words, reasons why large predators have always been relatively rare are the same reasons why they are vulnerable today. Whatever rarity existed prior to human interactions has obviously increased in recent centuries.

Human Impacts and Sliding Predator Baselines

The first human records of a terrestrial ecosystem show an intact carnivore and herbivore megafauna. Cave drawings from Chauvet, France, 35,000 years ago depict now extinct bears, lions, and leopards. The lions are shown stalking mega-herbivore ungulates such as bison, rhinoceros, and elephants. The cascading effects of those large predators to European biodiversity are not known. The data are absent or too spotty to conclude much for any region of the world. Nevertheless, the emerging global picture implicates humans as a catalyst for the extinction of the Pleistocene megafauna.

The correlation between the arrival of humans and the extinction of North American Pleistocene megafauna was first advanced by Martin (1973). Subsequent studies were consistent with that interpretation and suggested that most of the extinctions occurred during the first 2500 years after the human colonization of the New World 10,000 to 13,000 years before the present (Alroy 2001). The possibility that climate or other factors caused the mass extinctions of megafauna was rendered less likely when other southern hemisphere studies found elevated extinction of all terrestrial megafauna (mammals, reptiles, and birds) weighing more than 100 kg and 85% of those 45 to 100 kg around the time of first human contact. This also occurred in Papua New Guinea and Australia between 51,200 and 39,300 years ago, which was synchronous with human arrivals but asynchronous with climate changes (Roberts et al. 2001).

Despite the growing consensus of the correlation between humans and megafaunal extinctions, there is no consensus about causes in the terrestrial realm. Many believe hunters simply "overkilled" the predators and/or their prey (e.g., Flannery 1994 for Australia, Flannery 2001 for North America). Others believe large-scale human-induced changes to the ecosystem from fire, forest clearing, and direct hunting of megaherbivore prey contributed to the loss of megacarnivores (Grayson and Meltzer 2003; D. Corbett, pers. comm. 2001). The cause of the decline of large carnivores is immaterial to the bigger point that the carnivore component of the uppermost trophic levels irrevocably changed at the time that humans became significant components of many ecosystems.

Recently, ecologists have come to recognize that higher trophic levels are at greater risk of extinction (Duffy 2002). Further, ecological extinctions in which a species loses its interaction strength due to rarity (*sensu* Estes et al. 1989) can have the same effect of weakening top-down control (Duffy 2002) and increasing the importance of bottom-up forces. A careful study in many ecosystems would find the baseline interaction strength from large predators has shifted. Studies of historical ecology suggest that predator baselines began to shift thousands of years ago in some ecosystems (Jackson et al. 2001; Pandolfi et al. 2003), and these shifts have accelerated in recent centuries (Steneck et al. 2004).

Overkill in the Sea

The notion that humans hunted and gathered exclusively terrestrial resources during their colonization of the New World has recently been challenged (Erlandson 2001). New evidence from several sites suggests that maritime peoples were the first to colonize the Americas (Erlandson 2001). Impacts from these early settlers are apparent in archaeological deposits, as evidenced by a decline in size and frequency of large predatory sheephead fishes and a resultant increase in their invertebrate prey (such as sea urchins and abalone) (Erlandson and Rick 2002). Similar evidence from the Aleutian Islands of Alaska showed an early depletion of sea otters and an increase in their sea urchin prey (Simenstad et al. 1978). The megafauna overkill may have extended to the sea with the early extirpation of the whale-sized Steller's sea cow (*Hydrodamalis gigas*) and its final extinction soon after first European contact in the 1700s (Domning 1972; Clementz 2002). Of course the megafauna overkill continues today. Whaling may have led to declines in the northern right whale *Eubalaena glacialis*, contributed to the extinction of the North Atlantic gray whale *Eschrichtius robustus*, and reduced other whale species to a fraction of their former population size.

Apex predators in the marine realm are perennially the targets of fisheries because predatory fish are what people like to eat. Consumers prize fish such as tuna, cod, bluefish, grouper, and kingfish, and many are important players in ecosystems. Most have been extirpated by fishing (Jackson et al. 2001; Myers and Worm 2003). At a global scale fisheries are literally fishing down food webs (Pauly et al. 1998, 2001). The ecosystem consequences of this systemic loss of apex predators have been profound. Serial extirpations of apex then mesopredators in the marine realm has resulted in increases in lower trophic levels that are entirely consistent with HSS's green-world or top-down trophic cascades (see Steneck and Sala, this volume). Some of the best-known examples, such as the sea otter or cod from the North Pacific and North Atlantic, respectively, have been well studied. However, there is a growing list of examples of marine apex predator extirpation followed by population flushes of herbivores, resulting in large-scale denuding of all vegetation over expansive areas (Steneck et al. 2002). Such fishing down of marine food webs is thought by some to be one of the most serious threats to ecosystems of the world (Jackson et al. 2001).

Conserving Biodiversity

Large carnivores have been going extinct or decreasing in number for millennia. Extinction rates accelerated after humans arrived on the scene, but declines in abundance continue today. If predators played important functional roles in the past, then theory suggests today's community structure may be altered at all trophic levels. Thus many ecosystems that had been under strong top-down control may be bottom-up controlled today due to the extirpation of key predators. The ecosystem still functions, but it functions differently than it did in the past.

Humans have entered terrestrial and aquatic ecosystems as the ultimate apex predator. This could have the effect of changing the classic HSS three-level trophic cascade in which plants are abundant, into a four-level cascade in which herbivores dominate the system. In some cases, hyperabundances of herbivores have denuded landscapes. In other cases they have also been prone to epizootic diseases. Such large-scale instabilities appear to have escalated since humans became strong interactors in the ecosystem (Alroy 2001; Steneck and Carlton 2001).

Conserving biodiversity requires both a scientific understanding of the problem and the political will to act. The scientific linkage between biodiversity and ecosystem function is becoming clearer with several recent studies (reviewed by Duffy 2003). Specifically, functional loss of apex predators weakens top-down control, and the loss of species at lower trophic levels reduces functional redundancies, thereby reducing stability (Duffy 2002; Steneck et al. 2002). The political will to restore ecological function is more difficult because of common misperceptions that many parks and woodlands today look pristine. Often there is little public comprehension about how much the entire community has changed. Ecosystems dominated by abundant herbivores or mesopredators are commonly perceived as natural because the predator baseline shifted long ago or so slowly that the changes were not noticed.

Preserving ecological functions such as carnivory is much more difficult to explain than, say, extinctions, because their impacts are diffuse in space and time. In most cases, predators are extant but their population densities have fallen below levels where they limit their prey or effect lower trophic levels. Such "trophic level dysfunction" (*sensu* Steneck et al. 2004) results when an entire trophic level has so few consumers that their impact to other organisms or lower trophic levels is undetectable. In some cases trophic level dysfunction is similar to kidney dysfunction in humans. A loss of over 90% of a kidney's filtering capacity is asymptomatic. Similarly, when fishing pressure reduced most herbivorous fishes in the Caribbean, herbivory was maintained by a single sea urchin species without any obvious systemwide change in vegetation. Only when the sea urchin succumbed to a disease did vast areas of the Caribbean phase-shift to a highly vegetated alternate state (Knowlton 1992).

For some groups of consumers, ecological function and population densities are nonlinear. Thus threshold population densities exist, across which small changes in abundance result in alternate community states (Knowlton 1992) or "catastrophic shifts in ecosystems" (*sensu* Scheffer et al. 2001: 591). For keystone species, their impact will be high at relatively low population densities but may be undetectable at modestly lower levels. By definition, threshold densities for other predators will be higher, but these densities may change in different ecosystems (Soulé et al. 2003). Such real-world complications make conservation goals difficult both to define scientifically and to explain to policy makers and the public.

Conservation biologists have begun to address this problem by proposing to make ecological function a goal in conservation for highly interactive consumers such as predators (Soulé et al. 2003). This chapter attempts to outline some of the relevant ecological theory and predictions based on those ecological functions. The remainder of this book provides the case studies examining the range and depth of ecological effects of large carnivores on biodiversity.

Summary

In 1960 Hairston, Smith, and Slobodkin proposed that predators can structure entire natural communities. They asserted that carnivores limit herbivore abundance, releasing plants to attain abundances at which they become resource-limited. In this way, predators affect community structure at most, or all, trophic levels. In this chapter I use this seminal paper as a starting point to discuss the effects of large carnivorous animals on ecosystem biodiversity within the context of contemporary ecological theory. I then lay out a set of conclusions that have emerged from the ecological research that has resulted from the HSS paper, including concepts such as "keystone predators" and "trophic cascades." I discuss how differing numbers of trophic levels seem to affect how trophic cascades influence community structure and how omnivory, predator switching, trophic levels, and weakened interaction strengths due to high species diversity can make cascades less distinct. I additionally consider how all these factors vary between terrestrial and marine environments, pointing out that large-scale changes in vegetation resulting from predator impacts have been reported only for shallow coastal marine ecosystems in several of the world's oceans and seas with no comparable deforestation known from terrestrial ecosystems.

Large predators have always been relatively rare. They are vulnerable both because they are usually trophically specialized and because they are large in size. Human impacts also seem to have been important, with extinction having occurred on all continents except Antarctica and in both terrestrial and marine environments. In many places, predators have been absent or rare for so long that managers and scientists have never realized that they were ever important in the ecosystem. How should we manage ecosystems with such chronically sliding predator baselines? As food webs become increasingly depleted of higher-order consumers, top-down controls will likely give way to bottom-up or resource-limited community control. To minimize bias from such shifting baselines, managers and policy makers must consider an ecosystem's history to determine how it was structured and how it functioned before large carnivore effects were extinguished in the system.

ACKNOWLEDGMENTS

I wish to thank Kent Redford, Justina Ray, and Joel Berger for helping me develop this chapter, and Jim Estes, Justina Ray, and Kent Redford for their thoughtful reviews. Other helpful discussions came from Paul Dayton, Jim Estes, Jeremy Jackson, Bob Paine, and Blair Van Valkenburgh. Joanna Zigouris provided important support in the preparation of the manuscript. To all I am grateful.

Large Carnivorous Animals as Tools for Conserving Biodiversity: Assumptions and Uncertainties

Justina C. Ray

Vibrant growth in the disciplines of conservation biology and landscape ecology has been made possible through advances in species inventories, new approaches to data analysis, and changing perceptions in ecology. The growth and maturing of these fields have paved the way for the development and implementation of a variety of new conservation tools designed to combat the loss of biological diversity. As science continues to establish a strong role in conservation, such tools are being applied in tandem with traditional management techniques in the hopes of addressing the causes of species and ecosystem decline. As the biodiversity crisis deepens, conservation tools have been modified and developed, allowing a shift from reactive to proactive, ad hoc to strategic, emotional to science based, and small scale to large scale, with an increasing emphasis on process rather than pattern (Meffe and Carroll 1997; Margules and Pressey 2000; Groves 2003).

The birth of modern wildlife conservation in the early 1900s came in response to the plight of individual species. In contrast, increased ecological understanding during the latter part of the century led to the consideration of ecosystems and biodiversity in a broader sense as both conservation targets and operational units of focus (Franklin 1993; Norse 1993; Redford et al. 2003). There has been a growing realization that the traditional focus on single species with a blind eye to the whole may result in limited, if any, conservation gains (Simberloff 1998). However, a shift to the ecosystem scale may result in lack of sufficient protection for some biodiversity elements. Furthermore, knowledge of the biology of individual species within a system typically exceeds that of the processes driving that system (Schwartz 1999). Despite a voluminous literature, both biodiversity and ecosystem conservation remain relatively abstract concepts, especially to the public (Entwistle and Dunstone 2000; McNeely 2000). Indeed, such concerns have been the impetus for "coarse filter–fine filter" conservation strategies that combine both approaches in various ways (Groves 2003).

The robust use of a species-centered approach to biodiversity conservation requires the selection of appropriate target species. Large carnivorous animals are often chosen as centerpieces of conservation efforts in both terrestrial and marine domains for a variety of reasons. Various life history characteristics, such as their low population densities, space-demanding habits, and position at the top of the food chain, make members of this group potentially useful tools for conserving a broad array of coexisting biodiversity (where they persist or could feasibly be reintroduced). Like the canary in the coal mine, large carnivores are generally the first elements to disappear in a given system (Gittleman et al. 2001b; Pauly and Maclean 2003). Finally, because of the reliance of large carnivores on large spatial and temporal scales, efforts focused on their conservation may hold a key to reconciling the tension between coarse- and fine-scale conservation strategies. As such, they are often thought to provide a "useful entry to operationalize large-scale long-term conservation" (Clark et al. 1996: 396).

Historically, large carnivores have been viewed in negative or strictly utilitarian terms (Kellert et al. 1996; Lavigne et al. 1999). Gradually, the perception of top predators as impediments to conservation (e.g., through their wanton destruction of favored human prey) has given way to one in which the same animals are recognized for their broad public appeal. The nature of scientific inquiry regarding the role of predation has gone through a similar evolution, from a focus on the act of predation to one that seeks to understand the impact of top predators on ecosystems. Development of conservation approaches based on large carnivores, however, rests on a variety of assumptions that are untested. One of the most important of these assumptions is that carnivore conservation will necessarily result in the conservation of other biodiversity elements (Linnell et al. 2000; Hooker and Gerber 2004).

This chapter explores ways in which large carnivorous animals have been used by conservation practitioners as tools to increase the efficiency of attaining conservation goals. My specific objectives are to (1) review the types of tools and examples of conservation initiatives that have deployed large carnivores as tools, (2) provide a framework to categorize these tools, and (3) examine the rationale and underlying assumptions behind them. Rather than providing an exhaustive review, I analyze and discuss various examples that represent the range of carnivore-based conservation tools. Although these examples include both terrestrial and marine ecosystems, the terrestrial realm is emphasized. This imbalance stems in part from my expertise, but also from the fact that some of the tools reviewed here are not widely adopted in marine systems. This may be due either to the relative infancy of conservation biology in the marine realm or the fact that inherent ecological differences between marine and terrestrial systems preclude simple application of models developed in one system to the other (Carr et al. 2003; Sanderson et al. in prep.).

Large Carnivorous Animals as Conservation Tools

Conservation initiatives focusing on large carnivores can be divided into two major groups: those that focus entirely on such species with little attention paid to the system of which they are a part, or those that focus on carnivores as a tool for conserving the biodiversity in which the species is found. A large carnivore can serve simultaneously as both a target of and a tool for conservation action. This occurs when the specific objective of a conservation action is the protection of a particular carnivore, with an additional impetus the assumption that its conservation will achieve conservation for other biodiversity elements as well. The important distinction between these two objectives behind large carnivore conservation is in how conservation success is measured.

The deployment of large carnivorous animals as conservation tools can be grouped into six broad categories (Table 3.1). Many are typically associated in the literature with well-known identifiers (such as "indicator," "umbrella," and "flagship"; see Table 3.1). However, because the associated definitions are not consistent and multiple meanings are used for the same term (see Simberloff 1998; Caro and O'Doherty 1999; Leader-Williams and Dublin 2000), I focus here primarily on the nature of the tools themselves rather than the jargon or terminology with which they are generally linked (Caro and O'Doherty 1999).

Table 3.1

Categories and descriptions of conservation tools that employ large carnivorous animals for the purpose of achieving conservation of biodiversity

Tool Category	Common Identifier(s)	How Carnivore Is Used	Objective(s) of Conservation Action[1]
1. Ecosystem Conservation	Umbrella species[a] Keystone species[b]	Conserve or maintain large carnivore population(s)	Focused conservation of large carnivores to ensure conservation of biodiversity
2. Ecosystem restoration	Keystone species[b]	Reintroduce large carnivore	Restoration of biodiversity
3. Conservation symbols	Flagship species[c]	Use large carnivore as icon, emblem, or symbol of conservation initiative	Harnessing public support for biodiversity conservation
4. Identification of priority areas for conservation action and/or protection	Focal species[d] Umbrella species[a]	Large carnivore presence and/or habitat associations inform placement of priority conservation areas	Identifying areas for priority action to conserve biodiversity
5. Site-based conservation planning	Focal species[d] Umbrella species[a]	Large carnivore habitat and area requirements inform size and configuration of conservation areas	Manage or conserve existing areas for biodiversity conservation
6. Monitoring the status of biodiversity	Indicator species[e]	Routine evaluation of large carnivore presence or population status	Monitoring erosion of biodiversity, ecological integrity, or status of biodiversity

[1]The particular objectives of initiatives are not always stated.

[a]Roberge and Angelstam 2004; [b]Simberloff 1998; [c]Leader-Williams and Dublin 2000; [d]Lambeck 1997; [e]Caro and Doherty 1999

Category 1: Ecosystem Conservation

The first example of the use of large carnivorous animals as conservation tools spotlights these species, but has a larger goal of preventing biodiversity loss. The focus on carnivores takes many forms—from research projects to public relations campaigns—but all have in common the underlying supposition that the continued persistence of large carnivores will maintain or enhance the status of biodiversity.

As an example, the primary goal of the Large Carnivore Initiative for Europe is "to maintain and restore . . . viable populations of large carnivores as an integral part of ecosystems and landscapes across Europe" (LCIE 2004). However, the initiative also goes on to note that "[if] their habitats are successfully preserved, many other animals and plants, together with some of Europe's most important habitats and ecosystems, will benefit" (LCIE 2004). Campaigns and research projects built around large carnivores typically provide explanations of why such species are important, or why the public should care about their continued persistence. For instance, sharks "play vital roles in marine ecosystems. [Their] depletion or removal may lead to increases or declines in other species, with unpredictable consequences for ecosystems. It is likely that the removal of significant numbers of sharks will affect numerous species below them in the food chain" (Watts 2001: 11). The World Wildlife Fund–Canada Large Carnivore Conservation Strategy views the status of carnivores as a "useful point of entry to a broader set of conservation issues," emphasizing that "although our primary goal is to strengthen the position of carnivores in the natural community, the strategy is inclusive of all wildlife species" (Paquet and Hackman 1995: 1–2).

Other examples of large carnivore–focused campaigns that belong in the same tool category are initiated in reaction to predator control programs. In marine and terrestrial systems alike, the species that are viewed as threatened by predators are usually prey species of particular interest to humans, such as game species in terrestrial systems and valuable food species in marine systems. As public attitudes toward carnivores have shifted over time, proposals to control predators have sparked increasing levels of controversy, often being subjected to considerable public scrutiny and heated environmental campaigns (Dunlap 1983; Messmer et al. 2001). Frequently, opponents of predator control invoke broader ecosystem arguments to justify their stance (Dunlap 1983; Messmer et al. 2001). It was not until a clearer picture of the mechanisms of population regulation developed that scientists began to defend predators as an "essential element of nature," rather than "a constant drain on the prey population" (Dunlap 1983). Enhanced public awareness of marine conservation issues as well as scientific knowledge concerning the complex interrelationships among ecosystem components has likewise led to increasing calls for protecting large oceanic predators as an important part of protecting global marine biodiversity (Baum and Myers 2003). Concerns about beach netting programs to reduce interactions between sharks and swimmers, for example, are increasingly focused on the ecosystem and species-level costs of removing sharks from coastal ecosystems (Cliff and Dudley 1992).

Category 2: Ecosystem Restoration

The term "conservation" implies maintenance of a certain set of conditions, but in many areas, conservation goals can only be attained after restoration to a previous state. Such restoration often includes species reintroduction. Although only possible once a shift in general attitudes toward predators took place in the middle of the 20th century, many large carnivore reintroductions have been attempted in terrestrial environments since the 1940s, with mixed success (see Breitenmoser et al. 2001). Today, large carnivore reintroduction proposals figure prominently in conservation strategies for North America and Europe, where many top predators were exterminated over a century ago (e.g., DeBoer 2000; CREW 2004; Defenders of Wildlife 2004; LCIE 2004).

The primary motivations for reintroductions of large carnivores have been to restore a missing element of the original fauna and to correct an ethical wrongdoing (Warren 1997; DeBoer 2000; Breitenmoser et al. 2001; Nemethy 2002). Increasingly, however, reintroduction proposals are being justified as a scientific prerequisite for successful ecosystem restoration based on the presumed top-down influence of large predators (Warren et al. 1990; Soulé and Noss 1998; Terborgh and Soulé 1999; Noss 2001; Pyare and Berger 2003). Such a shift in reasons for reintroduction has come about not only because of research results documenting the ecological roles of large carnivores and the secondary ecological impacts that can result from their removal (Terborgh et al. 1999; Steneck, this volume) but also because opposition to such reintroductions may require a stronger argument than simply a desire to return the extirpated species (Warren et al. 1990). A prime example is the restoration of wolves (*Canis lupus*). In the Northeastern United States, burgeoning deer populations are seen by many as evidence of the impact of extirpation of large predators. In this light, the restoration of wolves would address not only the deer problem but associated ecosystem problems: "Restoring wolves would increase overall species diversity and help restore the balance of nature. Like many other ecosystems, those in the Northeast will not regain full ecological integrity until its [*sic*] top predator is restored" (Defenders of Wildlife 2004). Other large carnivores have likewise been introduced for reasons that reach beyond recovery of that species. Bobcat (*Lynx rufus*) introduction onto Cumberland Island, Georgia, from where they had been extirpated since the early 20th century, was in part initially justified from the standpoint of "restoring ecological control over several species of exotic and native herbivores by restoring a native predator to the island's ecosystem" (Warren et al. 1990: 582). Similarly, a proposal to reintroduce wolves to Japan asserted that the return of wolves would serve as a solution to wildlife pestilence (chiefly monkeys and deer), and a healing of Japan's moribund forests, which have been damaged by the disappearance of this top predator (Knight 2003).

In the marine conservation realm, arresting the decline of large predators such as sharks and marine mammals is more of an emphasis than is reintroduction. Methods for rearing and maintaining many marine species in captivity are as yet unknown, hindering scientists' ability to use reintroduction as a conservation tool for large marine carnivores and marine systems (Norse 1993). One exception is sea otters (*Enhydra lutris*), which have undergone a number of translocation attempts in coastal waters of western North America, with varying degrees of success (Jameson 1998). The impetus for such reintroductions has been focused more on replenishment of individual populations than ecosystem restoration.

Category 3: Conservation Symbols

Large carnivores are commonly used as symbols to promote conservation. Many species—including large cats, marine mammals, sharks, wolves, grizzly bears (*Ursus arctos*)—today have a global appeal that can be harnessed as effective symbols, especially in fundraising (Entwistle et al. 2000). They provide the "focus, rallying point and command centre" of conservation efforts promoted by many conservation organizations (Leader-Williams and Dublin 2000). The distinction between conservation symbols and other tools discussed here is not a clean one; in many situations, researchers and practitioners are drawn to these species for the same reasons as the general public. The difference, however, lies in the premise for the selection of large carnivores as conservation symbols, in this case being explicitly based on the species' high visibility or charisma, rather than for their ecosystem roles (as in Category 1 tools).

Large carnivores are used strategically as conservation icons by conservation organizations not only to promote conservation of the species in question but also to leverage protection for biodiversity or ecosystem protection. For example, a description of the World Wildlife Fund Global Species Programme (WWF 2004) notes the selection of flagship species that "act as ambassadors for a natural habitat, issue, campaign, or environmental cause. By focusing on, and achieving conservation of that species, the status of many other species which share its habitat—or which are threatened by the same threats—may also be improved." Large carnivores have also been deployed as "poster animals" to support broader regional conservation objectives. In the early 1990s a conservation initiative was founded to build a network of wildlife corridors and protected areas throughout the length of the Central American isthmus. The project was originally named "Paseo Panthera," to symbolize the wide range and space-demanding needs of mountain lions (*Puma concolor*) and hence "the need for continuity of natural environments throughout Central America if this region's stupendous biodiversity is to be preserved" (Coates and Carr 2001: xi). It is interesting to note that the conservation activities surrounding this enormous effort have not focused specifically on pumas beyond their symbolic value. In later years, after the goals of the project had shifted from preservation of biodiversity to the integration of sustainable development with ecological protection, it was renamed the Mesoamerican Biological Corridor (Illueca 2001). A further example is provided by license plate programs such as in Florida, where the public can pay more for a panther-bearing license plate to support statewide biodiversity conservation efforts (Simberloff 1998; Maehr et al. 2001b).

Category 4: Identification of Priority Areas for Conservation Action and/or Protection

Strategic planning has become a central tenet of conservation action that replaces the ad hoc approaches of the past with more proactive ones (Margules and Pressey 2000; Groves 2003). The goal of such planning is generally to maximize biodiversity protection in landscapes that are impacted by human activity. Conservation planning takes place in two stages: (1) priority setting and (2) the development of operational strategies (Ginsberg 2001; Redford et al. 2003). Large carnivores often figure prominently in both.

Establishing conservation priorities has become increasingly critical in the face of dwindling resources. Large carnivores often play a central role in such exercises by providing an objective means of determining location of conservation areas. Although some priority-setting exercises are specifically directed toward large carnivores themselves, such as tigers (*Panthera tigris*) (Wikramanayake et al. 1998), jaguars (*Panthera onca*) (Sanderson et al. 2002b), and African carnivores (Mills, et al. 2001), the added value of a focus on top predators as a means to achieve broader conservation goals is often mentioned. In the case of tigers, for example, "landscape level planning in representative habitat types also contributes greatly to biodiversity conservation at regional and continental scales . . . [and] reinforces the role of these mammals as umbrella species for conservation of diverse species assemblages associated with them" (Wikramanayake et al. 1998: 875).

Several conservation organizations have engaged in global priority-setting exercises for the purpose of identifying regions that are of particular value in conserving the planet's biodiversity (see Redford et al. 2003). In some cases, large carnivores have been used to aid in the selection of these areas. For example, an underlying premise of the WWF's ecoregional priority-setting initiative is the processes that sustain biodiversity, such as intact predator–prey assemblages (Olson and Dinerstein 2002). In one WWF planning exercise conducted in the Indo-Pacific region, ecoregions were assigned to five categories that reflected their current conservation condition, the descriptions of which all contained mention of the condition of top-predator communities (Wikramanayake et al. 2002). These ranged from "relatively intact" regions where top predators occurred at densities within the natural range of variation, to "critical" or "extinct" regions where top carnivores were absent or nearly absent. Another initiative that identified tracts of relatively undisturbed forests as priority conservation areas was the Frontier Forest Initiative of the World Resources Institute (Bryant et al. 1997). In this approach, one of the criteria identifying intact forested areas was that they were of "sufficient size to support ecologically viable populations of the largest carnivore and herbivores associated with that particular ecosystem, although they might not actually contain those species" (Bryant et al. 1997: 39). The rationale for use of large carnivores in this case was that their large area requirements would meet the habitat needs of most other co-occurring species.

Similar approaches have been used to pinpoint priority areas for conservation in marine environments. The identification of higher predator foraging distribution and hotspots of increased biological productivity has driven identification of marine protected areas (Hyrenbach et al. 2000; Price 2002; Hooker and Gerber 2004). For example, a recently conducted priority-setting exercise for the Bismarck-Solomon Sea included a focus on large pelagic predators in identifying key sites for conservation (WWF 2003). Increasingly, marine predators are being used as the basis for marine conservation areas; for example, whale and other marine mammal sanctuaries listed by Hooker and Gerber (2004).

Category 5: Site-Based Conservation Planning

Once landscapes are selected for protection and/or conservation action, site-specific planning is required (Noss et al. 1999; Margules and Pressey 2000; Noss 2000; Sanderson et al. 2002c) to answer questions of protected area size, configuration, connections with other important areas, and conservation of matrix (unprotected) areas. Where biodiversity conservation is a goal of conservation planning, it requires consideration of the biological requirements of resident species. Large carnivores are often regarded de facto as focal species (*sensu* Lambeck 1997; Sanderson et al. 2002c) for this purpose by virtue of their relatively large spatial requirements, presumed disproportionate ecological role, and public appeal (Noss et al. 1999). Examples where large carnivores have been used as focal species for planning in terrestrial habitats include habitat and distribution modeling (Carroll et al. 2001), conservation area design (Carroll et al. 2003), spatially explicit population viability modeling (Noss 2000; Carroll et al. 2003), and evaluation of landscape permeability (Singleton et al. 2002). A recent conservation planning exercise for the Patagonian Shelf Large Marine Ecosystem was based on the selection of a suite of seascape species and explicitly considered the protection of trophic relationships and the conservation of top predators along with intermediate predators and prey species (Sanderson et al. in prep.).

Category 6: Monitoring the Status of Biodiversity

The presence of large carnivores is often heralded as a sign of ecological integrity—the intact and resilient condition of a landscape (Groves 2003). Such a condition is meant to represent a standard from which to measure progress (or regression) relative to a baseline (Karr 2000). Underlying this is the recognition that human activity alters not just patterns of biodiversity but also ecological processes (Karr 2000; Noss 2000). Practically speaking, determining whether a system possesses its historic fauna and ecological processes is an important challenge for the conservation practitioner. One approach to addressing this challenge involves selecting a species whose population levels will help in determining how far the system has deviated from a chosen baseline (Noss et al. 1996; Karr 2000). Routine evaluation of traits of such species (e.g., presence, density, reproductive success, etc.) might represent an index of the ecological condition and biodiversity status of the landscape they inhabit.

If large carnivores are indeed ecologically "pivotal" species in that they are among the first to disappear in the face of human intrusion and are instrumental in structuring ecosystems (Noss et al. 1996; Pauly et al. 1998), it makes a good deal of theoretical sense to consider their status an indicator of integrity of a given system. If the frequency with which such animals are referred to as appropriate indicator species were used as a measure of their actual employment in this capacity, just about every monitoring program would have resident large carnivores heading its list of indicator species. Despite common reference to carnivores as appropriate indicator species, practical considerations usually mean that carnivores do not well suit this role (Hilty and Merenlender 2000).

Although the true practical value of large carnivores as indicators has rarely been shown, they are nevertheless used in some conservation monitoring programs in both marine and terrestrial systems. For example, grizzly bears, black bears (*Ursus americanus*), cougars (*Puma concolor*), wolves, and wolverines (*Gulo gulo*) are formal indicator species in national parks in Canada (S. Woodley, pers. comm. 2003; Parks Canada 1997). In this case, the basis for selection of these species as indicators (through monitoring mortality/natality, immigration/emigration, and population viability) relates to the fact that large carnivores are among the first species to be eliminated by humans from a system. Their decline can serve as early warning for current or impending threats to other elements of biodiversity (Woodley 1997). Monitoring is therefore a feedback system for park management, with ecological integrity considered a "management endpoint" (Woodley 1993). Sharks, whales, and seabirds are indicator species in various marine protected area management plans, including the Northwest Hawaiian Islands National Marine Sanctuary (sharks) and the Great Barrier Reef Marine Park (whales).

Assumptions Underlying the Use of Large Carnivores as Conservation Tools

The increasing interest in conservation activities reaching beyond individual species to serve the broader interests of biodiversity conservation has resulted in a need for strategies underpinned by strong science. The multiple rationales for using carnivores as conservation tools with their underlying, often unstated, assumptions make overall analysis difficult (Andelman and Fagan 2000).

I have reviewed the six categories discussed above and grouped the assumptions underlying them into three categories: value, efficiency, and functionality (Table 3.2). The first category pertains to the subjective worth of a species from the human point of view, whereas the latter two relate more to objective biological characteristics of the species and the ecosystems of which they are a part. The differences between the types of assumptions that govern use of large carnivores as conservation tools are rooted in the distinction between the strategic and ecological roles of these animals (Leader-Williams and Dublin 2000). Some pertain to characteristics of the species themselves, whereas others are concerned with the relationship between their conservation and that of other elements of biodiversity (see Table 3.2).

Table 3.2

Types of assumptions behind the use of large carnivorous animals as conservation tools

Type of Assumption	Assumptions Concerning Characteristics of Large Carnivores	Assumptions Concerning the Link between Large Carnivore Conservation and Biodiversity Conservation
Value	Large carnivores have inherent value as a part of biodiversity and deserve protection	Large carnivores will garner greater resources in support of conservation than other less popular elements of biodiversity
		Members of the public (and potential donors) value the rest of biodiversity as much as they do large carnivores so that they will be happy to support broader biodiversity if giving targeted support to large carnivores
Efficiency	The area required for persistence of large carnivore population is larger than that required by most other species	The ranges of large carnivores are large enough to encompass the habitats of other sympatric species, hence large-carnivore conservation will ensure the protection of the rest of biodiversity sharing the same range
	Large carnivores are among the most vulnerable of species to human perturbations and will be eliminated faster than other species	Focusing on large carnivores will provide an early warning signal of the erosion of biodiversity
Functional	Large carnivores play critical, functional roles in the system of which they are a part	The disappearance or removal of large carnivores will result in a degraded and simplified ecosystem
	Large carnivores affect the productivity, juvenile survival, and/or breeding density of prey populations	The return of large carnivores to a system will result in the restoration of that system

Value

The premise behind this most straightforward set of assumptions is that large carnivores have intrinsic value and hence deserve to be conserved on that basis alone. Several tools, however, go a step further by assuming that they have greater value than many or most other components of biodiversity. For example, an assumption that generally governs the use of large carnivores as symbols is that the public values carnivores disproportionately to other species and will thus be compelled to provide support, which can then be used for the general cause of biodiversity conservation.

Efficiency

A second set of assumptions is rooted in the premise that large carnivores can act as surrogates or representatives for biodiversity, such that focusing on their conservation will be the most efficient way to bring about conservation of all elements of biodiversity (Lambeck 1997; Andelman and Fagan 2000; Roberge and Angelstam 2004). Such species may be more sensitive to ecosystem deterioration or have more stringent ecological requirements, hence focusing efforts on this group should take care of the needs of other elements of biodiversity that are less sensitive and that have more modest requirements. Thus key focal species are often selected for efficient concentration of planning or monitoring as representatives for the remainder of biodiversity, under the assumption that this will indeed result in the conservation of other species (Lambeck 1997; Carroll et al. 2001; Groves 2003; Roberge and Angelstam 2004).

Functionality

There is increasing recognition that conservation efforts must be focused on the mechanisms or processes that maintain biodiversity (Smith et al. 1993). The overall assumption behind using large carnivores as tools in this sense relates to the importance of the ecological roles they are thought to play. For example, top-down forces may be exerted by top predators on species at lower trophic levels through myriad ecological and evolutionary interactions (Steneck, this volume), making

carnivores critical elements in system conservation. In cases where large carnivores have been extirpated, such forcing will be drastically weakened, and ecosystems will subsequently become degraded and simplified (Terborgh and Soulé
1999). As discussed previously, a corollary assumption is that a return of large carnivores to a system will restore degraded systems (e.g., Knight 2003).

Testing the Assumptions behind Using Large Carnivores as Conservation Tools

If the preceding assumptions are true, then countless other species would benefit
from conservation actions focused on large carnivores. In truth, however, such assumptions have seldom been tested, making the conservation benefits of a focus
on large carnivores in many cases speculative, even if the assumptions have a
strong theoretical basis. The remainder of this chapter examines the conditions
under which the three types of assumptions are most likely to be satisfied, and
where scientific input is most needed in guiding the employment of large carnivores as biodiversity conservation tools.

Meeting Value Assumptions

A basic tenet of conservation biology is the inherent value of all species residing
on this planet (Soulé 1985). That certain species—such as large-bodied, charismatic, and threatened mammals—are more attractive to the average person has
been well documented in polls. Marketing strategies used to attract donations
from the public to particular programs or conservation causes have been found to
enjoy more success if certain species serve as the focus of such campaigns (Leader-
Williams and Dublin 2000). Hence, assumptions behind the use of large carnivores
as conservation tools would seem to be supported.

There can, however, be considerable variation in what is perceived as charismatic and worth saving (Leader-Williams and Dublin 2000; Linnell et al., this volume), indicating that careful attention needs to be paid to whom the conservation
message is being directed. A global conservation symbol cannot be assumed to
work at the local level because large carnivores are often most attractive to those

residing in countries or regions that do not possess them (Kellert et al. 1996; Entwistle and Dunstone 2000). For example, it may be necessary to make use of different communication tactics targeting urban people than those used on individuals who live close to the human–wildlife interface (Leader-Williams and Dublin 2000). A case in point is provided by wolverines in northern Canada, which are being employed as conservation symbols to engage urban Canadians in northern boreal conservation initiatives. Meanwhile, to many individuals living in areas where wolverines persist, these animals are regarded as pests and threats to hunting and trapping success (Ray, unpubl. data).

Meeting Efficiency Assumptions

In contrast to value assumptions, efficiency assumptions are more easily evaluated. The utility of large carnivores as surrogates for biodiversity is often taken as a given because of the relatively large area requirements of populations, which theoretically would encompass even more individuals of co-occurring species (Roberge and Angelstam 2004). The extent to which other elements of biodiversity benefit from attention to large carnivores depends, however, on the biogeographic characteristics of the region and the extent to which carnivore habitats overlap with those that are rich in biodiversity or vulnerable species (Noss et al. 1996). Indeed, empirical evidence has not been encouraging in this regard. Several studies have concluded that the habitat associations of large carnivores were not particularly congruent with (1) areas of high endemism (Noss et al. 1996), (2) habitats of other taxa (Kerr 1997), and/or (3) species of concern (Andelman and Fagan 2000). In spite of their often vulnerable status, large carnivores have a remarkable capacity to persist in degraded landscapes, provided that they are secure from human conflict and have sufficient food (Linnell et al. 2000; Karanth et al. 2004). A growing recognition of the general unsuitability of large carnivores as biodiversity surrogates is leading either to use of other umbrella species that are more sensitive to habitat condition to complement large carnivores (e.g., Carroll et al. 2001) or to deploying a combination of tools rather than exclusive reliance on large carnivores in an umbrella role (e.g., Noss 2003).

Despite these problems, the utility of large carnivores as biodiversity surrogates can still be of considerable value during the protected area design and zoning phases of conservation. The area requirements of the widest-ranging species and those most vulnerable in the face of human activity can serve as inspirations for conservation area design (Noss et al. 1996; Hooker and Gerber 2004). If large carnivores can provide a mechanism to delineate a large area that by virtue of its size will be protected from other threats, such as overfishing (Gell and Roberts 2003) or habitat degradation (Maehr et al. 2001b), then their utility as focal species is quite likely well founded.

Meeting Functionality Assumptions

Determining the validity of assumptions of functionality necessitates an understanding of the ecological role of large carnivores in any given system. Particularly in terrestrial environments, it is enormously challenging to determine the ecosystem role of top predators, not only because of the difficulty of studying them but also because of the logistical challenges of experimental designs that can test for their effects relative to other factors (NRC 1997; Terborgh et al. 1999). To comprehend the effects of large carnivore predation on one trophic level, let alone any cascades that might result (Estes, this volume), has proven to be a difficult task, with evidence of the top-down effects being considerably stronger in aquatic than terrestrial environments (Pace et al. 1999; Steneck, this volume).

Despite this difficulty, increasing evidence points to the emergence of trophic cascades when top predators are removed from both terrestrial and aquatic ecosystems (Pace et al. 1999). This provides support for a focus on large carnivores for maintaining interactions in natural communities. However, for at least three important reasons discussed following here, being able to meet the assumption that predators are critical components of natural communities does not necessarily provide assurance for the corollary premise that maintaining or restoring this element will ensure conservation of other elements of the ecosystem of which it is a part.

Functional Viability

Conservation of top predator populations is usually defined on demographic and genetic parameters rather than on functional roles (Pyare and Berger 2003; Soulé et al. 2003), which are very difficult to quantify (Groves 2003). In many places where large carnivores persist, their densities are far lower than in the past, calling into question their continued ability to interact significantly with other species (Jackson 2001; Redford and Feinsinger 2001; Myers and Worm 2003; Steneck, this volume). As a result, even a demographically viable population may easily be functionally extinct (Redford and Feinsinger 2001) or ecologically ineffective (Soulé et al. 2003).

There are at least two important obstacles in the way of achieving large carnivore population levels that are ecologically effective, and hence achieving functionality goals. First, although much research and effort has been directed at measuring population viability, our ability to detect the point at which top predators cease to perform ecological functions that maintain biodiversity lags far behind (Redford and Feinsinger 2001; Soulé et al. 2003). Second, and perhaps more problematic, is that, in many localities, the focus is on maintaining populations of carnivores at socially acceptable, rather than ecologically functional, densities (Mech 1996; Linnell, this volume). Such constraints could compromise biodiversity goals, even if the conservation or reintroduction of that species were deemed to be successful by other measures.

Spatial and Temporal Variation in Functionality

Focusing on a top carnivore in one system for the benefit of biodiversity conservation as a whole may bring about more advantages in some places than others. Understanding the functional role of a carnivore is complicated by the fact that ecosystem-level interactions vary with ecological context, making it difficult to make broad generalizations regarding the roles of large predators. The reintroduction of wolves in the United States provides a provocative example of this phenomenon.

Various negative biodiversity impacts can be clearly attributed to the absence of this top predator, as one significant species missing from the otherwise intact Greater Yellowstone Ecosystem (see Smith et al. 2003). Hence, Yellowstone was considered to be a natural place for wolf restoration that would serve broader biodiversity goals. Bolstered by the apparent success of the Yellowstone reintroduction and recolonization, as well as the beginning signs of associated ecosystem restoration that have since taken place (Smith et al. 2003; Berger and Smith, this volume), conservation organizations have been eager to apply lessons learned from this experience to other areas. In the eastern United States, for example, wolves have been hailed as one remedy for restoring function to the degraded ecosystem characterized by overpopulated white-tailed deer (*Odocoileus virginianus*) (Soulé and Noss 1998). However, the Northeast has a significantly different climate, vegetation, prey base, and land use history from that of Yellowstone (Mladenoff et al. 1999). Perhaps more significantly, the region has suffered a host of anthropogenic impacts, including extirpations of some other large mammals, significant land clearing, forest fragmentation and edge effects, burgeoning human populations, and a long history of human occupation. Although the extirpation of large predators during the latter years of the 19th century presumably caused negative ecosystem impacts, many other factors related to the long history of human land use have similarly been responsible for the erosion of biodiversity (McShea, this volume). Such factors make it less likely that the return of wolves would, by itself, serve the same biodiversity goals that it has in the western United States. A significant additional factor is that areas of deer overabundance are typically associated with high human populations, which are the least likely places for wolves to recolonize (Mladenoff et al. 1997). This does not discount the importance of wolves as top predators in ecosystems in which they once flourished, but it does raise the question as to whether they can, by reestablishing trophic cascades that have been lost, provide a vehicle for biodiversity conservation in a highly derived system (Pace et al. 1999).

Complexity Breeds Unpredictability

Forecasting the future nature and extent of community changes that might occur following the loss (or reestablishment) of a certain large carnivore is an even greater challenge than assessing whether a top carnivore has, or once had, an important functional ecosystem role in the present or past (Micheli et al. 2000). Discerning and addressing root causes of biodiversity loss is complicated by the complexity and unpredictability of ecological interactions and food web dynamics. Time lags and unique ecosystem characteristics, such as species and habitat diversity, and even weather, can further impede the ability of scientists to make other than general predictions (Smith et al. 2003). As already noted, in regions with long and complex human histories, it is difficult to untangle ecosystem changes caused by the loss of predators and those due to the presence of humans, or both. The direction and magnitude of impacts are even harder to predict in the marine context, for which particularly complex food webs are characteristic (Northridge and Hofman 1999; Yodzis 2001). It stands to reason that if cascading impacts of predator removal are challenging to predict, the same would be true regarding the consequences of restoring predators to a dysfunctional ecosystem (Peterson 2001).

Conservation Implications: Large Carnivorous Animals in Conservation Practice

The past century has witnessed a profound shift in our perceptions of the roles of top predators in ecosystems, with an increasing weight of evidence indicating that large carnivores can have important functional roles within ecosystems. This has translated into a corresponding zeal on the part of the conservation community to employ them as centerpieces in conservation strategies. The choice of conservation tools to combat biodiversity loss can profoundly affect the efficiency and likelihood of success (Salafsky et al. 2002). Successful implementation, however, often depends on the veracity of both implicit and explicit assumptions. The urgency of the extinction plight facing both carnivores and biodiversity often does not, however, leave room for careful testing of such assumptions.

Uncertainty about or lack of evidence regarding the functional role of top predators in some cases may signify that assumptions concerning their conservation value are invalid. On the other hand, the true functionality of a large carnivore, and hence its potential to serve as a conservation tool, may be masked by the complexity and hence unpredictability of ecosystem interactions, the inherent unpredictability of natural ecosystems, the temporal scale of inquiry, and the complicating actions of humans. This becomes problematic not only with respect to formulating appropriate action but in communicating the rationale of conservation activities to the public. Conservation programs that involve large carnivores often seek public endorsement, either because of the need to fundraise for such high-profile expensive projects or because of potential human–carnivore conflicts. It is important that such initiatives recognize the potential stumbling blocks that are gradually being uncovered by scientific inquiry; otherwise they risk perception of failure by the public.

A case study of the hazards of overselling the functional roles of top predators to the public is offered by Warren et al. (1990), who examined a bobcat reintroduction project on Cumberland Island, Georgia. In this case, the original
justification for the reintroduction was for deer population control, which became
the focus of media coverage. After only one year, even though the project had
reestablished a bobcat population, it was deemed a failure by some media because
the deer problem had not been solved. The lesson the authors took from this experience was that public support for the return of bobcats themselves should not
have been underestimated. Indeed, the communication strategy subsequently was
shifted to one that emphasized the premise for reintroduction of bobcats to "restore one aspect of that island's original fauna." The attention span of the media
and public does not often align with time lags that might be required for the
restoration of a predator's functional role.

The pressure for conservation practitioners to articulate scientific rationales
behind conservation actions becomes awkward in the face of the uncertainties inherent in trophic relationships and ecosystem dynamics. In some cases, a desire to
emphasize certain points, or anticipate points of opposition, can unfortunately
lead to conflicting messages. This stems from the complicated fact that carnivores
mean "simultaneously different things to different people, and sometimes also to
the same people" (Macdonald 2001: 527). Conservation initiatives that oppose
predator control and those that advocate for predator restoration offer some interesting examples of this phenomenon. For example, background articles provided on one website (NoSnare.org 2003) set up to increase public awareness about
government agency policies on coyote (*Canis latrans*) control gave conflicting messages. One article emphasized the importance of coyotes as "part of the ecological web in various communities because they help to regulate species at different
trophic levels" (Bekoff 2003), whereas another noted that predator control should
not be used as a means of maintaining strong deer populations by arguing that
"coyote populations are naturally regulated by deer, not the other way around"
and that "never has it been proven that these kills effectively impact deer populations, even on the most local level" (Jenssen 2003). Similar discordant messages
can be heard with regard to proposals on the table to reintroduce large predators,
which are compelled to argue for the restoration of the predator function while
at the same time refuting the notion that such predators will impact game populations (Phillips and Smith 1996).

Many underlying assumptions that form the basis of conservation action will be seized upon, applied, and generalized long before the science is sound—a particular concern, given a tendency among conservation practitioners to apply general principles in a "simplistic and uncritical" fashion (Noss 2000). We have seen here, however, that although such assumptions do at times rest on solid ground, there are many cases when they cannot be taken for granted. This makes it necessary to diversify the conservation toolbox, so as to alleviate the risk of relying solely on large carnivores to maintain or restore biodiversity. Through the process of adaptive management (Walters and Holling 1990), which can be designed specifically to allow action in the face of uncertainty, implementation will be allowed to test assumptions and revise them accordingly.

Meanwhile, none of this discussion diminishes the reality that large carnivores are often important functional ecosystem elements and have intrinsic value as an element of biodiversity. It is not always necessary for conservation practitioners to shy away from intrinsic value claims that require no clear scientific evidence to sustain them (Freyfogle and Newton 2002) and when clearly understood and recognized, can form an unshakeable foundation of conservation strategy (Maguire 1994). Both rational and nonrational arguments are needed to effect conservation, and should be complementary, not conflicting, elements to successful conservation efforts.

Summary

Research focused on large carnivorous animals during the past half-century has revealed the broad scales at which they perceive their environment, their particular vulnerability in the face of human intrusion, and their propensity to impart top-down control or limitation of prey populations. Such characteristics make this species group attractive for employment as species-based conservation tools. It is therefore a natural leap to suppose that focusing on top carnivores as surrogates should be able to offer protection to other elements of biodiversity with which they share their range. This chapter reviews the types of conservation tools in which large carnivores have been deployed in both terrestrial and marine environments, provides a framework to categorize them, and examines the rationale and underlying assumptions that characterize their use. The deployment of large carnivores in conservation strategies can be grouped into six broad categories: (1) ecosystem conservation, (2) ecosystem restoration, (3) conservation symbols, (4) identification of priority areas for conservation action and/or protection, (5) site-based conservation planning, and (6) monitoring the status of biodiversity.

The successful implementation of conservation tools to combat biodiversity loss rests on both implicit and explicit assumptions, based on the value, efficiency, and functionality of large carnivores. Although some assumptions are more easily met than others, others remain untested, making the conservation benefits of a focus on top predators in many cases speculative, and perhaps ineffective. Functionality assumptions are particularly problematic. Being able to meet the assumption that predators have critical ecological roles in natural communities does not necessarily provide assurance for the corollary premise that maintaining or restoring this element will ensure conservation of other elements of the ecosystem of which it is a part. Not being able to meet or even test for functions behind such conservation tools becomes an issue not only with respect to formulating appropriate action but also in communicating the rationale of conservation activities to the public. This makes it necessary both to evaluate carefully the premise behind the use of large carnivores as conservation tools, and to diversify the conservation toolbox, so as to alleviate the risk of relying solely on these animals as a means to maintain or restore biodiversity.

ACKNOWLEDGMENTS

I am most grateful to Joel Berger, Josh Ginsberg, Jay Malcolm, Reed Noss, Kent Redford, and Bob Steneck for significantly strengthening this paper with their insightful comments. Special thanks to Liz Lauck for her great help in bolstering the marine component, and to Joanna Zigouris for important support provided in the preparation of this manuscript.

PART II

The Scientific Context for Understanding the Role of Predation

Tremendous strides have been made in the past half century in understanding the behavior and ecology of many large carnivore species. It is only recently, however, that this science has extended beyond questions focused on the species themselves, to their relationships with other faunal and floral elements with which they co-occur. The authors of the six chapters in this section provide important examples of the types of information conservation practitioners will need to address both the imperilment of large carnivorous animals and the conservation of other elements of biodiversity.

The first three chapters present several of the best-known research projects that have examined the ecosystem-structuring role of large carnivorous animals in systems in which local predator extirpations provide a basis of comparison against as-yet intact systems. As a pioneer investigator of questions of this nature, James Estes (Chapter 4) took the essential leap from exploring the role of predation in relatively simple ecosystems with small apex predators to complex environments with larger-bodied predators at the top of the food chain. Dr. Estes and his colleagues have investigated the ecosystem impacts following both the demise and the recovery of sea otter populations in various coastal marine locales. Although this story has been told in many different forms, the scale of inquiry keeps expanding. In this volume, Dr. Estes takes a bird's-eye view of the research to date, offering strong evidence that trophic cascades extend far beyond the system in question, temporally as well as spatially.

In the second contribution, John Terborgh (Chapter 5) describes the results of an experiment made possible by the creation of Lago Guri, a hydroelectric impoundment in Venezuela. Dr. Terborgh reveals how a decade of research has rejected multiple alternate explanations and provided support to the prevailing hypothesis that changes to biotic communities on the islands since their isolation were directly attributable to the loss of top predators. He uses this study as a centerpiece in revisiting a set of competing theories regarding the role of predation in structuring ecosystems.

Shifting to the North American Greater Yellowstone Ecosystem, Joel Berger and Doug Smith (Chapter 6) review the substantial body of research that has documented the myriad ecosystem effects following the extirpation and subsequent reintroduction of wolves. This situation, combined with the relative lack of confounding influences and systematic monitoring, has allowed a clear understanding of the ecological roles of this pinnacle predator, thereby permitting a clean and careful examination of the question, Do wolves matter?

Robert Steneck and Enric Sala (Chapter 7), review a lesser-known set of studies that offer equally compelling evidence regarding the role of large predatory fish in structuring coastal marine systems. As Dr. Steneck pointed out in his framing chapter in the introductory section of this book (Chapter 2), it is largely in marine systems that there has been consistent documenting of trophic cascades caused by the impacts of predators. Drs. Steneck and Sala review evidence from the world's tropical and temperate marine systems, providing multiple examples of the ecological roles of marine carnivores.

The last two chapters in this section examine a different dimension of the science of understanding the role of predation. The first of these, by William McShea (Chapter 8), examines one hallmark of ecosystem decay—the overabundance of ungulate populations. Some of the direct impacts of uncontrolled ungulate populations are clearly evident even to nonscientists; the more subtle, but equally damaging, indirect effects are further detailed in this chapter. Although it is tempting to attribute such changes directly to the loss of predators, Dr. McShea reveals the complexity of the situation, discussing the confounding role of human-induced changes in the landscape. This means that healing the system will be more challenging than might otherwise be expected (e.g., by recovering extirpated top predators).

Although most investigators study the impacts of one species at a time, Rosie Woodroffe and Joshua Ginsberg (Chapter 9) discuss "megapredator" guilds, examining for the first time how sets of top predators might collectively exert influence on ecosystems. Despite the paucity of available data, they assemble valuable information to examine important questions, including assessing the conservation implications of possible redundancy within guilds of top carnivores.

CHAPTER 4

Carnivory and Trophic Connectivity in Kelp Forests

James A. Estes

In what many people see as the wholesale unraveling of global biodiversity, large carnivorous mammals are often among the first species to disappear (Redford, this volume). It follows that human actions intended to preserve large carnivorous mammals should slow or even curtail this unraveling. The large carnivores can thus be thought of as de facto "umbrella species" for the conservation of biodiversity.

The preceding logic, which seems like a good starting point for imagining why large carnivores are especially important to biodiversity conservation, does not require that they occupy a position of functional significance in their associated interaction webs. A reserve or conservation unit conceived and designed around large carnivores as umbrella species need not even be populated with these creatures to maintain its biological integrity. And since large carnivores cause so many societal problems (like eating people and the things people eat), some might argue that their presence is not essential.

The question of carnivore essentiality depends on the nature of interspecies connectivity. If species are connected to one another and their physical environment principally through bottom-up forcing processes (Hunter and Price 1992), the nonessentiality argument is reasonable. If, however, top-down forcing processes are important in population regulation, that argument is at best overly simplistic and at worst wrong. Despite abundant evidence to the contrary, a strict bottom-up perspective of interspecies connectivity is probably close to the layperson's view of food web dynamics. This perspective also has strongly influenced the ways many ecologists have studied everything from population regulation to ecosystem dynamics. Thus it is not surprising that large carnivore conservation

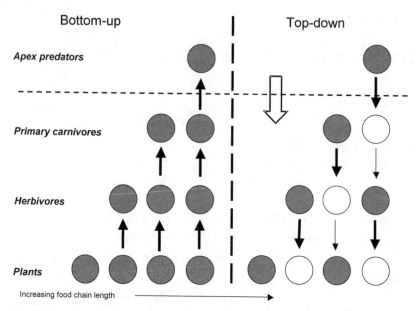

Figure 4.1

Schematic showing the interactions among trophic levels in food chains of varying length under bottom-up (left of vertical dashed line; upward pointing arrows) and top-down (right of vertical dashed line; downward pointing arrows) control. Bold and light arrows indicate strong vs. weak interactions, respectively. Filled and open circles indicate that species within that trophic level are limited by resources vs. disturbance by their consumers, respectively. Horizontal dashed line shows the consequences of removing the highest trophic level consumers by "fishing down" the food web (*sensu* Pauly et al. 1998). Hairston et al.'s (1960) Green World Hypothesis was built around a view of the world depicted by the three-tiered food chain to the right of the dashed line. Fretwell's (1987) expanded theory of food chain dynamics is also illustrated to the right of the dashed line by the alternating weak to strong plant–herbivore interactions as increasing food chain length alternates between an odd and even number of trophic levels. Strong nutritional connections between prey and their consumers under a strict bottom-up forcing scenario are depicted to the left of the vertical dashed line.

has focused primarily on the maintenance of minimum viable populations and, to a lesser extent, the value of these species as umbrellas for the conservation of other species.

Bottom-up linkages are clearly important in connecting consumers with their prey. However, the distribution and abundance of species and populations are not determined solely by bottom-up forcing. Unequivocal evidence for top-down forcing has existed for decades. Farrow's (1916) rabbit-proof cage experiments in England provided a dramatic early example. Yet it wasn't until 1960, when Hairston, Smith, and Slobodkin (HSS) advanced their famous Green World Hypothesis (Hairston et al. 1960)—a proposal that the immense abundance of plant biomass depends on predator limitation of herbivore populations—that the roles of top-down forcing and apex predators seriously entered the intellectual fore of community and ecosystem dynamics. Fretwell (1987) refined HSS by pointing out that, for systems under top-down control, plant–herbivore interaction strength should alternate between weak and strong depending on whether the number of trophic levels is odd or even (Fig. 4.1). Paine (1980) referred to these downward-reaching indirect interactions as "trophic cascades."

This chapter considers the ecological connections between large apex predators and other species. My focus will be on two related but largely unstudied issues. One is the relative importance of direct versus indirect food web pathways (Fig. 4.2). Other than via trophic cascades, ecology has paid very little attention to the question of how carnivory might influence associated species through indirect food web pathways. The other issue concerns the dimensions of spatial and temporal scale to food web connectivity. That is, how distantly in space and time must one look to properly understand how food webs work? In particular, to what degree do important interaction pathways transcend the periods of our field studies and the boundaries between what we perceive as distinct ecosystems? My treatment of these issues has been strongly influenced by the work my colleagues and I have done over the past three decades on the dynamics of trophic interactions between sea otters (*Enhydra lutris*) and kelp forest ecosystems in the Northeast Pacific Ocean. Our findings illustrate some of the complex ways in which other species are influenced by carnivory, thereby establishing the potential importance of carnivores to biodiversity conservation.

The Sea Otter–Kelp Forest Ecosystem

Sea otters and kelp forests provide unusual insights into the complex interplay between mammalian carnivory and food web connectivity. There are several reasons for this. One is that since the early 1970s when my colleagues and I first began

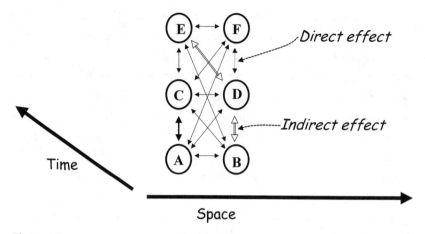

Figure 4.2

Schematic showing the array of potential interspecies linkages in a simple six species food web with two autotrophs, two primary consumers, and two predators. Indirect linkages (exemplified by the open arrows) are those that contain one or more intermediate species.

studying the ecological dynamics of the sea otter–kelp forest system, the questions we asked were formulated with top-down forcing very much in mind. In retrospect, this way of thinking resulted in part from HSS and in part from earlier field studies of lakes (Brooks and Dodson 1965) and temperate-latitude, rocky intertidal communities (Connell 1961a; Paine 1966) that provided empirical support for top-down control. Furthermore, Paine and Vadas (1969) had just demonstrated that sea urchin grazing could denude macroalgal stands. We knew that sea otters consumed sea urchins, so a significant indirect linkage from sea otters to sea urchins to algal stands was easy for us to imagine. Finally, history provided a natural experiment for exploring the sea otter's role in kelp forest ecosystems. This experiment was set in motion by the Pacific maritime fur trade, which reduced once-abundant sea otters to near extinction. The surviving remnants began to recover with cessation of hunting in the early 1900s. Moreover, recovery was asynchronous in space and time, due to the fortuitous locations of the remnant sea otter colonies and the species' limited dispersal ability. Ocean passes further inhibited the spread of sea otters between islands (Kenyon 1969). By the early 1970s, when I arrived on the scene, these processes had created a highly fragmented sea otter distribution across southwestern Alaska's Aleutian archipelago. We thus were able to observe the role of sea otter predation in kelp forest ecosystems by contrasting otherwise similar islands with and without sea otters, and by monitoring these systems through time as sea otters became reestablished and their numbers increased and later declined.

Sea Otter Predation—Direct Effects and Trophic Cascades

Shallow reef systems with and without sea otters differed markedly (Fig. 4.3). Where otters were absent, sea urchins were abundant and kelps were rare. Where otters were abundant, sea urchins were rare and kelps were abundant (Estes and Palmisano 1974). These patterns are now known for other parts of the North Pacific Ocean (Watson 1993; Estes and Duggins 1995). Additionally, phase shifts between kelp-dominated and urchin-dominated community states have been observed as otter numbers have waxed and waned over the past three decades. For instance, shallow reef habitats at Attu Island (westernmost of the Aleutians), Torch Bay (in southeast Alaska), and along the west coast of Vancouver Island all shifted from urchin barrens to kelp forests as sea otters repopulated these areas. Similarly, rising sea urchin populations devoured long-established kelp forests in parts of southwest Alaska as sea otter numbers plummeted in the 1990s, apparently because of increased killer whale (*Orcinus orca*) predation (Estes et al. 1998). These findings provide the main evidence for a sea otter–induced trophic cascade.

Indirect Effects

The sea otter's influence on coastal ecosystems does not stop with herbivore–autotroph interactions at the bottom of the trophic cascade. Kelp forests exert a variety of important influences on shallow reef systems. For instance, kelps are important photosynthesizers, fueling high rates of secondary production (Duggins et al. 1989); they create three-dimensional structure and habitat for other species (Dayton 1985; Foster and Schiel 1985); they reduce light penetration to the seafloor (Reed and Foster 1984); and they exert drag on the water column, thus attenuating coastal currents and the force of waves on shorelines (Jackson and Winant 1983). For these and other reasons, "spin-off" effects from the sea

Figure 4.3
Typical kelp (a) and urchin barrens (b) at Amchitka Island. The photographs are both of the same approximate site: (a) in 1971 (by P. K. Dayton) when sea otters were still abundant, and (b) in 1999 (by M. Kenner) following the sea otter population collapse.

otter–induced trophic cascade were easy to imagine. My colleagues and I have attempted to measure some of these effects by contrasting island systems with and without sea otters, and by monitoring change at particular sites as sea otter numbers waxed and waned through time. This approach is identical to that used in our earlier studies of the sea otter–induced trophic cascade, except our focus here has been on different parts of the food web. The potential number of interaction web pathways that could be influenced indirectly by sea otter predation is vast. We have studied but a few of these that appeared as though they might be linked to the otter–urchin–kelp cascade in some simple or obvious way.

Our most extensive and successful search for indirect effects has centered on differences in the source of organic carbon and rate of production between otter-dominated and otter-free systems. Not surprisingly, these differences are striking. Kelps and other macroalgae generate roughly three quarters of the organic carbon fixed by photosynthesis in coastal waters at otter-dominated islands (Duggins et al. 1989). Otter-free islands, in contrast, derive most of their production from phytoplankton and other microautotrophs. The well-developed kelp forests at otter-dominated islands generate a severalfold increase in total production, influencing in turn the abundance and growth rates of other species. For example, filter-feeding mussels and barnacles grow more rapidly at otter-dominated islands than at otter-free islands (Duggins et al. 1989). Reef fish populations, most of which rely on kelp forests for food, protection, or other habitat requisites, are similarly affected. Densities (measured as catch per effort) of rock greenling (*Hexagramos lagocephalus*), the most common shallow-water reef fish in the western and central Aleutian Islands, are 8- to 10-fold greater in kelp forests than they are in urchin barrens (Reisewitz 2002).

Other species are also influenced indirectly by sea otter predation. For example, the interaction web linkage connecting sea otters, sea urchins, kelp, and kelp forest fishes affects the foraging behavior of glaucous-winged gulls (*Larus glaucescens*), one of southwest Alaska's most abundant coastal seabirds (Fig. 4.4). This species is mostly piscivorous where sea otters are abundant, yet it feeds largely on intertidal invertebrates where otters are absent (Irons et al. 1986). Complex behaviors result from these differing diets. Gulls forage more strongly in concert with the tidal cycles in otter-free systems, and the species composition and profitability of their foraging efforts vary between spring and neap tides. This is because

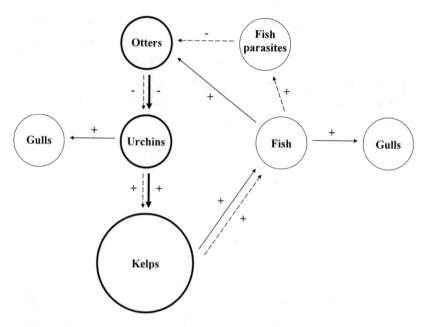

Figure 4.4

Food web pathways of several indirect interactions stemming from the otter–urchin–kelp trophic cascade (circles and arrows in bold) that are described in the text. The solid arrows connecting otters, urchins, kelps, and fish illustrate how the consumption of kelp forest fish by sea otters can create a positive feedback loop that strengthens the cascade. The dashed arrows illustrate how parasitic infections carried by the fish might prevent this positive feedback loop from becoming a runaway process. This partial food web also illustrates how the pathways leading to glaucous-winged gulls (*Larus glaucescens*) differ when otters are present or absent in the system.

they are only able to gain access to the relatively unprofitable mussels and barnacles during neap low tides, whereas during extreme spring low tides they are able to feed lower in the intertidal zone on more profitable sea urchins (Irons et al. 1986). However, the largest and potentially most profitable urchins cannot be easily opened or swallowed intact. For these large prey, the gulls employ an air-drop tactic in order to break open the urchin's test and expose its internal viscera, in turn attracting other gulls that utilize a pirating strategy to quickly gobble up the exposed viscera before the air-dropper can land and consume it.

The sea otter–urchin–kelp trophic cascade has feedback effects on the otters themselves (see Fig. 4.4). By driving coastal reef systems from urchin barrens to kelp forests, sea otters elevate total production, thereby altering the abundance and species composition of their potential prey. In some parts of the Aleutian Islands, this has caused the sea otters to supplement their diet with fish (Estes et al. 1978), creating in turn a positive feedback system by increasing the number of otters, thereby further reducing sea urchin numbers and the intensity of herbivory and increasing the abundance of kelp and hence the abundance of fish (Estes 1990). But some fish also carry parasites that are pathogenic to otters (Rausch 1953), and high parasite loads lead to elevated mortality and otter population declines (Kenyon 1969). This linkage between otters, fish, and fish parasites may prevent the positive feedback cycle among otters, urchins, kelp, and fish from becoming a runaway process (see Fig. 4.4).

In the preceding section of this chapter, I have briefly described several known or suspected interaction web pathways that are connected to the sea otter–urchin–kelp trophic cascade via behavioral or demographic effects on the recipient species. The majority of potential indirect effects are unstudied. However, these examples suffice in demonstrating the complex influences of sea otter predation on coastal food web dynamics.

Oceanic Connectivity

The previously described patterns and processes were discovered from a priori hypotheses and field studies designed to test those hypotheses. Our findings rarely departed significantly from initial expectations. The first real surprise came in the early 1990s when the sea otter population at Adak Island, where my colleagues and I were working at the time, began to decline rapidly. This decline led us to wonder about its cause, its geographical extent, and its effects on the kelp forest ecosystem.

The geographical extent of the sea otter decline was easily determined. We began by resurveying sea otters at three other islands (Amchitka, Kagalaska, and Little Kiska) for which we had measures of abundance from before the decline. Similar declines were seen at all of the islands (Fig. 4.5). In 2000 the U.S. Fish and Wildlife Service conducted an aerial survey of sea otters throughout the Aleutian

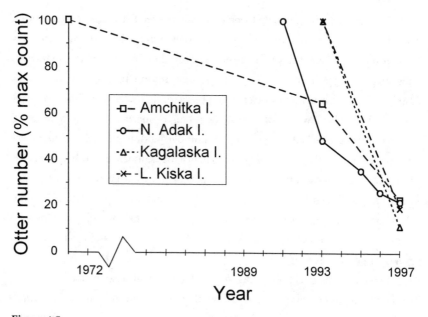

Figure 4.5

Sea otter population trends at Adak, Amchitka, Little Kiska, and Kagalaska islands. Data (from Estes et al. 1998) are plotted as proportions of maximum counts for each island. Survey methods described by Estes (1990) and Doroff et al. (2003).

archipelago. Similar surveys had been done in 1965 and 1992, and the 2000 results showed that sea otter populations had declined to low densities across this region (Doroff et al. 2003).

Effects of the sea otter decline on sea urchin and kelp populations were also easily determined because we had been monitoring kelp forest habitats at various islands since the 1970s, well before the decline started. The data from Adak and Amchitka islands were especially useful because these islands supported large otter populations and dense kelp forests throughout the period of our monitoring program. By the late 1990s, the habitats around both islands had shifted from kelp forests to urchin barrens (Estes et al. 2004).

The cause of the sea otter decline was more difficult to ascertain. Field studies of tagged and radiotelemetered sea otters quickly revealed that elevated mortality was the general cause. Weakened or sick sea otters typically come ashore to die, thus producing numerous stranded carcasses (Kenyon 1969; Bodkin et al.

2000). The near absence of strandings during the decline seemed inconsistent with food limitation or disease as the cause of the decline. Increased sightings of killer whales very near shore (something we had not seen previously), and several observed attacks by killer whales on sea otters (also something not previously seen—Hatfield et al. 1998) led us to suspect that the decline was caused by increased killer whale predation. This suspicion was supported by the fact that, throughout the decline, high sea otter densities persisted in Clam Lagoon on Adak Island, an apparent refuge from killer whales. In addition, every other possible explanation that we could imagine for the decline was inconsistent in one important way or another with the available evidence. However, a vast discrepancy between the large number of required losses and the small number of observed kills by killer whales was also difficult for us to reconcile.

We explored this paradox by first computing the number of additional sea otter deaths required to generate the decline. We began with an estimate of the pre-decline population abundance of sea otters from Kiska Island in the west to Seguam Island in the east—that region of the central Aleutian archipelago within which we had been working. Using that abundance estimate and life-table statistics for a stationary sea otter population, we created a matrix population model to estimate the number of additional deaths (about 40,000, beyond those expected in a stationary population) needed to force the population downward at the observed rate and magnitude. We then estimated the proportion of these deaths that would have been observed by members of our field team, assuming that their occurrence was randomly distributed in space and time and that the team members sampled them in a representative manner. Our estimate was accordingly based on the proportion of the area/time sample space that was "sampled" over the course of the decline, given that all of the added deaths were caused by killer whale predation. The numbers matched closely (six kills observed; five observed kills predicted from our model—Estes et al. 1998). The small number of observed kills was thus easily reconciled.

Initially, it seemed that an unrealistically large number of killer whales would have been required to eat so many sea otters over such a short time period. However, by combining the loss estimates with measures of killer whale nutritional requirements and sea otter nutritional content, we calculated that fewer than four killer whales feeding solely on sea otters could have eaten all of the 40,000 sea otters. These various observations and analyses led us to conclude that killer whale predation was a reasonable and likely explanation for the decline.

What caused the killer whales to suddenly eat so many sea otters? We began exploring this question by assuming that either increased numbers of killer whales or a change in their behavior caused it to happen. Since a rapid population increase by killer whales seemed unlikely, we focused on the behavioral ecology of killer whales and the history of their food resources within the area of the sea otter decline.

Three killer whale ecotypes have been described for the North Pacific Ocean: those that feed mostly on fish (the "residents"), those that feed mostly on marine mammals (the "transients"), and those that feed in offshore oceanic waters (the diet of these animals is poorly known). A strong potential for cultural evolution created by the species' distinct matrilineal social structure and a very long period of association between mothers and their young are thought to be important in the generation and maintenance of these ecotypes (Baird and Dill 1996). Two inferences emerged from the preceding view of killer whale behavior. One was that transient killer whales were probably responsible for the sea otter declines. The other was that the prey base of these transients must have changed in some way to cause this to happen. The well-documented collapse of pinniped populations, including Steller sea lions (*Eumetopias jubatus*), northern fur seals (*Callorhinus ursinus*), and harbor seals (*Phoca vitulina*)—all of which are eaten by killer whales—in the western Gulf of Alaska and Aleutian Islands during the 1970s and 1980s (National Research Council 2003) provided a reasonable explanation. We hypothesized that some of the pinniped-eating killer whales expanded their diets to include sea otters as the dwindling numbers of pinnipeds were no longer able to sustain them. This hypothesis is supported by the fact that the sea otter decline began as the most precipitous phase of the Steller sea lion decline in the Aleutian Islands was ending (Fig. 4.6).

If our hypothesis is true, then a search for the ultimate cause of the sea otter collapse rests squarely on the question of why the pinnipeds declined. Several lines of evidence and reasoning led my colleagues and me to suspect that increased killer whale predation had also caused the pinniped declines. For one, the sea otter and sea lion declines were similar in pattern and geographical range. If killer whale predation caused the sea otter decline, it seemed reasonable to us that this process

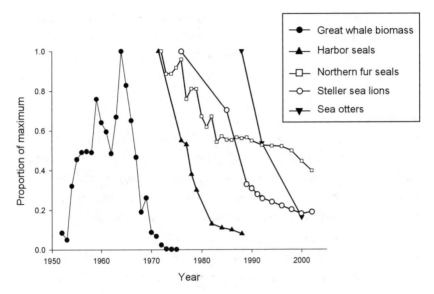

Figure 4.6

Reported great whale landings (biomass) and trends in numerical abundance of pinnipeds and sea otters from the Aleutian archipelago and nearby regions of southwest Alaska. Figure modified from Springer et al. (2003).

might have figured prominently in the sea lion declines as well. Demographic and energetic analyses, broadly similar to those described previously for sea otters, have demonstrated that relatively small changes in killer whale behavior here too could have driven the sea lion declines (Springer et al., 2003; Williams et al. 2004). We estimate that as few as 26 killer whales could have generated all of the necessary mortality in the Aleutian Islands if these animals fed exclusively on sea lions. Alternatively, a dietary shift of less than 1% (based on source of caloric input) by the region's entire killer whale population (estimated at about 3800 individuals—Springer et al. 2003) also could have driven the decline. These extreme-case scenarios establish the ease with which some intermediate possibility could have caused the sea lion population decline. Finally, we grew to favor the killer whale predation hypothesis in part by default. So far, all other possible explanations seem to be inconsistent with the available information.

The insights gained from these analyses prompted us to further explore the history of marine mammals in the Aleutian Islands. In addition to the pinniped and sea otter declines, we knew that whaling had destroyed the great whales (National Research Council 1996). However, it was not until we put all of the data together that the sequential nature of these declines became evident (Fig. 4.6), and we began to consider how they might be interrelated.

How could whaling have caused the pinniped and sea otters to decline? Killer whales prey on all or most great whale species (Matkin 1994). Therefore, industrial whaling could have substantially altered prey availability for these predators. My colleagues and I have proposed that this change caused whale-eating killer whales to begin feeding on pinnipeds and sea otters (Springer et al. 2003). We hypothesize that these species were unable to sustain the elevated mortality rates that would have resulted from such a dietary shift, and thus that the sequential decline of harbor seals, Steller sea lions, and sea otters resulted from a growing dietary breadth of killer whales to include increasingly less profitable prey as they serially depleted this component of the marine food web.

Summary of Food Web Dynamics

We have documented striking changes in the distribution and abundance of sea urchins and kelp as food chain length effectively increased (Fig. 4.7) from two (sea urchins and kelp) to three (sea otters, sea urchins, and kelp) to four (killer whales added as the apex predator). These changes are consistent with Hairston et al.'s (1960) and Fretwell's (1987) theoretical predictions. That is, herbivores are rare and plants abundant in odd-numbered food chains whereas the converse is true when food chain length is even-numbered (see Fig. 4.7). A variety of indirect effects from this trophic cascade influence the distribution, abundance, and behavior of other kelp forest species. Although uncertainty remains concerning the ultimate reasons for the entry of killer whales into this system, it is likely that the causal factors operated on large spatial and temporal scales, involving linkages with oceanic ecosystems.

Implications for Other Species and Ecosystems

Although the sea otter's story may seem unique, the interactive processes influencing the dynamics of this species and its associated ecosystem occur widely in nature. There is now abundant evidence for top-down forcing and trophic cascades

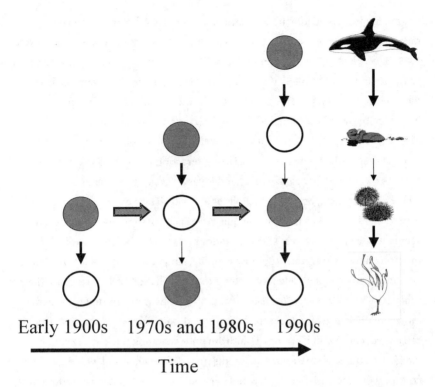

Figure 4.7

Diagram showing how plant–herbivore interaction strength in kelp forest ecosystems of the Aleutian archipelago changed over the course of the 20th century from (1) early in this period when sea otters were rare or absent, to (2) the middle–late decades when otter populations were recovering, to (3) the 1990s when killer whales entered the system and again reduced the number of sea otters. Bold and light arrows indicate strong vs. weak interactions, respectively. Filled and open circles indicate that species within the indicated trophic level are limited by resources vs. disturbance by their consumers, respectively (as in Fig. 4.1). Plant–herbivore interaction strength shifts from strong to weak to strong as the growing food chain alternates between an even and odd number of trophic levels, as predicted by Fretwell (1987).

from diverse ecosystems (Pace et al. 1999). Predator removal and additional experiments (some fortuitous, others purposeful) have established the importance of trophic cascades in lakes (reviewed in Carpenter and Kitchell 1993), whereas experiments on river food webs confirm Fretwell's prediction of alternating variation in plant–herbivore interaction strength with food chain length (Power 1990). Other examples of trophic cascades in aquatic food webs include those originating from predation by sea stars and oystercatchers in rocky intertidal communities (Paine 1966; Wootton 1995); fishes and lobsters in temperate reefs (Cowen 1983; Babcock et al. 1999; Steneck et al. 2002) and tropical reefs (McClanahan and Shafir 1990; Sala et al.1998; McClanahan et al. 1999b; and even various fishes in the open sea (Verity and Smetacek 1996; Shiomoto et al. 1997). Trophic cascades have been more difficult to demonstrate in terrestrial ecosystems although the list of examples is growing (Schmitz et al. 2000). Some trophic cascades also have more widespread influences on associated interaction webs (Terborgh et al. 2001).

Most ecologists recognize that trophic cascades occur, and are now debating where and why (e.g., Strong 1992; Polis et al. 2000). This debate does not seem to have resulted in a productive line of inquiry, especially for the conservation and management of large carnivores. There are various reasons for this. First, as discussed in the preceding paragraphs, top-down forcing and trophic cascades have been demonstrated in enough different systems to establish their wide occurrence (Pace et al. 1999; Schmitz et al. 2000; Estes et al. 2001). A few more examples, even many more, are unlikely to alter the substance of the debate over where and why trophic cascades occur. Second, ecology is far from understanding the limiting conditions (i.e., key features of species or ecosystems) that define when and where top-down control will occur (Menge 2003; Schoener and Spiller 2003). Arguably, this quest would be better served by a search for systems that do not fit the HSS/Fretwell model. Such examples, if they can be found, will be tremendously useful in the development of a more comprehensive theory. Finally, numerous obstacles (listed following here) stand in the way of assessing the generality of top-down control and trophic cascades across large carnivore species and their associated ecosystems.

1. Large carnivores are now missing from many systems in which they once were plentiful. Gray wolves (*Canis lupus*) and grizzly bears (*Ursus arctos*) are all but gone from temperate North America (Soulé and Terborgh 1999); large predators have been depleted from the coastal oceans worldwide (Jackson et al. 2001); and whales and large predatory fishes have been depleted throughout the world's oceans (Clapham and Baker 2002; Pauly et al. 2002; Baum et al. 2003; Myers and Worm 2003). Scientists have wondered about the functional importance of these species, but answers have been slow in coming because those processes are difficult to understand where they no longer occur.

2. Even where large carnivores still exist at normal densities, their potential ecological roles are hard to ascertain. This is because the distributions and abundances of these animals are typically almost stationary over the time scales that we study them, and the dynamics of stationary systems are difficult to understand (May 1973). In fact, much of what we know about the ecological roles of large carnivores comes from a few studies of perturbed or changing carnivore populations. Examples include the reintroduction or reestablishment of gray wolves (McLaren and Peterson 1994; Ripple and Larsen 2000; Berger et al. 2001a), the local extinction of large cats and harpy eagles (*Harpia harpyja*) from recently created land-bridge islands (Terborgh et al. 2001), the extinction and recolonization of chaparral fragments by coyotes (*Canis latrans*—Crooks and Soulé 1999a), and post–fur trade recovery and the recent collapse of sea otters (Estes and Duggins 1995; Estes et al. 1998—described earlier in chapter). By and large however, ecologists have been hard pressed to design studies that place the ecological roles of large, apex predators into dynamic perspectives.

3. Plant generation times vary immensely across taxa, from hours or days for phytoplankton to centuries or millennia for some forest trees. The rate of response by plants to predators should vary accordingly, which probably explains to some degree why the best experimental evidence for top-down forcing and trophic cascades comes from lakes, streams, and coastal marine systems, and why similar processes have proven so difficult to observe and understand in terrestrial systems (Estes 1995; Duffy 2002).

4. Many ecologists continue to view food web dynamics from a strictly bottom-up perspective. Not surprisingly, this is especially true of those who work in systems where top-down forcing has been difficult to demonstrate experimentally. The result is a positive feedback strengthening intellectual inertia. Scientists, like most people, tend to look for what they believe is true. Recent work on the food web dynamics of cordgrass marshes in the southeastern United Stated demonstrates the insidious nature of this phenomenon. For decades variation in nutrient availability and bottom-up forcing processes were thought to drive the salt marsh system. However, experimental manipulations of the herbivore (snails) and predator (blue crabs—*Callinectes sapidus*) populations demonstrated not only that top-down forcing and trophic cascades are essential to the maintenance of

this system but that increasing fishery pressure on blue crabs was likely responsible for recent cordgrass habitat reduction (Silliman and Bertness 2002).

5. Philosophically, science operates on the premise that something does not exist until otherwise demonstrated. For this reason, many ecologists seem to think that the examples of top-down forcing and trophic cascades are exceptions rather than the rule. But there is almost no evidence for the converse—that top-down forcing and trophic cascades are not important. Nor is such evidence likely to be forthcoming any time soon, for the reasons already discussed.

6. The standards of scientific inference are too strict to put the question of top-down forcing and trophic cascades to a broad general test. Here, science is trapped in the tyranny of Type I error (rejecting the null hypothesis when it is true). What many scientists fear most is concluding that something exists when in fact it does not. To avoid that mistake, they establish decision criteria such that the probability of a Type I error (in this case, concluding that predators are important when in fact they are not important) is very small. This inevitably makes the probability of a Type II error (accepting the null hypothesis when it is false—in this case, concluding that predators are not important when in fact they are important) relatively large. Key effects of carnivores and carnivory may be discounted as insignificant for that reason alone. It is the age-old problem of the burden of proof.

It will be impossible to evaluate the importance of top-down forcing and the existence of trophic cascades for every predator in every ecosystem because there are too many species and too many kinds of ecosystems. At some point inferences must be made about these species and systems from what has been learned elsewhere. I believe that time is now.

Implications for Conservation and Management

The conveners of this symposium asked us, among other things, to imagine how the methods and philosophy of biodiversity conservation might be recrafted to take a more "carnivorecentric" perspective. My thoughts are based largely on what

I have learned or come to believe about sea otters and kelp forests. Hence, it is worth considering whether the workings of the sea otter–kelp forest system are unique or more general.

The HSS/Fretwell model provides a unifying theme that connects apex carnivores with other species. We now have enough evidence at least to know that this unifying theme occurs widely in nature. Thus, in my view, there is relatively little to be gained by spending our precious time and money in an effort to fill the remaining gaps in our empirical knowledge of predator–herbivore–plant interactions (which admittedly are many). Furthermore, the HSS/Fretwell model defines the influences of carnivores too narrowly. Numerous food web pathways are linked to the sea otter–induced trophic cascade, thus connecting the demography and behavior of many other species to the presence or absence of sea otters, often in complex ways. I see no reason why similar interactions should not accompany trophic cascades in other systems. Hence, one broader lesson for conservation and management seems clear. Carnivores can influence functional biodiversity across a wide array of associated species.

Even this expanded view of carnivore functionality fails to capture the scale and complexity of their role in biodiversity conservation. Our findings show or suggest that carnivores connect species over very large spatial and temporal scales. We have seen that kelp forests do not function in isolation of other ecosystems. Coastal kelp forests appear to be linked with the open sea via predator–prey interactions between large, high-level carnivores. I have attempted to show that a sufficiently long view of history is necessary to see and understand these interactions. This realization leads to the question, What defines critical habitat for sea otters? Despite the fact that these animals spend all or most of their lives living in shallow water close to shore, something much more is needed to protect sea otters and kelp forests. Almost certainly, any successful effort to conserve sea otters and kelp forests in the Aleutian Islands must also consider the open sea. The conservation of other carnivore species and their associated habitats might equally depend on such large-scale interecosystem connectivity.

Our studies of sea otters and kelp forests provide three general lessons for conservation biology and resource management. The first is that the minimum viable population concept is deeply flawed as a strategy for carnivore conservation (Soulé et al. 2003). Although a minimum viable population may be sufficient to prevent the target species from becoming biologically extinct, it often will be insufficient to prevent ecological extinction. The second lesson is that the temporal and spatial scales of effective conservation must be large enough both to include linkages across ecosystems and to redress the course of recent history. These may not be welcome messages to an increasingly fragmented world whose human societies seek short-term and small-scale solutions to conservation problems. A final lesson concerns the scientific uncertainties that are sure to surround our understanding of these rare and secretive species, and of processes of such scale and complexity. I doubt that we will ever understand these species and processes with enough certainty to satisfy the skeptics. Conservation planning must proceed in the face of this uncertainty, based on the weight of available evidence.

Summary

Although top-down forcing processes and trophic cascades have now been demonstrated in a diversity of species and ecosystems, the ecological roles of large, apex carnivores are difficult to demonstrate and thus remain poorly understood. The sea otter–kelp forest ecosystem provides unusually clear insights into these processes because (1) a history of overexploitation and recovery fragmented the species' distribution, thereby permitting contrasts between otherwise similar habitats in which the species was present or absent, and (2) the subsequent recovery and more recent collapse of sea otter populations afforded us the opportunity to chronicle co-occurring community and ecosystem-level changes through time. These contrasts demonstrate that sea otters initiate a trophic cascade by feeding on herbivorous sea urchins, which in turn feed on kelp. This trophic cascade indirectly influences the distribution, abundance, and behavior of numerous other kelp forest species. Sea otter populations in southwest Alaska have recently declined precipitously because of increased killer whale predation, thus causing sea urchins to increase and the kelp forest to decline. The sea otter decline was the most recent event in a sequential megafaunal collapse that includes northern fur seals, harbor seals, and Steller sea lions. Although still speculative, there is growing evidence that post–World War II industrial whaling caused killer whales, believed by some to be the great whales' foremost natural predator, to expand their diet to include pinnipeds and sea otters, thereby driving populations of these smaller and less abundant species downward. The sea otter's coastal marine food web thus appears to be interconnected with events acting on large spatial and temporal scales. Many of the same kinds of processes that structure the sea otter–kelp forest system, although difficult to observe and understand, may very well occur elsewhere in nature. These findings suggest that large, apex carnivores figure prominently in the maintenance of biodiversity, and that conservation and management strategies aimed at preserving these species and systems must be planned on large spatial and temporal scales.

ACKNOWLEDGMENTS

The number of people who have collaborated with me over the past three decades in studies of sea otters and kelp forests are too numerous to list individually. I am deeply indebted to all for their insights, friendship, and hard work. I thank Joel Berger, Karen Phillips, Justina Ray, Kent Redford, Robert Steneck, and one anonymous reviewer for comments on earlier drafts of the manuscript. I am especially grateful to Robert T. Paine for inspiration and encouragement over the course of my career. Financial support for my work has been provided by the U.S. Department of the Interior, the U.S. National Science Foundation, the U.S. Navy, the U.S. Air Force, and the Pew Fellows Program in Marine Conservation.

CHAPTER 5

The Green World Hypothesis Revisited

John Terborgh

In a landmark article published four decades ago, three distinguished ecologists, Hairston, Smith, and Slobodkin (HSS), proposed a model for trophic interactions that has been dubbed the "Green World Hypothesis" (Hairston et al. 1960). The world is green, the trio argued, because herbivores are kept in check by predators. Herbivores therefore do only minor damage to vegetation. Implicit in this scenario is the notion that modern plant communities evolved in predator-regulated ecosystems that maintained a rough proportionality among interactive guilds such as seed dispersers, seed predators, and seedling herbivores.

At about the same time, phytochemists were discovering a vast pharmacopoeia of what were then referred to as plant secondary compounds. These included simple and condensed phenolics, alkaloids, terpenoids, glycosides, and many others, few if any of which could be assigned roles in the major pathways of plant metabolism. Through the pioneering work of Fraenkel (1959) and others, it became established that the primary function of plant secondary compounds was the deterrence of herbivory. It was later demonstrated that such deterrence can be costly, comprising as much as 30% or more of the dry weight of leaves (Coley et al. 1985). It thus stands to reason that the more plants invest in antiherbivore defenses, the less they can invest in roots, shoots, and seeds.

Recognition that plants were making such costly investments to deter herbivory quite logically led to an alternative worldview in opposition to HSS (Murdoch 1966). According to what we shall term the "Plant Self-Defense Hypothesis," the world is green because plants defend themselves to the degree necessary to prevent herbivores from doing excessive damage to their foliage. Under the Plant Self-Defense Hypothesis, predators are regarded as an epiphenomenon, and herbivore densities are regulated by limited availability (in space and/or time) of ingestible foliage.

Which view is correct? the Green World Hypothesis, in which predators are paramount? or the opposing view of plant prevalence in chemical warfare? The alternatives appear so distinct that there should be no difficulty in distinguishing them. But here appearances are deceptive, for distinguishing the two hypotheses has stymied ecologists for decades. Nevertheless, the question of which theory better describes nature is of paramount importance to conservation, because the HSS scenario implies that predators are of vital importance to ecosystem stability, whereas the plant self-defense scenario implies that ecosystem stability is largely independent of the function of predators.

If HSS is correct, then predators are crucial to controlling consumers (animals that feed on plants or their products, such as flowers, fruits, and seeds). In the absence of predators, consumers are predicted to increase with ensuing impacts on vegetation in what is known as a top-down trophic cascade (Paine 1980; Oksanen et al. 1981; Carpenter and Kitchell 1993). The term refers to the alternating positive and negative effects of a perturbation at one trophic level on the organisms at successively higher and lower levels. Thus elimination of predators is expected to result in higher levels of consumers with consequent severe damage to vegetation, as has been so elegantly demonstrated in the case of sea otters (*Enhydra lutris*), sea urchins, and kelp forests (Estes and Palmisano 1974).

Large animals have been eliminated over much of the contemporary world to the extent that intact ecosystems replete with top predators and large ungulates have effectively been reduced to the status of museum exhibits in parks (Redford 1992; Flannery 2001; Wright 2003). Under the HSS scenario, increases in consumers in the absence of predators could lead inexorably to the destabilization of ecosystems and a contraction of biodiversity. If, on the other hand, the Plant Self-Defense Hypothesis were to be confirmed, ecosystem stability would not be seriously threatened by a loss of predators. Under such a scenario, predators are presumed to exert only a minor influence on consumer numbers, and trophic cascades would be weak or absent in the terrestrial realm (Power 1992; Strong 1992; Persson 1999; Halaj and Wise 2001).

In attempting to draw conclusions about which of these two diametrically opposed worldviews more closely describes the working of nature, I shall proceed as follows. First, I shall consider a number of natural and artificial situations that test the Plant Self-Defense Hypothesis. I will then describe results from a large-scale predator exclusion experiment. Along the way it will become apparent that neither of the simplistic worldviews projected by the two hypotheses holds up to scrutiny. Instead, the available evidence will suggest that nature functions through a more complex web of interactions in which both bottom-up (plant productivity) and top-down (predation) forces play decisive roles.

A vital bottom-up role in regulating consumer numbers is affirmed by data sets from arctic, temperate, and tropical regions showing that consumer biomass correlates positively with plant productivity (Oksanen and Oksanen 2000). Superimposed on this pattern is the action of predators that prevents many (but not all—to be discussed) consumers from becoming even more abundant—so abundant that they alter the structure and composition of vegetation. Available evidence points to a crucial stabilizing role of predators in many terrestrial ecosystems. Loss of predators through human persecution and habitat fragmentation thus threatens the future of much of the earth's biodiversity.

Tests of the Plant Self-Defense Hypothesis

Casual observations challenge the notion that plants are adequately protected by their own defenses, but mitigating circumstances and lack of controls leave open other possible interpretations.

For example, crop plants routinely suffer from pest damage, requiring farmers to invest heavily in artificial chemicals to supplement those produced by the plants themselves. But crop plants have been selected for traits such as high yield and synchronous ripening, which may be achieved by plant breeders at the expense of constitutive chemical defenses.

Numerous examples attest to the vulnerability of island vegetation to introduced domestic stock or game animals (Bramwell 1979; Cronk 1980). Vertebrate herbivores introduced to islands lacking such animals typically explode in number until an island's vegetation is severely damaged or denuded, whereupon the herbivore population often crashes (Klein 1968; Coblentz 1978). The explosion of herbivores could be explained by a lack of predators, or, alternatively, by low investment in defenses on the part of plants evolved in an environment naturally devoid of vertebrate herbivores. Investigation of endemic island plants has shown that they invest substantially less in chemical defenses than mainland counterparts, a fact that supports the latter interpretation while not refuting the former (Bowen and van Vuren 1997).

Natural plant communities are occasionally observed to undergo wholesale defoliation by an insect that is normally present but uncommon (Mattson and Addy 1975). Entomologists often attribute such outbreaks to a rare conjuncture of physical conditions that favors the defoliator, for example, by allowing high overwinter survival, or by synchronizing the hatching of larvae in the spring with the emergence of leaves by the host plant. Because outbreaks occur suddenly, plants are unable to react through increased investment in defense. The insect is typically so abundant that naturally occurring predators are unable to control it. Although constituting obvious violations of the Self-Defense Hypothesis, the rarity of such outbreaks makes them easy to dismiss as mere curiosities of nature. Nevertheless, the fact that outbreaks do occur is compelling evidence that at least some arthropod defoliators are able to overcome the defenses of a wide range of plant species.

Apart from stringent meteorological requirements, outbreaks may be rare because induced defenses can retard larval or nymphal development, thereby precluding the rapid-fire series of generations that results in a population explosion. Moreover, once induced by herbivore damage, plants often maintain high levels of defense for prolonged periods (Edelstein-Keshet and Rausher 1989; Zangerl and Rutledge 1996). Investment in defense tends to wane during periods when herbivore attack is infrequent or light.

Most insect herbivores are small and lack the ability to ingest the tough foliage of mature leaves. Leaves consequently suffer greatest damage from insect herbivores when they are young and expanding (Coley 1980; Coley and Barone 1996). Once expansion is complete, leaves increase greatly in toughness, and the concentration of essential components, such as proteins, may fall (Kursar and Coley 1991). Mature leaves are thus not favored by insect herbivores.

Mature foliage is the province of larger arthropod herbivores such as late instar Lepidoptera and adult Coleoptera and Orthoptera. However, the heaviest consumers of mature foliage are vertebrates. Like insects, most vertebrate herbivores, including domestic livestock, prefer fresh young foliage. Unlike insects, however, most vertebrate herbivores live several to many years and are obliged to survive through periods of low productivity by subsisting on mature foliage, twigs, buds, and even bark (Mclaren and Peterson 1994).

Vertebrate herbivores can have devastating impacts on vegetation (Alverson et al. 1988; McShea et al. 1997). A century of overgrazing by cattle has converted most of the high desert grassland of the southwestern United States to a spiny thicket of agaves, yuccas, and prickly pears (Valone et al. 2002). Is not this evidence enough that herbivores could reduce vegetation to tatters if left to their own devices?

No, would reply the skeptics. Herds of livestock are a creation of ranchers who artificially elevate population densities through such practices as digging wells, sheltering and immunizing newborn calves, and providing supplemental feed during winters and droughts. Populations of wild ungulates and other large herbivores are not observed to overgraze to the extent of transforming whole landscapes (but see Berger and Wehausen 1991).

Native herbivores in postcolonial North America are but a faint echo of an earlier North America that was replete with mammoths, mastodons, giant ground sloths, tapirs, camels, bison, horses, and literally dozens of other large herbivorous mammals. All but a handful of these are now extinct (McDonald 1984; Lange 2002). What was the impact of this formidable assemblage of herbivores on the vegetation of the Pleistocene? Unfortunately, we shall never precisely know. But what we do know is that, where such assemblages of herbivores remain extant, such as in parts of Africa and India, their impacts can be profound (Owen-Smith 1988; Calenge et al. 2002).

Parts of East Africa (Tsavo, Selous) have alternated between open savanna and closed thorn scrub as elephant (*Loxodonta africana*) populations were episodically decimated by ivory hunting, first in the late 19th century and again in the late 20th century. When elephant numbers are high, open savanna habitat dominates because elephants eat trees and by doing so create grassland. But when elephant numbers are low, thorn scrub quickly takes over and shades out the grassland, with profound consequences for other herbivores (Owen-Smith 1988). Grazers are favored during periods of high elephant numbers, whereas browsers thrive when densities of elephants are reduced. These are not fairy tales; these are historical observations documented by photographs and estimates of animal numbers.

I have been fortunate to see first hand the impact of high elephant densities in the woodlands of the Moremi Reserve and Chobe National Park in Botswana.

Hardly a tree was unscathed, and most were in tatters, with their limbs reduced to ragged stubs. Trees of many species were damaged and few or none appeared to be immune. The thought kept returning to me that this must have been how parts of North America looked 12,000 years ago before the mastodons and ground sloths were driven to extinction.

"Megaherbivores" (defined as those weighing more than 1000 kg as adults) have been a key part of the earth's ecology for the last 200 million years. But due to "overkill," only a few remain in the living menagerie of our time (Martin and Klein 1984; Flannery 1994). Megaherbivores are relatively free of the dietary limitations that restrain insects and lesser mammals from defoliating the landscape. Their huge size implies a slow metabolism, and a slow metabolism in turn implies that food can be processed in the gut at a leisurely pace. Slow passage through the gut provides time for breakdown of a high fiber diet and permits chemical processing of plant constituents via microbial activity. Elephants are thus able to eat almost anything green, and do so wherever their numbers temporarily exceed the productive capacity of the environment. Such circumstances arise naturally during droughts and unnaturally when poaching or human encroachment causes elephants to concentrate in national parks.

The same trait—huge size—that enables megaherbivores to eat mature foliage also allows them to escape predation, for there are no predators that regularly attack and kill adult elephants, hippos, or rhinoceroses. The prevalence of megaherbivores throughout evolutionary history invalidates both the Green World and the Plant Self-Defense hypotheses, the former because it (tacitly) assumes that all herbivores have predators, and the latter, because plant defenses fail to deter megaherbivores in some, but not all, circumstances.

Megaherbivores roamed over most of the earth's terrestrial habitat before the Late Pleistocene and Holocene overkill. They are widespread in Africa from the tops of Mt. Kenya and Kilimanjaro to the lowland rainforests of the Congo Basin and from the fringes of the Sahara through the Namib desert to the Cape of Good Hope. Where megaherbivores persist today, their collective biomass tends to constitute roughly half of the biomass of vertebrate herbivores (Owen-Smith 1988).

Woolly mammoths lived in Siberia and Alaska in the harshest vegetated landscapes on the planet (Owen-Smith 1988; Flannery 2001). Obviously, cold was no barrier. Before overkill, elephants and their proboscidian relatives lived on all continents except Australia and even some islands. Large islands lacking pro-boscidians supported other giant herbivores, like elephant birds in Madagascar, moas in New Zealand, diprotodons and giant kangaroos in Australia, and giant tortoises in the Galapagos, Seychelles, and Aldabara (Martin and Klein 1984; Flannery 1994). Thus, before humans disrupted ecosystems on a global scale, mega-herbivores, or their smaller (predator-free) insular counterparts, occupied nearly every substantial landmass on Earth.

Predator Exclusion Experiments

Why, in the more than 40 years since HSS published their landmark article, has science not resolved the question, Why is the world green? The answer is embarrassingly simple. It is because ecologists, with the scant resources available to them, are unable to conduct experiments on a sufficiently large scale. In principle, a test of the Green World Hypothesis would be easy: construct large hectare or km^2 scale exclosures, remove all predators (but not consumers), and document the ensuing dynamics of consumers and vegetation. But, for lack of resources, most of the predator-exclusion experiments conducted by North American and European scientists have been done at the scale of a few square meters, and have generally not included herbivores larger than grasshoppers or mice (Marquist and Whelan 1994; Englund 1997; Schmitz et al. 2000). These small-scale predator-exclusion experiments typically demonstrate increases in the numbers of herbivores inside predator-proof exclosures, but, for a variety of technical reasons, most of the documented increases in herbivore density have been unimpressive (Schmitz et al. 2000).

The one notable exception to the few-square-meter scale of predator-exclusion experiments is the experiment (1 km^2 scale) carried out by Charles Krebs and his team in the Yukon (Krebs et al. 1995; Hodges et al. 1999). A single strand of electrified wire kept out coyotes (*Canis latrans*) and lynx (*Lynx canadensis*) but allowed lemmings and snowshoe hares (*Lepus americanus*) to come and go at will. Access by other potential predators of hares and lemmings, such as mustelids and raptors, was unimpeded (Rohner and Krebs 1996). The experiment included controls plus three treatments: lynx/coyote exclusion, provisioning of food for hares, and both. Over a decade that included both the rising and falling portions of the hare cycle, hares were on average several times more abundant in the predator exclusion and provisioning treatments and an order of magnitude more abundant in the predator exclusion plus provisioning treatment (Krebs et al. 1995). A concerted effort to examine alternative hypotheses has left predation as the only plausible interpretation of the differences in hare abundance between predator exclusion treatments and controls (Krebs et al. 2001).

The Krebs et al. (1995) experiment points to a strong role of coyotes and lynxes in regulating the numbers of hares, but leaves open the question of what would happen to vegetation in an entirely predator-free world. The question remains open because the experiment achieved only partial predator exclusion while allowing herbivores to move freely in and out of the exclosures. A more conclusive experiment would entail complete predator exclusion and would deny herbivores the option of density-dependent emigration from predator-free areas.

Such an experiment can be constructed by building on one of the most consistent "rules" in ecology; namely, that predators require more living space than their prey. A fortuitous "experiment" that created the desired conditions was initiated in 1986 in the Caroni Valley of Venezuela with the completion of the giant Raul Leoni dam. The resulting impoundment, Lago Guri, flooded 4300 km² of hilly terrain, thereby creating hundreds of so-called land-bridge islands. The term "land-bridge" refers to the fact that the islands were formerly part of the surrounding mainland and thus offer representative samples of the mainland tropical dry forest ecosystem.

Many of the islands in Lago Guri are too small to support predators of vertebrates but are large enough to sustain populations of consumer species. It was the prospect of studying such islands that drew me to Lago Guri in 1990 to begin a project that has so far lasted more than a decade. Because many of the results obtained by our research group have already been published, I shall skip over all but the findings most relevant to testing the Green World Hypothesis.

Surveys of a dozen islands and the mainland revealed early on that three-quarters or more of the vertebrates present on the mainland were already absent from small (> 0.5, < 1.5 ha) and medium (≥ 5, < 12 ha) islands by the early 1990s. Rapid species loss from land-bridge islands and forest fragments is a well-documented phenomenon (Ferraz et al. 2003). What is special about Lago Guri is that its myriad islands are effectively sealed from uncontrolled immigration and emigration of nonflying animals, and afford satisfactory levels of replication. The species that persisted on small and medium Lago Guri islands constituted a highly nonrandom subset of the preinundation fauna. There were no predators of vertebrates and few to no frugivores and seed dispersers, whereas seed predators (several genera of rodents) and herbivores were overrepresented (Terborgh et al. 1997b). Meanwhile, the faunas of large islands remained intact or nearly so.

Populations of vertebrates that persisted on small islands nearly all demonstrated hyperabundance, referring to the fact that their densities were far higher than those estimated on large islands and the mainland. Thus the animal "communities" present on small islands, and to a lesser degree on medium islands, were strongly imbalanced in that some ecological functions were underrepresented or absent, whereas others were overrepresented.

Most importantly for a test of HSS, predators of vertebrates were entirely absent, although predators of invertebrates were often hyperabundant, such as birds, lizards, toads, and spiders (Terborgh et al. 1997a). The most noteworthy feature of these imbalanced faunas was a vast hyperabundance of generalist herbivores that included small rodents of several genera, porcupines, iguanas (*Iguana iguana*), tortoises, howler monkeys (*Alouatta seniculus*), and leaf-cutter ants (*Atta* spp., *Acromyrmex* sp.). The densities of some of these species were 10- to 100-fold higher than on the mainland (Terborgh et al. 1997b; Rao 2000; Terborgh et al. 2001). If such exaggerated densities were simply a response to release from predation, the response was vastly greater than had ever been observed in small-scale exclosure experiments (Schmitz et al. 2000).

Hyperabundance of herbivores in the absence of predators is anticipated by the Green World Hypothesis but is in conflict with the Plant Self-Defense Hypothesis. Moreover, the persistence of hyperabundant herbivore populations on a number of islands from the early 1990s through 2003 implies that any herbivory-induced increases in plant defenses were ineffective. Although vertebrate herbivores are well-known for their capacity to denude oceanic islands, evolved loss of defenses is not a factor in the vegetation of Lago Guri islands. That is because the individual trees that compose the forest of these islands all began life on the South American mainland long before inundation stranded them, and are thus genetically representative of the mainland flora.

What are the impacts of herbivore hyperabundance on the vegetation of Lago Guri islands, and what are the long-term consequences of these impacts? These are the questions we set out to answer in 1996 when we began to establish sample vegetation plots at all our sites: small ($N = 6$), medium ($N = 4$), and large ($N = 6$). Every tree on the small islands was tagged, measured, mapped, and identified. Similar numbers of trees (300 per site) were incorporated into sample plots on medium and large landmasses. In addition, we delimited 15 × 15 m subplots within the adult tree stands in which we marked, measured, mapped, and identified smaller stems in two size classes, small saplings (≥ 1 m tall, and < 1 cm dbh), and large saplings (≥ 1 cm dbh to < 10 cm dbh). Overall, we marked, mapped, and identified 4771 trees and 7027 saplings in 32 sample plots at 16 sites.

The islands were 10 or 11 years old when we established the plots and change was already apparent. The numbers of small saplings on small islands were depressed to only a third of that on large landmasses, suggesting that plants of the smallest size class were being selectively impacted (Terborgh et al. 2001). However, the numbers of large saplings and adult trees were similar at all sites.

Several years previously, colleagues at the Smithsonian Tropical Research Institute in Panama had suggested that reproduction of trees might fail on small islands if young seedlings or saplings were exposed to desiccating dry season winds (Leigh et al. 1993). To investigate this possibility we sampled saplings in pairs of plots, one located on the windward and one on the leeward slope of each site where we had installed an adult tree plot. We hypothesized that if exposure to prevailing winds was impeding plant establishment, there would be greater mortality of stems, and/or reduced recruitment on the windward slopes of our sites.

We conducted recensuses of all sample plots in 2001 and 2002. The results provide conclusive evidence that both the structure and the composition of the vegetation of small and medium islands are gradually being degraded. Over the five years, the numbers of small saplings on the small islands dropped from a third to a quarter of control (large landmass) levels as a consequence of mortality rates averaging 46%, coupled with extremely low recruitment, amounting to only 20% of the rate recorded on large landmasses. Mortality of nearly all plant species exceeded recruitment. Those species that were the most resilient tended to possess thick and fibrous evergreen leaves. Larger-stem classes also suffered elevated mortality rates on small and medium landmasses. Adult trees, for example, died on small islands at a rate 50% higher than on the large landmasses. We failed to detect any effect of windward versus leeward exposure on mortality, recruitment, or survival.

High mortality and low recruitment, concentrated especially in the small sapling class, result in long-term consequences that can be projected into the future via stage-based matrix models. These models project a complete collapse of the forest on both small and medium islands within the next few decades as new saplings fail to replace adult trees that die (Fig. 5.1). Complete collapse of the forest is predicted to occur, first on the small islands, where the densities of herbivores are most elevated, and a decade or two later on medium islands. In contrast, the tree stands of large islands and the mainland are predicted to persist indefinitely.

Evidence of the degradation of small island forests forcefully strikes the eye of even a casual visitor. The scant understory vegetation allows a person standing in the middle of one of these islands to see light streaming in around the entire perimeter. In place of a carpet of leaf litter, the soil surface is mostly bare, displaying the bright orange-red color of subsoil excavated by leaf-cutter ants. Dead twigs, branches, and vine stems, by-products of canopy dieback, litter the ground and in places lie in heaps. It is a desolate scene.

Medium islands, in contrast, appear as green havens, covered in leaf litter and replete with saplings. In 1997, when we established the sapling plots, the numbers of small and large saplings on medium islands were indistinguishable from those on the mainland, so we thought that these islands were somehow immune to the progressive degradation that was so evident on the small islands. Our recensus results revealed something the unaided eye could not detect: that the numbers of new saplings recruiting on the medium islands were depressed to almost the same degree as on the small islands. We now attribute this to the appearance of new leaf-cutter ant colonies on the medium islands during the five-year period between censuses (see following).

What are the forces driving vegetation change on Lago Guri islands? Hyperabundant herbivores are the prime suspects, but there are several candidates. Arboreal species such as porcupines, iguanas, and howler monkeys could be contributing to elevated tree mortality and canopy dieback, but are doubtfully of much consequence to the survival of seedlings and saplings near the ground. Tortoises were initially common on several islands but are avidly harvested by poachers and have become scarce. Capybaras (*Hydrochaeris hydrochaeris*) use the forest for daytime resting but feed in the grassy drawdown zones surrounding the islands and not in the

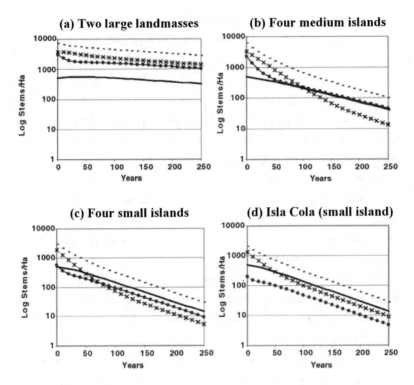

Figure 5.1
Projections of tropical dry forest stands on large, medium, and small landmasses:
(a) composite data for stands at two large landmass sites, (b) four medium landmasses, (c) four small landmasses, and (d) Isla Cola, the worst-case example. Canopy
trees only. The ordinate represents numbers of stems in three size categories and
total number of stems, standardized to 1 hectare. Legend: ——— Adult Trees
✕✕✕✕ Large Saplings ●●●● Small Saplings ------ Total

forest itself. By elimination, the circumstances point to leaf-cutter ants, which can
attain 100 times normal densities on small Lago Guri islands (Rao 2000). Leaf-cutter
ants swarm over the small islands every night and no plant escapes their scrutiny.

Seedlings set out under ant-proof exclosures survive well even on small
islands where unprotected seedlings sometimes fail to survive even one night
(unpubl. data). In a few instances, seedlings set out under ant-proof exclosures in
1999 were still alive and healthy in 2003. The strong inference of these experiments
is that leaf-cutter herbivory on seedlings and small saplings is the principal factor
responsible for the high mortality and low recruitment of saplings.

Nevertheless, there are other possible contributory mechanisms that so far have not been adequately assessed. One is that hyperabundant rodents are suppressing recruitment by consuming many of the seeds produced by the forest. Another possibility is that repeated defoliation is stressing trees and inducing them to allocate resources to vegetative growth at the expense of reproduction. Still another possibility is that fragmentation has impeded pollination and has thereby reduced the size of seed crops (Aizen and Feinsinger 1994). The first two possibilities entail "top-down" effects of hyperabundant consumers transmitted via interaction pathways distinct from that of herbivory, while the third reflects a possible but still undocumented type of ecological imbalance resulting from fragmentation. Further research will be required to evaluate these possibilities quantitatively, but overall, one can hardly doubt that hyperabundant herbivores are having major impacts on the vegetation.

The creation of Lago Guri constitutes the first comprehensive predator removal experiment conducted with replicates and controls on a large enough scale of both time and space to allow the responses of a whole ecosystem to be observed. Although there is more research to be done, we are satisfied that we have conducted a full and fair test of the Green World versus the Plant Self-Defense hypotheses, and that the Green World Hypothesis has won hands down. Hairston, Smith, and Slobodkin were right, but only in the aberrant context of a world lacking megaherbivores.

Nature Reduced to an Artifact

Megaherbivores have disappeared from most of our planet, and so they have largely been ignored in the Green World debate. Megaherbivores have had the most dramatic impacts in seasonally dry climates where trees grow only to modest heights and produce relatively short-lived leaves. In both Africa and India, vertebrate herbivore biomass peaks at 15,000 to 20,000 kg/km^2 in seasonal climates supplying around 1000 mm of rain (Owen-Smith 1988; Karanth and Nichols 1998; Sukumar 2003). Herbivore biomass is linearly proportional to rainfall up to $\pm$ 1000 mm, indicative of a strong bottom-up influence. In less seasonal evergreen forests that receive > 1000 mm rainfall, herbivore biomass is reduced to intermediate levels.

One can presume that evergreen forests support lower herbivore biomass than seasonal forests because long-lived leaves are chemically better defended than short-lived leaves (Coley et al. 1985; Coley and Barone 1996). Elephants and other large herbivores fail to destroy wet forests in Africa and Asia because much of the accessible foliage must be unpalatable to them. Thus plant defenses do deter herbivory, especially in wet and/or nutrient-poor sites where leaf lifetimes exceed one year. Regardless, elephants can have major impacts on the structure and regeneration of evergreen forests (Kortlandt 1984; Struhsaker et al. 1996).

Megaherbivores constitute one of two classes of vertebrate herbivores neglected by HSS; the other is herd-forming migratory ungulates (Terborgh et al. 1999). Whereas megaherbivores escape predation through huge size, migratory ungulates are able to reduce predation by aggregating in large herds (Sinclair and Norton-Griffiths 1979; Fryxell et al. 1988). Herds have the effect of concentrating an overwhelming number of prey within the territory of just a few predators, whether wolves (*Canis lupus*), lions (*Panthera leo*), or hyenas (*Hyaena, crocuta*). Predators, with their altricial young, are unable to migrate, and hence are ineffective at regulating the numbers of herd-forming ungulates (Bergerud 1988; Fryxell et al. 1988). Thus Serengeti wildebeest (*Connochaetes taurinus*) are clearly bottom-up regulated and most individuals die of natural causes (including, prominently, malnutrition; Mduma et al. 1999). Almost certainly, bison were similarly bottom-up regulated in precolonial North America.

In the absence of megaherbivores and migratory herd-forming ungulates (a condition characterizing much of the world today, and the case considered by HSS), predators appear to regulate the numbers of prey to levels well below those at which the prey begin to transform the vegetation. Plants, in turn, invest in antiherbivore defenses to the degree necessary to restrain the damage done by insects and small vertebrates.

So long as their numbers are low, lesser herbivores, such as the ones stranded on Lago Guri islands, will pick and choose the most tender and nutritious foliage, with most vegetation escaping unscathed (Glander 1981). Under such conditions, forest regeneration proceeds in a fashion we have come to consider "normal." But plants are not adapted to resist hyperabundant herbivores whose feeding is far less selective (unpubl. data). In the presence of hyperabundant herbivores, evolved levels of investment in defenses are manifestly inadequate and vegetation suffers severe damage. Prolonged herbivore hyperabundance eventually drives a substitution of heavily defended plants for less defended, faster-growing species (de Mazencourt and Loreau 2000). For example, in parts of the United States where white-tailed deer (*Odocoileus virginianus*) have exploded in the absence of wolves and mountain lions (*Puma concolor*), declines of herbivore-sensitive plant species are already widespread (Alverson et al. 1988; Miller et al. 1992; Garrott et al. 1993).

Plant defenses are most effective in conjunction with predators and in the absence of megaherbivores and herd-forming ungulates. Our results suggest that the defenses of most plants—in themselves—are inadequate to protect them from herbivores, especially during the vulnerable stage of leaf expansion. Forests around the world are able to regenerate because herbivores, both large and small, are prevented by predators and/or human hunters from attaining hyperabundance, except in unusual circumstances such as those documented in this chapter.

Putting together the various pieces of the puzzle, it is evident that neither the Green World nor the Plant Self-Defense hypothesis provides a complete picture of how nature works. Instead, both contribute to a larger picture, which must necessarily include megaherbivores and migratory ungulates. The world is green wherever the climate is warm and wet enough to support vegetation, and would be so even if the earth's megafauna were still intact (Oksanen and Oksanen 2000). Megafauna do not convert lush forests to dustbowls, but they can convert woodlands to savannas and maintain high rates of disturbance in forests, with little-known effects on plant composition.

We humans have greatly altered the natural balance of forces that regulate herbivory, first by eliminating most of the world's megafauna and herding ungulates, and then by systematically persecuting large predators. The greatly diminished state of nature that now prevails over perhaps 90% of the globe is thus an aberration produced by our own species. Even scientists tend to take this aberrant world as their frame of reference, and by doing so overlook key pieces of the picture. Such culturally determined blind spots have distorted the thinking behind both the Green World and the Plant Self-Defense hypotheses, because the explanatory power of these hypotheses is greatest in the context of the radically altered world in which most of us live.

The last fully natural terrestrial ecosystems on Earth amount to a few scattered remnants of landscapes that filled whole continents just a few generations ago. What is to become of these few precious remnants? No one can say, for their future is far from secure. If we lose the last elephants, lions, and rhinos, what other aberrations of nature will become the basis of future ecological theories? In the absence of megaherbivores, migratory ungulates, and top predators, ecology will by default become the science of human artifacts.

Conservation Recommendations

Results from the tropical dry forest ecosystem at Lago Guri, Venezuela, affirm that an absence of predators of vertebrates can trigger a massive, communitywide trophic cascade that leads to a catastrophic transformation of the habitat. The clear conservation message to emerge from this work is that predators are essential for maintaining the stability of many, if not most, terrestrial ecosystems. Thus the restoration of wolves and mountain lions to as much of the North American continent as practical should be a conservation imperative.

Terrestrial ecosystems from the early Mesozoic onward have supported both large carnivores and megaherbivores (those large enough as adults to escape predation). Today, such animals occupy less than 10% of the earth's terrestrial realm. What are the effects on other species, both plants and animals, of losing these two leading components of ecosystems? The question is an extremely important one, but strangely, it has hardly been investigated. In a more speculative vein, one wonders whether we should restore elephants, camels (*Camelus* spp.), horses, cheetahs (*Acinonyx jubatus*), and lions to the North American continent. Would the impacts of doing so have positive or negative consequences for sustaining native biodiversity? It would be nice to have answers to questions such as these, for they are highly relevant to conserving biodiversity on Earth.

Summary

A trio of distinguished ecologists proposed more than 40 years ago that "the world is green" because predators regulate the numbers of herbivores and thereby prevent herbivores from destroying vegetation. Their argument was soon rebutted by advocates of the opposing view that the world is green because plants defend themselves with toxic chemicals (the Plant Self-Defense Hypothesis). Decades have now passed with no clear resolution of this debate.

Lack of progress in resolving the issue is largely attributable to the formidable difficulty of conducting large-scale predator removal experiments with appropriate controls. Here I report results from such an experiment that was initiated in 1986 with the filling of Lago Guri, a 4300 km² hydroelectric impoundment in Venezuela. Rising water created hundreds of islands, many of which are too small to support predators of vertebrates, but large enough to sustain populations of consumers.

A number of Lago Guri islands have now been under study for more than a decade. Islands lacking predators of vertebrates support bizarrely imbalanced ecological communities containing a paucity of dispersers (and perhaps pollinators), but a pronounced hyperabundance of predators of invertebrates, seed predators, and, especially, generalist herbivores. A five-year study of the dynamics of small and large saplings and canopy trees on small, medium, and large landmasses revealed high mortality of stems of all sizes and low recruitment of small stems on predator-free islets. A matrix model paramaterized with the results predicted complete collapse of the forest of predator-free islets within a few decades because tree recruitment fails to keep pace with mortality. Thus, in the absence of predators, plant defenses do not deter herbivory to the degree necessary to prevent the destruction of vegetation. At face value, our results support the Green World Hypothesis and reject the Plant Self-Defense Hypothesis. The results from Lago Guri should properly be interpreted, however, in the context of a prehuman world that abounded with megaherbivores capable of transforming vegetation. In the presence of megaherbivores, neither the Green World nor the Plant Self-Defense Hypothesis is valid because megaherbivores are not regulated by predators and are only conditionally deterred by the defenses of many plants. The explanatory power of both these hypotheses is greatest in the highly artificial world occupied by most scientists, a world lacking both megafauna and top predators. If humans complete the process of exterminating these guilds worldwide, ecology will irrevocably become the science of human artifacts.

ACKNOWLEDGMENTS

Warm thanks go to the administration of EDELCA (Electrificación del Caroní), and to Luis Balbas in particular, for longstanding support of our project at Lago Guri. I am extremely grateful to the many individuals who helped with the tedious work of installing and censusing plant dynamics plots. The John T. and Katherine D. MacArthur Foundation and the National Science Foundation (DEB-97-07281, DEB-01-08107) provided financial support.

CHAPTER 6

Restoring Functionality in Yellowstone with Recovering Carnivores: Gains and Uncertainties

Joel Berger and Douglas W. Smith

For those interested in the conservation of biological diversity, the word "natural" often conjures up systems that may be remote and also are likely to be pristine. To many, such regions exist where human influences are minimal and ecosystem processes operate today as they have in the past. To ecologists, however, such systems are rare or nonexistent. Nevertheless, these are precisely the images that jump to the forefront of the public mind when natural ecosystems are mentioned.

In reality, no one really knows how best to creatively define "natural"—a difficulty that arises in part because of variation in dimensions that include time and space. If one were to imagine a vast boreal landscape devoid of modern human activity, the claim could easily be made that the system is more natural than one filled with gas pads, dammed rivers, or clear-cuts. But the conundrum is obvious when dealing with large carnivores, whether wolves (*Canis lupus*), bears (*Ursus* spp.), or lynx (*Lynx canadensis*), which are still abundant at some sites but not others. So could "natural" systems be those with a continued presence of top predators (Anderson 1991; Pritchard 1999)?

The area in and around Yellowstone National Park provides a useful place to explore the relationship between carnivores and the definition of "natural." The last wolves from the area outside the park were removed in the 1930s and from Yellowstone in 1926 (Weaver 1978), and wolves were not reestablished until 1995 and 1996 (Bangs and Fritts 1995; Phillips and Smith 1996).

In this chapter we summarize results on how the return of wolves is reshaping the 8991 km² Yellowstone National Park (YNP), and, in particular, how these apex carnivores are altering biological diversity (Smith et al. 2003). Continuous monitoring over periods of time when wolves were both present and absent offers a unique opportunity to examine the link between large carnivores and bio-diversity in this North American temperate ecosystem. YNP is part of a broader ~60,000 km² ecosystem, so we also describe what changes wolves have effected beyond the park's borders. This latter point is relevant scientifically and from a con-servation perspective, particularly because there are many areas in the United States and other parts of the world where wolf or other large carnivore reintro-ductions will never be possible. In this case, however, gauging how wolves mod-ify landscapes lends itself to an understanding of what can and cannot be achieved for the conservation of biological diversity, especially when predation is lost from other systems and where ecological dynamics can unfold with minimal current in-tervention by humans.

From a conservation perspective it is also necessary to know how important large carnivores can be, especially wolves, in ecosystem functioning. Some say they are not important at all, that ecosystems can do without them, that their return only makes things seem "perceptually" healthier (Theberge 2000). But on the other hand, we know that wolves bring about dramatic change to the animals and plants present in the system. So is it really true that we don't need wolves? Or have we lived long enough without them to know the long range impact of not having them? We also know that human hunters do not serve as surrogate wolf hunters (Wilmers et al. 2003b; Berger, this volume), even though this has been one ra-tionale for massive global predator extermination. So where are we? Do wolves matter? Is carnivore conservation even relevant to anything of importance to human beings or ecosystems?

Ecology in Yellowstone National Park with and without Wolves

An understanding of the prior management of YNP is fundamental to interpret-ing the role of wolves in restructuring ecological communities. Volumes have been written about the history of YNP (Haines 1977; Schullery 1997; Pritchard 1999). Here we draw only the most rudimentary historical sketch of species and processes relevant to assessing wolf-related effects.

During the latter part of the 19th century, just after YNP had been established, poaching resulted in the killing of large numbers of elk (*Cervus elaphus*) and other big game (Haines 1977). Bison (*Bison bison*) dropped to fewer than 50 animals (Meagher 1973). Subsequently, the existing herd was supplemented by the introduction of bison from elsewhere, and, within Yellowstone, all were fed and virtually farmed (Pritchard 1999). The population increased and such practices terminated in the 1960s (Meagher 1973). Elk within and beyond YNP were also managed. From the 1930s to 1968, elk from the northern Yellowstone herd, the park's largest, were controlled by the Park Service. In 1968 the herd numbered 4000 animals. On the northern range of YNP animals were reduced by some 75% (Houston 1982), and the animal populations within the northern YNP boundaries were limited to around 4000. Once human harvest terminated, by the late 1980s, and then again after a crash, in the early 1990s the numbers within the park grew to approximately 19,000. Like bison (which increased to about 4000 animals in 1997, crashed, and then increased again to 4000 in 2004), there was little apparent significant predation on adults of either species (Houston 1982). Since animals move beyond park boundaries to nonprotected areas, they can be shot. Control actions beyond park boundaries limit numbers within the park because the removed animals cannot, obviously, return (Meagher 1989; Lemke et al. 1998; Smith et al. 2003).

Although other species, including cougars (*Puma concolor*) and white-tailed deer (*Odocoileus virginianus*), have also changed in population size and distribution within and beyond the park, much attention has focused on abundance of the large herbivores such as elk and bison. During the 70-year period between wolf extirpation and reintroduction, strikingly different viewpoints have been expressed about changes in native biodiversity, often centering on what is natural and what is not. Among notable changes in wildlife since the establishment of the park has been colonization by moose (*Alces alces*) (Tyers 2003); the virtual elimination of beavers (*Castor canadensis*)(from the northern range; YNP, unpubl. data) and white-tailed deer; lack of recruitment in, and reduction of, aspen and cottonwood communities; diminished willow communities; and reduction of songbirds (Hansen and Rotello 2002). Such modifications have variously been attributed to human mismanagement (Chase 1987), loss of indigenous human hunters (Kay 1994), extirpation of wolves, weather cycles and fire, erosion, and complexities that have yet to be unraveled (Singer et al. 2002, 2003). Although undoubtedly there is much merit in both the empirical support and the opinions that underlie these varied po-

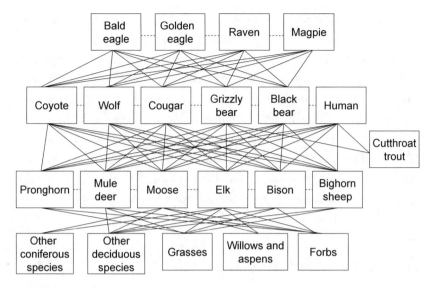

Figure 6.1

Overview of simplified trophic level interactions involving scavengers, terrestrial carnivores, fish, ungulates, and plants in Yellowstone National Park. Dotted lines indicate possible competitive pathways. (Modified from Smith et al. 2003).

sitions, what is now clear is that, since their reintroduction within the park, wolves have mediated a ripple of effects throughout the food chain (Fig. 6.1). Among the major changes are both direct and indirect effects on processes, some with visible influences on species and landscapes.

Predator–Prey Relationships

Of the seven native ungulates within the Yellowstone ecosystem—mule deer (*Odocoileus hemionus*), pronghorn (*Antilocapra americana*), white-tailed deer, elk, bison, bighorn sheep (*Ovis canadensis*), and moose—all but white-tailed deer have been killed by wolves within YNP; and this may be due to the fact that their populations are low due to restricted habitat. The major prey of YNP wolves both numerically and by biomass is elk, representing 92% of 1582 kills between 1995 and 2001. Calves are taken disproportionately relative to their availability (43% vs. 15%), adult females reflect the opposite (28% killed, 60% available), and adult

₄ies are killed in proportion to their abundance (21% vs. 25%, respectively; Smith et al. 2004). Condition of elk as determined by marrow fat indicated that wolves may be selecting for elk in poor condition. An increasing number in early winter killed by wolves are in poor shape and virtually all of them in late winter show marrow fat depletion, the last fat reserve utilized by ungulates (YNP, unpubl. data).

Among other large herbivores, moose and bison combined represented less than 2% of the total kills, although bison calves, like elk, were killed in numbers disproportionate to their abundance (Smith et al. 2000). Other species killed by wolves include beavers, coyotes (*Canis latrans*), badgers (*Taxidea taxus*), Canada geese (*Branta canadensis*), ground squirrels (*Spermophilus* spp.), and other small mammals (Ballard et al. 2003).

Relationships Involving Individual Species

At least 16 vertebrates have been shown to be affected in ecologically important ways by wolf predation (see Fig. 6.1). For instance, not only have prey species such as elk been reduced in total population size (and for reasons not entirely due to wolves; other factors, such as female harvest of prime-aged elk, multiple carnivores, and multiyear drought are also important), but coyote densities have also been diminished at local scales by at least 50%, a change brought about through both spatial avoidance and intraguild predation (Crabtree and Sheldon 1999). For example, coyote population size in one region of the park declined from 80 to 36 individuals (Crabtree and Sheldon 1999).

Wolves have also facilitated the availability of carrion for grizzly bears, not simply because wolves kill large items such as elk but also because grizzly bears now have more carrion available (Ballard et al. 2003). Indeed, grizzly bears tend to dominate wolves at carcasses (Smith et al. 2003). Influences of wolf populations on other carnivores have yet to be documented, although cougars tend to be driven from their kills by wolves more than the converse, and several cougar kittens plus one adult female were killed by wolves (Ruth et al. 2003; Smith et al. 2003). On the other hand, cougars have also killed wolves (YNP, unpubl. data).

In addition to possible or demonstrated effects on fish, carnivores, and ungulates, the carrion made available by wolves also attracts at least 12 vertebrate scavengers, 5 of which visit nearly every wolf kill: bald (*Haliaeetus leucocephalus*) and golden (*Aquila chrysaetos*) eagles, coyotes, ravens (*Corvus corax*), and magpies (*Pica pica*) (Wilmers et al. 2003a). For at least one of these species, ravens, a change in their patterns of food detection and acquisition has been noted since the return of wolves; ravens arrive at wolf-killed meals on average within one minute of death and actually *follow* wolves to locate food before they make a kill (Stahler et al. 2002). The greatest number of ravens ever recorded on a wolf kill, 135, comes from a report in YNP (Stahler et al. 2002), which recorded an average number of 29 ravens per wolf kill. It is hard to overstate the importance of wolf-provided carrion to the scavenger guild; indeed, this may be one of the most significant impacts of wolf recovery because no other animal provides the food to this group in the manner that wolves do. Cougars, the other primary large carnivore in YNP, cover their kills, making them unavailable to scavengers. Wolves, on the other hand, try to out-eat the masses of scavengers and can lose significant amounts of food to them (Vucetich et al. 2004).

Landscape-Level Change

As is well appreciated from theory and a plethora of ecological studies, biological communities are often dynamic. Such dynamism has characterized a 100,000 ha region of northern YNP where aspen (*Populus tremuloides*) recruitment has declined during the past 100 years—a reduction that coincides with decades of high elk densities (Houston 1982; Ripple et al. 2001; Larsen 2002). However, abiotic and biotic factors have also been implicated in the decline of aspens and other mesic or riparian plant communities (Romme and Despain 1989; Kay 1994; Singer et al. 2002, 2003). Regardless of the cause of the decline, aspen stands are biodiversity hotspots across the West, especially for songbirds (Turchi et al. 1995).

If wolves affect the foraging behavior and/or distribution of elk (as is suggested for numerous vertebrate predators; Brown et al. 1999), then, assuming that elk are structuring agents themselves, wolves may indirectly affect plant growth and recruitment. Elsewhere in this book, evidence is presented that large herbivores, especially at relatively high densities, have strong ecological impacts (see McShea; Terborgh; Berger, this volume). In the Greater Yellowstone Ecosystem, some evidence of a wolf-mediated pathway with ultimate effects on plants is based on work on aspen sucker growth. Ripple et al. (2001) found sucker height to be significantly greater in areas with high wolf densities and where elk dung counts were lowest. These results, however, were ephemeral and not long term, suggesting multiple causes for the reported response (YNP, unpubl. data). Willow growth is perhaps a better example, but is equally complicated and also likely to be interactive between factors. Although willow growth is significant in some places, it is important to note that this response has not been uniform across the northern range (YNP, unpubl. data). However, we feel the most parsimonious explanation is that elk have reduced their use of areas where predation pressure was high and, as a consequence, the intensity of browsing on aspens and willows was relaxed (Ripple and Larsen 2000; Ripple et al. 2001).

Beyond vegetation and elk, wolves may possibly affect diversity at other levels. For example, the nesting density of songbirds such as the common yellowthroat (*Geothlypis trichas*), Lincoln's sparrow (*Melospiza lincolnii*), warbling vireo (*Vireo gilvus*), Wilson's warbler (*Wilsonia pusilla*), and yellow warbler (*Dendroica petechia*) in willow stands varied in a nonlinear fashion with the severity of browsing, suggesting a threshold effect (Jackson 1992). Similarly, where both grizzly bears and wolves were extirpated to the south of YNP, avian diversity was less than in areas where humans controlled another large cervid, moose (Berger et al. 2001a). Although not completely tested in YNP, both to the south (Grand Teton National Park; Berger et al. 2001a) and to the north (Banff National Park; Nietvelt 2001), a link between wolves, ungulates, vegetation, and songbirds has been established. We expect that, with further examination, the same will hold true in YNP. Evidence to include beavers in this mix is also accumulating: their return to the northern range in the past six years is correlated with wolf recovery and appears to be linked with increased availability of willow (Nietvelt 2001; Smith, unpubl. data). Hence, if the threat of predation by wolves on elk or moose redistributes these large-bodied ungulates away from willow communities, then herbivory is likely to be reduced, which may subsequently influence biological diversity (Berger et al. 2003). However, the relationship may not be so straightforward since many factors other than herbivory per se also affect the density and growth of willows and avian reliance upon these plant communities (Cody 1981; Finch 1989; Singer et al. 2000).

It is important to note that grasses drive the northern Yellowstone system (and are not suppressed by ungulate grazing; Frank and McNaughton 1992) because only 2 to 4% of the entire area is deciduous woody vegetation (Smith et al. 2003). But there is evidence that the importance of woody vegetation is disproportionate to its abundance. Therefore, any increase may increase biodiversity; for example, as we are beginning to see with beavers responding to increased willow, which in turn provides aquatic habitat for other plants and animals (Baker and Hill 2003; YNP, unpubl. data).

Ecology beyond Yellowstone National Park with and without Wolves

Although YNP is the one of the world's premier locations, both for understanding ecological processes and because it has galvanized carnivore restoration globally (Clark et al. 1999), the park itself represents only ~15% of the entire Greater Yellowstone Ecosystem. As a consequence there is much to learn not only by temporal contrasts of processes prior to and after wolf reintroduction but also spatially through assessment of ecological effects of variation in wolf densities within and beyond park boundaries. Such contrasts of course are never as clean as purists hope because in addition to carnivore densities, sites vary in topography, history, structural community components, human management, and many other factors. Additionally, there have been only limited scientific study of areas beyond (or even within) YNP prior to wolf extirpation. Nevertheless, much can be gleaned by contrasts between wolf-related effects in the park itself relative to those outside the protected boundaries.

Wolf densities on the northern range of YNP (~50 per 1000 km^2) are some of the highest known in North America (Smith et al. 2003), and these are supported by a prey biomass that is similarly high (Singer and Mack 1999). Importantly, effects of current wolf densities can be contrasted between the park and regions beyond where densities are much lower due to conflicts with livestock.

As apex carnivores, wolves would be expected to exert effects outside the park that might be similar to those within, assuming other factors to be equal. Given that lands outside YNP receive differing levels of protection, none as strict as within (where there is no hunting and grazing of livestock), opportunities exist to determine what lack of effects may be occurring because wolves are at reduced densities. For instance, in lands adjacent to the park, about 100 wolves have been killed due to conflicts with people and livestock. Indeed, wolf densities in some areas of the Greater Yellowstone ecosystem may be 1% or less than those within YNP (see Fig. 16.4 in Berger, this volume). Human tolerance in the Yellowstone ecosystem, despite relatively higher economic well-being, is lower than tolerance for carnivores in other areas (e.g., Africa or India) where human population density is higher and economic well-being orders of magnitude lower (Creel and Creel 2002; Jhala and Giles 1991). This disparity is interesting and counterintuitive. Clearly, where there are no or few wolves, they will play no or little role in directly shaping ecosystems and biodiversity, except perhaps through a series of cascading processes where ungulate densities are unregulated or modified by humans.

Understanding why wolves are not playing an ecologically functional role beyond YNP boundaries but still within the much larger Greater Yellowstone Ecosystem is relevant more from a conservation than an ecological perspective. This is because wolves in this region of the United States are soon to be de-listed from federal protection since population sizes have achieved a level at which demographic viability is expected (Smith et al. 1999). Once this occurs, the American public may believe that wolves are playing functional ecological roles. However, this is unlikely to be the case because wolves are killed beyond protected boundaries, and ungulates and other ecological responses may be derived more as a consequence of the human milieu than human hunting or wolves (Pyare and Berger 2003; Soulé et al. 2003). Nevertheless, the fact that wolves are currently playing an apex role in YNP offers unprecedented opportunities to understand basic ecology while minimizing the role of anthropogenic disturbance that characterizes so much of the terrestrial world, and allows us to better answer the question, Do wolves and other large carnivores matter?

Summary

Following a 70-year absence, wolves were reintroduced to Yellowstone National Park (YNP) in 1995 and 1996 where they now play a key role as apex predators. Continuous monitoring of this north temperate ecosystem during the absence and return of wolves has provided an opportunity to examine the relationship between wolves and biodiversity. Not only do wolves define ecological relationships with ungulates and vegetation but they are initiating subsequent effects on biodiversity through interactions with other predators and scavengers. Although wolf density on the northern range of YNP is one of the highest reported, effects of wolves on ecosystem processes and biological diversity beyond park boundaries are likely to remain less obvious. This is because beyond YNP boundaries there is a confounding array of other processes that include landscape-level changes and associated human activities, and in general human tolerance for wolves is low. Although the goal of wolf reintroduction into YNP was to support a self-sustaining population throughout the entire region, less attention was initially focused on functional ecological relationships. The YNP portion of the overarching U.S. Rocky Mountain wolf population therefore enables a fundamental understanding of how wolves affect biodiversity unfettered by direct anthropogenic manipulations. Beyond park boundaries, however, wolves are not likely to play critical long-term ecological roles.

CHAPTER 7

Large Marine Carnivores: Trophic Cascades and Top-Down Controls in Coastal Ecosystems Past and Present

Robert S. Steneck and Enric Sala

The ways in which marine carnivores affect biodiversity are conspicuous. Studies documenting some of these effects have shaped modern ecology. For example, concepts such as "keystone species," in which carnivores at relatively low abundance structure natural communities by mediating competition (Paine 1966), and "trophic cascades," in which predators regulate herbivores allowing edible plants to be limited only by resources (Paine 1980; Carpenter and Kitchell 1993), were both first observed and described in the marine realm. These examples were top-down (consumer-dominated) effects driven by marine carnivores (Menge and Sutherland 1987; Scheffer et al. 2001; Steneck, this volume). These examples came from the marine realm in part because shallow marine benthic ecosystems are accessible and scaled conveniently in space and time for ecological studies and also because documented predator impacts in these systems appear to be particularly strong (Schmitz et al. 2000; Shurin et al. 2002; Steneck, this volume).

Oceans cover two-thirds of Earth, but most are nutrient-poor and populated with hyperdispersed and often highly migratory organisms that are difficult to study. Our review focuses on a subset of the marine realm: the coastal zones, shallow continental shelves where marine life concentrates at or near the seafloor. In this chapter we illustrate how widespread and globally important carnivores, particularly large carnivores, have been, and are, to the structure of marine ecosystems. We do not suggest that carnivores heavily influence all shallow benthic marine ecosystems, but they do influence many (or they have in the past) and thus should be considered for managing the biodiversity of those ecosystems. Large carnivores and top-down controls are also important in some pelagic and deep-sea realms, but they are studied less and are outside the scope of this review.

Predation Theory and Evidence of Effects

Carnivores are predators that consume animals; herbivores are predators that consume plants. The effect of consumers on prey populations is measured as "interaction strength" (Paine 1980, 1992), which depends on more than just its abundance. We define per capita interaction strength as the effect of an individual predator on the population of its prey (interaction strength hereafter), and population interaction strength as the effect of a population of predators on the population of their prey (ecological impact hereafter). We focus primarily on strong interactions among marine carnivores that limit prey abundances. The communitywide consequences of such interactions depend on whether the prey of top, or apex, predators are themselves strong-interacting predators. Such "mesopredators" can be either carnivores if they eat animals or herbivores if they eat plants. Typically they are smaller than apex predators. Trophic cascades result when carnivores affect at least three trophic levels, such as apex predators controlling herbivorous prey that are strong interactors with the plants they eat (see Steneck, this volume for a review).

Predator effects can be direct or indirect (Menge 1995). Direct interactions are characterized by those in which predation reduces prey populations. However, evidence of direct predator control can be elusive. This is not because such predator effects are subtle, but because predators have become rare and/or their body sizes small due to overfishing (Pauly et al. 1998; Jackson et al. 2001; Duffy 2002). The secondary consequences of prey reductions, or indirect effects, consist of increases in subordinate competitors following predator-induced reduction of competitively dominant prey. Undoubtedly, ripple effects from indirect predation are widespread, but our focus will be primarily on direct predator interactions across several trophic levels.

We review the literature for evidence demonstrating direct carnivore impacts in marine ecosystems. We begin by considering the decline in marine carnivores over the past few centuries. We then review case studies for which archaeological and historical research suggests large marine carnivores were important. By integrating those temporal trends in predator abundance and size with contemporary research on interaction strengths we make the case for the probable effects these carnivores once had on entire communities.

We will highlight a few modern examples of predator effects on benthic marine community structure where the strongest evidence comes from controlled experiments. Although these are relatively easy to perform with slow-moving invertebrate predators, they are less effective or impossible to apply to large, mobile carnivores. For them, we focus on three categories of evidence: (1) experimental evidence such as marine reserves in which fishing has been curtailed over a sufficiently large area to increase predator abundances; (2) gradient analyses in which stratified samples over a gradient in predator abundance allows their impacts to be assessed; and (3) strong inferences based on multiple lines of evidence such as documented diets, past abundances, and size.

We Eat Large Marine Carnivores: Fisheries-Induced Declines in Predator Abundance

Globally the abundance and body sizes of predatory fish have declined (Jackson et al. 2001; Dayton 2003). Archaeological evidence from Alaska (Simenstad et al. 1978), California (Erlandson and Rick 2002), Gulf of Maine (Steneck et al. 2002), and coral reefs worldwide (Pandolfi et al. 2003) all suggests these fishing-induced changes began centuries to millennia ago because large carnivores are often targeted first. Unfortunately, many of these large species are very susceptible to overfishing because they are often long-lived with relatively low birth rates, they readily attack baited hooks, and/or they aggregate for spawning, making them easy to overharvest (Sala et al. 2001; Ames 2003). Overall landings records can mask fisheries depletions due to the opportunistic switching of fishing effort to the remaining most abundant species, and increasing effort to land the same biomass of fish. Nevertheless, since the late 1980s the reported global marine fisheries landings have begun an apparently steady decline (Fig. 7.1; Pauly et al. 1998, 2001). Although fisheries statistics have been collected globally in a more or less rigorous way over the last 50 years, the depletion of other large carnivores such as seals,

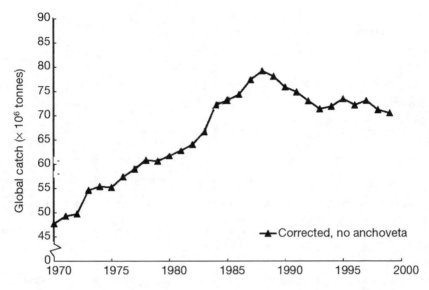

Figure 7.1

Temporal trend of global reported marine fisheries landings excluding the highly
variable Peruvian anchoveta from 1970 to 2000. (Reprinted with permission from
Nature [Watson and Pauly 2001], Copyright (2001), Macmillan Magazines Limited.)

turtles, and crocodiles has gone mostly unreported. In any case, the sequence of
changes appears to be universal, with large mammalian carnivores being targeted
first, followed by large fishes, smaller fishes, and finally invertebrates. Only large herbivores such as sea cows (*Dugong dugon*) and green turtles (*Chelonia mydas*) have been
targeted intensely as soon as or earlier than large carnivores (Pandolfi et al. 2003).

Carnivorous fish are declining most rapidly in recent times. Because they can
be caught with hook and line, their catch rate is a measure of abundance (called
catch per unit effort). Since 1960, the abundance of carnivores declined precipitously (Fig. 7.2; Myers and Worm 2003). Once apex predators are reduced, fishing effort often shifts to second- or third-order consumers such as mesopredators
and herbivores, which is known as "fishing down food webs" (Pauly et al. 1998).
Daniel Pauly and coworkers (Pauly et al. 1998, 2001) demonstrated this by as-
signing numbers for each trophic level, starting with plants at the bottom as
trophic level number one, up to apex predators such as the great white shark (*Carcharodon carcharias*) with the highest number. The mean trophic levels of fish

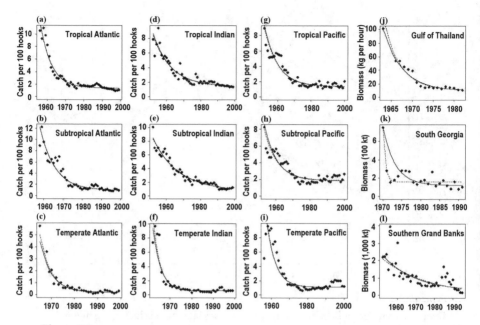

Figure 7.2

Time trends of fish community biomass in oceanic (a–i) and shelf (j–l) determined from Japanese long-line fishing efficiency (i.e., catch per 100 hooks fished). (Reprinted with permission from *Nature* [Myers and Worm 2003], Macmillan Magazines Limited.)

landed globally show a general decline over the last half-century (Fig. 7.3a), with different ocean systems showing different starting trophic levels and rates of decline. For example, the Mediterranean declined more gradually than has the western North Atlantic (Figs 7.3b,c). The dramatic post-1970 decline in North Atlantic predators (Fig. 7.2c) resulted in the steep drop in mean trophic level (Fig. 7.3c). The ubiquity of such declines suggests that large carnivorous apex predators are being, or have been, extirpated globally.

Sharks are the largest predatory fish. Unfortunately, we know very little about their impact because by the time we sought to address that question they were absent from most marine ecosystems. Sharks have been fished for hundreds of years, and even supported fisheries in the Caribbean during the colonial period for consumptive uses, or they were simply killed because of fear (Sánchez Roig and Gómez de la Maza 1952). However, they have been caught, killed, and discarded

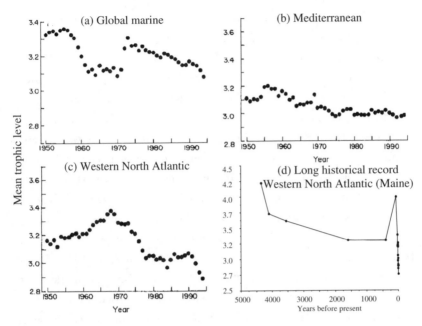

Figure 7.3

Trends of mean trophic level of fisheries landings for (a) global, (b) Mediterranean, (c) western North Atlantic for 1960–1994 (after D. Pauly et al. 1998) and (d) for coastal western North Atlantic over past 4500 years (after Steneck et al. 2004, figure 2a with permission, Copyright, Springer-Verlag). Top carnivores have the highest numbers, plants have the lowest. Vertebrate predators usually range from 3.5 to 4.6, invertebrate mesopredators 2.5 to 3.5, herbivores 2.0 to 2.4, and plants 1.0.

as bycatch, without records, for centuries (Baum et al. 2003). Their low fecundity, slow growth rate, and relatively long life make them vulnerable to, and slow to recover from, depletion (Schindler et al. 2002). Recent estimates suggest that since the 1980s large shark abundance has declined 50 to 95% from earlier, already low population densities (Baum et al. 2003). It is exceedingly difficult, therefore, to estimate the ecological impact of sharks as apex predators in marine ecosystems prior to human effects. In some coastal ecosystems, sharks must have been important apex predators, but we can only guess what effect they may have had on those ecosystems. For instance, a simulation model suggested that the removal of tiger sharks from the coral reefs of French Frigate Shoals in the northwest

Hawaiian archipelago would produce a ninefold increase in the abundance of marine turtles and reef sharks and a threefold increase in bottom fish abundance (Stevens et al. 2000).

Our interpretation of the structure and functioning of marine ecosystems may be skewed by historical loss of large carnivores. Many ecosystems today are thought to be primarily under bottom-up (resource driven) control because consumers have been removed by fishing (Jackson et al. 2001). However, there is ample evidence that large carnivores were more important in the past. For example, one of the first illustrations of marine life in the New World comes from a painting of the Chesapeake Bay made in the 1500s, which depicts indigenous people spearing and trapping fish in weirs. Beneath the canoes filled with large fish, we see through the clear water hammerhead sharks (*Sphyrna* spp.) and other large fish (see cover of *Science,* Volume 293, Number 5530, 2001). Today there are neither hammerhead sharks nor other large predatory fish in the murky Chesapeake Bay ecosystem. The most economically important species, such as blue crabs (*Callinectes sapidus*), may well have been prey of predators past (interestingly, they were not depicted in the early New World painting). Many modern ecological studies point to bottom-up forces as the cause of the current crisis in this ecosystem. However, the Chesapeake may be one of myriad such systems that were once top-down structured but, through fishing down of the local food webs, have become bottom-up structured. This does not mean that human disturbance has eliminated top-down control, since humans are exerting the strongest top-down forcing we have known. What this means is that humans have simplified, homogenized, and accelerated the turnover of marine food webs in a way that makes marine communities much more susceptible to bottom-up factors (Sala and Sugihara 2005).

Evidence for Past and Present Top-Down Predator Effects Altering Trophic Cascades in Major Benthic Marine Ecosystems

The clearest evidence of top-down predator effects comes from coastal marine communities dwelling on hard substratum (Shurin et al. 2002; Steneck, this volume). Globally two fundamentally different types of ecosystems develop on hard substrata: coral reefs in the tropics and kelp forests at higher (nonpolar) latitudes. These two ecosystems differ biological and geologically. They have almost no higher taxa in common, yet both show evidence of predator effects.

Kelp Forest Ecosystems of North America

Some of the best examples of trophic cascades are found in kelp forests. In well-studied systems of the eastern North Pacific and the western North Atlantic, predators control sea urchins, thus providing evidence of recent or past strong top-down control. Following here we describe those and other kelp forests apparently structured by trophic cascades.

Alaska

Alaska's Aleutian kelp forests were likely well developed before human contact be-cause sea otters (*Enhydra lutris*) were probably abundant and their predation on sea urchins prevented overgrazing on kelp (Simenstad et al. 1978; Estes et al. 1998; Steneck et al. 2002). This story, detailed elsewhere in this volume (Estes), merits some degree of repetition here because it remains one of the best-documented trophic cascades in any system. Aboriginal Aleuts locally diminished sea otters be-ginning around 2500 BP, with a corresponding increase in the size of sea urchins (Simenstad et al. 1978). European and North American fur traders subsequently hunted the remaining otters to the brink of extinction in the 1700s and 1800s, caus-ing the collapse of kelp forests because they were grazed away by sea urchins released from sea otter predation (Steneck et al. 2002). Legal protection of sea ot-ters in the 20th century reversed their decline and reestablished their trophic cas-cade in the 1970s and 1980s. Beginning in the 1980s, however, sea otters began declining again—this time due to predation by killer whales (*Orcinus orca*) (Estes et al. 1998; Estes, this volume).

Killer whales are the largest living apex predator. They travel over large areas and are capable of consuming the largest animals on Earth—the great whales. Estes (this volume) suggests that the sea otter decline was just the last in a series of predator-induced extirpations caused by killer whales, precipitated by diet shifts to harbor seals (*Phoca vitulina*), then Steller sea lions (*Eumetopias jubatus*), and now sea otter populations, all populations of which eventually collapsed throughout the Aleutians. Following this decline of sea otters, the population of sea urchins increased, causing almost complete deforestation of kelps (Estes, this volume).

Although trophic cascades occur over vast areas, the Aleutian kelp forest ecosystem has relatively few strong-interacting species. It is clear that, in this system, the sea otter is a true keystone species. Estes and Duggins (1995) documented the remarkably strong associations between otters and kelp at numerous islands with this predator, and between sea urchins and grazing-resistant coralline algae at sites without otters. The results are consistent in space and time because at the time of the study, otter populations had been increasing since the 1970s when strong interactions on their herbivorous prey were first described (Estes and Palmisano 1974). This progression stopped and reversed after otters were eliminated by predation from killer whales (Estes et al. 1998). This is one of the few observed marine examples of top-down impacts resulting from large and highly migratory predators. It illustrates well Hairston, Smith, and Slobodkin's (1960) theory that predators regulate herbivores that regulate edible plants in three-trophic-level systems. However subsequent theory suggested that if an even higher-order apex predator representing a fourth trophic level enters the system, it will effectively control the predators of herbivores, allowing herbivores to overgraze plants (Fretwell 1977, 1987; Oksanen et al. 1981; Steneck, this volume). With the unprecedented attacks of killer whales on sea otters beginning in the 1990s, sea otters lost their status as the system's apex predator and the Aleutians once again lost their kelp forests to grazing sea urchins (Estes; Steneck, this volume). This example shows that entire communities can change with even subtle changes in apex predator populations. It also illustrates the context-dependent nature of top-down controls (Pace et al. 1999).

Western North Atlantic

Marine carnivores of the western North Atlantic were both more abundant and larger in the past. In Maine, archaeological evidence indicates that coastal people subsisted on Atlantic cod (*Gadus morhua*) for at least 4000 years (Steneck 1997; Jackson et al. 2001). Cod constituted up to 85% of the bone mass in middens and they averaged about a meter in length (Jackson et al. 2001). Significantly, prey species such as lobsters and crabs were absent from the excavated middens in the region, perhaps because large predators had eaten them (Steneck et al. 2004).

Today cod are ecologically extinct from coastal zones. Although they are not biologically extinct, they are so rare and small (average size is less than 40 cm), that they have no measurable ecological impact in coastal zones (Steneck 1997). This stands in stark contrast to estimates of the past when cod's abundance, size, and, significantly, ecological impact were all great. Interaction strength scales with predator size but very large cod such as those exceeding 90 kg (200 pounds) are no longer found (the last one of that size was caught in the late 1800s; Collette and Klein-MacPhee 2002). However, cod stocks were sequentially extirpated, first from coastal zones in the 1930s, and finally from offshore banks in the 1990s (Steneck 1997). Modern ecological studies in coastal zones found that crab, sea urchins, and adult lobsters tethered to the sea floor were no longer vulnerable to attack by vertebrate predators (Witman and Sebens 1992; Vadas and Steneck 1995; Steneck 1997, respectively).

Where small relict populations of large cod and other predators remain in less fished, offshore, kelp-forested, submarine pinnacles (Witman and Sebens 1992), predation rates and ecological impacts were much greater than those recorded in coastal zones (Steneck and Carlton 2001). Where large predators remained, lobsters, crabs, and herbivorous sea urchins were rare, kelp was abundant (Vadas and Steneck 1988, 1995), and attack rates on adults of all three invertebrate groups were high (Witman and Sebens 1992; Vadas and Steneck 1995; Steneck 1997). Over the past half-century in predator-free coastal zones, sea urchin populations expanded dramatically. In fact, in the 1980s, meetings were held to discuss what could be done about the "plague" of sea urchins in coastal New England and the Canadian Maritimes (Pringle et al. 1980). At that time, vast carpets of sea urchins grazed all marine algae, leaving "barrens" that were widespread until the 1990s (Steneck et al. 2002). Thus, as with Alaska's sea otter, in the absence of apex predators, herbivore abundances increased until their grazing controlled biodiversity by denuding coastal zones of their kelp forests.

The abundant lobsters and sea urchins that had formerly been the prey of apex predators became the primary target of local fisheries. By 1993, the value of sea urchins harvested in Maine (whose roe is a highly valued food in Japan) was second only to lobsters. Because this species is also the *only* strong-interacting herbivore in the western North Atlantic, as its populations declined, so too did communitywide rates of herbivory (Steneck 1997). In less than a decade, sea urchins became so rare that they could no longer be found over large areas of the coast (Andrew et al. 2002; Steneck et al. 2004) and as a result, kelp forests and other macroalgae came to dominate the coast once again. Accordingly, the fractional trophic level of harvested species from Maine's coastal zone continued its very recent and very rapid decline toward lower trophic levels (Fig. 7.3c; Steneck et al. 2004).

The urchin-free phase shift back to kelp forests and other macroalgae superficially looked like the ecosystem's initial state, although this time devoid of large vertebrate predators. The combination of abundant algae without large predators was ideal for a population increase of large predatory Jonah crabs (*Cancer borealis*) (Leland 2002). The rise in abundance of crab mesopredators de facto makes them the new apex predators because there is no longer any higher-order predator present in this system. This was well illustrated when 36,000 adult urchins were relocated to six widely spaced patches over a two-year period to an area that had been an urchin "barren" a decade earlier, only to observe and record all urchins being eaten by this newly abundant and highly migratory crab (Leland 2002).

The changes in coastal zones of the western North Atlantic illustrate a shift in controlling factors from a strongly top-down to a more bottom-up structure. Today larval settlement and available nursery habitat controls the demography of lobsters (Steneck and Wilson 2001), crabs (Palma et al. 1999), and sea urchins (Vavrinec 2003) in the Gulf of Maine. In all cases, nursery habitats have become the limiting resource (i.e., bottom-up) as opposed to predation on adults (top-down) that probably regulated abundances in the past. Specifically, the extirpation of large carnivorous fishes compressed the "demographic bottleneck" (Wahle and Steneck 1992), controlling each of these three invertebrates. Large and abundant fish predators eat these invertebrates from the time of settlement usually through adulthood. Based on prey growth rates, relative to measured size-specific attack rates (Witman and Sebens 1992; Steneck 1997), lobsters were vulnerable to predators for at least a decade and sea urchins and crabs were vulnerable their entire life. In the absence of large predatory fishes, small fish feed on very small juveniles that have recently settled out of the plankton. This effectively compresses the demographic bottleneck to a relatively brief period of time, and population densities of these invertebrates correspond to the abundance of unoccupied nursery habitat (Wahle and Steneck 1991 for lobsters and Ojeda and Dearborn 1991 for sea urchins).

Obviously, the present is not the key to the past in coastal ecosystems of the western North Atlantic because the once abundant large carnivorous predators and their structuring effects are now absent. The resulting chain reaction altered biodiversity by allowing invertebrates that were formerly prey to large carnivorous fish to dominate the system. Specifically, the rise in abundance of lobsters and crabs (both of which were absent from ancient Indian middens) is impressive and continues in Maine to this day with unknown consequences.

Global Patterns and Processes in Predator-Induced Changes in Kelp Forest Biodiversity
There are convergent characteristics in the community structure among most of the world's kelp forests. Of course they are now, or have been, dominated by kelp. Globally, sea urchins constitute nearly 70% of the strong-interacting (i.e., potentially deforesting) herbivores in this system (Table 7.1; Steneck et al. 2002), but other herbivores include fish and gastropods. Apex predators in kelp forests are phyletically diverse, with just over half being fishes, 30% invertebrates such as lobsters and crabs, and the remaining nonfish vertebrate predators such as the well-studied sea otter (e.g., Estes, this volume) and seabirds (see Table 7.1).

The tri-trophic structure of kelp forests (apex predator, herbivore, and kelp), as already described for North America, is ubiquitous (Table 7.1). However, relatively few of these systems have been evaluated experimentally due to constraints of space and time necessary for such manipulations. An exception exists in northern New Zealand where no-take marine reserves were established in 1978 so that predatory fish stocks and other organisms could recover from overfishing (Shears and Babcock 2003). Over the subsequent 25 years, the size and population density of predatory fish increased, sea urchins populations declined, and kelp and other macroalgae abundances increased in abundance relative to fished control sites (Shears and Babcock 2003). This remains one of the best experimental studies to demonstrate predator-induced changes in kelp forest biodiversity.

There are two ways predators control herbivory, thereby affecting biodiversity throughout the food web: (1) direct predation (most examples so far, including the previous one from New Zealand), and (2) modifying the foraging behavior of the herbivores. An excellent example of the latter is found in the United Kingdom where large predatory crabs (*Cancer pagurus*, *Liocarcinus puber*, and *Carcinus maenas*) drive a trophic cascade by modifying the behavior of the grazing purple sea

Table 7.1

Comparison of subtidal kelp forest ecosystems of the world. Numbers in parentheses denote number of ecologically important species in subtidal kelp forests for specified taxa (after Steneck et al. 2002).

Site and Latitude	Dominant Kelps	Deforesting Herbivores	Predators or Diseases of Kelp Herbivores	Regional Distribution	Depth Range	Local Distribution	Duration Deforested	Refs.
Western North Atlantic								
Nova Scotia 43–45° N	Laminaria (1), Agarum (1)	Echinoid (1)	Fishes (2) Urchin disease	Widespread	Broad	Homogenous	Decades	1
Maine 43–44° N	Laminaria (1), Agarum (1)	Echinoid (1)	Fishes (2) (crabs)	Widespread	Broad	Homogenous	> Decades	1
Eastern North Atlantic								
North Iceland 65° N	Laminaria (1)	Echinoid (1)	?	Widespread	Broad	Homogenous	?	1–2
North Norway 65–71° N	Laminaria (1)	Echinoid (2)	Seabirds	Widespread	Broad	Homogenous	Decades	3–5
South Norway 55–64° N	Laminaria (1)	Echinoid (1)	?	Restricted	Broad	Patchy	Decades	4
Britain and Ireland 52–55° N	Laminaria (3)	Echinoids (2)	Crabs	Restricted	Broad	Patchy	?	6–8

Table 7.1 Continued.

Site and Latitude	Dominant Kelps	Deforesting Herbivores	Predators or Diseases of Kelp Herbivores	Regional Distribution	Depth Range	Local Distribution	Duration Deforested	Refs.
E. North Pacific								
Alaska (Aleutians) 50–55° N	Alaria (1), Laminaria (3), Thalassiophyllum (1) Agarum (1)	Echinoid (1)	Sea otter	Widespread	Broad	Homogenous	>Decades	1
Southern California 30–35° N	Macrocystis (1), Laminaria (1), Pterygophora (1) # other spp.	Echinoids (3), Gastropods (8), Fishes (2)	Sea otter (1) Fish (1) Lobster (1)	Restricted	Broad	Patchy	<Decade	1
W. North Pacific								
North Japan (SW Hokkaido) 39–46° N	Laminaria (2)	Echinoid (1–3)	Crabs Urchin disease	Widespread	Broad	Homogenous	Decades	9–10
South Japan (W. Honshu) 36–38° N	Undaria (1), Eisenia (1), Ecklonia (1),	Echinoid (3) Fish (1)	?	Restricted	Broad	Patchy	?	9–10

Table 7.1 Continued.

Site and Latitude	Dominant Kelps	Deforesting Herbivores	Predators or Diseases of Kelp Herbivores	Regional Distribution	Depth Range	Local Distribution	Duration Deforested	Refs.
E. South Pacific								
North Chile 18–42° S	Lessonia (1) Macrocystis (1)	Echinoids (2) Fishes (1) Gastropods (2)	Asteroids (3) Fishes (3)	Widespread	Shallow	Patchy	Decades	11–13
South Chile 46–54° S	Macrocystis (1) Lessonia (2)	Echinoids (1) Gastropods (1)	Asteroids (1)	Restricted	Shallow	Patchy	Decades	14
Southernmost Chile 55° S	Macrocystis (1) Lessonia (2)	Echinoids (4)	Asteroids (1)	None				15–17
Argentina 42–55° S	Macrocystis (1) Lessonia (1)	Echinoid (1)	?	None				18
W. South Pacific								
Australia (New South Wales) ~32–35° S	Ecklonia (1)	Echinoids (1) Fishes (1)	Fishes (2)	Widespread	Mod. deep	Patchy	?	19–22
Australia (Tasmania) ~43° S	Macrocystis (1) Ecklonia (1)	Echinoid (1)	Fish (1) Lobster (1)	Restricted	Broad	Homogenous	Years	23

Table 7.1 Continued.

Site and Latitude	Dominant Kelps	Deforesting Herbivores	Predators or Diseases of Kelp Herbivores	Regional Distribution	Depth Range	Local Distribution	Duration Deforested	Refs.
New Zealand (North Island) 34–37° S	*Ecklonia* (1) *Lessonia* (1)	Echinoids (2) Gastropods (2)	Fishes (1) Lobster (1)	Widespread	Mid-depth	Homogenous	Decade	24–31
New Zealand (South Island) 41–47° S	*Ecklonia* (1) *Lessonia* (1) *Macrocystis* (1)	Echinoids (1)	?	Restricted	Broad	Patchy	?	27, 32
E. South Atlantic								
South Africa ~30–35° S	*Ecklonia* (1) *Laminaria* (1) *Macrocystis* (1)	Echinoids (1) Gastropods (1)	Lobster (1) Fish (?)	Widespread	Deep only	Patchy	?	33, 34
E. Indian Ocean								
W. Australia 28° S	*Ecklonia* (1)							35

1. Steneck et al. 2002; 2. Hjorleifsson et al. 1995; 3. Hagen 1983; 4. Sivertsen 1997; 5. Bustnes et al. 1995; 6. Kain 1975; 7. Kitching and Ebling 1961; 8. Ebling et al. 1966; 9. Fujita 1998; 10. D. Fujita, pers. comm.; 11. Vasquez 1993; 12. Vasquez and Buschmann 1997; 13. Ojeda and Santelices 1984; 14. Dayton 1985; 15. Castilla and Moreno 1982; 16. Santelices and Ojeda 1984; 17. Vasquez et al. 1984; 18. Barrales and Lobban 1975; 19. Andrew 1993; 20. Andrew and Underwood 1993; 21. Andrew 1994; 22. Andrew and O'Neill 2000; 23. Edgar and Barrett 1999; 24. Choat and Schiel 1982; 25. Andrew and Choat 1982; 26. Choat and Ayling 1987; 27. Schiel 1990; 28. Cole and Babcock 1996; 29. Babcock et al. 1999; 30. Cole and Syms 1999; 31. Willis et al. 2003; 32. Schiel et al. 1995; 33. Anderson et al. 1997; 34. G. Branch, pers. comm. 2001; 35. Hatcher et al. 1987.

urchin (*Paracentrotus lividus*) (Kitching and Ebling 1961; Ebling et al. 1966). The three species of urchin-feeding large crabs are nocturnal, and their foraging activity forces the urchin to be restricted to shelter-providing habitats where they remain at night and feed during the day. Algal abundances were shown to vary inversely with urchin populations. Similar predator-induced behavioral limitations on herbivory have been described for nonkelp benthic marine ecosystems by Carpenter (1984) for the Caribbean, and Sala et al. (1998) and Hereu (2004) for the Mediterranean.

Certainly, historical fisheries-induced changes in predator distribution, abundance, and size can skew our impression of the importance of large carnivores. For example, a classic study of Kitching and Ebling (1961) in the British Isles describes three trophic-level interactions among crabs, sea urchins, and algae. It was one of the earliest publications to describe subtidal predator–prey interactions. But is it reasonable to assume that crabs have always been the dominant predator in this ecosystem? In a recent global review of trophic cascades in benthic marine ecosystems, Pinnegar et al. (2000) concluded that there are no important predators or trophic cascades in the eastern North Atlantic, including the British Isles. Further, we know of no studies that suggest that large predatory fish were ever important in coastal zones of the eastern North Atlantic (see Table 7.1, Steneck et al. 2002). However, past studies may not have considered the important historical impacts of fishing in Europe where large predatory fish had undoubtedly once lived but just as certainly had been depleted centuries ago. Woodcarvings from 1415 depict cod fishing along the coast of England, and more than a century later, the "Cod Wars" were fought on the same shores (Kurlansky 1997). In 1556 in the Netherlands, Pieter Brugel the Elder's fanciful painting *Big Fish Eat Little Fish* depicted fish, crabs, and lobsters spilling out of the dissected stomach of a house-size cod caught near a coastal village. Nearly 500 years ago, large carnivorous fish were part of Europe's coastal ecosystem. The depletion of large predatory fish from Europe may have promoted the development of distant fisheries in Iceland and eventually North America (Kurlansky 1997). Thus it is possible that the very early extirpation of apex fish predators in Europe's coastal zones centuries ago stimulated a mesopredator release (*sensu* Crooks and Soulé 1999b)—just as has recently occurred in the Gulf of Maine (Leland 2002).

In many ecosystems, the presence of abundant lobsters or crabs may indicate a general absence of large vertebrate carnivores. These invertebrates have high food value for large predators because of their high fat content. However, lobsters and crabs are relatively large and usually only large carnivores are capable of eating them and controlling their populations. This could explain why two regions with a long history of fishing—Japan and the United Kingdom—both have benthic ecosystems dominated by crabs as the dominant predators of sea urchins (see Table 7.1, Steneck et al. 2002).

Tropical Coral Reefs

Coral reefs are defined as tropical shallow marine communities dominated by stony corals. There is almost no geographic or biodiversity overlap between coral reefs and kelp forest ecosystems. Coral reef ecosystems are in grave trouble. Their distribution and abundance have declined globally for several reasons that include increased thermal stress, bleaching, disease, and loss of critical trophic levels (Hughes et al. 2003; Bellwood et al. 2004). The decline is global but is most advanced in the Caribbean (Pandolfi et al. 2003; Bellwood et al. 2004). Central among the causes of this decline is the functional loss of consumers that perform critical ecological services.

In coral reefs, more than in most coastal ecosystems, the role of large carnivores is poorly understood. One by now familiar reason may be that they were more important long ago before being fished out. We suggest this for two reasons that have come to light only recently. First, that even low levels of artisanal (nonindustrial) fishing can reduce the abundance of the largest predatory fishes on reefs was demonstrated in Fiji on Indo-Pacific reefs (Dulvy et al. 2002) and at several locations in the Caribbean (Hawkins and Roberts 2004). Second, in some locales declines in predator abundance and size began more than a millennium ago, even before first European contact, although the rate of decline was higher after European arrival (Wing and Wing 2001; Pandolfi et al. 2003). In the following sections, we expand on these and other predator-related relationships for the biodiversity of Caribbean and Indo-Pacific coral reef ecosystems.

Caribbean

Currently, the biodiversity implications of altered food webs on Caribbean reefs are most conspicuous among the primary producers. Macroalgal abundance is the single best indicator of poor conditions for coral reefs (Lirman 2001, Kramer 2003). A primary *proximate* cause for increased macroalgae on Caribbean reefs is the loss of herbivory, especially following the mass mortality of the abundant sea urchin (*Diadema antillarum*) (Hughes 1994 for Jamaica; Steneck 1994 for St. Croix; Miller et al. 1999; Hughes et al. 1999; see also Knowlton 1992). However, the *ultimate* reason for the sudden collapse of coral reef ecosystems following the *Diadema* decline may have been the urchin's response to the long history of fishing down Caribbean reef food webs (Hughes 1994). Trophic cascades may be important to Caribbean coral reefs but they are difficult to discern. This is primarily because the ecological *function* of high-order (i.e., apex) predators in these coral reef ecosystems is poorly understood (Hixon 1997; Pennings 1997). It is possible that predators are simply not strong interactors in Caribbean reefs, or that the most important large carnivorous apex predators on reefs were reduced or eliminated long before the debut of modern science. The strongest evidence for the latter comes from the recent findings that prehistoric fisheries existed at numerous locations throughout the Caribbean (Wing and Wing 2001). This archaeological study reported that between 1500 and 500 years ago, the abundance and body size of carnivorous fishes declined. Further, the mean trophic level index declined at all sites, suggesting that the highest-order carnivores were becoming relatively rare over that interval as prehistoric fishers turned to herbivorous fishes (Wing and Wing 2001). Thus the predator baseline may already have slid by the time the first Europeans arrived in the Caribbean. Against this baseline, recent fishing has caused further declines of large carnivorous fishes such as sharks, groupers, and snappers (Munro 1996; Williams and Polunin 2000; Hawkins and Roberts 2004).

The recent collapse of Caribbean reefs, therefore, could have been the direct result of long-term fishing down food webs in these low-diversity reefs with very few functional redundancies (Pandolfi et al. 2003; Bellwood et al. 2004). In most areas of the Caribbean, the predatory fish such as groupers, snappers, and triggerfish are targeted for their high food value. Hook and line fishing was commonly used until population densities declined. Following predator declines, spearing and trapping of fish became more widespread (Munro 1996). This allowed herbivorous fish such as parrotfish and surgeonfish (*Acanthurus* spp.) to be harvested. Herbivorous fish and *Diadema* sea urchins compete for algal resources (Hay and Taylor 1985). Thus the sea urchins could have expanded due to the absence of both carnivorous predators and herbivorous competitors (Fig. 7.4) (but see Jackson 1997). By the early 1980s, the entire herbivore trophic level was maintained by the single

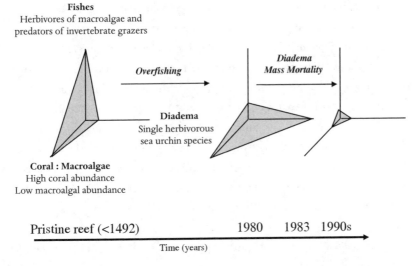

Figure 7.4

Changes over time in the coral reef ecosystem in Jamaica indicated by coral to macroalgal abundance, herbivorous and invertebrate predatory fish functional groups, and the herbivorous sea urchin *Diadema antillarum*. After predatory and herbivorous fishes were extirpated by overfishing, the high coral, low macroalgal abundance phase remained because the increased grazing effect of *Diadema* compensated for the loss of fish species. At this point, the "ecosystem insurance" was absent. The reef ecosystem collapsed soon after the mass mortality of the urchin (see text). Source: Done (1995), figure 4d with permission—Copyright, Springer-Verlag, modified after Jackson (1994).

sea urchin species *Diadema antillarum*. Such hyperabundances are rife for epizootic disease, and that is exactly what happened throughout the Caribbean in 1983 and 1984—effectively creating herbivore trophic level dysfunction (*sensu* Steneck et al. 2004) on all reefs that had come to rely on grazing from this urchin, and this contributed to the collapse of reefs throughout the region.

Other evidence supports the thesis that fishing pressure on reefs contributed to the rise of *Diadema* sea urchins (Levitan 1992). When this urchin is abundant, it consumes most edible algae, leaving grazer-resistant calcareous coralline algae (Steneck 1994), causing urchins to develop more robust mouthparts. Using museum collections, Levitan (1992) found that in areas with high fishing pressure on reefs the mouthparts were most robust. Further, as local human populations adjacent to reefs grew, they arguably increased fishing pressure and thus reduced the predators and competitors of this urchin. Other ecological studies conducted prior to and following the mass mortality concluded that urchin abundance was highest where fishing pressures were highest (Hay 1984). Thus circumstantial evidence suggests that predators control biodiversity on Caribbean reefs through a trophic cascade of predatory fishes, grazing sea urchins, and macroalgae.

Only a few studies have employed manipulative techniques to demonstrate cascading change from changes in apex predator abundance. One by McClanahan (this volume) showed two direct predator effects in protected areas in Belize. As predator densities increased there was a slight but significant decline in herbivorous sea urchins. Importantly, there was also a significant decline in the abundance of damselfish.

Damselfish may contribute to trophic cascades initiated by apex predators on reefs. Highly territorial damselfish are so aggressive that they effectively exclude most other grazing fishes (Brawley and Adey 1977) creating patches of elevated algal biomass. Predators have been shown to control the abundance of these damselfishes (Hixon and Beets 1993; Hixon 1997; McClanahan, this volume). If damselfish and their gardens increase on reefs as a result of predator declines, then the effects could be the same as if predators were directly limiting herbivore populations. Thus this trophic cascade results from predators indirectly affecting the process of herbivory without necessarily affecting the population density of the herbivores themselves.

The diversity of species within each trophic level is effectively a type of insurance against trophic level dysfunction (Bellwood et al. 2004). The low species diversity of Caribbean reefs leaves them with few functional players at each trophic level. The apex predators such as sharks and large groupers and snappers have been extirpated or are locally extinct on many reefs. Additionally, widespread use of fish traps removed many of the herbivorous fish, leaving only a single species of sea urchin as the primary herbivore in this tri-trophic system. The disease-induced mass mortality of this sea urchin was a natural experiment of gargantuan proportions. The widespread and rapid phase shift to algal domination may have been one of the world's most rapid and widespread shifts in biodiversity ever documented.

It should be noted, however, that this scientific evidence is another example of shifting baseline, since we describe mechanisms involving trophic cascades involving only three species, such as the triggerfish, sea urchin, and corals/algae.

In the past, as it still happens in a couple of remote and well-protected reefs, the apex predators were large carnivores such as sharks and the goliath grouper (*Epinephelus itajara*). This indicates the existence of longer trophic chains in intact coral reef food webs. The historical removal of the apex predators may have set the conditions for the recent collapse of Caribbean reefs. The best way to know about past events would be to shift our focus from popular but seriously degraded reefs where damselfish and wrasses are the dominant species, to those few reefs where we can still witness nature in all its splendor.

Indo-Pacific
The vast Indo-Pacific region contains most of the world's coral reefs and has, by far, the highest marine species biodiversity. For example, there are at least 350 species of corals in the Indo-Pacific compared to the Caribbean's 64, and there are 3000 species of reef fish compared to the Caribbean's 750. If species diversity within trophic levels is ecosystem insurance, then the Indo-Pacific reefs appear to be well insured (but see Bellwood et al. 2003). Many reefs in the Indo-Pacific still have significant populations of large predatory fish. Nevertheless, these predators remain very susceptible to overfishing (Dulvy et al. 2002). Although Indo-Pacific reefs are not as degraded as Caribbean reefs, there are signs that fish (Graham et al. 2003) and coral (Hughes et al. 2003) abundances are declining (Jones et al. 2004). In marine protected areas (MPAs) in the Philippines where fishing is not allowed, large predatory fish abundance increased by more than an order of magnitude within a decade (Russ and Alcala 2003).

With changes in the abundance of large reef predators, their prey abundances have changed as well. For example, strong interactions of predators on their herbivorous prey were recently shown for MPAs on the Great Barrier Reef (Graham et al. 2003). Abundance of the largest predator in that system, the coral trout (*Plectropomus leopardus*), was inversely correlated with the aggregate of prey species biomass. Several of the prey species with the strongest declines were herbivorous damselfish and parrotfish and mesopredators such as wrasses. Further, in the expansive northwest Hawaiian Islands, where fishing pressure is relatively low, large predators, including sharks and jacks, remain abundant (Sudekum et al. 1991; Friedlander and DeMartini 2002). There, the trevally jacks that can attain sizes exceeding 100 kg, are known to prey heavily on some of the dominant reef herbivores such as scarid parrotfish. In fact, some reefs in the Northwest Hawaiian

Islands with abundant large sharks (*Carcharhinus* spp.) and jacks (*Caranx ignobilis*) have so few smaller herbivores that the reef is dominated by macroalgae (Parrish and Boland 2004). In Fiji, fishing pressure directly relates to a reduced biomass of piscivorous predators and an increased biomass of invertebrate carnivores (Jennings and Polunin 1996). In that system, there was no significant relationship with herbivore biomass. Thus some of the prey responses to predators are muted in the diverse tropical reefs of the Indo-Pacific, perhaps due to diffuse impacts from predator switching among highly diverse prey (see Steneck, this volume).

Elsewhere, small to medium-sized predatory fish have been shown to control populations of invertebrate herbivores. For example, the abundance of herbivorous sea urchins in East African reefs were shown to be inversely related to the abundance of local predators (McClanahan and Shafir 1990; McClanahan et al. 1999a, most notably the red-lined triggerfish (McClanahan 2000).

In general, the trophic structure of coral reef ecosystems throughout much of the Indo-Pacific remains less degraded than those in the Caribbean. Thus it follows that macroalgal phase-shifts are less common on Indo-Pacific reefs than they are in the Caribbean. Nevertheless, when large herbivore fish exclusion cages were erected at a site on the Great Barrier Reef, macroalgal abundance steadily rose over the two-year period of the study (Hughes et al. 2003; Bellwood et al. 2004). This shows that herbivorous fish control macroalgal biomass as other herbivores do elsewhere (e.g., in the Caribbean).

All of this suggests that diffuse trophic cascades operate in highly diverse ecosystems such as the coral reefs of the Indo-Pacific, with the largest predators able to limit the abundance of mesopredators and herbivores, and herbivores able to limit the abundance of their algal prey. However, none of the examples are as clean as those that have been observed in low-diversity ecosystems.

Evidence from Other Marine Ecosystems

One of the best examples of a trophic cascade is found in the Mediterranean (Sala et al. 1998). Sala and coworkers conducted research in a no-take marine reserve in the northwest Mediterranean that has been in place since 1983 and compared this relatively unfished system with fished control sites. This reserve had both higher population densities of predatory fishes and larger body sizes than those in

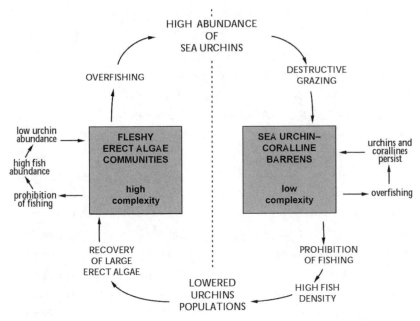

Figure 7.5
Fishing-mediate trophic cascade controls algal community structure in the western Mediterranean (from Sala et al. 1998, with permission—Copyright, Blackwell Publishing).

the fished control sites (García-Rubies and Zabala 1990). From these two predator states, they tested the hypothesis that predators control the population densities of their sea urchin prey, and subsequently the abundance of macroalgae, which are eaten by the sea urchins. Tethered urchins were attacked by fish predators within the protected area at rates that were five times greater than in the fished site (Sala and Zabala 1996). Where predator attack rates were high, urchins were relatively rare, and macroalgae were abundant. In the predator-free control sites, the herbivorous urchins were abundant and all erect edible algae were grazed away leaving mostly calcareous coralline algae (Fig. 7.5).

The linkage between the predatory fish, the sea urchins, and algal communities was not always straightforward in the Mediterranean, according to Sala and coworkers (1998). The effectiveness of resident predators was mediated by the architecture of their habitat. Where shelters were abundant (i.e., there was high spatial heterogeneity), urchins coexisted with predators, but in habitats with the same density and size of predators but fewer hiding places, urchins were rare (Hereu 2004). These findings were similar to those of Kitching and Ebling (1961) in the United Kingdom, Carpenter (1984) in the Caribbean, and Andrew (1993) in South Australia. In all cases, urchins were safe in their shelters when predators were active and they grazed in the vicinity of those shelters at times when the predators were rare or absent. Regardless of the existence of other regulating factors, the abundance of predators generally appears to be a necessary condition for the existence of macroalgal communities.

As we have seen for many other benthic marine ecosystems, size of predatory fishes in the Mediterranean system is an important factor to consider. This is because only the largest fish (i.e., three species of sparids) can attack adult sea urchins (Sala 1997; Hereu 2004). Because fishing often removes larger predators and leaves the small and often commercially less important species, the vulnerability to and demographic impacts of fishing are not at all the same as those of predation. If fishing selectively removes larger fish, it skews the population toward juveniles and species that are smaller as adults. This changes the nature of predation and moves the ecosystem from top-down to bottom-up control. Many of the best examples of the biodiversity impacts of large carnivores are found in benthic marine ecosystems (see Steneck, this volume). This is not to say that trophic cascades cannot or will not occur in pelagic ecosystems. In fact, in the Black Sea a clear fisheries-induced trophic cascade exists (Daskalov 2002). In that system, fishing on pelagic predatory fish, including bonito (*Sarda sarda*), mackerel (*Scomber scombrus*), bluefish (*Pomatomus saltator*), and dolphin (*Coryphaena hippurus*), caused increases in the system's planktivorous fishes. This resulted in a sharp decline in zooplankton abundance and an increase in phytoplankton biomass. It also led to a massive population explosion of jellyfish during the 1970s and 1980s, thereby causing a further decline in zooplankton abundance.

General Consequences of the Loss of Large Carnivores, and Implications for Conservation

Large marine carnivores are increasingly less frequent and smaller (Figs. 2, 3d, Steneck 1998; Jackson and Sala 2001; Steneck et al. 2004). If body size drives the intensity of predation by controlling how much biomass can be consumed in a bite or a swallow, then population densities drive the frequency of predation by controlling the rate at which attacks occur. Evidence from tropical reefs (Pandolfi et al. 2003), estuaries (Jackson et al. 2001), and kelp forests (Steneck et al. 2002) suggests that past predators were big and their attack rates were probably frequent, but both began to decline long ago due to overfishing. In effect functional baselines have slid so far that both the structure and the functioning of the ecosystem bear little resemblance to the one that was assembled over the last tens of thousands of years from components evolved over the past several million years. The baseline has shifted so much even among scientists, that the "large predators" in the literature are not really large. For instance, recent studies carried out in Australian coral reefs dealing with "large predatory fishes" considered "large" a size larger than only 20 cm (Connell 1998). This is one order of magnitude smaller than really large carnivores such as sharks and large groupers. Based on stomach contents and energetic studies (Froese and Pauly 2004) we can estimate that a 200 kg goliath grouper, absent in most reefs, would eat 310 kg of lobsters, 55 kg of crabs, 25 kg of turtles, and 60 kg of fish per year. These phenomenal ecological impacts are all but unknown to most modern scientists, who began to study marine ecosystems long after strong shifts in vertebrate:invertebrate biomass ratios occurred (Jackson and Sala 2001). We should admit current marine predators are not very big and their predatory effects have slowed to below detectable levels. It is indeed remarkable that, in places where there used to be sharks and large groupers, the only keystone species we can identify today are medium-sized fishes such as triggerfish and sparids.

With few exceptions, the strength of the top-down control exerted by marine carnivores has diminished due to overfishing (Steneck 1998; Jackson et al. 2001; Hughes et al. 2003; Pandolfi et al. 2003). As nonhuman top-down forces diminish, the likelihood of large fluctuations in species abundance, mainly of lower trophic levels, increases. This happens because humans exert a radically different kind of top-down control on marine ecosystems than nonhuman predators do (Sala and Sugihara 2005). Careful study of many marine communities today may fail to realize that large marine carnivores once structured the ecosystem (e.g., in the British Isles and Japan). Such "sliding baselines" (Dayton et al. 1998) can give managers of marine ecosystems skewed management goals. Perhaps if managers better understood past functional attributes of marine ecosystems, they would be better able to remediate or restore them. Without considering what was there, current and future managers may assume most marine systems are controlled by the bottom-up and try to manage them that way. Without a solid understanding of the role of multispecies trophic interactions of complete communities we cannot have rigorous ecological expectations, or even be able to evaluate the success of conservation actions.

The available data on trophic cascades suggest that the long-term preservation of marine ecosystems must happen through the preservation of their complexity, including all strong trophic interactions. The removal of functional trophic levels has cascading effects on the stability and productivity of entire ecosystems. But managing predator populations in marine ecosystems is especially challenging. Large predators are still hunted for an apparently insatiable global appetite for fish. The sea and its fish are common property that too often suffers from the "tragedy of the commons" (*sensu* Hardin 1968). As the notion of the sea as an inexhaustible resource fades, more traditional stakeholders are coming to realize that sustaining their marine predators can be the equivalent of job security for them. Indeed, the large carnivores we eliminated early on might have to be restored in order to ensure the stability of marine communities. As we learn more about how fragile marine ecosystems are, the wall between environmentalists interested in preserving biodiversity and fisheries managers interested in preserving fish stocks crumbles. Increasingly we are seeing large marine carnivores as essential for preservation of intact food webs. Nevertheless, few fisheries management plans consider the ecological *function* of managed stocks, only what rate of harvest a given population can sustain. The question now is how to achieve this goal. Will carnivores be able to regulate and maintain complex ecosystems only in places where humans are absent? Or can we exploit marine communities in a truly sustainable way such that other predators are not undesirable competitors but insurances against catastrophes and wild fluctuations? A challenge for future marine conservation will be to identify and manage for the ecological function of strong interactors such as large predators (Soulé et al. 2003).

Summary

Large marine carnivores can have striking impacts on marine biodiversity by controlling herbivores and thus indirectly controlling vegetation in nearshore, bottom-dwelling communities. In our review, we provide evidence that the distribution and abundance of kelp forests in the western North Atlantic, eastern North Pacific, and New Zealand are strongly controlled by the abundance of large predators. Similarly, many other areas such as coral reefs and Mediterranean marine ecosystems are controlled by medium-sized fish predators, although larger carnivores likely played this role in the past. Unfortunately, because fishing has targeted these predators for literally thousands of years, the carnivore fauna we have today (still under intensive fishing with rapidly declining stocks) is a fraction of that in existence prior to human impacts. By integrating historical ecology with experimental, process-level research, we can estimate what the ecological impact of large carnivores must have been. Where large areas have been put off limits to fishing, the results have been impressive. However, such area management is much more difficult in the marine realm because of the lack of property rights. It is our hope that through this chapter more people will consider historical trends and ultimately learn from the ghosts of marine predators past.

ACKNOWLEDGMENTS

We thank Justina Ray, Kent Redford, and Joel Berger for their help and encouragment for us to write this chapter. We discussed aspects of the paper with Jim Estes, Bob Paine, and Justina Ray. We received helpful reviews from Tim McClanahan, Kent Redford, and Elizabeth Stephenson. Justina Ray and Joanna Zigouris edited and assisted in the final assembly of the paper. To all we are grateful.

CHAPTER 8

Forest Ecosystems without Carnivores: When Ungulates Rule the World

William J. McShea

A corollary to the premise that removal of large carnivores from some ecosystems has increased prey densities and led to severe changes in biodiversity because of loss of plant species and subsequent trophic cascades is that reestablishing large carnivores will restore biodiversity. In this chapter, I address this issue primarily in forested ecosystems where large carnivores are missing. Specifically, I examine the phenomenon of high-density prey populations (i.e., ungulates), the effect of those populations on biodiversity, and whether subsequent reductions of ungulates restore biodiversity. Because most of the world's biodiversity resides in forested systems (Mittermeier et al. 1998), predators may serve a critical role as biodiversity "managers" —a role that is not often obvious because large predators have been missing from many of these systems for decades or in some cases centuries. Although the scope of this chapter is limited to ungulates, high densities of other mammal species, including rodents (*Castor canadensis,* Naiman et al. 1994; *Rattus* spp., Cabin et al. 2000), leporids (*Lepus americanus,* Boutin, this volume), and elephants (*Loxodonta africana,* Dublin et al. 1990), can also influence biodiversity or nutrient flow. I focus on whether forest biodiversity can be conserved by managing ungulate populations, and whether large predators can be an effective means to limit ungulates.

Extent of the Problem of High-Density Ungulate Populations

Not every ungulate species occurs at high densities. There are reports of high densities for wild boars and feral pigs (*Sus scrofa*), but these almost always involve introduced populations (Van Dreische and Van Dreische 2000). There are about

140 species of bovids, yet reports of high-density populations are limited to only a handful. High densities of domestic cows, sheep, and goats, through food supplementation and predator protection by humans, can cause significant reductions in biodiversity (Cabin et al. 2000; Van Dreische and Van Driesche 2000; Hobbs 2001). Most other reports of high density are limited to introduced populations, such as chamois (*Rupicapra rupicapra*, Homolka and Heroldova 2001), tahr (*Hemitragus jemlahicus*, Caughley 1970), mountain goats (*Oreamnos americanus*), or feral goats and sheep (Coblentz 1990; Van Dreische and Van Dreische 2000).

Among species of deer (Cervidae), the problem is likewise not universal. There are approximately 47 species worldwide, with reports of high density limited to 11 species. This list becomes further reduced to 8 species if only native populations are considered. Elk (*Cervus elaphus*, Putman and Moore 1998; Howell et al. 2002), white-tailed deer (*Odocoileus virginianus*, Butfiloski et al. 1997; Martinez, et al. 1997), mule deer (*O. hemoinus*, McCullough et al. 1997), fallow deer (*Dama dama*, Putman and Moore 1998), roe deer (*Capreolus capreolus*, Fuller and Gill 2001), sika deer (*Cervus nippon*, Takatsuki et al. 1994), caribou (*Rangifer tarandus*, Messier et al. 1988; Couturier et al. 1990), and moose (*Alces alces*, Brandner et al. 1990; Berger et al. 2001a) have been reported at densities high enough to cause substantial effects on vegetation. Introduced populations of caribou (Klein 1968; Leader-Williams et al. 1989), sika deer (Smith 1998), Chinese water deer (*Hydropotes inermis*, Putman and Moore 1998), and muntjacs (*Muntiacus reevesi*, Cooke and Farrell 2001; Flowerdew and Ellwood 2001) have also been implicated in similar ways.

Although the number of deer species involved is not impressive, it is more sobering when one considers that these eight species occupy most temperate and boreal forests in the northern hemisphere. When introduced populations are considered, the issue becomes equally relevant for temperate forests in New Zealand (Coomes et al. 2003) and South America (Veblen et al. 1992). With the possible exception of tropical forests, the issue of high-density ungulate populations is important for all forest managers.

Ungulates that maintain high densities are mostly temperate or boreal species (Geist 1998; Renecker and Schwartz 1998). Some tropical deer, such as sambar (*Cervus unicolor*) in New Zealand (Coomes et al. 2003) and Australia (Bentley 1998), and Reeve's muntjac in the British Isles (Cooke and Farrell 2001; Fuller and Gill

2001), have increased rapidly when introduced into temperate systems but have never exhibited high numbers in their native habitat. High-density populations of ungulates tend to be in developed, temperate or boreal countries with effective controls over hunting (Wemmer 1998).

Populations Near Carrying Capacity

I have used "high density" as a shorthand way to denote populations that are above the carrying capacity (i.e., K) of their habitat. The density of animals is not as important as their number relative to the productivity of the habitat. There are two profiles of ungulate population growth that result in significant damage to plant communities: (1) populations that irrupt beyond the carrying capacity of the habitat and then crash when disease or food limitations lead to increased mortality, and (2) populations that fluctuate at or above K without severe crashes in numbers. Both of these population types can lead to changes in vegetation diversity and structure (McCullough 1979, 1997), and are the ones that introduced predators would be expected to regulate by keeping prey numbers below K.

Ungulate populations that increase exponentially beyond carrying capacity and then crash, such as the infamous Kaibab Plateau population in northern Arizona (Leopold 1943) or many introduced populations (Caughley 1970, 1981), were originally of prime concern for wildlife managers because this type of abundance profile was supposed to reduce the carrying capacity of the environment. Empirical evidence, however, has not supported this hypothesis (Caughley 1970; McCullough 1997), and the long-term consequences of population crashes on future measures of carrying capacity appear to be minimal (McCullough 1997). Research focus has shifted to populations chronically above K, where most of the evidence for ungulate effects on biodiversity has been reported.

There are three common scenarios for populations that chronically approach or exceed K: (1) when animals have been introduced to islands or refuges where both significant competitors and predators are missing (McCullough 1979; Leader-Williams et al. 1989; Putman and Moore 1998); (2) after predators have been removed or reduced from previously stable systems (Brandner et al. 1990); and (3) around suburban or agricultural areas, decades after predator removal, but following habitat changes (Butfiloski et al. 1997; McCullough et al. 1997; Fuller and

Gill 2001). All three scenarios involve the absence of predators, but for those populations within the third category, the loss of predators did not immediately trigger increases in ungulate populations. In North America, the late 1800s through the 1970s was a period of extensive shifts in land-use, hunting habits, forestry practices, and game management. In these human-dominated systems, there is an interaction between the reduced role of hunters and predators and the increased input of nutrients and disturbance to the habitat, which combine to boost ungulate numbers.

Some ungulate populations persist for years above what is supposed to be the carrying capacity (Underwood and Porter 1997). Several authors (Schmitz and Sinclair 1997; Stromayer and Warren 1997) have described the potential for alternative stable states in forested systems, where ungulates might cause a plant shift toward more grass and herbaceous species that can persist in heavily browsed ecosystems, essentially increasing K. Grazing systems also have degraded states that are relatively stable under high densities of livestock (Friedel 1991; Laycock 1991). The persistence of ungulate populations at or above K might produce a shift toward a less diverse, but possibly more productive, plant community (Augustine and McNaughton 1998).

Although ungulate numbers in grasslands might rise and fall with annual rainfall and grass productivity (Augustine and McNaughton 1998), the connection between ungulate abundance and available plant biomass is more complex in many forested systems. Deer populations can persist above K when there is no short-term feedback between ungulate browsing and plant productivity (Augustine and McNaughton 1998). For example, consumption of seeds or seedlings in forested systems does not affect the seed production of mature trees until the lack of recruitment decreases the number of trees in the adult age class—something that might not occur for many decades in temperate forests. Peterken and Tubbs (1965) noted that recruitment once every 50 to 100 years was enough to sustain New Forest in southern England in a mature state since 1650, despite chronically high densities of deer. Agricultural crops are renewed with little regard for the density of ungulates feeding on them. Forest fragments in the agricultural midwestern United States support deer densities well beyond those seen in nearby contiguous forests (Hansen et al. 1997). The highest densities of moose occur in forested landscapes where timber management keeps a large proportion of the forest in an early seral stage (Karns 1998). The diet of ungulates in suburbs consists of plants subsidized with both nutrients and water by landowners. The result of all of these scenarios is that ungulate populations persist in numbers that appear to be beyond K because of extrinsic inputs of energy.

Whereas K is set by the ability of the habitat to support a species, erosion of biodiversity occurs when that species affects the ability of the habitat to maintain other species. The severity of this loss is certainly linked to K, because the closer an ungulate population is to K the broader its foraging patterns and the more obvious its impact on vegetation. Biodiversity, however, can be negatively impacted at levels far below K, when the forage species is preferred (Alverson et al. 1988). From a conservation perspective, the focus should be on biodiversity loss and not K.

Ungulate Effects on Biodiversity

The direct and indirect effects of ungulates on various plants and animals have received an increasing amount of research attention over the past decade or so, allowing for a more comprehensive picture of the consequences of high-density herbivore populations to the state of biodiversity.

Plant Communities

Browsing by ungulates impacts the species composition of both woody and herbaceous plant species in forested systems (Hobbs 1996; Waller and Alverson 1997). For herbaceous plants the impact is primarily on reproduction, with fewer flowering and fruiting plants within forests with high ungulate densities (Augustine and Frelich 1998; Webster and Parker 2000; Fletcher et al. 2001a,b). With the exception of woodlands in England (Kirby 2001; Morecroft et al. 2001), most data for impacts on herbaceous plants come from white-tailed deer in North American forests. Miller et al. (1992) reported 98 rare forbs browsed by white-tailed deer, with wildflowers in the lily and orchid families making up the bulk of the reports (Bratton 1979; Anderson 1994; Balgooyen and Waller 1995; Augustine and Frelich 1998; Webster and Parker 2000; Fletcher et al. 2001a,b). Aside from rare plants, high densities of deer also result in the reduction in bramble (i.e., *Rubus*) at forest sites (deCalesta 1997; McShea and Rappole 2000, Kirby 2001; Morecroft et al. 2001). Increased abundance of grasses and sedges in forests are obvious following introductions of deer (Cooke and Farrell 2001), or after decades of high densities (Stromayer and Warren 1997). Similarly, elk populations maintain grasses and sedges within spruce–hemlock forests (Schreiner et al. 1996), which converted to ferns after elk numbers were reduced.

Although some herbaceous species disappear under heavy browsing pressure, others often proliferate in forests with high ungulate densities. deCalesta (1997) noted increases of hay-scented fern (*Dennstaetia punctilobula*) within understories with high densities of deer. This shift is important because dense cover of ferns results in regeneration failure for several woody species in eastern North American forests (Horsley et al. 2003). Invasive species often fill the void left by browsed species. Honeysuckle (*Lonicera* sp.) is spread through the feces of white-tailed deer (Vellend 2002). Garlic mustard (*Allaria officinalis*) is associated with high-density populations of deer in the Southern Appalachians (McShea, pers. obs.), and its propagation and spread may be enhanced by soil disturbance during deer grazing (Anderson et al. 1996). Furthermore, most invasive and exotic plant species benefit from high levels of disturbance and degradation (Hobbs 2001) and would probably thrive under high densities of ungulates. Chinese privet (*Ligustrum sinense*), for example, forms dense thickets in disturbed habitat, and can persist under heavy browsing pressure from deer (Stromayer et al. 1998). Such a combination can lead to common associations of privet and high densities of deer in the southeastern United States.

For species that reach maturity in the forest understory, such as hobblebush (*Viburnum alnifolium*, Hough 1965) and Canada yew (*Taxus canadensis*, Balgooyen and Waller 1995), high densities of ungulates result in their elimination or severe reduction in density because all life stages of the plant are vulnerable to browsing. Severe impacts due to deer are also noted in forests that do not grow beyond the browsing height of deer, such as conservation coppices in England (Putman and Moore 1998).

In the case of trees that reach maturity in the canopy, there are abundant studies that demonstrate that survival of seedlings and saplings is impacted by deer browsing (Anderson and Loucks 1979; Alverson et al. 1988; Tilghman 1989; Healy 1997; Gill and Beardall 2001; Horsley et al. 2003). Similar results have been found for both moose (Brandner et al. 1990; Reneker and Schwartz 1998) and elk (Singer et al. 1994). Lower densities of saplings and adults for many late successional species such as hemlock (*Tsuga canadensis*), white pine (*Pinus strobus*), and red oak (*Quercus rubra*) are attributed to ungulate browsing (Alverson et al. 1988; Healy 1997; Healy et al. 1997; Rooney 2001; Horsley et al. 2003).

Once canopy trees grow beyond the reach of deer (approximately 2 m), only damage from antler rubbing, scent marking, or bark stripping can be considered significant to the tree's survival (Bowyer et al. 1994; Putman and Moore 1998). An important question is whether ungulate damage to seedlings and saplings converts into real changes in the composition of mature forests. Those working in managed systems maintain the importance of deer browse in shaping future forests (deCalesta 1997; Healy et al. 1997; Horsley et al. 2003). Successional models, however, do not always support these predictions. Mladenoff and Stearns (1993), for example, found few significant impacts of browsing intensity on the abundance of eastern hemlock due to the predominant influence of life history traits, land use patterns, environmental stochasticity, and site quality in determining successional patterns. Seagle and Liang (1997) predicted the composition of bottomland hardwood forests would only be affected under low browsing intensity, where deer browsing is more selective. Old field succession into forest was also more influenced by plant life history traits then deer browsing pressure (Bowers 1997). Occurrence of herbaceous species in British forests was more accurately predicted by abiotic factors than deer browsing intensity (Watkinson et al. 2001). In most forest models, life history traits (i.e., shade tolerance, moisture requirements, annual growth rates) are the predominant factors in determining which trees reach the canopy.

Differences in the importance of ungulate browsing to composition of mature forests could be due to real differences in forest ecosystems. In managed forests, preharvest levels of seedlings are critical to the successional path of the forest following harvest (Dey 2002). In mature forests, young trees exhibit slow growth for decades while they wait for canopy gaps to occur (Runkle 1990). In mature forests, browsing pressure may not be as significant as shade tolerance. Although browsing by ungulates does change the proportion and abundance of species among seedling and sapling age classes (Horsely et al. 2003), disturbance rates may determine if this impact is significant at the stand level.

Ungulates regulate nutrient flow in forested ecosystems by conversion of plant material into fecal pellets and urine, and by changing the composition of plant material over time through selective browsing (Hobbs 1996). In boreal forests, ungulates can retard (Brandner et al. 1990), or change (Pastor and Naiman 1992; Pastor et al. 1993) the successional pathway of forests. For example, selective browsing by moose on aspen (*Populus* sp.) can shift boreal forests toward a climax community composed predominantly of spruce (Pastor and Naiman 1992). This shift from aspen to spruce has ramifications for nutrient flow, with available nitrogen levels in boreal spruce forests being significantly lower after decades of heavy browsing by moose. Although there is evidence that succession is retarded by heavy browsing in the more nutrient-rich temperate forests (Tilghman 1989), the predicted changes in species composition of mature forests (deCalesta 1997; Healy et al. 1997) are still to be demonstrated, perhaps because high ungulate densities are such a relatively new phenomenon.

An additional complication to the impact of ungulates on forest succession is the patch size of disturbances (Hobbs 1996; Pastor et al. 1998). The plant growth within small openings in the canopy created by small disturbances such as windfalls and lightning strikes can be exploited rapidly and thoroughly by ungulates. Large disturbances, such as fire or hurricane, are unevenly exploited and create a mosaic of successional patches. Systems characterized by smaller disturbances, such as mature mesic deciduous forests (Runkle 1990), may be more severely impacted by ungulate browsing than more xeric or grassland systems that experience disturbance at a larger scale.

Forest Birds

Dramatic changes in forest composition and density do have ramifications for other vertebrates. The best example of trophic cascades involving ungulate browsing is within forest bird communities, primarily because the distribution of birds is closely tied to vegetation (MacArthur and MacArthur 1961; Urban and Smith 1989). deCalesta (1994) reported migratory birds that foraged within the intermediate canopy on a hardwood forest in Pennsylvania to increase when deer densities dropped below 8/km². The distribution of Kentucky warblers (*Oporornis formusus*) in a Virginia forest shifted over a 10-year period with a shift in the distribution of deer (McShea et al. 1995). Abundance changed for 11 species of migratory birds within 9 years after deer were excluded from Virginia forest sites (McShea and Rappole 2000).

As with vegetation, there is some debate as to the relative importance of deer browsing in influencing bird population levels. In England, nightingale (*Luscinia megarhynchos*) populations appear to decline in the presence of high densities of deer, wheras other species persist as long as suitable habitat is maintained (Fuller 2001). Perrins and Overall (2001) concluded a review of bird population studies in England by stating that deer numbers probably play a minor role in species declines compared to other anthropogenic factors. Again, the degree of disturbance probably determines the impact of deer browsing. In protected habitats, deer browsing may be the main agent of change, but their role is diminished in forests already susceptible to fragmentation, timber harvest, livestock grazing, and the like.

Other Vertebrates

There is evidence that high densities of ungulates impact organisms other than birds. Voles and rodents were at higher densities within deer exclosures in an ancient woodland in England (Putman et al. 1989). The evidence was strongest for bank voles (*Clethrionomys glareolus*) that rely on moist, dense understories (Flowerdew and Ellwood 2001). Forest salamanders and amphibians also rely on the density and diversity of understory plants to structure moisture and pH of soil and leaf litter (deMaynadier and Hunter 1995) and may be similarly affected, although this may not hold for redback salamanders (*Plethodon cinereus*, Brooks 1999).

It would be inaccurate to label all herbivory as detrimental to biodiversity. Rare species of butterflies in England are maintained in forests that are browsed sufficiently to open the canopy, but not to the extent that the herbaceous host plants are removed (Feber et al. 2001). The concept of biodiversity linked to intermediate levels of herbivory has received repeated attention from researchers (Crawley 1983; Hobbs 1996; Augustine and McNaughton 1998).

The role of ungulates in complex ecosystems is best illustrated with oak (*Quercus* sp.) forests. All oak forests in North America are highly modified by human activities (McShea and Healy 2002), and both white-tailed and mule deer can reach high densities within these systems. High densities of deer in oak forests do affect small mammal abundance through competition for seed crops, but only when annual mast production is low (McShea 2000). High seed production (> 200 kg/ha in eastern oak forests), on the other hand, can support high densities of rodents, regardless of deer densities. Since the effect of deer is relative to seed production, events that reduce seed production, such as disease or insect irruptions, become important cofactors.

Measuring the impact of high-density populations of white-tailed deer in oak forests should not be confined to counting species, because trophic interactions are also evident. Deer serve to facilitate Lyme disease (*Borrelia burgdorferi*) transmission to humans, with ticks (*Ixodes scapularis*) as the vector (Ostfeld 1997, 2002). White-footed mice (*Peromyscus leucopus*) serve as an intermediate host for the disease, and increases in rodent species diversity reduces disease transmission by having ticks feed on less effective hosts (Ostfeld and Keesing 2000). One consequence of ungulates reducing rodent diversity might be increased disease transmission to humans (Ostfeld 2002).

Is There Evidence That Reducing Ungulate Density Restores Biodiversity?

If high densities of ungulates have reduced biodiversity measures, how should managers respond? Manipulation of predator densities does have measurable effects on ungulate populations (Gasaway et al. 1992; Boertje et al. 1996; Hayes et al. 2003). An Alaskan moose population increased threefold, while the caribou population doubled, within seven years of predator control (Boertje et al. 1996). Messier (1991) concluded wolf (*Canis lupus*) densities have a major influence on both moose and white-tailed deer populations on Isle Royale, Michigan. Similarly, Hayes et al. (2003) determined wolf predation in southwest Yukon affected both recruitment of caribou and moose populations and survival of adult moose, to the extent that wolves were limiting both species. Although there is endless debate on whether populations are controlled by top-down or bottom-up processes (e.g., Messier 1991; Boertje et al. 1996; Schmitz and Sinclair 1997), predator populations can reduce prey densities, particularly when they are already significantly below carrying capacity (Bowyer et al., this volume). The important question is whether reductions in ungulate prey density are sufficient to restore biodiversity measures.

Most of what we know about ungulate impacts on biodiversity is derived from deer exclosure (Putman et al. 1989; McShea 2000; McShea and Rappole 2000) or inclosure (Tilghman 1989; deCalesta 1994; Horsely et al. 2003) studies. Other studies involve correlations between deer density and biodiversity measures (Hough 1965; Anderson and Loucks 1979; Alverson et al. 1988; Balgooyen and Waller 1995; Fletcher et al. 2001a,b). Changing the density of ungulates does change the biodiversity of a system. There are few examples, however, of forest/ungulate systems where biodiversity was measured at the same time that predators were being introduced or excluded. One exception is the natural experiment of the reintroduction of wolves into the Yellowstone ecosystem (Berger et al. 2001a,b). In this case, the reintroduction of predators has resulted in increased biodiversity of avian fauna within riparian areas. The potential exists for similar impacts to be documented in the northern Midwest with the resurgence of wolf populations, and the planned reintroduction of large carnivores into the northeastern United States.

Hunters versus Large Carnivores

The most direct means to influence plant biodiversity may be to manipulate herbivore density directly by controlling hunter access to a site. A valid question, therefore, is whether the restoration of biodiversity mandates the use of large carnivores to control ungulate density, or can hunters fill that ecological role?

The first argument against the need for large carnivores to regulate ungulate numbers is that forest systems without large carnivores do not lack predators. In North America, black bears (*Ursus americanus*), coyotes (*Canis latrans*), lynx (*Lynx canadensis*), bobcat (*Lynx rufus*), and red fox (*Vulpes vulpes*) can be significant predators of ungulate fawns (Linnell et al. 1995). The expansion of black bear and coyote ranges in recent years could be attributed partially to regional increases in the abundance of ungulate prey. Automobiles may function as effective large predators on adults. Conover (1997) estimated > 725,000 deer are killed by automobiles annually. Pierce et al. (2000b) compared the demography of mule deer killed by automobiles and those killed by mountain lions (*Puma concolor*) and coyotes in the Sierra Nevada of California. They reported few significant differences, although deer killed by both predators were older than those killed by vehicles, and coyotes were more likely to kill female deer than were automobiles. Likewise, legal hunters annually remove approximately 200,000 deer in Virginia (Knox 1997); and over two million deer across the United States (M. Knox, pers. comm. 2003). Hunting, automobile strikes, and smaller predators, therefore, pose a significant source of mortality for a population of 15 million deer (McCabe and McCabe 1997).

A second argument that might cast doubt on the utility of using large carnivores to restore biodiversity is that regulating ungulate density through manipulation of predator numbers is cumbersome (Boertje et al. 1996; Hayes et al. 2003), and management might be better directed closer to the source of the problem (i.e., ungulates). Increased harvest by hunters or professional sharpshooters can reduce ungulate numbers in small areas, such as a township (Butfiloski et al. 1997) or historical park (Frost et al. 1997). At a larger scale, manipulation of hunting permits between regions of a state can also achieve some control over harvest levels (Knox 1997). In countries with effective law enforcement and hunter management, ungulate numbers may be managed more easily through use of hunters.

Arguments against the reintroduction of large carnivores, however, ignore the point that ungulate foraging has two components, numerical and functional (Hobbs 1996; Pastor and Cohen 1997; Augustine and McNaughton 1998). When modeling predator effects on ungulate populations, functional responses are as important as numerical responses (Eberhardt et al. 2003). The same probably holds for ungulates and their forage; the functional response of ungulates to changes in plant density and availability is critical to predicting their impact on biodiversity measures.

A functional response of ungulates at lower densities means that reduced numbers may not cause a rebound in biodiversity, unless ungulates are forced to shift their feeding preferences (Coomes et al. 2003). One ungulate foraging at ease could eliminate an entire localized population of an endangered species, if that species is preferred forage. Mechanisms that force ungulates to shift or avoid feeding sites have the potential for modifying any functional response. Affecting functional responses is more easily achieved with predators than with hunters that have both temporal and spatial constraints.

There is evidence that natural predators affect the functional response of ungulates. The reintroduction of wolves into the Yellowstone ecosystem has provided a good test of how predators change ungulate behavior. Berger and colleagues (Berger 1998; Berger et al. 2001b; Pyare and Berger 2003) have shown that ungulates increased movement and restricted feeding in certain habitats in response to the reintroduction of predators. These changes impacted their ability to forage on preferred browse species and changed the structure and composition of plant communities (Berger et al. 2001a). The consequences of these behavioral shifts were increased abundances of several bird species that are sensitive to vegetation characteristics.

It is a false dichotomy to advocate either hunting or predators. Bowyer et al. (this volume) discuss how ungulate populations are more likely regulated by predators when densities are low than when densities approach carrying capacity. Populations at carrying capacity (e.g., the problem populations discussed in this chapter) are driven by habitat productivity. Predators introduced into high-density prey environments may not seriously decrease overall browsing pressure. Hunters work on a different dynamic from predators and it may be difficult to entice hunters into areas with deer densities well below carrying capacity. In North America, state wildlife agencies have a difficult job of convincing hunters that populations should be reduced further through harvests of female deer (Diefenbach et al. 1997). A short-term solution would be for hunters to reduce chronically high populations below half of carrying capacity and then introduce large carnivores to maintain these lower densities at an alternative steady state that has high measures of biodiversity. Around suburban or agricultural areas, of course, reintroduction of large predators may prove unfeasible, and managers may be left with professional hunters to achieve numerical goals. The scenario of hunters and predators would be preferable to either option alone, since the goal is to change both the numerical and the functional response of the ungulates.

Conservation Recommendations

It is clear that ungulate populations impact plant communities, and that this can have far-ranging implications for forest biodiversity. This is not the same as saying that managers can always reach their biodiversity goals through reduction in ungulates. The forest modeling referenced earlier (Mladenhoff and Stearns 1993; Seagle and Laing 1997) indicated life history traits play a large role in determining which trees reach the forest canopy. Plant productivity is influenced by temperature, light, nutrients, and moisture (Crawley 1983), none of which are regulated in the short term by ungulates. Some sites lack the qualities needed for a measurable response in biodiversity to lower densities of ungulates. An obvious example is light; reductions of ungulate densities within mature or old growth forest have not always resulted in dramatic changes in plant biodiversity because of insufficient seed banks and light reaching the forest floor (Putman et al. 1989; Coomes et al. 2003). In those circumstances, ungulate reduction should be accompanied by other management activities that set the stage for increased plant productivity (e.g., selective harvest, controlled burning).

Managers should also be aware that sites do not start from a pristine state following the reduction of deer. Some species of herbaceous and woody plants persist under heavy browsing pressure due to unpalatable tissues or rapid growth rates (Augustine and McNaughton 1998; Horsley et al. 2003). The removal of deer within these systems likely would lead to rapid increases in already established species (Coomes et al. 2003). Any hope of rapidly restoring biodiversity to these systems must deal with control of exotic or invasive species prior to release of browsing pressure. If alternative stable states exist within plant communities (Schmitz and Sinclair 1997; Stromayer and Warren 1997), a release of plants from browsing pressure may not restore the original suite of species but, rather, a "new" community that has no historical precedent.

There is clear justification for severely reducing introduced populations of ungulates (Coblentz 1990; Garrott et al. 1993), but most of my discussion has centered on native ungulate populations. The role of herbivores in forest communities is to consume plant material and convert plant tissue into animal tissue and waste products, which will be consumed by other organisms (Crawley 1983; Hobbs 1996). Evidence that ungulates consume woody or herbaceous plants should not be shocking to ecologists or be used as justification for their control. Deer impact is enhanced when they occur at densities close to K, but their removal is not necessarily the path to restoration of biodiversity. The intermediate browsing hypothesis (Crawley 1983; Augustine and McNaughton 1998) states that an intermediate degree of herbivory maintains maximum diversity, and either extreme results in biodiversity loss due to overbrowsing (near K) or competitive exclusion between plants (far below K). An intermediate level of herbivory will be a difficult goal for most managers to achieve but can be obtained through focus on both numerical and functional responses of ungulates.

There is no single ungulate density that would provide a metric for managers managing for increased biodiversity because effects depend on K and on the availability of preferred forage. White cedar (*Thuja occidentalis*) and eastern hemlock are affected by > 3 deer/km² (Alverson et al. 1988), whereas other woody seedlings in a hardwood forest rebound when densities drop below 11 deer/km² (Tilghman 1989). We do not know the densities needed to restore migratory birds, but 8 deer/km² increased bird diversity in Allegheny hardwood forests (deCalesta 1994). It is relatively simple to pick a focal species and manage deer for that target species. It is orders of magnitude harder, however, to manage a whole forest ecosystem, and we have few successful examples (McShea and Healy 2002). There must be tight management parameters that result in densities of ungulates low enough to stimulate restoration of biodiversity but high enough to prevent competitive exclusion and maintain the attention of large carnivores or hunters. One solution to this problem is to increase the spatial scale of management activity. At the larger scale not all places will have the desired deer or biodiversity measures, but rare species will be maintained across the region through the dynamic movement of ungulates, hunters, and predators. In most forested systems ungulates need to be managed to achieve biodiversity goals, and large predators are a preferable means to regulate deer because they impact both the numerical and the functional properties of deer foraging.

Summary

In this chapter, I examined evidence for ungulate impacts on forest biodiversity in the absence of large carnivores. Wild populations from eight species that are distributed worldwide in temperate and boreal forests have been shown to have deleterious effects on biodiversity. Plant species are most often directly affected by ungulates, but additional trophic levels, composed of birds and mammals, are likewise affected. The closer ungulate populations are to carrying capacity the broader their impact. Biodiversity in forests can be restored following ungulate reduction if conditions are favorable. The reintroduction of large carnivores is one component in a restoration project that must take into account the potential of the site, invasive species, life history traits of target species, and disturbance regimes.

Hunters are not a direct equivalent to large carnivores because they often affect numerical, but not functional, responses in the prey. Large carnivores should not be expected to regulate ungulate populations that are at or above carrying capacity. The culling of ungulate populations prior to carnivore introductions might enhance the ability of carnivores to act as biodiversity "managers." It is preferable to produce large spaces that are variable for both the density of ungulates and plant biodiversity, because microhabitats within the landscape will differ in both their potential for biodiversity and their ability to "shield" ungulates from predation. We have the opportunity to conduct large-scale natural experiments, and these should be used to test the ideas presented in this volume. Forest communities will exhibit a large diversity of responses when placed back under the influence of large carnivores. Whereas it is difficult to predict the response of any one site to reduced ungulate density, it is not hard to predict that the impact over the landscape scale will be significant.

ACKNOWLEDGMENTS

I want to thank all the field assistants, graduate students, and colleagues who have helped develop these ideas over the years. Thank you to the editors, Terry Bowyer, and one anonymous reviewer for greatly improving the manuscript and to WCS for inviting me to a very informative workshop at White Oak.

King of the Beasts? Evidence for Guild Redundancy among Large Mammalian Carnivores

Rosie Woodroffe and Joshua R. Ginsberg

Across the vast majority of the modern terrestrial realm, the largest predators are mammals. On each continent, a guild of such species has evolved—specialists at hunting large-bodied ungulates (Fig. 9.1). Constrained by the need to fuel their own large bodies and provision their energetically demanding young, these species are dependent on large prey (Carbone and Gittleman 2002). Today, these guilds include members of the cat, dog, and hyena families, as well as bears in some areas (e.g., Fig. 9.1b).

Large mammalian predators (hereafter, "large carnivores") may have profound impacts on the ecosystems they inhabit by influencing both the density and the behavior of other species they kill. Multiple studies have demonstrated that predation by large carnivores clearly has the capacity to limit or regulate the density of ungulate prey (reviewed in Ballard et al. 2001; Sinclair et al. 2003). The extent to which this occurs varies substantially among systems, according to a variety of factors addressed throughout this volume. More subtly, the presence of large carnivores may influence the behavior of large herbivores in ways that can affect habitat quality for other species, often incurring beneficial influences for local biodiversity (Berger et al. 2001a; Ripple et al. 2001). Additionally, "mesopredator release" is a well-documented phenomenon associated with the localized extinction of large carnivores, which has marked impacts on populations of small-bodied species such as birds and rodents that are preyed upon by mesopredators but not by larger predators (e.g., Crooks and Soulé 1999a).

With few exceptions (Sinclair et al. 2003), studies of these ecological impacts of large carnivores tend to focus on evidence for a single keystone species. However, it is rare to find a natural system containing only one large carnivore species.

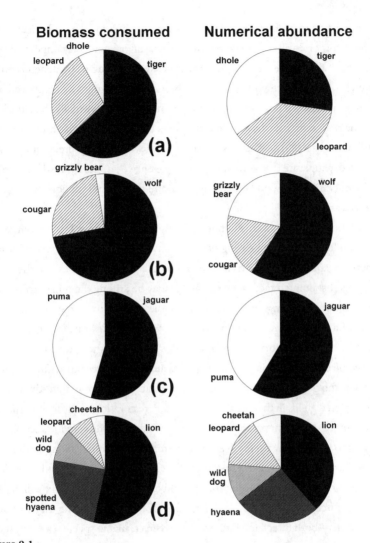

Biomass consumed   **Numerical abundance**

Figure 9.1

Relative consumption of mammalian biomass by large carnivores, and their relative numerical abundance, in (a) Nagarhole National Park, India (Karanth 1993); (b) the Northern Range of the Greater Yellowstone Ecosystem, USA; (c) Manu National Park, Peru (Emmons 1987); and (d) Kruger National Park, South Africa (Mills and Biggs 1993). Data for Yellowstone are based on predator numbers given in Smith et al. (2003) and prey consumption rates from Bjorge and Gunson (1989), Ross and Jalkotzy (1996), and Mattson (1997). Estimates of large carnivore diet are based upon a combination of kills and scats.

This raises two questions. First, can a single species assume the ecological role of top predator? or do all species within the large carnivore guild play important parts in structuring ecosystems? And, second, if ecosystem structuring depends on the existence of intact carnivore guilds, how best can intact guilds be conserved or, if necessary, restored? These are important questions because virtually all of the large mammalian carnivores have suffered range collapses in the last hundred years (Gittleman et al. 2001a). This has left some areas with no large predators and others with depauperate faunas retaining only one or two species from what were formerly much richer assemblages. Only a handful of sites still support intact guilds of large carnivores.

With this background, an understanding of the ecological role of large carnivore guilds has a direct bearing on how conservation priorities are established. If there are strong interactions between multiple predator species and their prey, then sites sustaining intact carnivore guilds may be among the most ecologically pristine areas remaining on Earth, retaining the majority of ecological processes and with a consequently high priority for conservation. Conserving such sites would be a higher priority than an overall strategy that might ensure the persistence of individual large predator species at different sites but without conservation of the guild as a whole. Moreover, if intact guilds are manifestly different from depleted guilds in their ecological function, restoration of the former may become a more significant priority. However, as we discuss, the very interactions one is trying to restore may themselves make restoration of complete guilds difficult.

In this chapter we make a first attempt at addressing these difficult questions. The ecological impacts of losing some members of a carnivore guild will depend upon the forces structuring the guild itself. Only by understanding the horizontal interactions—both direct and indirect—between carnivore species can we predict the vertical effects of guild collapse. Hence, we begin by briefly characterizing the world's major intact large carnivore guilds and reviewing the factors underlying their structure. We then characterize "partial" guilds by outlining the processes whereby large carnivore assemblages collapse under human pressure. With a better understanding of the differences between intact and partial guilds, we discuss whether partial guilds may be able to fulfill the ecosystem functions of intact guilds and review whether conservation of intact guilds is necessary. Finally, we discuss how interactions within large carnivore guilds may generate surprising (and sometimes counterproductive) effects of management activities aimed at preserving or restoring intact guilds. Many of our discussions are hindered by the limitations of available data, but we hope that they will at least spark further study.

Large Carnivore Assemblages and the Forces That Structure Them

Historically, every continent that supports native, large-bodied, herbivorous mammals also supported large carnivores. It is worth bearing in mind that, with the exception of Africa, all of these continents lost a significant proportion of their large mammal faunas during the late Pleistocene: Australia no longer supports marsupial lions (*Thylacoleo carnifex*), and North America no longer supports dire wolves (*Canis dirus*) or saber-toothed cats (*Smilodon fatalis*). Nevertheless, with the exception of Australia, each continent retains a more or less distinctive assemblage of large carnivores. Africa has the largest number of extant, sympatric, large carnivores (five; see Fig. 9.1d). A large number of large carnivores still occur in Asia, but some, like lions (*Panthera leo*) and cheetahs (*Acinonyx jubatus*), are restricted to isolated, remnant populations and are no longer part of intact guilds, whereas the geographic ranges of others do not overlap those of all the other species. South America supports the smallest guild of carnivores dependant on large-bodied prey, with just two large cats. Figure 9.1 shows that, at least in the systems illustrated, predation pressure—expressed very simply here as the biomass of meat consumed per unit time—is often dominated by one species: the tiger (*Panthera tigris*) in Asia, the wolf (*Canis lupus*) in North America, and the lion in Africa. The relative impact of different species may vary between sites, with the "dominant" role sometimes shared between two species, such as lions and spotted hyenas (*Crocuta crocuta*) in some parts of Africa, but there is a general pattern that ungulate biomass predation is not shared equally among species.

Feeding Competition

Species within large carnivore guilds are in competition with one another. The extent of dietary overlap is typically high (Karanth and Sunquist 1995; Creel and Creel 1996; Kunkel et al. 1999; Scognamillo et al. 2003). Some partitioning of prey may occur related to the body size of the prey being hunted, with the very largest carnivores (e.g., lions, tigers) or those that hunt socially (wolves, lions) able to exploit very large-bodied prey not accessible to smaller or less social guild members. Conversely, some species, such as leopards (*Panthera pardus*), puma (*Puma concolor*), and wolves in Europe, while preferring larger prey, may be able to subsist on smaller prey (Iriarte et al. 1991; Boitani 1992; Ramakrishnan et al. 1999). However, two important patterns emerge. First, medium-sized ungulates form the mainstay of the diets of most large carnivores (Mills and Biggs 1993; Seidensticker et al. 1999; Sinclair et al. 2003), creating the opportunity for exploitation competition among carnivore species. Second, because predator diets tend to be nested within one another rather than strictly partitioned, smaller-bodied prey are subject to predation by a larger number of carnivore species (Sinclair et al. 2003). From the point of view of the prey, this suggests that top-down regulation will more likely occur in smaller species (Sinclair et al. 2003). From the predator's point of view, it suggests that competition among carnivore guild members is likely to be asymmetric: larger bodied or social carnivores, dependent on larger prey species, may also contribute to suppressing the density of small prey by opportunistic predation. Hence, they may to some extent deplete food supplies for smaller carnivores while being themselves supported by larger prey not available to their smaller carnivorous competitors.

The ways in which large carnivores partition prey may not just be related to body size. Competing species often have different methods of hunting (e.g., stalking vs. coursing, Kunkel et al. 1999; Karanth and Sunquist 2000), which may lead to changes in prey selection related to the ecology and behavior of the prey (FitzGibbon 1990; Laundré et al. 2001). Also, competitors frequently hunt at different times within the diurnal cycle, with the largest species tending to be nocturnal while smaller species are crepuscular or even diurnal (Mills and Biggs 1993; Karanth and Sunquist 2000).

Intraguild Predation

Indirect competition over prey may have a less direct effect on the structure of predator guilds than direct, intraguild predation. Perhaps the greatest insight into the ecology of large mammalian carnivores in the last generation has been the recognition of the role that intraguild predation plays in structuring large carnivore assemblages (Palomares and Caro 1999). Deliberate killing of small carnivores by larger carnivores has been recognized for some time (e.g., Fuller and Keith 1981), but the ecological importance of killing of large carnivores by other species in the same guild has been appreciated only recently (Palomares and Caro 1999). Particularly profound effects have been documented in Africa's intact carnivore guilds, where predation by lions appears to limit the densities of both cheetahs (Durant 1998) and wild dogs (*Lycaon pictus*) (Creel and Creel 1996; Mills and Gorman 1997) to well below ecological carrying capacity. In both species, predation by lions is the most substantial cause of natural mortality (Laurenson 1995; Woodroffe et al. 2004). Observations of interactions between dhole (*Cuon alpinus*) and big cats in Asia suggest a similar ecological relationship, with tigers in particular behaviorally dominant over dhole and known to have killed dhole on several occasions (Venkataraman 1995; Karanth and Sunquist 2000). The outcome of competitive interactions between jaguars (*Panthera onca*) and pumas, tigers and leopards, lions and hyenas, and wolves and cougars have been less well studied, but intraguild killing has been recorded in most cases (Kruuk 1972; White and Boyd 1989; Karanth and Sunquist 2000). Given the number of competing predator species with broadly overlapping geographic distributions, intraguild predation has the potential to be a very powerful force determining the density and distribution of carnivore species (Caro and Stoner 2003).

Mammalian predators' attempts to avoid being killed by other members of the same guild may have as profound an ecological impact as the predation itself. Two strategies of avoidance emerge. Some species avoid larger, more successful competitors by placing their home ranges along the margins of their competitors' territories, in areas not favored by the larger guild member. Thurber et al. (1992) described this behavior among coyotes (*Canis latrans*) avoiding wolves on the Kenai peninsula, Alaska, with the same pattern occurring, on a smaller scale, among red foxes (*Vulpes vulpes*) avoiding coyotes in Yellowstone National Park (Crabtree and Sheldon 1999). Similar behavior has been described several times among leopards avoiding tigers (Schaller 1967; Seidensticker et al. 1990; Støen and Wegge 1996), which appears to exclude leopards from the cores of protected areas, forcing them into human-dominated areas on reserve boundaries (Seidensticker et al. 1990). In contrast, African wild dogs and cheetahs adopt a somewhat different strategy by avoiding areas frequented by lions (and, to a lesser extent, spotted hyenas) and seeking out areas of low prey density (Mills and Gorman 1997; Durant 1998; Creel and Creel 2002). In wild dogs, this leads to an unusual phenomenon: as the density of wild dogs' preferred prey increases, wild dogs' home range size also increases (Woodroffe in prep). As a result of this behavior, both wild dogs and cheetahs occupy home ranges substantially larger than predicted on the basis of their metabolic requirements (Gittleman and Harvey 1982). Radiotelemetry studies have still not been carried out on dhole; however, a minimum estimate of home range size based on visual observations (Venkataraman et al. 1995) indicates that they, too, occupy larger-than-expected home ranges (Woodroffe and Ginsberg 2000). This, coupled with observations of prey-stealing and predation by big cats (Venkataraman 1995; Karanth and Sunquist 2000) suggests that dhole may also experience intraguild predation, and, like African wild dogs and cheetahs, range widely to avoid it.

Overall Patterns of Guild Structure

In conclusion, while most intact guilds consist of several large mammalian carnivores, in most systems one species has a disproportionately large ecological impact. Such species are typically large bodied and/or social, hence capable of killing comparatively large prey not exploited by competing species. These predators are often behaviorally dominant over other species in the same guild, which they are able to displace and even kill. Through their impacts on both prey and competing predators, the loss of these dominant species—lions in Africa, tigers in Asia, and wolves in North America—might be expected to have the greatest impact on ecosystem function. We discuss this hypothesis in a later section.

When Guilds Collapse: Rules for the Disassembly of Large Carnivore Assemblages

If, as we have hypothesized, the loss of some species will have a particularly marked impact on ecosystem function, it is important to consider which species typically *are* lost from predator guilds. Following major contractions in geographic range of virtually all large carnivore species over the past hundred years, very few sites still support intact large carnivore guilds. The factors underlying these local extinctions are common to virtually all large carnivores: persecution by people as a result of human–wildlife conflict, habitat fragmentation, prey loss, and disease. However, species vary considerably in their ability to persist in the face of these threats, with the result that some are more extinction prone than others (Woodroffe 2001). Consequently, the partial large predator guilds that remain in many places do not comprise a random selection of the original assemblage. Rather, they tend to contain a fairly predictable array of species depending upon the site's size, habitat type, and degree of protection.

Research on the correlates of local extinction indicates that wide-ranging species tend to be most vulnerable. Mammalian carnivores with large home ranges require larger reserves for population persistence, tend to go extinct at earlier dates, and are more likely to be classified as threatened under Red Book criteria than are related species with smaller home ranges (Woodroffe and Ginsberg 1998, 2000; Woodroffe 2001). This association between wide-ranging behavior and extinction-proneness, independent of population density, is robust among large mammalian carnivores and has also been observed in other taxa, from primates to butterflies (Harcourt 1998; Thomas 2000). Among mammalian carnivores, wide-ranging behavior seems to be associated with extinction because it increases exposure to people, who constitute the overwhelming dominant agents of mortality, both inside and outside reserves (Woodroffe and Ginsberg 1998).

A species' relative vulnerability also depends upon the degree of protection from human influence. This is a function not just of its legal status, or the protection status of the land it inhabits, but also its own behavior. Some species (e.g., lions, jaguars) appear to respond well to protection and can sustain viable populations in comparatively small reserves (Woodroffe and Ginsberg 1998, 2000). Outside reserves, however, these species are often the first to become extinct (Woodroffe 2001). Although no quantitative data are available, this variation in vulnerability may reflect species' large body size, and hence the (real or perceived) threat that they pose to livestock and people (Woodroffe 2001).

In fragmented landscapes, species' vulnerability to local extinction may be influenced, both directly and indirectly, by intraguild predation. First, interspecific killing might force poor competitors to such low densities that populations inhabiting small habitat patches are simply too small to be viable (Vucetich and Creel 1999). Alternatively (or additionally) intraguild predation could force poor competitors to range into areas where they are forced into contact (and, hence, conflict) with human activities (Woodroffe and Ginsberg 1999; Woodroffe 2003). Woodroffe (in prep.) attempted to disentangle these twin effects of intraguild predation by investigating correlates of local extinction in African wild dogs. Although wild dogs were indeed more likely to become extinct from areas supporting high densities of lions, these areas tended to be small reserves where the degree of contact between wild dogs and people could also be expected to be high. Hence, it was not possible to determine whether local extinction of wild dogs was due to high lion density or small reserve size.

Parallels between lion–wild dog and tiger–dhole relations remain somewhat speculative. Nevertheless, we draw attention to a possible association between the local population density of tigers and local extinction of dhole. Dhole have experienced marked declines, at least in India, in recent years (Johnsingh 1985), and require larger reserves for persistence than do tigers (Woodroffe and Ginsberg 1998). Densities of both species may be estimated using camera trapping (Carbone et al. 2001). At 19 sites in Indonesia, Myanmar, and Thailand, all within dholes' historic range, dhole are absent from areas where tigers are photographed more frequently; (t_{17} = 2.23, p < 0.05; Wildlife Conservation Society unpubl. data; a total of over 28,000 camera trap-nights). More intensive ecological studies of tiger–dhole interactions are likely to be highly informative.

Whatever the underlying ecological reasons for their vulnerability, the tendency for wide-ranging carnivores to be most extinction-prone means these species are usually the first to disappear from large carnivore assemblages. This is an interesting finding because, as already discussed, the most wide-ranging species (e.g., African wild dogs, cheetahs, perhaps dhole) are often species that consume comparatively small amounts of mammalian biomass, and hence might have an equivalently small impact on the density and dynamics of ungulate prey populations.

Guild Redundancy and Compensation

The presence of several competing species within a large carnivore guild could entail a degree of redundancy. That is, the loss of one species from the guild could cause populations of the remaining species to increase in size, alter their hunting behavior to exploit prey species formerly monopolized by the missing guild member, hunt at favorable times of day from which they had hitherto been excluded, or expand the range of habitats in which they occur. If mammalian carnivores are able to compensate in this way for the loss of other guild members, then the ecological role of large carnivores might be conserved even if some species are lost from the guild.

A quasi-experimental study of the impact of removing part of a carnivore guild—carried out in East Africa—provides no support for the contention that predators can compensate for one another (Sinclair et al. 2003). Poaching and poisoning greatly reduced the densities of lions, hyenas, and jackals (*Canis* spp.) in part of the Serengeti ecosystem from 1980 to 1987 (Sinclair et al. 2003). It is unlikely that wild dogs or cheetahs would have been so severely affected, since both species scavenge only very rarely and are hence less sensitive to poisoning than other members of their guild. Loss of part of the predator guild from this region was associated with significant increases in the densities of small and medium-sized ungulates. No such increases were recorded in a neighboring area where predators remained undisturbed (Sinclair et al. 2003). This suggests that the remaining predators did not compensate for the loss of other members of their guild, thereby allowing prey populations to expand (Sinclair et al. 2003). It is worth commenting that the subsequent complete extinction of wild dogs in the Serengeti ecosystem has not been linked to any change in prey numbers.

Although this one quasi-experimental study provides no evidence for compensation within large carnivore guilds, redundancy within guilds might be inferred by comparing the density, diet, and behavior of large carnivores living in intact and partial guilds.

Effects on Population Density

Evidence that populations of large carnivore species increase in size following the extinction of competing guild members is mainly anecdotal. Perhaps the best information comes from Miquelle et al. (this volume) who showed that wolf numbers declined steadily in the Russian Far East as conservation efforts led to the recovery of tiger populations. Likewise, the distribution of leopards in Java and Malaysia is reported to have expanded following localized extirpation of tigers (Seidensticker 1986). Spotted hyenas apparently increased in number in Amboseli

National Park, Kenya, following local extinction of lions (C. Moss, pers. comm. 2002). By contrast, restoration of wolves to Yellowstone National Park appears not to have led to a reduction in the density of cougars, although it has substantially reduced the density of a local mesopredator, the coyote (Smith et al. 2003).

Snapshot data on large carnivores studied in the presence and absence of competitors are equally inconclusive. Cheetahs living on Namibian farmland, in the absence of lions and hyenas, occurred at densities no higher than observed elsewhere (Marker 2002). Likewise, wild dogs studied under similar circumstances of drastically reduced numbers of competitors lived at densities no higher than those recorded in similar ecosystems where lions and hyenas have remained (Pole 2000). However, such snapshot data must be interpreted with caution. Lions appear to achieve higher densities in areas where wild dogs have become extinct, but examination of time sequences shows that this is a spurious correlation generated by the pattern that wild dogs are more extinction-prone in areas of high lion density (Woodroffe in prep). Similarly, although we have interpreted the preceding data on the relative abundance of tigers in areas with and without dhole as indicating that dhole have become extinct where tigers (a larger competitor) are more abundant, in the absence of time-series data it is impossible to be confident that tigers have not, in fact, been able to increase in number following local extinction of dhole.

Negative correlations between the densities of competing predators at different sites have been discussed (e.g., Schaller 1967; Sunquist and Sunquist 2002) but rarely demonstrated. Indeed, the local densities of presumably competing predators are often *positively* correlated because densities of most predators tend to increase with prey density (Stander 1991). Laurenson (1995) found a negative association between lion and cheetah density after controlling for the confounding effects of prey density. Creel and Creel (1996) found a negative correlation between the densities of wild dogs and lions across sites, but Woodroffe (in prep.) found a positive correlation using a larger data set. The highest wild dog densities have been recorded in areas where lions are subject to sport hunting (Creel and Creel 2002), but cause and effect have not yet been demonstrated. Interestingly, wild dogs studied in the absence of larger competitors occupied home ranges only slightly smaller than those of packs living alongside lions and hyenas, and home range size remained unrelated to the density of their principal prey, impala

(Pole 2000). Wild dogs appear to range widely with only slight regard for the regional density of either predators or prey.

Effects on Diet

We were not able to find evidence to suggest that large carnivores altered their diets in response to the absence of other guild members. African wild dogs living alongside lions and spotted hyenas in the Kruger National Park, South Africa, preyed on the same species, and in the same proportions, as those inhabiting the Savé Valley Conservancy, eastern Zimbabwe, part of the same ecosystem where lions and hyenas had been eradicated (Fig. 9.2a). Curiously, lions persisting in wildlife areas from which wild dogs had been extirpated exploited medium-sized antelope (wild dogs' preferred prey) to a *lesser* extent than those still living alongside wild dogs (Fig. 9.2b). The reason for this is not clear. The diet of leopards living alongside tigers in Royal Chitwan National Park, Nepal, was indistinguishable from that of leopards living in the absence of tigers (but with the same guild of ungulate prey) in Wilpattu National Park, Sri Lanka (Fig. 9.3).

Our failure to find evidence of dietary compensation is perhaps not surprising. Carnivores' diets typically span the range of prey species they are able to kill: larger and/or more social predators can kill larger prey, but all species can and do kill small prey. Thus carnivore diets tend to be nested within one another (Sinclair et al. 2003). Within this size range, however, predators tend to specialize on those species they can kill most efficiently and profitably (Sinclair et al. 2003), which may not necessarily reflect the relative abundance of those prey species. Losing tigers (for example) from a guild might perhaps increase the abundance of larger ungulates on which tigers specialize, but it is unlikely to affect leopards' ability to subdue and kill those prey—hence it may not affect leopards' diet choice. Indeed if, as Sinclair et al. (2003) suggest, larger ungulates are unlikely to be predator-limited (but see Mills et al. 1995) while smaller ungulates are suppressed by a broad array of predator species, a leopard might be more likely to exploit an increase in the abundance of its preferred (smaller) prey as a consequence of tiger loss, than to switch to hunting larger ungulates. Conversely, loss of a smaller competitor such as a leopard would be unlikely to have much impact on the availability of the larger prey favored by larger predators such as tigers.

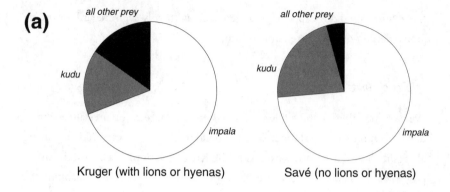

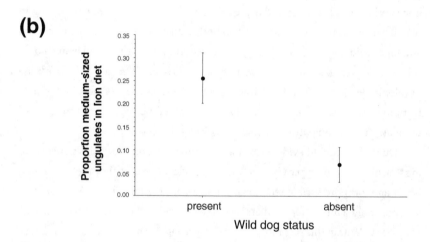

Figure 9.2

Diet of African wild dogs and lions living sympatrically and apart. (a) Wild dog diet
in the Kruger/Gona re Zhou ecosystem in the presence and absence of lions. Data
from Mills and Biggs (1993), and Pole (2000). There is no significant difference in
wild dog diet between the two sites. (b) Proportion of lion diet consisting of
medium-sized ungulates in five ecosystems where wild dogs were present at the
time of the study (Chobe, Manyara, Kafue, Mana Pools, and Kruger) and four
ecosystems where wild dogs had become extinct by the time of the study (Laikipia,
Ngorongoro, Etosha, and Nairobi). Lions take fewer medium-sized ungulates where
wild dogs are extirpated (Mann Whitney $U'_{4,5} = 18$, $p = 0.05$). Data from Elliott and
Cowan (1978), Stander (1992), Dunham (1992), Funston et al. (1998), R. Woodroffe
(unpubl. data), Schaller (1972), (Viljoen 1993).

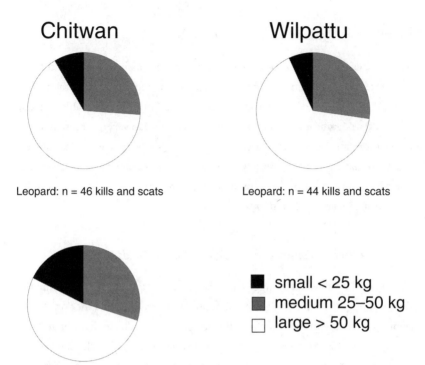

Chitwan

Leopard: n = 46 kills and scats

Wilpattu

Leopard: n = 44 kills and scats

■ small < 25 kg
▨ medium 25–50 kg
☐ large > 50 kg

Tiger: n = 135 kills and scats

Figure 9.3
Prey composition of leopard diet in the presence (Chitwan National Park, Nepal) and absence (Wilpattu National Park, Sri Lanka) of tigers, with an identical array of ungulate prey species available. Leopard diet at the two sites is indistinguishable ($\chi^2 = 0.12$, df = 2, $p > 0.9$). Data from Seidensticker et al. (1990), Schaller (1967), Eisenberg and Lockhart (1972), Sunquist et al. (1999), and Seidensticker (1976).

Effects on Activity Patterns

Large carnivore species might also respond to the local extirpation of another species in the same guild by altering the manner in which they hunt (e.g., by becoming more nocturnal or diurnal). However, we are not aware of any studies that have investigated this possibility. Although carnivores often become more nocturnal in human-dominated landscapes (e.g., Stander 1990; Vilà et al. 1995), it is difficult to know whether this is due to the absence of competing predators, or the presence of people.

Overall Evidence for Redundancy within Guilds

Taken together, these data provide little convincing evidence to suggest that large carnivore species respond to the local extirpation of competing guild members by increasing in numbers or even altering their behavior. A few available anecdotes suggest that significant effects may sometimes occur. However, at present, the evidence for redundancy within large carnivore guilds is poor. We caution, however, that systematic data are extremely restricted, and that many factors—particularly complex interactions between predators and their prey—could obscure subtle, or even significant, guild redundancy.

So, Do We Need to Conserve Intact Large Carnivore Guilds?

The limited data available suggest that large carnivores retain roughly the same diet, hunting behavior, and, perhaps, population densities, whether or not competing species are present. If this is the case, it could indicate that not all species are equal, and loss of one species from a large carnivore guild could alter the structuring impact that that guild has upon the ecosystem it inhabits. If predator species within a guild do not compensate for one another, it is fair to assume that, when a species is removed, the complex web of direct or indirect top-down interactions will be disrupted. This would tend to argue for conservation—and, where necessary, restoration—of intact large carnivore guilds where possible.

The impact of losing part of a guild will likely depend upon which species are lost. Clearly, losing a dominant predator population that consumes a significant amount of ungulate biomass will have the greatest effect: studies of the loss of lions and hyenas (Sinclair et al. 2003) and the restoration of wolves (Smith et al. 2003) show this very clearly. Such cascading effects would, of course, only be expected where the "dominant" predator is lost because of direct removal (e.g., through being killed by people, or through disease). If declining prey populations underlie the loss of such species—as has been shown for some tiger populations (Karanth and Stith 1999; Ramakrishnan et al. 1999)—any top-down structuring imposed by the predator must already have been swamped by anthropogenic impacts on prey.

The loss of species that occur at lower densities and consume a smaller biomass of prey might be expected to have a proportionately smaller impact on ecosystem function. In Africa at least (and perhaps also in Asia), these species appear to be the most extinction-prone. Wild dogs, cheetahs, and (perhaps) dhole are the first to be lost from protected areas; they also naturally occur at low densities, probably as a result of top-down structuring *within* large carnivore guilds, and hence have a comparatively small impact on total prey biomass (see Fig. 9.1).

Is it possible, therefore, that (in protected areas at least) the first to be lost are those that can best be spared? Might the earliest stages of large carnivore guild collapse (when only the most extinction-prone species are lost) have comparatively few implications for the structure of the ecosystem as a whole? This simplistic conclusion ignores the important indirect role that predation plays in structuring ecosystems; for instance, through its impact on ungulates' antipredator behavior. Predators can influence habitat structure (and, hence, a multitude of smaller species) simply by altering the way ungulates move within the landscape (Berger et al. 2001a; Augustine 2002; Ripple and Beschta 2003). The presence of multiple competing large carnivores, all adopting different hunting strategies, means that in intact ecosystems ungulates face a risk of predation at all times of day and night, in virtually all habitats and at virtually any population density. This is particularly the case for smaller ungulates, which are prone to predation by a broad array of carnivore species (Sinclair et al. 2003). The effect of this chronic risk upon ungulates' antipredator behavior has not, to our knowledge, been studied, but it could have implications for habitat structure and, hence, for biodiversity in general. In the absence of any data, it would be premature to conclude that less abundant predator species play no role in structuring ecosystems. Although this might not in itself provide a strong argument for restoring these species to the many landscapes from which they have disappeared, it does suggest that such species may have an ecological value, and that permitting further local extinctions of these species could have ecological consequences beyond the loss of their intrinsic value.

Conserving Complete Large Carnivore Guilds in Fragmented Landscapes: Some Complex and Surprising Predictions

That large predator guilds have collapsed around the globe, even where comparatively large areas of habitat remain, demonstrates that conserving carnivores is difficult. As we have reviewed here, the impacts of guild collapse, or even guild simplification, are poorly understood and have only recently become a focus of serious study. With scant evidence for redundancy within such guilds, we argue that sites that still support intact guilds should be priorities for conservation efforts. However, the conservation and restoration of these guilds are complicated by the finding that the very largest predators can, through intraguild predation, influence the extinction probabilities for smaller species.

Conserving intact ecosystems, including complete large predator assemblages, may entail some serious challenges for conservationists. One salient, and recurring, theme in our discussions has been that not all large carnivores are equal. Within the guild, dominant carnivores (whether lions and hyenas in Africa, tigers in Asia, or perhaps jaguars in South America) influence the behavior and ecology of other members of their respective large predator guilds. These intraguild interactions, which are themselves ecological processes worthy of preservation, argue strongly for both the conservation and the restoration of intact guilds. Achieving these goals, however, is likely to require a focus not on the Kings of the Beasts, but on the other less competitive guild members that are more extinction-prone. In this section we discuss conservation strategies for large mammalian carnivores and draw attention to three perhaps surprising predictions generated by an understanding of how intraguild predation structures carnivore assemblages.

The Largest Predators Are Not the Best Umbrella Species

The principal threat to large mammalian carnivores is deliberate or accidental killing by people (Woodroffe and Ginsberg 2000; Woodroffe 2001). Given the inherent difficulty for local people of coexisting with large, potentially dangerous animals, protected areas will continue to play an absolutely crucial role in carnivore conservation. Large carnivores place the greatest trophic demands on an ecosystem and therefore have the greatest area requirements. This suggests that they should be effective umbrella species (Ray, this volume), in that a protected area big enough to sustain large carnivores should also be big enough to sustain a multitude of other, smaller, species with lesser area requirements. Taking this trophic argument a step further, one might expect that the very largest carnivores, with the greatest energetic demands, should need the largest areas and therefore provide the most appropriate model for the design of protected areas. This argument ignores, however, the finding that smaller-bodied, less effective predators (such as cheetahs, wild dogs, and perhaps dhole) range far more widely than their larger guildmates, apparently as a way of avoiding intraguild predation. Such species thus have far greater area requirements: for example, lion populations in East Africa have tended to persist in reserves of 291 km^2 or larger, whereas wild dogs require protected areas in excess of 3,600 km^2 to have even a 50% chance of persistence (Woodroffe and Ginsberg 2000). Hence, a conservation area designed, even quite conservatively, to be large enough to protect lions would almost certainly be too small to conserve wild dogs effectively. As an example of this sort of logic in action, in 1990 Ginsberg and Macdonald mentioned that the recent establishment of a network of reserves in India under the auspices of Project Tiger would probably provide adequate protection for dhole (Ginsberg and Macdonald 1990). Subsequent analyses have shown that dhole have substantially greater area requirements (723 km^2 for a 50% chance of persistence) than do tigers (135 km^2) (Woodroffe and Ginsberg 1998).

In Praise of Low Prey Density

Some species are threatened by prey loss (e.g., tigers, Karanth and Stith 1999), suggesting that measures to augment prey densities could benefit predators. Moreover, in many species home range size declines as prey density increases (Gompper and Gittleman 1991; Grigione et al. 2002), suggesting that improved prey availability might mitigate the wide-ranging behavior that takes so many predators beyond the boundaries of protected areas and into conflict with people. However, this prediction is based upon an oversimplistic view of the relationship between some large carnivore species and their prey. Once again, the most extinction-prone species are often those species that are displaced from areas of high prey density by more effective competitors. Increasing prey density might therefore benefit the largest carnivores but have surprisingly negative consequences for species that are poorer competitors. As an example, the establishment of boreholes in Kruger National Park between 1930 and 1979 led to an increase in the density of resident ungulates. This measure may have inadvertently caused the local extinction of the brown hyena (*Hyaena brunnea*, which disappeared from Kruger in the same time period), by increasing the density of the spotted hyena, which is behaviorally dominant over the (smaller) brown hyena (Mills and Funston 2003). A second example comes from the Serengeti ecosystem, where the eradication of rinderpest led to rising densities of ungulates. This was associated with an increase in the local populations of both lions and hyenas; however, wild dogs declined over the same period (Hanby and Bygott 1979) and eventually disappeared from Serengeti in 1991 (Burrows 1995; Ginsberg et al. 1995).

Restoration May Not Always Be Attainable

The substantial area requirements of large carnivores mean that it may not always be possible to restore complete guilds. Complete assemblages were able to coexist in the past, in geographically extensive areas including different habitat types and prey densities. By contrast, in today's human-dominated landscapes, wildlife is often compressed into habitat fragments too small to contain the habitat mosaic needed for coexistence of multiple predator species. Restoration of complete, self-sustaining, large carnivore assemblages to such areas is probably not an attainable goal in most circumstances, although recent successes in the Greater Yellowstone Ecosystem suggest that some opportunities do exist for such action (Fritts et al. 1997).

Conclusions and Conservation Recommendations

The data presented here, although limited, support the idea that the largest carnivores in an ecosystem can have a significant impact on ecosystem function and biodiversity. Such species contribute the greatest "predation force" to a system: they extract the greatest amount of energy from lower trophic levels (see Fig. 9.1), and also, often being most numerically abundant, probably have the greatest impact on ungulates' antipredator behavior, which in turn influences habitat heterogeneity and local biodiversity. The "lion's share" is well named. In addition to their impacts on the density, distribution, and behavior of their prey, such species may have equally profound impacts on the ecology of competitors within the large predator guild. Indeed, the evidence suggests that top-down structuring within large carnivore guilds, although discovered more recently than top-down effects that occur between trophic levels, may be just as widespread. Conservation of such species, many of which face serious extinction risks, clearly has a high conservation priority.

Although intraguild predation appears to limit, either directly or indirectly, the abundance and distribution of somewhat smaller-bodied, less successful large carnivore competitors, there is little evidence to suggest that these species are redundant within ecosystems. Such species may contribute comparatively little to the total energetic offtake from ungulate populations (see Fig. 9.1). However, in comparison with their larger and more successful guildmates, such species may have adapted their behavior to mitigate direct competition. They often hunt at different times of day, in different habitats, and at different prey densities, suggesting that they may influence the antipredator behavior of ungulates in ways that could potentially affect habitat structure and, through this, local biodiversity. Such effects have rarely been detected—or even sought—but their existence should at least be considered before dismissing species such as cheetahs, African wild dogs, and perhaps dhole as ecologically irrelevant.

We also acknowledge that in assessing the conservation and ecological value of intact large carnivore guilds and of less successful competitors in the large carnivore guild, a precautionary approach should be observed. The lack of evidence for redundancy may be real, or may reflect insufficient data and an inability to conduct large-scale experiments in predator reduction/exclusion. Classic studies of compensation among carnivores in the marine intertidal zone show that identifying redundancy may depend on habitat and/or prey availability and structure (see papers in Kareiva and Levin 2003). Moreover, these species and the interactions that shape the guilds they occupy are important components of biodiversity worthy of conservation irrespective of their impact on lower trophic levels. Hence, conservation of intact predator guilds has a higher priority for the preservation of biodiversity than protection of guilds' constituent species, but separately at a range of different sites with little or no sympatry.

Although conservation of intact guilds is a high priority, direct competition between large carnivore species leads to some complex ecological interactions that may confound attempts to maintain or restore complete guilds. Intraguild predation has two major impacts: it may depress the densities of inferior competitors through simple offtake, and it may also displace them into less preferred areas of low prey density. Seeking out these areas often entails long-range movements, which in fragmented landscapes can increase the probability of conflict with local people. This means that, against expectation, the most numerically abundant large carnivores, with the greatest ecological impact, may not have the largest area requirements and would not be the most appropriate species on which to base protected area designs if one is looking to select an umbrella species (Sanderson et al. 2002b). Moreover, management activities that seek to augment prey densities may lead to further displacement of inferior competitors.

Clearly, conservation of intact large carnivore assemblages demands the protection of large areas of habitat, supporting a mosaic of high and low prey densities. Such areas might include private or communal lands, as well as legally designated protected areas. Preservation of small habitat fragments, sustaining high densities of ungulates and high densities of the most numerically dominant predators, was the fashion in the past and is reflected in the small size of many of the world's reserves. That this approach has failed to capture important elements of biodiversity is well-established (e.g., Mace et al. 1999). However, that it has also failed to capture the complete large carnivore assemblage, with consequential loss of a potentially vital ecological structuring force, is only now being realized.

Summary

Across the vast majority of the terrestrial realm, the largest predators are mammals. Through the predation pressure that they exert, these animals influence both the density and the behavior of their prey, with profound impacts on the structure of ecological communities. In most systems, one or a few of the largest-bodied predator species have the greatest impact on prey populations. This occurs partly because these species have the greatest energetic requirements but also because they dominate and kill smaller members of the same guild, which are consequently less abundant. Despite this predominance of a small number of carnivore species, two lines of evidence suggest that intact predator guilds, which can contain up to five sympatric species, may be required to maintain ecosystem function and to promote biodiversity. First, there is little evidence for redundancy within large carnivore guilds; if one species is lost, the others rarely compensate by increasing in density, broadening their diets, or altering their hunting behavior. Second, and more speculatively, the maintenance of ungulate antipredator behavior, a very powerful force influencing browsing patterns and hence habitat structure and biodiversity, may require a complete assemblage of large carnivores that hunt at all times of day and night, in all habitats and at all prey densities. Unfortunately, interactions between competing predator species can make it difficult to preserve or restore complete guilds. Conservation will be most effective in very large areas, containing a mosaic of high and low prey densities; hence such sites have a very high priority for the conservation of ecological processes.

ACKNOWLEDGMENTS

This chapter was greatly improved by comments from the editors, along with three anonymous reviewers. We would also like to thank the staff of the Wildlife Conservation Society in Indonesia, Thailand, and Myanmar for providing us with unpublished camera trapping data on tigers and dhole.

From Largely Intact to Human-Dominated Systems: Insight on the Role of Predation Derived from Long-Term Studies

Due to the difficulties inherent in developing proper experiments it is enormously challenging to distinguish the influence of predation on determining patterns of biodiversity from that of other factors. More often than not, the questions pertaining to the relationship between top predators and biodiversity have simply not been asked. In this section, five case studies are presented by ecologists who have worked on a long-term basis in a diverse array of ecosystems. In each of these case studies, the authors present new information on how the role of large carnivorous animals might be influenced by key ecosystem properties, such as productivity, predator community structure, and human perturbations. The ecosystems they examine range from relatively intact to heavily influenced by humans. They provide information essential for determining whether the functional importance of carnivores necessarily means that focusing conservation efforts on them will achieve conservation of biodiversity.

In the first chapter, Dale Miquelle and coauthors (Chapter 10) present results from a relatively pristine northern temperate ecosystem in the Russian Far East containing intact biological communities. They continue the discussion of functional redundancy among top predators started by Woodroffe and Ginsberg (Chapter 9) by examining the competitive relationship between tigers and wolves, and what this means for conserving biodiversity in the region. Gus Mills (Chapter 11) next turns to four relatively well studied and intact savanna ecosystems of Africa, where long-term predator–prey studies have provided data on the interplay between habitat, climate, behavior, and predation by resident carnivores. Dr. Mills provides a window into how predation, even in cases where it is not a dominant structuring force, might still be an important component of conservation strategies.

The remaining case studies focus on systems where humans have been more influential in shaping patterns of biodiversity. Bogumila Jędrzejewska and Wlodzimierz Jędrzejewski (Chapter 12) explain that, although the Białowieża Primeval Forest represents the last remaining intact forest in north-temperate Europe, it is small and embedded in modified landscapes. Additionally, resident carnivores and herbivores have experienced human-generated population fluctuations over the past few centuries. Nevertheless, unrivaled datasets documenting population cycles of the major players in this story, combined with long-term ecological research, allow them to explore questions pertaining to the ecosystem role of top predators and the implications for biodiversity conservation. Turning to the marine realm, Tim McClanahan (Chapter 13) compares two coral reef parks following the cessation of fishing in limited areas that primarily targeted predatory fish. Through detailed analyses of functional groupings within the extraordinary biodiversity of these areas, Dr. McClanahan evaluates the relative roles of predator-induced cascades and abiotic factors in bringing about the documented changes in these systems. In the final chapter of this section, Andrés Novaro and Susan Walker (Chapter 14) describe current conditions in the Patagonian Steppe—an ecosystem that has undergone sustained changes through the combined introduction of exotic herbivores and habitat modification. Although top predators have still managed to repopulate their former range under such conditions, Drs. Novaro and Walker explore why this has not translated into accompanying recovery of other key components of regional biodiversity.

CHAPTER 10

Tigers and Wolves in the Russian Far East: Competitive Exclusion, Functional Redundancy, and Conservation Implications

Dale G. Miquelle, Philip A. Stephens, Evgeny N. Smirnov,
John M. Goodrich, Olga J. Zaumyslova, and
Alexander E. Myslenkov

Abundant evidence indicates that predators can have profound indirect effects on many aspects of the diversity of a region through their direct effects on prey species. It follows that conserving predators is likely to be an essential component of conserving intact floral and faunal complexes. However, if its role can be ful-filled by an alternative species, the extinction of a predator need not lead to radi-cal changes in the biota. In particular, in areas where multiple members of the predatory guild are present, the loss of one predator may be compensated for by other members of the guild. Similarly, where one predator has historically ex-cluded a competitor, extinction of the former may permit the latter to colonize the area, once again compensating for the loss. This idea of functional redundancy (e.g., Walker 1992) among the predator guild relies heavily on close similarities in the effects of predators on their prey. If different species of carnivores are not func-tionally equivalent (i.e., if they have varying impacts on ungulate populations), then competitive exclusion, or any replacement of one large carnivore by another, could have important cascading effects on community structure.

Differences in body size, morphology, life history, and predation behavior among large carnivores can all potentially lead to differences in their impacts on prey populations and each other. For instance, comparisons of felid and canid life history strategies, social structures, and hunting techniques suggest that selection of prey and impact on prey populations can vary substantially (Kleiman and

Eisenberg 1973; Eisenberg 1984; Kunkel et al. 1999; Husseman et al. 2003). At the same time, within-guild competition among predators, expressed directly through avoidance, food usurpation, and outright killing, and indirectly by reducing prey abundance (Palomares and Caro 1999; Creel et al. 2001) can lead to competitive exclusion of one or more carnivore species by another. Unfortunately, although there are many comparisons of within-guild food habits of predators, comparisons of the relative impact of different species of predators on ungulate populations within an ecosystem (Jędrzejewska and Jędrzejewski 1998; Sinclair et al. 2003) are fewer. Similarly, although examples of competitive exclusion among large carnivores exist (Fuller and Keith 1981; Creel and Creel 1996; Durant 1998; Tannerfeldt et al. 2002), explorations of the impact of this phenomenon on prey species are rare.

In this chapter, we present a case study of how two predators, the Amur tiger (*Panthera tigris altaica*) and the gray wolf (*Canis lupus*), interact competitively and exert top-down pressures in the Sikhote-Alin ecosystem, Russian Far East. Specifically, we use a combination of historical records, more recent ecological data (from Sikhote-Alin and Białowieża Primeval Forest, a similar, well-studied area in Poland), and modeling techniques to address two principal questions: (1) What is the relationship between tigers and wolves in Sikhote-Alin, and is there evidence of competitive exclusion? and (2) Do these two species have similar impacts on prey, suggesting some degree of functional redundancy in this system? We provide evidence that tigers do exclude wolves from this system but that human persecution can lead to a reversal of that process. Our analyses suggest the direct effects of these two carnivores on prey species are different and, consequently, their roles in influencing ecosystem structure are unlikely to be equivalent.

Study Areas

The southernmost Russian Far East is dominated by the Sikhote-Alin Mountains, a coastal range that parallels the coast of the Sea of Japan from Vladivostok 1000 km north to the mouth of the Amur River (Fig. 10.1). The principal forest type of the original ecosystem is a mixed-composition forest dominated by Korean pine (*Pinus koraiensis*) and a variety of broadleaf species. In disturbed areas, Mongolian oak (*Quercus mongolica*) is dominant. As with plant communities, the faunal complex

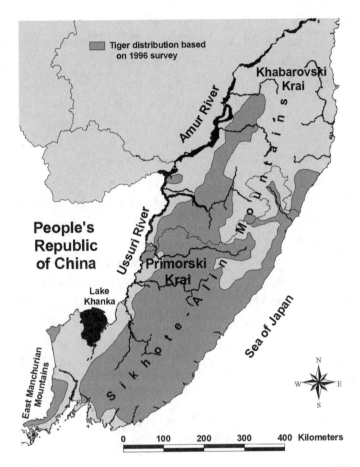

Figure 10.1
Primary features of the Southern Russian Far East, and tiger distribution based on a
1996 survey (Matyushkin et al. 1999).

is a mixture of Asian, Himalayan, and boreal species. The ungulate complex is represented by seven species: red deer (*Cervus elaphus*), wild boar (*Sus scrofa*), sika deer
(*Cervus nippon*), Siberian roe deer (*Capreolus pygargus*), Manchurian moose (*Alces
alces cameloidus*), musk deer (*Moschus moschiferus*), and ghoral (*Nemorhaedus caudatus*). In addition to tigers and wolves, Far Eastern leopards (*Panthera pardus ori-
entalis*), lynx (*Felis lynx*), brown bears (*Ursus arctos*), and Himalayan black bears
(*Ursus thibetanus*) occur in the region. Of these, tigers and wolves are the only large

carnivores that are widely distributed and depend upon the larger ungulates. Leopards are restricted to the East Manchurian Mountains, bears have not been shown to be important predators of ungulates, and lynx focus on roe deer and occasionally red deer calves (Okarma et al. 1997; Goodrich et al., unpubl. data).

Sikhote-Alin Zapovednik (SAZ), an IUCN Category I protected area, is situated in the central Sikhote-Alin Mountains, and extends from the coast of the Sea of Japan across the divide to the western side of the range (see Fig. 10.1). When established in 1935 it covered 11,570 km², but its size has varied dramatically. Reaching a nadir of 990 km² in 1951, it was restored to 3100 km² in 1960, and is presently 4000 km². Red deer are the most abundant ungulate in SAZ followed by roe deer (Stephens et al., in press) Wild boar numbers fluctuate widely, but the species is generally common.

Where necessary (see following), we compare information on prey selection in SAZ primarily to Białowieża Primeval Forest (BPF) in Poland, where prey composition is similar to SAZ, where wolves are the dominant large predator, and where an extensive database of both predators and prey exists (Jędrzejewska and Jędrzejewski 1998, this volume; Jędrzejewski et al. 2002). BPF retains essentially the same ungulate complex as SAZ, with red deer, roe deer, and wild boar the dominant species.

Data Analysis and Modeling Methods

To assess the relationship between tigers and wolves, and their impact on prey, we reviewed long-term monitoring data from SAZ and across the Sikhote-Alin Mountains and then modeled predation by the two species using data from both SAZ and BPF. We outline these approaches below.

Relationship between Tigers and Wolves in the Sikhote-Alin Ecosystem

We derived the relative abundance of tigers and wolves in the Sikhote-Alin ecosystem from anecdotal and historical accounts of their distributions and abundances, as well as from data derived from the archives of the Primorski Krai Department of Hunting Management and "Chronicles of Nature" from SAZ. Track counts (including expert assessment of tracks to derive absolute estimates) on standardized survey routes in winter in SAZ provide a basis for assessing changes in abundance over time (Smirnov and Miquelle 1999). Biases may exist in such data but we assume that these remain relatively constant from year to year, allowing an assessment of changes in relative abundance over time.

Impacts of Wolves and Tigers on Prey

The degree of similarity between the impacts of these two predators on prey depends on what they eat and the extent to which they limit prey populations. The first of these can be determined by looking at empirical data on diet breadth and prey selection. Unfortunately, there are no comparable estimates of the number of prey taken by the two different predators in SAZ. Consequently, we used data from the literature to parameterize models of prey removal by these species.

Diet Breadth and Prey Selection of Tigers and Wolves

We compared data on food habits from SAZ for two periods: 1962 to 1972 when both tigers and wolves were present, using data from Gromov and Matyushkin (1974); and 1992 to 2002 when wolves were absent and tiger numbers were considerably higher (Smirnov and Miquelle 1999). To supplement these data, we also used information on wolf diets from BPF for the period 1986 to 1996 (Jędrzejewska and Jędrzejewski 1998).

In SAZ, data on prey selection from 1992 to 2002 were obtained by locating kills made by both radiocollared (Goodrich et al. 2001) and uncollared tigers (Miquelle et al. 1996). We combined data for both collared and uncollared tigers after finding no significant variation in the ratio of prey species found (Miquelle et al. 1996). Kills were identified to species and, where possible, categorized into sex–age classes as adult males, adult females, yearlings, or young of the year.

We compared diet selection using Shannon's diversity (H) and equitability (E) indices. We also estimated Horn's index of diet overlap between: (a) tigers and wolves in SAZ (1962–1972); and (b) tigers in SAZ (1992–2002) and wolves in BPF (1986–1996) (all formulae available in Krebs 1989). Indices for wolves and tigers in SAZ from 1962 to 1972 were based on number of kills, but for tigers in 1992 to 2002 we converted data to biomass, using weights from Bromley and Kucherenko

(1983), Danilkin (1999), and Jędrzejewska and Jędrzejewski (1998) to derive comparable dietary diversity indices for tigers in SAZ and wolves in BPF (% biomass in Table 4.8, autumn–winter diets in western part, Jędrzejewska and Jędrzejewski 1998). For wolves in BPF we used data only for wild prey items and allocated "undetermined deer" in Jędrzejewska and Jędrzejewski (1998) to proportions of red deer and roe deer in wolf diets in western BPF (1984–1984) (as reported in Fig. 4.10 of Jędrzejewska and Jędrzejewski 1998).

Estimates of relative abundance of red deer and wild boar from both study periods in SAZ were determined from winter counts of fresh tracks (less than 24 hours) adjusted by the relative daily travel distance of each ungulate species (Stephens et al. in prep.) found along permanent routes within SAZ (Stephens et al. in prep.). Population composition of red deer and wild boar (the two most common prey) was determined by trained observers (scientists and forest guards) recording sex–age composition of all observed groups and individuals on a year-round basis. We considered data on predation and population structure for the winter period only (November–April). We used a multinomial test to compare selection by tigers and wolves for the three most important prey species (red deer, wild boar, and roe deer) with estimates of relative prey abundance as the expected ratios, and used chi-square analyses to compare kill selection for each sex–age class within a species (red deer and wild boar) to the proportion of that class found in the population.

Following Karanth and Sunquist (1995) and Kunkel et al. (1999), we used Chesson's (1978) index of selectivity (also know as Manly's alpha, Krebs 1989), to compare dietary preferences of wolves and tigers. We compared dietary responses of tigers to changes in red deer density by using relative abundance estimates of red deer and tiger kill composition (Miquelle et al. 1996), both averaged over approximately five-year intervals from 1962 to 1999.

Tiger and Wolf Predation on Prey Populations
The impact of predators on prey is dependent on three factors: (1) the density and productivity of the prey population, (2) the amount that each predator kills, and (3) the density at which the predators occur. Neither the productivity of prey nor the daily requirements of predators are known for wolves and tigers in SAZ; however, we were able to derive estimates of both from the literature on conspecifics

in similar systems. Productivity of the prey population varies with population density (because this affects population growth) and also depends on body mass. Red deer (and elk) population dynamics are well studied (Clutton-Brock et al. 1982; Houston 1982) and these represent the most important prey species for tigers and wolves in SAZ. Consequently, we limited our assessments of predator impacts on prey to this prey species. Our conclusions, however, should be qualitatively similar for a multiprey system. The dynamics of the red deer population were assumed to be of a ramped density-dependent form (e.g., Fowler 1987; McCullough 1992), with a simple, linear decline in population growth rate only above a threshold at $0.6 K$ (where K is the environmental carrying capacity in the absence of predation). Below this threshold, mean population growth, r, was assumed to be constant. Biomass production was estimated by assuming that the average adult weighed approximately 180 kg (this assumes a male to female ratio of 0.66, with mean adult masses of 149 kg for females and 224 kg for males, Bromley and Kucherenko 1983). Where necessary, we estimated standing biomass of prey assuming that an adult female would represent an individual of approximately average mass.

The density at which predators occur (the third factor underlying the impact of predators and prey) is complex, relying on prey density, predator population growth rates, and, potentially, predator social structure. Prey density is itself a product of the extent of predation, leading to potentially circular logic. Furthermore, flexibility in social structure (for example, the rate at which territory size changes) is difficult to quantify and may also be a function of prey density. As a result of these complications, we used two modeling approaches to assess the potential impact of predators on prey: an energy balance model and a simulation model.

Energy Balance Model. The energy balance model combined empirical relationships between predator and prey densities (hereafter, "numerical responses"), with estimates of prey productivity (see earlier) and predator kill rates. From these it is possible to estimate, for any given prey carrying capacity, the prey density at which prey productivity and prey consumption by predators are in balance. This, in turn, provides an estimate of the proportion by which the prey population will be depressed below its potential carrying capacity.

Predator–Prey Simulation Model. We generated simple matrix models to describe the dynamics of predators and prey. Productivity of the prey (red deer) was based on a stochastic version of the dynamics already described. Parameters used for tigers and wolves are summarized in Appendix 10.1. Predator dynamics were linked to prey availability through energetic constraints on survival and reproduction. Because estimation of functional responses is problematic (Marshal and Boutin 1999), energetic constraints were modeled using a simple depletion approach (Sutherland 1996). Specifically, we assumed that during each time step, predators could remove all required prey (red deer) from the environment down to some critical threshold density (below which predation is no longer energetically viable). Predator consumption rates were the same as those used for the equilibrium model. Predators that could not obtain their requirements during any time step were assumed to die or disperse. Social group sizes, reproductive behavior, dispersal behavior, and presence of transient animals were all modeled on the basis of empirical data (e.g., Mech 1974; Hayes and Harestad 2000a; Sunquist and Sunquist 2002; Kerley et al. 2003; Goodrich, unpubl. data).

A key factor determining the ability of predators to respond to changes in prey availability and, thus, to fully exploit prey populations, is their flexibility regarding territory size. Predator territory size is known to vary with prey densities across habitats for both wolves (Fuller et al. 2003) and tigers (Miquelle et al. 1999). However, flexibility (i.e., the rate of change) of territory sizes in response to changes in prey availability within a single area is poorly understood. For the wolf at least, Fuller (1989) provided evidence that territories may expand, contract, disappear, or be established in response to changes in prey availability or distribution, but the rate at which these changes occur is unclear. In the absence of definitive knowledge, we derived territory size from empirical predator–prey relationships, and retained a stable territory size through simulations, with the assumption that large carnivores will be conservative in adjusting territory size to changes in prey density.

Research Findings

Analyses of the competitive relationship of tigers and wolves in the Sikhkote-Alin ecosystem provide the necessary background for considering their relative influence on prey populations in temperate forest ecosystems.

Relationship between Tigers and Wolves in the Sikhote-Alin Ecosystem

Considered common throughout the region in the late 1800s and early 1900s, Amur tigers were driven to historical lows in the 1940s due to human persecution (Kaplanov 1948; Kucherenko 2001). With hunting of tigers outlawed in 1947, recovery of the tiger population continued for approximately 40 years, reaching an apparent peak in the late 1980s and early 1990s.

Distribution and abundance of wolves follows an inverse pattern to tigers. Wolves were absent or exceedingly rare in the southern Russian Far East at the end of the 19th and beginning of the 20th centuries. Abramov (1940) believed that wolves appeared in the Sikhote-Alin only after the beginning of the 20th century, coincident with the abrupt range reduction of tigers. Yudin (1992) suggested that wolves arrived earlier by infiltrating human-dominated regions (the Ussuri Basin and Lake Khanka regions) where tigers were largely eliminated during colonization by Russians in the late 1800s and early 1900s. In the broken forests and meadows of these regions, wolves survived on a combination of roe deer and domestic livestock (Yudin 1992). When tiger numbers dropped in the Sikhote-Alin and East Manchurian Mountains in the 1930s and 1940s, this peripheral population acted as a source for expansion and colonization. Wolf populations across the region decreased coincident with recovery of the tiger population since the 1940s. Today wolves are rare across the range of tigers, being found in scattered pockets, and usually as solo individuals or small groups.

Numerous anecdotal accounts suggest an inverse correlation between wolf and tiger numbers. For instance, in the absence of tigers, wolves apparently survived (at least intermittently) in relatively large numbers in the Pogranichniy Raion (district) of the Lake Khanka region (Yudin 1992) but disappeared from this fragment of habitat with the arrival of tigers (Matyushkin et al. 1999). Tigers have since disappeared again from this fragment, with reports of wolves returning (D. G. Pikunov, pers. comm., 2002).

Accounts of historical shifts in the abundance of tigers and wolves are especially well documented in the Zapovedniks, where long-term monitoring has been conducted. For instance, in Lazovski Zapovednik, Bromley (1953) reported that, although wolves were formerly absent, 105 wolves "had to be destroyed" in the 1940s, coincident with the low density of tigers. Wolf numbers declined

consistently from the 1960s through the 1980s, at the same time as tiger populations were recovering in the reserve. Wolf tracks were rarely observed in the 1990s,
and no tracks were registered in 1992 and 1993, whereas tiger numbers were high
and stable (Khramtsov 1995).

The population dynamics of tigers and wolves are best documented in SAZ.
Bromley (1953) noted that elder native people on the eastern slopes of the Sikhote-
Alin Mountains (including Terney Raion, where the Zapovednik is located) had
no recollection of wolves occurring in that region prior to the 1930s (Abramov
1940), coincident with a depleted tiger population (Kaplanov 1948). Although
rangewide tiger numbers began increasing in the 1950s, tigers were still virtually
absent in SAZ in the early 1960s (Matyushkin et al. 1981; Smirnov and Miquelle
1999), whereas wolves remained common despite efforts to control their numbers
even within the Zapovednik (Gromov and Matyushkin 1974) (Fig. 10.2). Restoration of reserve size and better protection led to recolonization by tigers in 1963,
and recovery of the tiger population through the mid-1990s (Fig. 10.2) (Matyushkin et al. 1981; Smirnov and Miquelle 1999). Based on fieldwork in SAZ through
the early 1970s, Gromov and Matyushkin (1974) argued against the perception,
common in Russia, that wolves are driven to low densities or extinction in the presence of tigers. In retrospect, it is clear that their observations were made during
a period of disequilibrium, when both species coexisted in SAZ in moderate numbers. As tiger numbers continued to increase, records of wolves in the Zapovednik decreased and became rare (see Fig. 10.2).

Despite the clear inverse correlation between wolf and tiger numbers, the
mechanism driving population declines of wolves is unclear. Gromov and Matyushkin (1974) reported both usurpation of wolf kills by tigers and scavenging of tiger
kills by wolves. The former has probably not been a primary factor influencing wolf
distribution. Although usurpation has been documented in cougar (*Puma concolor*)–wolf systems (Murphy 1998; Kunkel et al. 1999), it is not as common in these
forest habitats as in open savannas (Fanshawe and Fitzgibbon 1993; Caro 1994; Creel
2001). Gromov and Matyushkin (1974) believed that tigers did not prey directly on
wolves, and along with others (Yudin 1992) proposed that wolves actively avoid
areas used by tigers, resulting in spatial separation of the two species as tiger abundance increases, with wolves remaining only in peripheral areas. However, simple
displacement and avoidance seem unlikely to explain the dramatic decrease in wolf
numbers across such large areas. Although there are only four records of a tiger

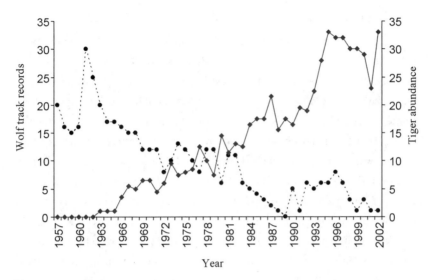

Figure 10.2

Estimates of wolf abundance (------) (based on the total number of tracks reported/year) and tiger abundance (—) (derived from an expert assessment based on number and distribution of tiger tracks/year) in SAZ 1957–2002. Data taken from the SAZ's "Chronicles of Nature" database.

killing a wolf (Miquelle et al. 1996; Makovkin 1999), Amur tigers are notorious for killing dogs (Makarov and Tagirova 1989; Miquelle et al. in press); official records indicate 104 dogs killed by tigers in or near SAZ (where dogs are illegal) between 1957 and 2002. Tiger predation on another canid, the dhole (*Cuon alpinus*), has also been reported on several occasions (Venkataraman 1995; Karanth and Sunquist 2000). Although rarely observed, direct killing of one predator by another is suspected to play an important role in limiting many predator species (Palomares and Caro 1999; Woodroffe and Ginsberg, this volume). Thus, despite lack of clear evidence, we propose that direct killing of wolves by tigers has likely been an important element in reducing wolves to a functionally insignificant role in the Sikhote-Alin ecosystem.

Impacts of Wolves and Tigers on Prey

Selection of prey species by tigers and wolves appears surprisingly similar both within and across study sites, but selection for sex–age classes varies. Our models also suggest that the impact of tiger and wolf predation on prey populations will also vary.

Diet Breadth and Prey Selection of Tigers and Wolves

We located 389 remains of 15 species of wild mammals and birds killed by tigers in SAZ between 1992 and 2002. During a similar span (1986–1996) in the western part of BPF, Jędrzejewska and Jędrzejewski (1998) reported 15 wild species (and plant material) taken by wolves, based on scats and kills. During a shorter time frame with smaller sample sizes, Gromov and Matyushkin (1974) reported that co-occurring tigers and wolves used five and six species, respectively, in SAZ (Table 10.1). Diversity of tiger and wolf diets appeared similar based on comparisons of data from Gromov and Matyushkin (1974) and, again, when based on more recent data on tigers in SAZ and wolves in BPF (Table 10.2). Shannon's diversity and equitability indices put greater emphasis on the diversity of more common species and, hence, large numbers of species taken very infrequently do little to suggest increased diversity. Indeed, the highest diversity score was calculated for wolves during 1962 to 1972 (see Table 10.2), in spite of the narrower range of species recorded (see Table 10.1). Available evidence thus suggests that diet breadth is similar for these two predators.

Overlap in species preyed upon by wolves and tigers was high (see Tables 10.1, 10.2). In all the studies, red deer, wild boar, and roe deer constituted 79 to 97% of kills made, with both tigers and wolves relying on red deer as their primary prey (57–65%) in both SAZ and BPF (see Table 10.1). Wild boar represented the majority of the remaining prey taken by tigers, but wolves relied more on roe deer (BPF) or musk deer (SAZ) than wild boar (see Table 10.1).

Despite a reliance on red deer as their major prey, important differences in diets of tigers and wolves existed. Tigers showed prey selectivity in both study periods (1962–1972: $\chi^2 = 51.52$, df = 2, $P < 0.001$; 1992–2002: $\chi^2 = 563.4$, df = 2, $P < 0.001$), taking red deer in proportion to their relative abundance but showing a strong selection for wild boar and against roe deer (Fig. 10.3a,b; see Table 10.2). This pattern was consistent across both time periods, irrespective of whether wolves were present. For the short time that data on both species were collected in SAZ, wolves appeared to take a slightly greater percentage of red deer than were available, took a smaller percentage of wild boar than were available, and clearly selected against roe deer (see Fig. 10.3a, Table 10.2). Wolf predation in BPF (Jędrzejewska and Jędrzejewski 1998; Jędrzejewski et al. 2002) showed similar patterns to the limited data that exist in SAZ, but with a stronger preference for red deer, and greater avoidance of wild boar (Fig. 10.3c). Wolves in SAZ used a surprisingly high percentage of

Table 10.1

Food habits of coexisting tigers and wolves in Sikhote-Alin Zapovednik, Russian Far East, 1962–1972, for tigers in Sikhote-Alin 1992–2002, and for wolves in an ecosystem with a similar ungulate complex in Białowieża Primeval Forest, Poland

| | % Occurrence | | % Biomass | |
| | Sikhote-Alin, 1962–1972[1] | | Sikhote-Alin, 1992–2002 | Białowieża 1986–1996[2] |
Prey Species	Tigers ($n^3 = 40$)	Wolves ($n^3 = 77$)	Tigers ($n^3 = 389$)	Wolves ($n = 528.2$ kg)
Red deer	57.5	64.9	63.9	65.4
Wild boar	27.5	7.8	24.0	12.8
Roe deer	5	6.4	0.9	19.2
Sika deer			4.9	
Musk deer		14.4	0.04	
Bison				1.8
Goral			0.4	
Brown bear	7.5[4]	1.3[4]	2.4	
Asiatic black bear			1.8	
Harbor seal			0.6	
Moose	2.5	5.2	0.6	
Badger			0.4	
Raccoon dog			0.03	0.2
Red fox			0.02	
Hedgehog				0.2
Brown hare				0.1
Red squirrel				0.01[5]
Bank vole				0.01[5]
Vole (undet. species)				0.01[5]
White-tailed sea eagle			0.01	
Ural owl			0.001	
Great spotted woodpecker				0.01[5]
Nuthatch				0.01[5]
Tawny owl				0.01[5]
Reptile (undet. species)				0.01[5]
Amphibian (undet. species)				0.01[5]
Plant material				0.10
Total	100	100	100	100

[1] From Gromov and Matyushkin 1974.
[2] From Jędrzejewska and Jędrzejewski 1998.
[3] n = number of kills or scats
[4] Brown and Asiatic bears combined
[5] Recorded as present by Jędrzejewska and Jędrzejewski 1998. Minimum value is presented here to estimate biomass.

Table 10.2

Indices of dietary diversity (Shannon index H), equitability (E), diet overlap (Horn's index), and preference for prey (Chesson's or Manly's alpha) for tigers and wolves in Sikhote-Alin Zapovednik, Russian Far East, and Białowieża Primeval Forest, Poland

	Tigers Sikhote-Alin 1962–1972	Wolves Sikhote-Alin 1962–1972	Tigers Sikhote-Alin 1992–2002	Wolves Białowieża Forest 1986–1996[a]
n (# kills or scats)	(40)	(77)	(389)	(344)
Dietary diversity, H	1.60	1.65	1.58	1.40
Dietary equitability	0.60	0.64	0.40	0.35
Diet overlap[b]	0.89	0.89	0.86	0.86
Preferences (alpha)				
Red deer	0.19	0.46	0.15	0.62
Wild boar	0.79	0.48	0.83	0.22
Roe deer	0.02	0.05	0.02	0.16

[a] Data from Jędrzejewski and Jędrzejewska 1998.
[b] Overlap comparisons between (1) wolves and tigers in SAZ, 1962–1972, and (2) tigers in SAZ 1992–2002, and wolves in BPF, 1984–1994.

musk deer, which are most common in higher-altitude, coniferous forests, a habitat that tigers rarely use, according to our more recent studies (Miquelle et al. 1999).

Jędrzejewska and Jędrzejewski (1998) suggested that the density of red deer was the primary factor driving wolf selection of ungulate prey, with proportion of red deer in the diet increasing with density (Fig. 10.4d). Our data suggest no such pattern for tigers: the contribution of red deer did not change (Fig. 10.4a), and percentage of wild boar and roe deer in the diet did not decrease with increasing densities of red deer (Fig. 10.4b,c), as was apparently the case in BPF (Fig. 10.4e,f) (Jędrzejewska and Jędrzejewski 1998: 203). Interpretation of data in SAZ is confounded by increasing densities of a number of prey species over time (Stephens et al. in prep.). However, the fact that red deer abundance does not appear to drive prey selection may also be due to the fact that tigers, in contrast to wolves, show a strong preference for wild boar, a species that fluctuates greatly in abundance and might consequently confound any potential relationships.

Tiger selection of red deer sex–age classes closely mirrored herd composition of red deer, averaged over the study period from 1992 to 2002 ($\chi^2 = 1.92$, df = 2,

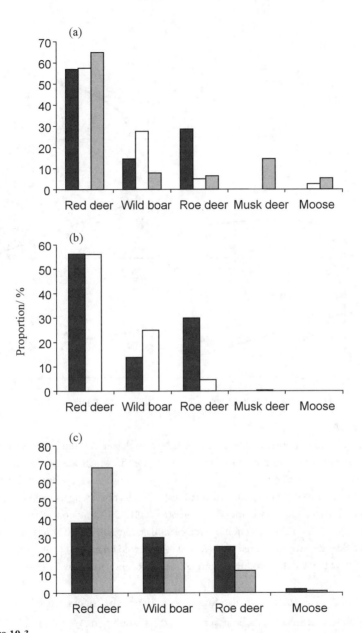

Figure 10.3

Relative abundance of ungulate species (■), and selection of those species by tigers (□) and wolves (■): (a) tigers and wolves in Sikhote-Alin Zapovednik (1962–1972); (b) tigers in Sikhote-Alin Zapovednik (1992–2002); and (c) wolves in the western part of Białowieża Primeval Forest, 1984–1994.

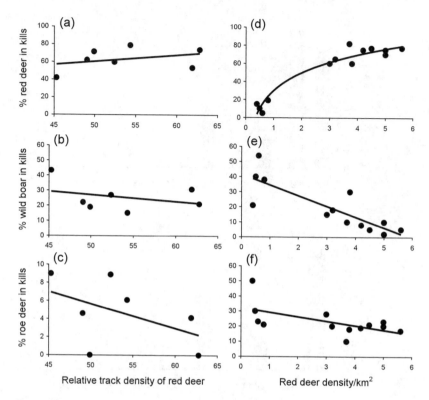

Figure 10.4

Diet selection responses to changes in densities of red deer. All trend lines were fitted by least squares regression and are linear, except for (d), which was best described by a logarithmic function. The R^2 values are given in parentheses following here, to give an indication of the strength of the relationships. Selection by tigers in SAZ, in relation to red deer abundance, for (a) red deer ($R^2 = 0.13$), (b) wild boar ($R^2 = 0.10$), and (c) roe deer ($R^2 = 0.23$); selection by wolves in Białowieża Primeval Forest, in relation to red deer density, for (d) red deer ($R^2 = 0.99$), (e) wild boar ($R^2 = 0.89$), and (f) roe deer ($R^2 = 0.91$) (d–f reproduced from Jędrzejewska and Jędrzejewski 1998).

$p < 0.382$) (Fig. 10.5a). We also found no evidence that tigers preyed selectively on sex–age classes of wild boar across all years combined ($\chi^2 = 2.31$, df $= 3$, $p = 0.5$) (Fig. 10.5b). Sex–age composition of the wild boar population fluctuated more dramatically than that of red deer (see Fig. 10.5b), and though there may have been more subtle within-year changes in selection of wild boar sex–age classes, sample sizes of boar kills limit meaningful yearly comparisons.

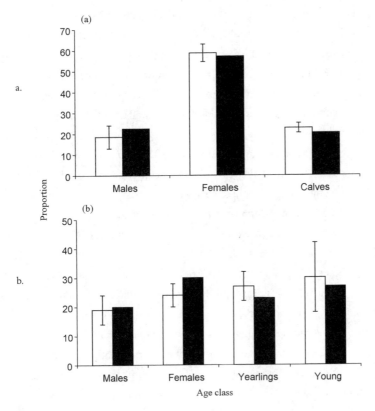

Figure 10.5

Sex–age composition (□) and tiger kill composition (■) of (a) red deer and (b) wild boar populations, SAZ, 1992–2002. Error bars show 95% confidence intervals.

In existing studies of wolf predation on red deer or elk, wolves preferentially fed on calves (Okarma et al. 1995; Mech et al. 2001; Husseman et al. 2003), but selection of adult males and females apparently varied over years in association with vulnerability (Mech et al. 2001).

Tiger and Wolf Predation on Prey Populations

Parameters underlying the two modeling approaches were largely derived from the literature. Demographic parameters for red deer (e.g., Clutton-Brock et al. 1982; Houston 1982) suggested an approximate mean population growth rate of $r = 0.3$ in the absence of density constraints. Above 0.6 K (where density constraints begin to act) this was assumed to decline linearly (to $r = 0$ at K). Estimates of kill rates were 5.1 kg wolf^{-1}d^{-1} (based on live prey eaten, Jędrzejewski et al. 2002) and approximately 8 kg tiger^{-1}d^{*1} (based on a food requirement of 5–6 kg of meat per day, Sunquist et al. 1999). The wolf numerical response has been subjected to considerable scrutiny but, most recently, Eberhardt et al. (2003) have shown that, within the range of available data, it is well represented by a linear regression through the origin. The gradient of this line (Eberhardt and Peterson 1999) is equivalent to approximately 0.1 wolves km^{-2} for each 4.9 red deer km^{-2} (adjusted for the size of red deer in SAZ). For tigers, we took data from 13 sites to construct the numerical response. This was found to be a Type II response, best represented by a Michaelis–Menton function of the form $T = aP / (b + P)$, where T is tiger density, P is prey density, and a and b are constants (see Fig. 10.6). The fit of this function was very highly significant ($R^2 = 0.484$, $F_{12,11} = 10.32$, $p < 0.001$). These parameters were all used to develop the two modeling approaches.

Energy Balance Model. Biomass production and requirement curves were constructed for a range of habitat carrying capacities, from 2 to 10 km^{-2} deer. Biomass production increases linearly up to 0.6 K, because mean growth rate is constant below this point. Above 0.6 K, production declines curvilinearly (Fig. 10.6a). Combining the numerical responses with energetic requirements of individual animals provides energy requirement curves for tiger and wolf populations (Fig. 10.6c). These results suggest constant prey depletion by wolves, but the curvilinear numerical response of tigers results in a relative reduction in energy offtake with increasing prey carrying capacity. Taken together, the production and requirement curves allow predictions of predator–prey equilibria (Fig. 10.6d). Due to the linear nature of both the wolf numerical response and the initial slope of the prey biomass production curve, prey depletion by wolves is predicted to be constant (with prey populations reduced by slightly over 28% of carrying capacity), irrespective of the initial carrying capacity of the prey population. By contrast, the Type II numerical response indicated for tigers suggests that the role of prey in limiting tiger densities declines with increasing prey density. As prey density increases, reduction of prey by tigers becomes gradually less significant, declining from approximately 23% of very low prey carrying capacities (2 km^{-2} deer) to 15% when prey carrying capacity is 10 km^{-2}.

Predator–Prey Simulation Model. Using existing estimates of prey biomass and

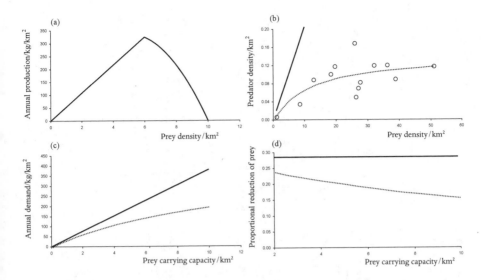

Figure 10.6

(a) Production curve for a red deer population when $K = 10$ km^{-2}. (b) Numerical response for the tiger ($\bigcirc$) (data from Karanth 1991; Karanth and Nichols 2000; Miquelle et al. 1999; Schaller 1967; Støen and Wegge 1996; Tamang 1982; Thapar 1986); the solid line shows the fitted regression (see text); the dashed line shows the predicted response for wolves, using data from Eberhardt and Peterson (1999). (c) Demand curves for wolf (———) and tiger (-------) populations, predicted from their numerical responses and food requirements. (d) Prey depletion by wolves (———) and tigers (-------) as a function of prey carrying capacity, K. Depletion, as used here, indicates the proportion by which the prey population is reduced below its carrying capacity in the absence of predation.

tiger density, we derived territory size of tigers based on the assumption that each territory of an adult resident tigress contains 3.3 tigers (a female, a third of a male, one to three cubs or a young daughter, and one transient) (Fig. 10.7a). For wolves, we derived a relationship between prey availability and territory size (Fig. 10.7b), based on data from Fuller et al. (2003). Using the predicted territory sizes, simulation models of predation suggested that wolves could deplete prey to a greater extent than tigers. This result was consistent in every scenario of prey availability and environmental stochasticity (Fig. 10.8). The proportion by which prey populations were reduced below K by tigers was typically in the range of 18 to 25% and

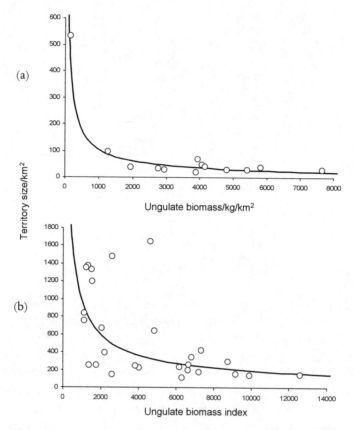

Figure 10.7

Relationships between prey availability and predator territory size.
(a) Tiger territory size, S (assuming 3.3 individuals per territory), and prey biomass, B; the solid line shows the least-squares regression, $S = 22270\ B^{-0.7764}$, $F_{11} = 42.5$, $p <$ 0.001. Mass data sources as for Figure 6 and Sunquist (1981). (b) Wolf territory size, S, and ungulate biomass index, I; the solid line shows the least squares regression, $S = 134138\ I^{-0.7126}$, $F_{25} = 17.7$, $p < 0.001$, data from Table 6.3 in Fuller et al. (2003).

never exceeded 30%. Tiger impact on prey decreased with higher prey density due to constraints on tiger density imposed by territoriality (see Fig. 10.7a), except when stochasticity was high. In this case, tigers were better able to limit prey when initial prey abundance was high, because this reduced the possibility of prey be-

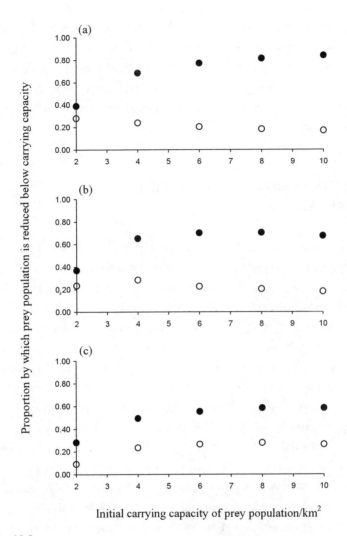

Figure 10.8

Prey depletion (proportion by which prey are reduced below carrying capacity) by wolves (●) and tigers (○) when territory sizes are set according to empirical relationships between prey availability and predator territory size. The amount of environmental stochasticity to which the prey population was subject varied from (a) none, to (b) moderate (prey population coefficient of variation, CV = 0.15), and (c) high (CV = 0.35). Rates of depletion are derived from simulations at the territory size predicted for the given level of prey abundance.

coming so scarce that tigers could not hunt effectively. By contrast, wolves typically reduced prey density to approximately 50% below carrying capacity, and by as much as 80% when prey were abundant and stochasticity was low. Wolves also showed a tendency to deplete prey more when the prey were initially abundant, because high rates of depletion in low prey availability scenarios led to frequent crashes in prey numbers.

Competitive Exclusion and Functional Redundancy in Tigers and Wolves

Mech (1974) suggested that few animals compete successfully with the wolf, but the tiger appears to be an exception. Wolves do not occur across most of the range of tigers in Southeast Asia. Whether this negative correlation is causal is unknown, but evidence from the Russian Far East strongly suggests that tigers depress wolf numbers either to the point of localized extinction or to such low numbers as to make them a functionally insignificant component of the ecosystem. Wolves appear capable of escaping competitive exclusion only when human persecution decreases tiger numbers. Although there are now many studies documenting the impact of intraguild competition among large carnivores (Creel et al. 2001), examples of one large carnivore driving another to localized extinction are relatively rare.

Woodroffe and Ginsberg (this volume) argue that the evidence for intraguild redundancy is poor but warn that data are extremely restricted. Our analyses suggest that the top-down influences of tigers and wolves (and, therefore, their broader impacts on biodiversity) are likely to differ substantially. Although diet breadth was similar, and both tigers and wolves depended on red deer as the primary prey species, there were important differences in prey selection. Jędrzejewska and Jędrzejewski (1998) argued that wolves strongly selected for red deer, and that changes in red deer density determined percentages of other prey that were taken by wolves. We found no such evidence to support this effect with Amur tigers. In contrast, tigers demonstrated a very strong preference for wild boar. Variation in preference, which may be due to the relative vulnerability of boar to ambush and cursorial predation, could alter the relative impact of tigers and wolves on ungulates in a multiprey system. In particular, when boar are common, tiger predation on red deer might be even lower than our models suggested, while wolf predation on red deer would remain high.

All available evidence suggests that wolves select for vulnerable age classes of *Cervus elaphus* (Jędrzejewska and Jędrzejewski 1998; Kunkel et al. 1999; Mech et al. 2001; Husseman et al. 2003). In contrast, data from SAZ suggest tigers exhibit virtually no selectivity among sex–age classes of red deer. Our findings are consistent with observations that solitary, ambush predators are not selective for vulnerable sex–age classes (Okarma et al. 1997; Kunkel et al. 1999), although such results are by no means universal (e.g., Karanth and Sunquist 1995).

Although these results reaffirm both theoretical and empirical indications that cursorial and ambush predators will differentially select prey (Kleiman and Eisenberg 1973; Husseman et al. 2003), we propose that absolute levels of prey depletion by predators may be even more important in determining differences between the top-down influences of tigers and wolves within terrestrial ecosystems. Both of our modeling approaches suggested that, for the majority of conditions, limitation of prey populations by wolves is likely to be considerably higher than by tigers. Our equilibrium approach relies partially on the accuracy of the numerical responses underlying it. Due to the fact that data on predator–prey systems may have been collected at a point in time when predators and prey were not in equilibrium, caution should be used in interpreting these results (Eberhardt et al., 2003). Nevertheless, there are three reasons why we have confidence in the general differences in the numerical responses of tigers and wolves used for this model. First, the numerical response used for wolves is the result of very thorough analyses of the underlying data (Eberhardt and Peterson 1999). Second, several of the data points underlying the tiger numerical response represent mean predator and prey densities averaged over a long period, decreasing the potential error associated with estimates from a single point in time. Third, evidence for density predictions of the numerical responses is borne out by the analysis of Carbone and Gittleman (2002), in which wolves fall substantially below the predictions of numbers per unit prey biomass. This is in agreement with the numerical response derived by Eberhardt and Peterson (1999), which indicates that 10,000 kg of prey supports only 48 kg of wolf (approximately half of the biomass predicted by Carbone and Gittleman 2002). Indeed, for the same prey biomass, the numerical responses underlying our equilibrium model predict higher biomasses of tigers than of wolves over the range of prey density values that exist for wolf–prey systems. This suggests that differences in predator density do not explain our predictions of higher depletion by wolves. Rather, the results arise from differences in food consumption.

Estimates of energy requirements made using standard allometric relation-ships (Nagy et al. 1999) suggest that wolves eat more than would be expected for their body mass, whereas tigers eat less than would be expected. Although it is pos-sible that there are some differences in assimilation efficiencies between these two predators, it is likely that most of these differences can be explained by two aspects of their life history: hunting mode and sociality.

As ambush predators, tigers hunt by stalking followed by very short rushes (Yudakov and Nikolaev 1992), with high success rates (Hornocker 1970). Chases by tigers rarely extend beyond 150 m (Miquelle et al., unpubl. data). Yudakov and Nikolaev (1992) reported 54% and 38% success of Amur tigers hunting wild boar and red deer, respectively. Because tigers are solitary, intervals between kills are high (six to nine days) (Sunquist and Sunquist 2002; Miquelle and Goodrich., unpubl. data). Collectively, high success rates, short chase distances, and long intervals between kills result in low energy expenditures for tigers. By contrast, wolves have low success rates, averaging 14% (based on individual prey) over 14 studies (Mech and Peterson 2003). As cursorial predators, they may chase prey for kilometers (Husseman et al. 2003; Peterson and Ciucci 2003). Although living in groups provides wolves the capacity to kill larger prey and obtain greater biomass per kill (Gittleman 1989), group living also has costs. In particular, shar-ing kills results in less energy acquired per individual per kill. Consequently, kill rates must be considerably higher for wolves, leading to greater travel distances as packs search for prey. An additional cost of group living is the time and energy ex-pended on social interactions, a cost that solitary species like tigers largely avoid.

In summary, higher kill rates (a consequence of sociality) and the associated greater travel distances, as well as greater energy expended in cursorial hunting (greater chase distances) and social interactions, all result in greater energy de-mands for wolves. These factors likely explain much of the difference in food con-sumption between wolves and tigers. As additional support for this argument, it is worth noting that the African wild dog (Lycaon pictus), another cursorial, social predator for which good data exist, is also renowned for its greater than predicted energy demands (e.g., Gorman et al. 1998).

Due to limitations in our understanding of prey dynamics, predator selectivity, and the flexibility of large predators to adapt territory sizes to reflect changing prey availability, our simulation modeling approach was necessarily coarse. Nevertheless, our simulations also suggest that wolves can limit prey to a much greater degree than can tigers. These results are supported by empirical data. Although few direct estimates of harvest rates have been made for tigers, in the high-ungulate-biomass systems of the Indian continent, offtake has been estimated at less than 10% (Schaller 1967; Støen and Wegge 1996). Our model, which predicted 10% depletion when $K = 20$ km^{-2} deer (equivalent to a prey biomass density of 3000 kg km^{-2}), is in agreement with these estimates. For wolves, offtake ranges from minimal (less than 10–15% of the standing biomass of red deer removed each year) (Glowacinski and Profus 1997; Jędrzejewski et al. 2002) to much more significant (35% reduction in moose populations, Messier 1994, 15–30% reduction in elk populations, Eberhardt et al. 2003).

In conclusion, it appears that, despite the fact that wolves and tigers have historically replaced one another as the top carnivore in the Sikhote-Alin ecosystem, they are extremely unlikely to be functionally equivalent. Not only are there differences in prey preferences but there are substantial differences in the extent to which these two predators impact prey populations.

Conservation Implications

In Russia, where there is a strong hunting tradition that is based on maximum sustainable yield, large carnivores are primarily viewed as competitors to human hunters. The impact of wolves on ungulate populations in Russia has received at least as much, if not more, attention than in North America (Filonov and Kaletskaya, 1985; Yudin, 1992). The general conclusion in Russia is that wolves can significantly depress ungulate populations and should be controlled to maintain high ungulate yields for hunters. A similar conclusion concerning wolf impact on ungulate populations has been reached by many in North America, but, whereas wolf control is practiced across much of Russia, in North America it has been controversial (Gasaway et al. 1992; Van Ballenberghe and Ballard 1994).

Our conclusions that tiger predation is unlikely to limit ungulate populations to the same extent as wolf predation concur with the opinions of local biologists

(Kucherenko 1974; Dunishenko 1987). Nonetheless, the relationship between tigers and Russian hunters is less than amicable, and "intraguild" killing of competitors extends to the hunter–tiger relationship. Abundant evidence suggests that competition killing is one of the primary motivations for tiger poaching in Russia (Miquelle et al. in press). Due to this inimical relationship, and because new regulations provide nongovernmental hunting groups with wide-ranging responsibilities to manage game species, hunters will be key stakeholders in determining the future of tigers in the Russian Far East (Miquelle et al. in press). Finding common ground with hunters must therefore be a primary task for those wishing to conserve the Amur tiger. Paraphrasing one argument for tiger conservation, local conservationists and biologists have proposed to local hunters that, "while tigers may not be desirable, they prevent wolves from becoming abundant . . . and we all know that wolves are worse than tigers in depressing prey numbers, so it is to your advantage to tolerate the tiger." Our models support local perceptions of the relative capacity of these two predators to impact prey populations and suggest that this "backdoor" rationale for tiger conservation has a biological basis. This argument does little to foster a more balanced perspective on carnivores, and further enforces a negative perspective on wolves. However, it appears that if Amur tiger conservation is a priority in the Russian Far East, maintaining high numbers of tigers in the Sikhote-Alin ecosystem will come at the cost of low wolf numbers due to competitive exclusion. Although promoting tiger conservation as a mechanism to control wolves may not encourage an appreciation of large carnivores or the intricacies of ecosystem processes, it provides a message understandable at the local level, appears to have a real biological basis, and is likely to elicit a more effective and rapid response than other approaches.

The absence of functional redundancy has implications not only for tiger conservation strategies but for biodiversity conservation as well. Because large carnivores are not necessarily functionally similar, saving "a" large carnivore is not equivalent to saving ecosystem integrity or ecosystem processes. The definition of "integrity" and the types of processes saved will depend on the carnivore retained in the system. Understanding the potential differences in the way large carnivore species structure communities is therefore a necessary prelude to defining biodiversity conservation strategies.

Conservation of large carnivores is obviously not synonymous with biodiversity conservation. For example, it provides no guarantee that other rare species, hotspots, or centers of endemism will be retained. Nonetheless, we believe that conservation of carnivores can help achieve these other conservation objectives. This chapter (and many others in this book) demonstrates that large carnivore ecology is largely driven by their relationship with prey species. Tigers and wolves, and indeed most large carnivores, are habitat generalists, and as such, minimum prey density is perhaps the key habitat parameter determining their presence. Large area requirements are not an intrinsic characteristic of large carnivores but a consequence of their need for adequate prey (Karanth and Stith 1999; Miquelle et al. 1999). Carnivore prey requirements therefore help define the minimum suitable area needed for biodiversity conservation but not necessarily the specific locations. Carnivore habitat is not spatially fixed but can be created by managing the prey base, a well-understood process that largely requires adequate protection from human harvest. Hotspots, areas of high endemism, and habitat for rare specialists can also be carnivore habitat because the exact parcel of land is less important than the presence of suitable prey for large carnivores. The charisma and large area requirements of large carnivores thereby provide a mechanism for achieving other conservation objectives. In this context, carnivore conservation is not synonymous with biodiversity conservation but as a mechanism to define ecosystem processes, identify minimum area requirements, and generate public interest, one of the necessary tools to achieve it.

Summary

Through their direct effects on prey species, predators can have profound indirect effects on many aspects of biodiversity. Where functional redundancy exists within the predator guild, however, conservation of a particular carnivore may not be essential to maintain the wider biodiversity of an area. Unfortunately, few studies have evaluated functional overlap of large carnivores. Here, we consider the impact of two predators, the Amur tiger and the gray wolf, on each other and on prey populations in the Sikhote-Alin ecosystem of the Russian Far East. Using historical data, we show that wolves do replace tigers as the top predator when anthropogenic influences depress tiger numbers; however, recovery of tiger numbers leads to competitive exclusion of wolves by tigers. Proportions of prey species in diets are similar, but, whereas wolves show a preference for red deer and select more vulnerable age classes, tigers exhibit a clear preference for wild boar and no apparent selection for any sex–age class. Two modeling efforts support the contention that wolves are more likely to deplete prey populations to a much greater extent than tigers. Local conservationists have used evidence for competitive exclusion and the apparent differential impact on prey to convince local hunters of the need to conserve tigers (as a means of reducing wolf numbers and wolf impact on ungulates). Although large carnivore conservation is not synonymous with biodiversity conservation, the charisma, large area requirements (related directly to prey requirements), and plasticity in habitat requirements of most large carnivores provide a mechanism for achieving other conservation objectives.

ACKNOWLEDGMENTS

Maurice Hornocker and Howard Quigley of the Hornocker Wildlife Institute conceptualized and initiated the Siberian Tiger Project, and we are grateful for their expertise, support, and dedication to research and conservation of tigers. The Siberian Tiger Project is funded by The National Fish and Wildlife Foundation's "Save the Tiger Fund," 21st Century Tiger, The National Geographic Society, The National Wildlife Federation, Exxon Corporation, The Charles Engelhard Foundation, Disney Wildlife Fund, Turner Foundation, Gary Fink, Richard King Mellon, and the Wildlife Conservation Society. Director A. A. Astafiev and former Assistant Director of Science M. N. Gromyko of Sikhote-Alin State Biosphere Zapovednik provided the logistical, administrative, and political support necessary to conduct fieldwork in Sikhote-Alin Zapovednik, and the Russian State Committee for Environmental Protection provided permits for capture of tigers. We thank I. Nikolaev, B. Schleyer, N. Reebin, A. Reebin, A. Kostirya, I. Seryodkin, V. Melnikov, A. Saphonov, V. Schukin, and E. Gishko for their assistance with data collection. P. A. Stephens was supported by a grant by the U.S. Forest Service, International Programs, to participate in this work. One anonymous reviewer provided valuable comments, as did all editors of this book, who, more importantly, had the patience and tolerance to guide us through formulation of this chapter.

Appendix 10.1
Life-history parameters used in the predator population models

Parameter	Tiger	Wolf	Sources
Survival			
Maximum age	25	15	(Mech 1974; Mazák 1981)
Background survival rate[a]	0.95 (females > 1 yr) 0.90 (adults > 2 yrs) 0.75 (cubs < 1 yr)	0.90 (males > 1 yr) 0.80 (yearlings) 0.75 (pups < 1 yr)	(Peterson and Page 1988; Hayes and Harestad 2000a; Kerley et al. 2003)
Fecundity and birth			
Age at first reproduction	4 yrs	2 yrs	(Mech 1974; Mazák 1981; Kerley et al. 2003)
Annual probability of female reproduction	0.55[c]	1.00[d]	(Mech 1974; Fritts and Mech 1981; Mech 1981; Fuller 1989; Kerley et al. 2003)
Mean (± SD) litter size[b]	2.38 (± 1.15)	6.00 (± 0.50)	(Mech 1974; Mazák 1981; Mech 1981; Jędrzejewska et al. 1996; Kerley et al. 2003)
Sex ratio at birth (males per offspring)	0.41	0.50[e]	(Kerley et al. 2003)

[a] Background survival rates reflect mortality from causes other than food limitation. The figures used were selected to reflect mortality in the absence of anthropogenic causes. Survival rates are expressed as annual equivalents.

[b] Litter sizes in the model were drawn from normal distributions described by these parameters but were reduced if food was limiting.

[c] The territories of male and female tigers are known to overlap. However, it was assumed that one male could mate with no more than three females in any one year.

[d] Only the alpha female could breed in any wolf pack (e.g., Hayes and Harestad 2000a).

[e] Assumed, in the absence of detailed information.

CHAPTER 11

Large Carnivores and Biodiversity in African Savanna Ecosystems

M. G. L. Mills

Large carnivores have attracted much attention and interest from zoologists and conservationists. Because they have the potential to exert a strong influence on other species and have often clashed with human interests, they have been at the center of many wildlife management issues and have been eradicated from much of their former range (Kruuk 2002). At the same time these striking and spectacular animals have attracted wide public attention and have a high ecotourism value.

Biodiversity is a broad concept incorporating compositional, structural, and functional attributes of ecosystems at four levels of organization—namely, landscapes, communities, species, and genes (Noss 1990). It should not be interpreted simply as a measure of species richness or species diversity. Because of the spatial scale at which relationships between large predators and their prey take place, it is only in a few pristine areas that the extent of the role of large carnivores in ecosystem dynamics and therefore their importance in the maintenance of biodiversity can be fully explored. These are areas of crucial importance because they can serve as baselines against which to reconcile arguments about the maintenance of biodiversity, the natural state of biotic communities and ecosystems, and the range of variation that can take place (Arcese and Sinclair 1997).

In this chapter I analyze patterns and processes of predator–prey relationships involving large carnivores in some of Africa's most pristine savanna areas. I concentrate on examples that demonstrate the spatial and temporal variation and dynamic nature of these processes, such as the differences between systems with migratory and sedentary prey and the influence of rainfall cycles. I show that predator–prey relationships are variable and may be driven by top-down or bottom-up processes, depending on ecological conditions, and that predation may also shape aspects of prey behavior and population structure. Another important component of large carnivore dynamics is the relationships between carnivores and the manner in which they influence each other's populations through interference and exploitation competition. Against this background, I will end up by discussing the biodiversity implications of these ecological relationships and issues pertaining to the conservation of large carnivores.

Most of the data presented here come from four more or less intact African ecosystems: (1) the Kruger National Park (KNP) in South Africa; (2) the Serengeti Ecosystem, especially the Serengeti Plains, in Tanzania and Kenya; (3) the Ngorongoro Crater in Tanzania; and (4) the Kgalagadi Transfrontier Park (southern Kalahari) in Botswana and South Africa. Detailed studies on large carnivores that have generated much data of relevance to the subject under discussion have been conducted in these four areas, each of which provides a different ecological template and set of conditions. Some of the data presented are from studies of comparatively long duration—a data set from the KNP of 24 years is presented (Mills et al. 1995), and the Serengeti ecosystem has been monitored for a similar time period (Sinclair and Norton-Griffith 1979; Sinclair and Arcese 1995a, with one study analyzing data from a 40-year period (Sinclair et al. 2003)—so that variability on both the spatial as well as the temporal scale is addressed.

Study Areas

The four study areas (Table 11.1) vary from the relatively dense tree savanna habitat with sedentary prey of the KNP, through to the vast open grassland plains with the massive migratory system of the Serengeti and the grassland enclosed caldera of the Ngorongoro Crater, to the semiarid thorn scrub with nomadic prey of the southern Kalahari. Lion (*Panthera leo*), spotted hyena (*Crocuta crocuta*), leopard (*Panthera pardus*), cheetah (*Acinonyx jubatus*), and wild dog (*Lycaon pictus*) are the major large carnivores in the systems, although wild dogs are vagrants in Ngorongoro and the southern Kalahari and have recently become so in the Serengeti as well, and cheetahs are vagrants in Ngorongoro.

The Kruger National Park is a 22,000 km² north to south, oblong-shaped area along a south (800 mm) to north (400 mm) summer rainfall gradient in the

Table 11.1

Features of the four major study areas discussed

	Kruger	Serengeti Plains	Ngorongoro	Southern Kalahari
Size (km²)	22,000	25,000	250	36,000
Rainfall (mm)	400–800	500–800	750	220–310
Habitat	Tree savanna	Vast open plains	Caldera plains	Arid shrub and tree savanna
Major prey species	**Buffalo**[a] zebra, **wildebeest,** impala	**Zebra, wildebeest,** Thomson's gazelle	**Zebra, wildebeest,** gazelle	**Gemsbok, wildebeest, springbuck**
Prey movements	Sedentary	Migratory	Sedentary	Nomadic
Large carnivores: resident (vagrant)	**Lion, spotted hyena,** leopard, cheetah, wild dog	**Lion, spotted hyena,** leopard, cheetah, wild dog	**Lion, spotted hyena,** leopard, (cheetah, wild dog)	**Lion, spotted hyena, cheetah, leopard** (wild dog)

[a] Species in bold are the ones discussed here.

northeast of South Africa. It is underlain by basaltic and granitic basement rocks, which influence the vegetation. The eastern basalt areas are an open woodland savanna with a dense shrub and grass layer with mopane (*Colophospermum mopane*) dominant in the north and marula (*Sclerocarya birrea*) and knobthorn (*Acacia nigrescens*) in the south. The western granitic soils support denser woodland communities with *C. mopane* again dominant in the north and bushwillow (*Combretum*) species in the south (Gertenbach 1983). The entire park is fenced, and numerous boreholes and dams provide additional drinking water for wildlife. African elephant (*Loxodonta africana*) make up the greatest proportion of the predominantly sedentary mammal biomass, with impala (*Aepyceros melampus*), buffalo (*Syncerus caffer*), plains zebra (*Equus burchelii*), and blue wildebeest (*Connochaetes taurinus*) the most important prey species for the large carnivores. Throughout its more than 100-year history numbers of all the aforementioned populations, as well as the large carnivores, have been manipulated to a greater or lesser extent at one time or another, although in recent years this has become much reduced as the park has become less intensively managed in terms of population control and the provision of water for animals (Freitag-Ronaldson and Foxcroft 2003; Mills et al. 2003).

The Serengeti ecosystem, comprising the Serengeti plains, the area of most importance in the discussions in this chapter, and to the north the woodlands, is usually defined as that area influenced by the wildebeest migration and covers approximately 25,000 km^2 across northern Tanzania and southern Kenya. Rainfall is variable but seasonal, with the short rains usually being experienced in November and December, followed by the long rains from March to May, along a rainfall gradient from 500 mm in the southeast to 800 mm in the northwest. The soils on the 5200 km^2 Serengeti plains are volcanic. The eastern plains are characterized by short grasslands, which become progressively taller to the west as the soils deepen and rainfall increases. The herbivore biomass in the ecosystem consists mainly of migratory wildebeest, zebra, and Thomson's gazelle (*Gazella thomsonii*) that concentrate on the plains in the wet season where they calve, move to the western corridor in the early dry season, then north into the Mara region in Kenya for the late dry season. Apart from the devastating rinderpest outbreak in East Africa in 1890 and the subsequent spillover of the disease from cattle into wildlife through the early 1970s, mammal populations have not been greatly influenced by human activities (Sinclair 1979a).

The Ngorongoro Crater, is immediately east of the Serengeti plains. It is a caldera about 18 km in diameter with walls 400 to 500 m high, which are mostly covered with thick forest. The floor of the crater is an open grass plain with two small patches of fever tree (*Acacia xanthophloea*) forest, some marshes, and a lake. It receives about 750 mm of rain per year. Wildebeest, zebra, and gazelle, both Thomson's and Grant's (*Gazella granti*) gazelles, are the dominant ungulates in this virtually self-contained ecological unit (Kruuk 1972), although recently wildebeest numbers have declined and buffalo numbers increased (Runyoro et al. 1995).

The southern Kalahari study area is today known as the Kgalagdi Transfrontier Park. It covers an area of 36,000 km^2, about one-third of which is in the northwest corner of South Africa, and the remaining area is in southwestern Botswana. The western, southwestern, and southern boundaries are fenced off from neighboring stock farming areas. The area is covered with sand arranged in a series of long, parallel dunes, through which two dry riverbeds run and numerous pans are scattered. The rainfall gradient goes from 220 mm in the southeast to 310 mm in the northeast. The vegetation is an extremely open tree and shrub savanna with tall annual and perennial grasses. Scattered boreholes occur along the riverbeds and in the dune areas on the South African section and in the extreme eastern section in Botswana. The major ungulate species are gemsbok (*Oryx gazella*), wildebeest, and springbuck (*Antidorcas marsupialis*) (Mills 1990).

Predator–Prey Relationships in Various Systems

It is of course as predators that the impact of large carnivores on ecosystem dynamics and biodiversity is most obvious. However, predator–prey relationships are complex, and it is impossible to make many general statements about the impact of predation, especially in the multispecies ecosystems of Africa. Predators may affect prey numbers and density, and in turn prey may affect predator numbers and density (e.g., Van Orsdol et al. 1986). The behavior of the prey also affects the impact of predation (e.g., by being migratory or sedentary), and predators may influence aspects of the biology of their prey (e.g., sex ratios and social behavior). Ecological variables such as habitat and climate likewise may have a large influence on these relationships. The following examples illustrate the diversity and dynamic nature of predator–prey relationships.

The Impact of Predation on Prey Numbers and the Influence of Rainfall— the Kruger Example

The lion is the dominant predator in the diverse savanna woodland habitat of the KNP. Although spotted hyenas are numerically equal to lions in this system, they hunt less and scavenge more than they do in the other African ecosystems discussed here (Mills and Funston 2003). Mills and Shenk (1992) examined the role of lion predation in the dynamics of blue wildebeest and zebra populations through simulation models. The data used in the models were from intensive observations over four years in a 235 km² study area in the southeast of the KNP. Population estimates for the two prey species were made from aerial surveys, and sex and age ratios from ground counts. Lion numbers were determined from observations of marked and radio-collared individuals and predation from continuous direct observations of lions for periods of up to 336 h. The numbers of both prey species remained constant throughout the study. There were approximately 235 wildebeest at a density of $1/km^2$ throughout the year in the study area, whereas the zebra were semimigratory, fluctuating from approximately 150 at a density of $0.6/km^2$ in the dry season to 660 at $2.8/km^2$ in the wet season.

There was no discernable difference in the frequency with which lions killed these two species, with zebra making up 14.4%, and wildebeest 13.5% of kills. No selection by lions for sex or by season could be found for either prey species. There were, however, differences in how lions selected individuals of different ages from the population. Zebra foals (less than one year) were killed more frequently than expected from their occurrence in the population ($\chi^2 = 24.89$; df $= 1$; $P < 0.001$), whereas wildebeest adults (many of which were in their prime reproductive years) and calves were killed in relation to their occurrence in the population ($\chi^2 = 0.519$; df $= 1$; $P > 0.05$).

The models ascertained the number of killing lions (defined as adult females) that could be supported by each prey population in the study area while remaining stable. A single model was constructed for the sedentary wildebeest population, and a wet season and dry season model for the semimigratory zebra population. The models predicted that the wildebeest population would stabilize with 7.7 killing lions, which was close to the actual number of approximately 7.0 that used the study area. The winter zebra population stabilized with 6.8 killing lions, but the increased summer population with 19.4. The kill age structure for each prey species was then swapped to determine the number of killing lions that the altered prey selection parameters might support. Thus wildebeest predation was made selective toward calves, and zebra predation was made nonselective for age. With these parameters the wildebeest population was estimated to stabilize with 10.7 killing lions and the zebra population with 5.4 in winter and 15.1 in summer.

These results suggest that lion predation affected wildebeest more severely than zebra during the study. It was in fact seen as the major cause for keeping the wildebeest population stable, providing evidence for top-down regulation. This was through the manner in which lions selected their prey by age, and because of the sedentary behavior of the low-density wildebeest population, in contrast to the semimigratory behavior of the zebra. These data support the hypothesis of Fryxell et al. (1988) that predators can regulate resident herbivores at low population densities, whereas such regulation is rare for migratory herds. They also demonstrate that age selection of the prey by predators may be important in determining the extent of their influence on prey populations.

Sensitivity analyses of the model parameters showed that manipulation of kill rate followed by adult fecundity rate of prey had the greatest effect on population size of both prey species. This is relevant because the aforementioned study was completed during years of average to slightly above average rainfall. Earlier, Smuts (1978) had shown that wildebeest and zebra populations declined during years of exceptionally high rainfall, and speculated that high predation rate was the major cause due to increased hunting cover and fragmentation of herds caused by the proliferation of tall rank grass. This situation was exacerbated by the fact that wildebeest and zebra numbers had earlier been controlled by culling. In order to rectify this perceived imbalance, the largest systematic predator culling operation in the KNP's history took place when 445 lions and 375 spotted hyenas were removed from an area of approximately 4500 km^2 in the Central District of the KNP over a five-year period (unpubl. records). The operation was terminated when it was found that the reduction of the two predator species had no detectable influence on the population trends of the wildebeest and zebra (Whyte 1985).

Much of southern Africa experiences a 20-year rainfall oscillation consisting of approximately 10 years of above-average and 10 years of below-average rainfall (Tyson and Dyer 1975). Mills et al. (1995) examined the influence of this temporal variation in rainfall on predator–prey relations. Using rainfall, prey population trends, and lion kill returns from park rangers' records over 24 years, they constructed a series of spreadsheet models based on simple and multiple regressions in an attempt to calibrate the contribution of the relevant parameters in the dynamics of buffalo, wildebeest, and zebra populations in the Central District of the KNP. Zebra and particularly wildebeest numbers fluctuated inversely with rainfall, whereas buffalo showed the opposite trend. Lions preyed proportionally more on the resident buffalo and wildebeest than on the semimigratory zebra populations. Moreover, they preyed more frequently and had a bigger impact on the buffalo population in drought conditions, whereas the converse was true for wildebeest during above average rainfall periods (Fig. 11.1).

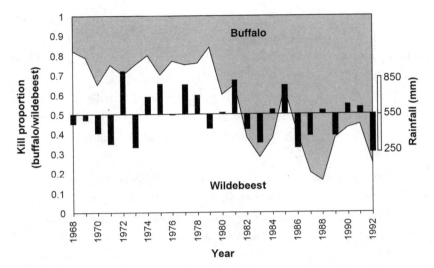

Figure 11.1
Lion kill proportions (the ratio of buffalo to wildebeest kills recorded by rangers per
year) for buffalo and wildebeest in wet and dry cycles from the Kruger National
Park. Note: The rainfall is the mean of the current and two previous years after
Mills et al. (1995).

In a related study, the buffalo population in the southeastern region of the
KNP also exhibited large variation over a 20-year period, which was also correlated
with environmental conditions (Funston 1999). A model developed by Starfield et
al. (1992) accurately predicted buffalo population trends, but only after the default
parameters had been adjusted by using predation data to calculate mortality. The
initial parameters were too conservative and did not influence the model sufficiently, particularly during severe drought periods, suggesting how important
increased lion predation rates on buffalo are under drought conditions. Spectacu-
larly, the buffalo population from the entire KNP dropped from 29,000 in 1991 to
15,000 in 1993 after the most severe drought on record (Whyte and Wood, unpubl.
report). Following several years of high rainfall the population had recovered to
24,000 by 2003 (I. Whyte, pers. comm. 2003.). The buffalo population, therefore,
was most heavily influenced by predation during population declines associated
with drought conditions. The lack of adequate forage for these tall grass feeders
that select for high protein and carbohydrates (Sinclair 1977a) resulted in a severe

and widespread loss of condition, rendering them susceptible to predation. The effects of environmental conditions on wildebeest and buffalo thus seem to be strongly mediated by lion predation. It would seem, therefore, that most lion predation on buffalo is compensatory rather than additive.

Although lions prey more heavily on zebra in higher rainfall periods than in drier periods, the predation rates even in wet periods do not seem to have a big effect on the population (Mills et al. 1995). It is still unclear what controls zebra numbers in the KNP.

The consequences of these findings for management suggest that control of both predator and prey populations should not be taken lightly and that fluctuations in prey numbers should be seen as part of natural ecosystem functioning. As a result, no population manipulation of these species is carried out anymore.

The Impact of Predation on Prey Numbers and the Influence of Habitat— Lessons from the Serengeti and Ngorongoro

In the geographically linked Serengeti ecosystem and Ngorongoro Crater, lions and spotted hyenas are numerically the top predators of the larger herbivores. They hunt predominantly wildebeest and zebra, but due to ecological differences between the two systems the impact of predation on these prey is strikingly different (Kruuk 1972; Schaller 1972).

During the time of Kruuk and Schaller's studies, there were estimated to be 2200 lions, 3000 spotted hyenas, 360,000 wildebeest, and 280,000 zebra in the Serengeti ecosystem. In the Ngorongoro Crater, these animals occurred at far higher densities than in the Serengeti. There were estimated to be 385 hyenas, 55 lions, 13,500 wildebeest, and 4500 zebra in the Crater (Table 11.2).

Not only did wildebeest and zebra occur at higher densities in Ngorongoro, but there was also a far higher density available to predators (see Table 11.2). Moreover, these ungulates are highly mobile in the Serengeti, leaving large areas on the Serengeti plains, where both Schaller and Kruuk made most of their observations, devoid of prey for several months of the year. In Ngorongoro where the prey are far more sedentary, there was a higher density of predators, mainly hyenas, and more predators per unit of prey than on the Serengeti plains. In the Serengeti, the ratio of lions to hyenas was closer to parity (see Table 11.2).

Table 11.2

Some relevant predator–prey parameters in the Serengeti and Ngorongoro Crater ecosystems[a]

	Serengeti	Ngorongoro
Size (km^2)	25,500	250
Hyena density(/km^2)	0.12	1.54
Lion density (/km^2)	0.09	0.22
Wildebeest density (/km^2)	14.1	54.0
Zebra density (/km^2)	11.0	18.0
Wildebeest / hyena + lion	69	31
Zebra / hyena + lion	54	10
Annual % of adult wildebeest population killed by hyena	1.6–2.6	11.0
Annual % of adult wildebeest population killed by lion	2.2–3.3	—
Annual % of adult zebra population killed by hyenas	0.8–1.0	9.0
Annual % of adult zebra population killed by lion	5.9–7.2	—

— Unknown, but considerably less than hyenas.
[a] Data from Kruuk (1972) and Schaller (1972).

In the Serengeti, although thousands of wildebeest calves died each season, the extent of this loss to predators was mitigated by the behavior of the wildebeest. They concentrated en masse at the calving grounds on the plains and synchronized the birth of the calves, thus making huge numbers of calves available to predators for a short period only. In fact, predation was not even the major calf mortality factor, most calves died from abandonment probably caused by malnutrition and disease (Sinclair 1977a). Many wildebeest became available to hyenas in the Serengeti not only as kills but also through lion predation and disease or starvation, so the hyenas there scavenged about half of their food. Additionally, their mobile behavior made it possible for the hyenas to track the migratory wildebeest far more efficiently than the much more sedentary lions could (Kruuk 1972). However, it was calculated that hyenas hunted only 1.6 to 2.6% of the adult wildebeest population, the majority being old animals. Lions removed a similar proportion (see Table 11.2), because the relatively sedentary lions on the plains had the highly mobile and abundant migratory wildebeest resident in their territories for only about four months a year. They were forced to rely on low-density sedentary prey species for the rest of the year. A recent detailed analysis of 40 years of data on the Serengeti wildebeest confirmed that food regulates this population (Mduma et al. 1999).

Zebra in the Serengeti were slightly more prone to predation by lions (see Table 11.2), primarily because the area over which they migrated was not as extensive as that of the wildebeest, so they were available to the plains lions for longer periods than the wildebeest were. Schaller (1972) found that at least a few zebra were available to most of the plains lions for two-thirds of the year. Schaller (1972) speculated that predation may have been more of a regulating factor on the zebra population than the wildebeest population, but that the depressing influence of disease and malnutrition was also necessary to keep the population from fluctuating widely.

Soon after the conclusion of Schaller and Kruuk's studies, the wildebeest population in the Serengeti underwent a dramatic increase so that by the late 1970s the population was calculated to stand at 1,320,000—a 360% increase. This was due to both the eradication of the disease rinderpest and an increase in dry season rainfall (Sinclair 1979b). As a result, the spotted hyena population more than doubled (Hofer and East 1995). The plains lion population, on the other hand, was not able to take advantage of this increase in wildebeest numbers to the same extent because there were still extended periods when the migrants were absent from lion territories. However, the ecological changes also benefited resident herbivores and so the lion population was able to increase to an extent through the availability of larger numbers of residents (Hanby and Bygot 1979).

In Ngorongoro, hyenas were the dominant predators, taking 11% of the wildebeest population and 9% of the zebra annually (see Table 11.2). Hyenas also killed a higher proportion of prime-aged wildebeest; the major mortality agent for wildebeest in Ngorongoro was hyena predation. The wildebeest responded to this by reaching puberty at an earlier age than their Serengeti counterparts and exhibiting a higher population turnover. Although the predation rate for zebra was far higher than in the Serengeti no effect on the zebra population composition could be shown (Kruuk 1972; Schaller 1972).

The differences in density and dispersion of food between the two systems impacted the two hyena populations differently, although both populations were to a certain extent controlled by their food supply. There was more competition for food between the hyenas in the Crater than in the Serengeti, and mortality of hyenas was found to be related to this competition. In the Serengeti the hyena population was limited by higher cub mortality because the mothers had to commute to the migrants in order to find food, leading to starvation of cubs at the den. With the increase in wildebeest numbers, however, recruitment of cubs into adulthood increased (Hofer and East 1995). In Ngorongoro the number of adult hyenas probably increased up to the limits of the food supply, whereas in the Serengeti this was prevented. This implies that Ngorongoro hyenas would be able to exert a regulating effect on the prey population whereas in the Serengeti they could not.

Taking this argument one step further, if hyena numbers were controlled by their prey numbers, they should not be able to limit that supply themselves—they should be in balance with one another. Kruuk (1972) concluded that the food supply to the prey population set the level. As food becomes limiting for the wildebeest population, more animals will lose condition, and hyenas, through their cusorial hunting style, will be very quick to select such individuals. However, should the condition of the prey improve, fewer would become vulnerable to predation until a new level of prey availability was reached and more weak animals became available once again.

Lion numbers in the Serengeti were limited by the migration. In the Ngorongoro Crater the population had crashed in 1962 due to an extraordinary outbreak of *Stomoxys* biting flies (Fosbrooke 1963) and was still recovering. By 1975 the population had reached 75 and maintained a level of 75 to 125 since then (Packer et al. 1991).

Taking a broader look at predation, Sinclair et al. (2003) analyzed a 40-year data set involving the entire large carnivore (over 10 kg) predator and prey community from the Serengeti region. This provided evidence that the preferences of carnivores impose the pattern of predation pressure on prey populations. As a result there exists a threshold in body size (around 150 kg) beyond which the cause of mortality switches from top-down in small prey species to bottom-up control in larger species, suggesting that, in this system, biodiversity allows both predation and resource limitation to act simultaneously to affect herbivore populations.

Other Impacts of Predators on Prey—Examples from the Southern Kalahari

In the southern Kalahari, herbivore concentrations occur along the two perma-
nently dry fossil river beds and pans during the wet season and disperse into the
dune areas during the dry season (Mills and Retief 1984; Van der Walt et al. 1984).
Springbuck are numerically the most common antelope along the 435 km of dry
riverbeds that run through the national park area. During the 1970s and 1980s
numbers fluctuated through the year between an average annual abundance of
4000 in the dry season to 8000 in the wet season. The range was from an absolute
low of 113 along the Auob riverbed in November 1982 and 816 along the Nossob
riverbed in December 1972 to highs of 6028 in April 1975 and 7350 in February
1974 along the two riverbeds, respectively (Mills and Retief 1984).

Feeding off these springbuck were about 100 lions, 60 leopards, and 60 cheetahs making up 13%, 65%, and 87% of the three predators' kills, respectively (Mills
1984, 1990). As for the migratory Serengeti wildebeest, the impact of predation
on springbuck numbers appeared to be quite small because the ratio between predator and prey numbers was low and the most common predator (lions) did not select springbuck. Additionally, the springbuck were nomadic so that prey numbers
were low during lean times, thus keeping predator numbers low. Further evidence
that predation had little impact in this population comes from the observation that
leopards and cheetahs selected older adult prey over younger ones (Fig. 11.2), thus
taking animals at the end of their reproductive lives. The data are particularly
strong for cheetah because of the large sample size ($\chi^2 = 19.36$, df $= 4$, $p < 0.001$,
$n = 82$ kills) (Mills 1984). Springbuck numbers were correlated positively with rainfall (Mills and Retief 1984), suggesting that food availability was the major determinant of springbok numbers.

Although they did not influence springbuck numbers, predators in the southern Kalahari did seem to affect the sex ratio of the springbuck population by selecting predominantly males. The adult sex ratio of springbuck in the living
population was 1 male to 1.70 females, yet the sex ratio of springbuck killed by
predators was 1 male to 1.06 females. Taking into account the sex ratio in the living population, all three predators showed a significant selection for males ($\chi^2 =
9.91$, $p < 0.01$; $n = 35$ for lion; $\chi^2 = 5.85$, $p < 0.02$; $n = 26$ for leopard; $\chi^2 = 35.50$,
$p < 0.001$; $n = 131$ for cheetah). However, the sex ratio of springbuck that died

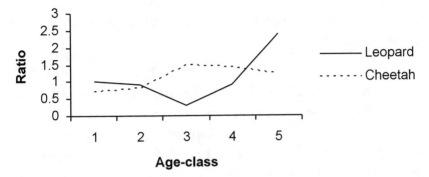

Figure 11.2
The ratio of the percentage of adult springbuck of each age class killed by cheetah ($n = 82$) and leopard ($n = 18$) to the percentage of springbuck of each age class randomly shot ($n = 191$) in the southern Kalahari.

from natural causes other than predation did not differ significantly from the expected ($\chi^2 = 1.2$, $p > 0.05$; $n = 32$) (Mills 1984). The preponderance of females in the springbuck population therefore may well have been due to the selection pressure exerted on males by all three cats.

Predation on an evolutionary time scale also may have had an influence on the form and behavior of the prey (Bertram 1979). The relationship between predator and prey is a delicate balance, an ongoing evolutionary struggle. Many of the attributes of both predator and prey have evolved to help the hunter catch the hunted and the hunted to escape. For example, the spotted hyena is a successful predator of gemsbok in the southern Kalahari, being particularly adept at selecting animals under one year of age (Mills 1990). Of 55 hunting attempts on gemsbok herds with calves, 40 (73%) were successful. Adult gemsbok, on the other hand, were far less susceptible to predation by spotted hyenas, with only 11 (7.6%) of 145 encounters ending in kills. The long rapier-like horns of the gemsbok, well developed in both sexes, are their main means of defense. Gemsbok horns also grow relatively quickly; by one year of age they have on average attained 74% of full length, compared to 39% for a selection of other antelope (Mills 1990). The shape of the horns, their lack of sexual dimorphism, rapid growth, and the successful use against spotted hyenas and also lions (Eloff 1964), all suggest that defense is an important function of gemsbok horns. This is in contrast to most other antelope where horns take on only a secondary sexual function (Packer 1983).

The Role of Predation in the Decline of a Rare Antelope Species—Another Kruger Example

The examples already cited deal with the relationship between carnivores and abundant prey species. What about the situation involving a rare species? Between 1986 and 1993 the roan antelope (*Hippotrgus equines*) population in the KNP, which mainly inhabits the northern basalt (*Colophospermum mopane*) plains, dropped from about 450 to 40 (Harrington et al. 1999). Competition from zebra and wildebeest, which moved into the roan's range after the introduction of water points in the late 1960s and 1970s, exacerbated by habitat deterioration due to an extended drought in the late 1980s and early 1990s (Grant et al. 2002), are believed to be important factors for this decline (Harrington et al. 1999; Grant and van der Walt 2000; Grant et al. 2002).

Harrington et al. (1999) noted that the population crash was associated with an increase in adult mortality, but little apparent change in calf survival, suggesting that nutritional factors were not the prime cause. They suggested that increased predation on adult roan due to a buildup of lion numbers following the wildebeest and zebra influx was the proximate cause of the decline, although recognizing that the ultimate cause was almost certainly the provision of numerous artificial water points. However, the wildebeest and zebra populations, and therefore presumably the lion numbers too, had increased several years before the roan decline. This suggests that if predation was the proximate cause of the decline, the roan became susceptible to predation after several years of increased lion numbers. Perhaps habitat conditions had changed due to a combination of heavy grazing and a number of dry years to make them more vulnerable to predation. Closure of some of the artificial water points resulted in the movement away from the prime roan habitat of many of the zebra and lions (Grant and van der Walt 2000; Grant et al. 2002). However, the expected positive response by the roan population did not materialize in spite of good rains (Grant and van der Walt 2000). This may have been because the roan numbers had dropped so low that even mild predation became a major block to population increase. It would appear from this example that through management policies, predator–prey relationships may become disruptive to biodiversity processes.

Intraguild Relationships

Early studies on African carnivores drew attention to the fact that both interference and exploitation competition are important in the dynamics of their ecological relationships (Kruuk 1972; Schaller 1972), however carnivore populations were generally believed to be limited by the food supply (Macdonald 1983). It is only more recently that the impacts of these relationships in some areas on the smaller and less dominant members of the large carnivore guild such as wild dogs, cheetahs, and brown hyenas (*Hyaena brunnea*) have become apparent. These relationships are discussed in detail by Woodroffe and Ginsberg (this volume).

Intraguild relationships are important and variable components of functional biodiversity, and also have consequences for management. Species negatively affected by these relationships are often sensitive to human-induced perturbations such as disease, as may have happened to wild dogs in the Serengeti (summarized in Creel and Creel 2002), and management actions, as may have happened to brown hyenas in the KNP (Mills and Funston 2003). However, like all these relationships, intraguild relationships are variable and influenced by ecological conditions. So, for example, the heavy predation by lions on cheetahs observed on the Serengeti plains and which was found to have had a major influence on the population dynamics of cheetahs in that region (Laurenson 1994, 1995) appears to be of less consequence to cheetahs in the more wooded KNP (Broomhall 2001).

Biodiversity Implications of Large Carnivore Ecology

The preceding sections have illustrated that large African carnivores affect their prey, and even one other, in a number of different ways in time and space. They are, therefore, an integral part of biodiversity, and their relationships with prey and one another influence population dynamics, behavior, and evolutionary processes. It is erroneous to consider biodiversity exclusively in terms of species richness and animal numbers. The intricate and dynamic relationships between animals and their environment, and the time scale over which these events take place are inextricably linked to the functional component of biodiversity. It is only where we are able to conserve intact ecosystems that all these interactions can be accommodated and the full role of large carnivores in ecosystem dynamics and therefore biodiversity conservation can be expressed.

The range of interactions that take place between predator and prey, and among predators, is testimony to the flexibility in behavior, social system, and ecological relationships of these animals. The variability in their behavior and their responses to ecological changes is a key component of biodiversity and evolution. The greater the range of ecosystems that can be conserved to accommodate large carnivores, the greater will be the number of opportunities for these variable interactions to be played out and for adaptations to changing conditions to evolve.

Humans have coexisted and competed with carnivores for millions of years (Kruuk 2002). The removal of humans as hunters from an African ecosystem constitutes the removal of an important predator, and therefore, it can be argued, has a negative impact on biodiversity. Modern weapons and high human population densities, however, would obviously give humans an unfair advantage over the other predators as to make their continued existence unsustainable—an important reason for the establishment of protected areas. Even limited hunting with modern weapons is probably very different from the ecological role that primitive humans once played because much modern human hunting is highly selective for trophy animals, which are often the most important contributors to the gene pool of the population. Traditional hunting such as practiced by the San Bushmen in the Kalahari would on the face of it seem more desirable and sustainable. However, I would argue that unless they were subject to all the environmental pressures of the Kalahari, which clearly today is impossible, they should not hunt in national parks because they would still exert an unfair advantage over the other predators. A new dynamic equilibrium, without humans, has been established in many African national parks, which appear to be sustainable. Wildlife management areas (in which sustainable hunting of both large carnivores and their prey takes place) surrounding large core protected areas such as are presently implemented in Botswana (Hunter 1993) and in some areas around the Serengeti National Park in Tanzania (Mbano et al. 1995; Leader-Williams et al. 1996) appear to be a sustainable and satisfactory compromise.

Conserving Carnivores and Carnivory in Ecosystems

It is often tempting to keep ecosystems as we know them and not to tolerate change. Long-term ecosystem studies in the Serengeti (Sinclair and Norton-Griffiths 1979; Sinclair and Arcese 1995a) and KNP (Du Toit et al. 2003), however, have demonstrated that fluctuations are indeed a part of ecosystem functioning and that intact ecosystems are generally resilient to change. Large carnivores are often important agents in ecosystem dynamics. The real difficulty arises when change is due to anthropomorphic causes and results in a loss of biodiversity. In more extensively managed ecosystems; for example, where water provision is extensive, this would seem to have a much heavier impact on the system than predators (Walker et al. 1987), or as has been discussed may disrupt predator–prey relations. Although there is probably no protected area in the world that is not influenced in some way by modern humans, the decision of when and how to intervene when ecological changes take place because of temporal variation or spatial constraints is often difficult to establish. Recognition and acceptance of the cyclic and dynamic nature of ecosystems are important and should always be taken into account before implementing management actions such as population control.

The history of management in the KNP is an example of a change of attitude from an intense management system (population control, the provision of artificial waterholes, and a rigid burning policy) to one that is more tolerant of natural fluctuations (Mills et al. 2003), and is based on the principle of conserving the system with all its facets and fluxes. This has come about partly because of the realization through long-term studies of the role of large carnivores in these fluctuations. Where interventions are implemented, an adaptive management approach should be taken (Biggs and Rogers 2003). It is impossible to guess what these systems may have looked like before the influence of postindustrial humans, nor is this really important. What is important is that the influence of modern humans, which can mask that of many other processes in a system, should be reduced to the minimum, so that the other species can play their role in ecosystem dynamics.

It is well known that the size of an area is crucial for maintaining biodiversity (Willis and Whittaker 2002), but one of the unknown factors in large carnivore dynamics is the spatial scale at which the ecological processes can be maintained without losing species. How large must an area be to allow these processes to be played out in some sort of equilibrium? When and how should intervention take place? An experimental approach to this conundrum is called for, but managers are not usually prepared to take the risks.

This question of scale is becoming more relevant as the area available for large carnivore conservation is shrinking. Moreover, in southern Africa a proliferation of small ecotourism reserves is taking place, most of which are fenced, and large carnivores are being introduced into the reserves because of their high ecotourism value. When do these operations make a contribution to biodiversity conservation and when are they of little more value than a safari park or captive breeding institution? Although the genetic issues of these small populations can in theory be fairly easily addressed through a metapopulation management approach whereby animals are moved between reserves (e.g., Mills et al. 1998), questions of carrying capacity and population management are far more complex (Vartan 2002).

Not only are the issues of predator–prey dynamics relevant, but intraguild dynamics need to be considered. The high predation pressure on cheetah cubs recorded by Laurenson (1994) on the Serengeti plains has led to a widely held perception that cheetah numbers are low in protected areas due to competition with other large predators (Nowell and Jackson 1996). As a result Marker-Kraus et al. (1996) suggested that protected areas cannot be relied upon to support the survival of viable cheetah populations, and that the best hope for the cheetah is on the extensive areas covered by commercial farmlands in Namibia where other large carnivores have been eliminated. However, this high predation pressure on cheetah cubs is a natural biodiversity process and may not have such an impact on cheetahs in savanna woodlands where the increased cover provided by trees and bushes makes it easier for females to hide their cubs (Broomhall 2001). Additionally, persecution and other pressures by farmers have apparently led to cheetahs occurring at lower densities in these areas than in conservation areas of similar habitat (Marker 2002). Obviously the conservation of carnivores in general and cheetahs in particular outside conservation areas is important. However, I submit that large protected areas in suitable habitat should still be regarded as the priority areas for cheetah conservation because viable cheetah populations still occur in many of them and this is where ecological processes with all their facets and fluxes, including interspecific competition, can take place. This is an example of a tendency in conservation to be more concerned with quantity (numbers of animals) than quality (the ecological processes and conditions under which the organisms exist).

Ecosystems without Carnivores and Other Conservation Implications

In light of what we know about predation in African ecosystems, what would a system look like in the absence of large carnivores? Much predation in the areas that have been discussed appears to be compensatory. Where the prey are migratory, such as in the Serengeti, they move out of reach of the predators for a large part of the year. In a woodland savanna like the KNP where the prey are mainly sedentary, the impact of predation is largely influenced by ecological conditions, particularly rainfall. In addition to the KNP predation studies cited in this chapter, during the second year of a severe drought in the KNP in the 1980s herbivore mortality, much of it due to starvation, was up to 35% for some species. However, these populations recovered to predrought levels within two years in spite of the full spectrum of large carnivores being present (Walker et al. 1987).

Without large carnivores, fluctuations in the densities of prey species would be expected to be more drastic; ungulate numbers would rise higher during favorable times and experience larger declines during poor times with resultant changes in vegetation structure and composition. This might have a cascading effect on a range of other ecological and behavioral interactions (e.g., Berger 2002). For example, more sensitive competitors such as roan antelope (*Hippotraggus equinus*) might decline in the absence of large carnivores because the more abundant species such as buffalo, wildebeest, and zebra would no longer be affected by predation and would therefore exert a heavier impact on their competitors through excessive utilization of the herbage layer. Additionally, other more drastic mortality factors such as disease might spread more easily through prey populations in the absence of large carnivores because they would be likely to weed out sick animals before the disease could spread. A novel way to manage these areas without large carnivores is to set off-takes with the aim of simulating the gross impacts of predation (Goodman and Hearn 2003).

Relationships between predators and their prey and within predator guilds allow for the expression of the full evolutionary potential of the species concerned and thereby contribute to the maintenance of biodiversity as defined by Noss (1990). As the opportunities for them to be expressed are limited to the few adequately conserved pristine areas left in Africa (and indeed in the rest of the world), they need to be conserved as a priority. In order for this to be achieved, interference by management, such as population control, should be minimal. The most important management actions should strive to maintain the integrity of the area and prevent disturbances by modern-day humans. Less pristine and smaller areas present particular challenges as to the best way of conserving large carnivores and their ecological relationships.

Summary

Predator–prey and predator intraguild relationships are complex and dynamic interactions in time and space. It is only in a few pristine areas that the extent of the role of large carnivores in ecosystem dynamics and biodiversity can be fully explored. Examples of this phenomenon are given from four well-studied African savanna ecosystems—Kruger National Park, Serengeti, Ngorongoro Crater, and southern Kalahari—that represent a broad spectrum of ecological and management conditions. All of these relatively pristine areas provide illustrations for how the impacts of carnivores on prey populations are variable and depend on a variety of factors, including habitat, climate, and behavior (i.e., the degree to which prey populations are migratory). Although predation may not necessarily be the dominant force influencing prey numbers, it is an integral component of these relationships, and the influence may be significant, not only in broader aspects of population dynamics (for example in influencing sex ratios), but also in shaping form and function. Moreover, interactions between carnivores may be powerful mechanisms in their population dynamics and in influencing their behavior.

The consequences of these findings for biodiversity conservation and evolution are significant. The intricate and dynamic relationships between animals and their environment, and the time scale over which these events take place are inextricably linked to the functional component of biodiversity. It is only where intact ecosystems can be conserved that we are able to accommodate all these interactions, such that the full role of large carnivores in ecosystem dynamics and therefore biodiversity conservation can be expressed. In order to conserve these processes, the conservation of suitable, usually large, pristine areas is essential. Interference by management, such as predator or prey population control, or the provision of additional water should be minimal.

ACKNOWLEDGMENTS

I thank South African National Parks for giving me the opportunity and providing the resources to work in two of Africa's wonderful large conservation areas, the Kruger National Park and the Kgalagadi Transfrontier Park. I also thank the Endangered Wildlife Trust and the Tony and Lisette Lewis Foundation for financial and logistic support, and WCS, the organizers and sponsors of the workshop, for inviting me and making it possible for me to attend. Finally, I am grateful to the editors and reviewers for their constructive and helpful comments.

Large Carnivores and Ungulates in European Temperate Forest Ecosystems: Bottom-Up and Top-Down Control

Bogumiła Jędrzejewska and Włodzimierz Jędrzejewski

A typical food chain found throughout the temperate and boreal forests of the Holarctic zone includes three trophic levels: forest plants, large herbivores (ungulates), and their predators (large carnivores). Whether these ecosystems are structured from the top-down or bottom-up is still an unresolved issue. If the food chain functions with bottom-up control, positive correlations should occur between density changes at all trophic levels. In the opposite model, when food chains are subject to top-down control, changes in abundance at the top trophic level (large predators) not only would cause the opposite change in ungulate density but should be further transmitted to plants.

In a bottom-up controlled ecosystem, both ungulates and carnivores would be food-limited, and the extinction of predators would not affect the abundance of lower trophic levels (ungulates and forest plants). Support for this interpretation from McNaughton et al. (1989) demonstrated a significant, global positive correlation between herbivore biomass and primary productivity. Positive correlation between the biomass or densities of large carnivores and biomass of ungulates has also been documented (Skogland 1991; Messier 1995). If, on the other hand, trophic cascades occur in forest ecosystems subject to top-down control, large carnivores would limit densities of ungulates and have a noticeable indirect influence on the abundance and regeneration of forest trees.

The concept of trophic cascade, first proposed by Paine (1980), was later generalized to food chains of one to five trophic levels in the hypothesis of exploitation ecosystems (HEE; Oksanen et al. 1981; Fretwell 1987). According to HEE, in ecosystems with three trophic levels, herbivores will be strongly limited by predators and, thus, their densities will not change along a productivity gradient. Empirical data for a forest plants–ungulate–carnivore food chain are relatively scant, not least because the adequate time scale for addressing such a question would be decades, and at spatial scales of hundreds of square kilometers. Nonetheless, there is a great and still growing interest in determining whether large predators limit ungulate numbers and thus indirectly shape the plant cover (e.g., Wright et al. 1994; Smith et al. 1999; Flueck 2000). If such top-down control operates in forests, it would have profound consequences for ecosystem biodiversity.

This chapter assesses the available evidence for top-down and bottom-up controls in a forest–ungulate–carnivore food chain in a specific temperate ecosystem to determine whether the cascading effects of carnivores on forest regeneration, mediated by herbivore density and behavior, occur. Based on ecological research conducted in Białowieża Primeval Forest (Poland and Belarus), we address the question of whether large carnivores limit ungulate densities in the multiple-herbivore community. Finally, we discuss how the removal of top carnivores affects the biodiversity of temperate and boreal forests.

Białowieża Primeval Forest: Study Area and Methods

Located on the Polish–Belarussian border, the 1500 km² Białowieża Primeval Forest (BPF) is regarded as one of the best-preserved lowland forests in the temperate zone of Europe. Set aside as a royal hunting ground until the early 20th century, it still contains a fair proportion of natural old-growth forests. BPF is composed of mixed coniferous and deciduous stands, with the dominant tree species spruce (*Picea abies*), pine (*Pinus silvestris*), oak (*Quercus robur*), hornbeam (*Carpinus betulus*), black alder (*Alnus glutinosa*), ash (*Fraxinus excelsior*), lime (*Tilia cordata*), maple (*Acer planatoides*), and birch (*Betula pendula* and *B. pubesces*) (Faliński 1986). The Polish part of BPF (600 km²) includes the protected Białowieża National Park (a 100 km² Man and Biosphere Reserve and World Heritage Site) and commercial forests with small nature reserves. The Belarussian part (900 km²) is currently designated as a state national park of the Belarus Republic. Since 1981, the Polish and the Belarussian parts have been separated by the wire fence (2.5 m high) built by the

Soviets along the state border. The fence serves as a barrier to ungulates, but wolves (*Canis lupus*) and lynx (*Lynx lynx*) are known to cross it in some places. However, BPF as a whole is well connected by forest corridors with other vast woodlands to the northeast, northwest, southwest, and southeast. The climate is transitional between Atlantic and continental types, with continental features prevailing. Mean daily temperature in January is –4°C and in July is 19°C. Snow cover persists on the ground for 50 to 150 days and its maximal depth in the 1990s reached 63 cm. Mean annual precipitation is 620 mm.

Five species of ungulates native to European lowland forests roam in BPF: wild boar (*Sus scrofa*), red deer (*Cervus elaphus*), roe deer (*Capreolus capreolus*), moose (*Alces alces*), and European bison (*Bison bonasus*) (Jędrzejewska et al. 1997). The population of European bison undergoes some culling by the services of the national parks, and the other ungulates are subject to hunting harvest. The original guild of large predators included three species: wolf, lynx, and brown bear (*Ursus arctos*), although bears were exterminated in the late 19th century (Jędrzejewska and Jędrzejewski 1998).

Long-Term Datasets

From 1991 to 1999, 12 wolves from four packs and 18 lynx were radio-collared and studied by telemetry techniques in the Polish part of the BPF. Their predation impact on ungulates was estimated by a combination of radio tracking, snow tracking, search for prey remains, analysis of scats, and surveys of ungulate abundance (Okarma et al. 1997, 1998; Schmidt et al. 1997; Jędrzejewski et al. 2000, 2002). To address the questions posed by this chapter, we supplement this work with a data set spanning more than 100-years on population trends of predators, ungulates, and annual temperature in the whole BPF (Jędrzejewski et al. 1996; Jędrzejewska et al. 1996, 1997; Jędrzejewska and Jędrzejewski 1998).

Using this long-term data set, we also analyze whether predation on ungulates has been affected by the long-term changes in the productivity of the forest ecosystem, as approximated by climatic records. Over this length of time, annual temperature is a good proxy of forest productivity for ungulates. In the mixed coniferous–deciduous forest located in the transitional nemoral–hemiboreal zone, periods of warmer climate are characterized by better regeneration of deciduous trees (which means a greater supply of preferred browse for deer and bison), more frequent and abundant seed crops of oaks (greater supply of acorns, the favorite food of wild boar, consumed also by bison and red deer), and better access to winter food due to less snow cover (important to all ungulate species) (Pucek et al. 1993; Okarma et al. 1995; Mitchell and Cole 1998). In this analysis, we used a 10-year moving average of annual temperature (years n-9 to n) to exclude large year-to-year variation and include the cumulative effect of changes in forest productivity over several years.

Wolf and Lynx Predation on Ungulates

The largest and the least numerous among Białowieża's ungulates—European bison and moose—were not attractive to predators. Lynx did not hunt them at all and wolves rarely killed moose and only occasionally seized bison. It was the two deer species that were strongly selected by large carnivores: red deer by wolves and roe deer by lynx (Table 12.1). Wolves supplemented their diet with wild boar and roe deer, and lynx also hunted for female and young red deer.

In the 1990s, the densities of red deer in the Polish part of the BPF were 3.6 to 6.1 inds/km² (mean 4.6) in late winter and 5.1 to 8.6 (mean 6.5) in summer, when the young were born. Annually, wolves killed 0.6 to 1.0 deer/km² (mean 0.8) and lynxes 0.4 to 0.7 deer/km² (mean 0.6) (Fig. 12.1). Their combined predation averaged 70% of the annual increase of red deer due to reproduction and nearly 40% of the recorded yearly mortality (Jędrzejewska and Jędrzejewski 1998; Jędrzejewski et al. 2002). During the study, the human hunting harvest was higher than in earlier years because of foresters' complaints about damage caused by ungulates to forest regeneration. Accordingly, hunters shot 1.3 to 1.4 red deer/km² annually. Since predation and hunting harvest were additive, the red deer population declined markedly during the 1990s (see Fig. 12.1).

Predation pressure was even heavier for the roe deer population, the densities of which varied from 2.9 to 4.9 inds/km² (mean 3.8) in winter and 5.0 to 8.2 (mean 6.4) in summer. Annual take by lynx was 1.1 to 1.8 deer/km² (mean 1.6) and wolves 0.2 deer/km² (see Fig. 12.1). The two predators removed nearly 75% of the roe deer yearly increase due to reproduction and were responsible for over

Table 12.1
Densities and biomass of the five species of ungulates, their percentage shares in the community, and characteristics of predator–prey relationships in Białowieża Primeval Forest, eastern Poland[a]

Parameter	European Bison (400)	Moose (200)	Red deer (100)	Roe deer (20)	Wild boar (80)
Mean density in winter (N/km^2)	0.5	0.2	4.6	3.8	3.0
Percent numbers in community	4	2	38	31	25
Mean biomass (kg/km^2)	184	32	461	76	241
Percent biomass in community	18	3	47	8	24
Percentage of wolf kills	+	1	68	19	12
Percentage of lynx kills	0	0	22	77	1
Percentage annual mortality caused by: wolves	+	5	21	6	13
lynxes	0	0	17	47	1
stray dogs	0	0	1	2	4
hunting/culling	82	66	46	32	56
other[b]	18	29	15	13	26

[a] Data for the 1990s. Numbers in parentheses are mean body masses of ungulates (kg). + below 0.5%. From Jędrzejewska and Jędrzejewski (1998).
[b] Disease, starvation, and cold during severe winters, poaching, traffic accidents.

50% of their annual mortality. Because hunters also took a heavy toll (1 ind/km^2), roe deer numbers declined during the 1990s (see Fig. 12.1).

Predation on wild boar (mainly by wolves) amounted to 10 to 17% of boar annual increase due to breeding but played a lesser role compared to other factors of mortality such as severe winter conditions and hunting harvest (Jędrzejewska and Jędrzejewski 1998, Jędrzejewski et al. 2000). Wolf predation was a noticeable but secondary factor of moose mortality (see Table 12.1).

The study of the wolf–red deer relationship lasted long enough (nine years) to examine how predation rate changed relative to prey density. Density dependence of predation is a crucial issue in the discussion on whether predators are capable of regulating ungulate densities (which happens when percentage predation increases with growing density of prey to maintain equilibrium prey densities) or whether they limit prey numbers and destabilize the population dynamics (predation rates not necessarily related to prey density; see Sinclair 1989; Messier 1991,

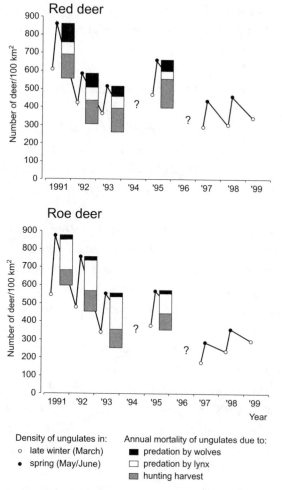

Figure 12.1

Wolf (*Canis lupus*) and lynx (*Lynx lynx*) predation and hunting harvest (1991–1996) in relation to red deer (*Cervus elaphus*) and roe deer (*Capreolus capreolus*) population dynamics (1991–1998) in Białowieża Primeval Forest, Eastern Poland. Modified from Jędrzejewski et al. (2000).

1995). In BPF, wolf predation rates were inversely density dependent, thus wolves limited numbers of red deer but were unable to regulate their prey. Stronger limitation occurred at low deer densities and was weaker at high density (Jędrzejewski et al. 2002). By eliminating a fair number of red deer compared to their annual production, wolves in BPF are capable of at least hampering population growth of deer and prolonging the time until they reach carrying capacity.

Long-Term Data on Large Carnivores and Ungulates

A first glance at the long-term dynamics of ungulates and large carnivores reveals a great variation in both of their abundances during the time period between 1890 and 2000 (Fig. 12.2). Wolf and lynx densities were significantly correlated in time (Kendall's coefficient of concordance tau = 0.264, $P = 0.05$) so we treated them jointly. Large predators had been exterminated twice: at the turn of the 19th century and during the 1960s. After the cessation of predator control, they quickly recovered due to immigration from Belarus. Generally, during times of war and political chaos wolves and lynx flourished, whereas during longer periods of stability they were subject to hunting and control. Extermination and recoveries of large carnivores created unintended experiments of predator exclusion from the ungulate community.

We recorded the opposite situation for ungulates. As a source of food to humans, they were overexploited during wars and times of economic downturn and political chaos but were otherwise managed sustainably or even protected. During the last 150 years, three species were temporarily absent from the community at one time or another. Red deer died off before 1800 (probably due to climate cooling) and were reintroduced in the 1890s (Jędrzejewska et al. 1997). Moose and European bison were extirpated by 1919 following the massive overkill of all ungulates during World War I. Moose recolonized the BPF in 1946 from its eastern refuges, and bison were reestablished by a national program that started in 1929 (Krasiński 1967; Pucek 1991). Following two decades of breeding in enclosures, European bison were released to the wild in 1952. Two other species, wild boar and roe deer, inhabited the BPF continuously, though their numbers also varied.

Generally, in the long term, the population dynamics of all five species of wild ungulates were markedly synchronized. Of ten pairwise comparisons, population trends of eight pairs of species were positively correlated (Kendall's coefficients of concordance tau = 0.272 to 0.664, $P = 0.0005$ to 0.05). The exception was the red deer–bison pair, where no correlation in the temporal variation in numbers was

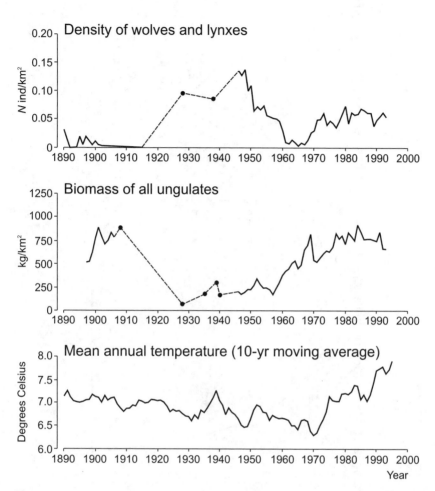

Figure 12.2

Density of large predators (wolves and lynxes combined), crude biomass of wild ungulates, European bison (*Bison bonasus*), moose (*Alces alces*), red deer, roe deer, and wild boar (*Sus scrofa*) combined, and mean annual temperature (10-year moving average) in Białowieża Primeval Forest, 1890–1995. Based on data from Jędrzejewska et al. (1996, 1997), and Jędrzejewski et al. (1996).

detected. The deer and the bison were subjected to the most frequent interference from humans (extirpations and reintroductions).

During the century covered by this analysis, productivity of the forest ecosystem also varied. (see Fig. 12.2). Abundance of ungulates (expressed as crude biomass of all species per unit area of BPF) was strongly positively correlated with

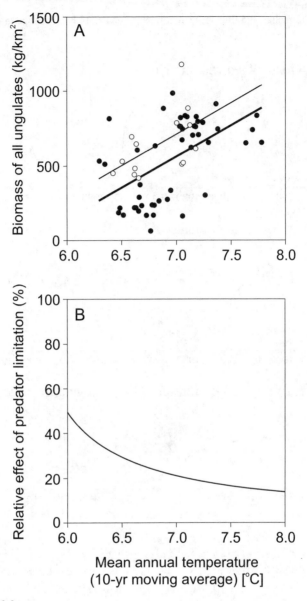

Figure 12.3

(a) Regression between mean annual temperature (*Temp;* a proxy of forest ecosystem productivity) and ungulate biomass (*B*) for years with predators exterminated or in low densities (open points, upper line; $B = -2230.48 + 420.21$; *Temp, R^2_{adj}* = 0.305, $n = 15$ years, $P = 0.03$), and years with predators present in moderate and high densities (black points, lower line; $B = -2349.87 + 415.93$; *Temp, R^2_{adj}* = 0.293, $n = 49$ years, $P < 0.0005$). Data points from Figure 12.2. (b) The role of predation on ungulates in relation to annual temperature: the limiting impact exerted by predators (the difference between the upper and the lower line in part (a) is presented as percent of the maximum ungulate biomass, shown by the upper regression line in part (a).

Table 12.2

The roles of climate (mean annual temperature in degrees Celsius; 10-year moving average including current year and nine preceding years) and large carnivores (combined densities of wolves and lynxes, N ind/100 km^2) in shaping the abundance of wild ungulates (crude biomass, kg/km^2) in BPF as shown by results of multiple regression analysis

Species	N Years	Total variation explained, R^2_{adj}	Contributions of Independent Variables to the Total Variation Explained, sr^2	
			Annual Temperature	Predator Density
European bison	70	0.300^c	(+) 0.315^c	(−) 0.030^{NS}
Wild boar	55	0.284^c	(+) 0.241^c	(−) 0.065^a
Moose	61	0.422^c	(+) 0.222^c	(−) 0.192^c
Red deer	55	0.149^b	(+) 0.088^a	(−) 0.089^a
Roe deer	56	0.314^c	(+) 0.055^a	(−) 0.279^c
All species pooled	54	0.495^c	(+) 0.303***	(−) 0.204^c

Data sources as in Fig. 12.1 (+) positive correlation; (−) negative correlation; $^a p < 0.05$, $^b p < 0.01$, $^c p < 0.001$, ^{NS} = not significant. Regression equation for all species pooled: *Ungulate biomass* $= -1940.80 + 384.96Temp - 33.50Pred$, $R^2_{adj} = 0.495$, $n = 54$ years, $P < 0.0005$).

annual temperature. Biomass of ungulates increased as the climate warmed (Fig. 12.3a). Interestingly, the positive correlation between temperature and ungulate biomass held in both situations: when predators were scarce or absent, and when they thrived. The periods of wolf and lynx control ($n = 15$ years) were characterized by very low density large predators (mean 0.75 inds/100 km^2, SE 0.13). In other years ($n = 41$), the combined density of wolves and lynx averaged 6.3 inds/100 km^2 (SE 0.41).

In the multiple regression analysis, mean annual temperature and predator density explained nearly 50% of the observed variation in ungulate biomass (Table 12.2). Semipartial correlations squared (Tabachnick and Fidell 1983) demonstrate the unique contributions of each independent variable to the total variation explained. For the community of ungulates as a whole, temperature was more important ($sr^2 = 0.303$) than predation ($sr^2 = 0.204$). However, the relative roles of these two factors differed greatly among species. Climate was crucial for bison and wild boar, and its role declined in moose, with even smaller effects in red and roe deer (see Table 12.2). The opposite trend was manifest in the role of predation. It ranged from negligible (and statistically insignificant) in bison to significant in red and roe deer. The interspecific differences in susceptibility to predation revealed by this analysis of long-term data are in agreement with our recent field research on wolves and lynx (Okarma et al. 1997; Jędrzejewski et al. 2000, 2002): roe and red deer appeared to be the ungulates most affected by predation.

The two regression lines in Figure 12.3a do not differ in slope (in fact they are nearly parallel), so the absolute value of the limitation by predators (about 120 kg of crude biomass of ungulates/km^2) seems to vary little within the whole observed range of climatic variation. Compared to the maximum densities attained by ungulates (upper regression line), the impact of predation was stronger during cooler, less productive periods than in the warmer, more productive times. In conditions of lower forest productivity, wolves and lynx were able to suppress ungulate biomass by 40 to 50% relative to the habitat carrying capacity, whereas in the most productive warm periods, suppression by predators was only 10 to 20% of the ungulate biomass (Fig. 12.3b). Thus, while higher productivity of a temperate forest ecosystem leads to increase in biomass of herbivores, limitation by predators apparently becomes weaker. Again, these results are in line with the conclusions of the wolf research (i.e., the rate of predation on red deer was negatively related to deer densities) (Jędrzejewski et al. 2002).

In conclusion, the long-term data set suggests that ungulates in the BPF have been affected simultaneously by bottom-up (climate-related food availability) and top-down (predation) forces. We suggest that food supply (shaped by climate) is the ultimate factor determining carrying capacity of the forest for ungulates. However, predators, if not controlled by humans, effectively limit density of ungulates below the level set by food resources. Moreover, top-down control is stronger in periods of cooler climate.

Top-Down and Bottom-Up Forces in Diverse Guilds of Predators and Prey

The limiting impact of wolf predation on ungulate (moose, elk) densities has been frequently reported in North American studies based on experimental reduction of wolves (e.g., Hayes et al. 2003, see also review in Boutin 1992) or observations of ungulate population response to wolf recolonization (Hebblewhite et al. 2002). Ballard et al. (2001) reviewed 30 studies on coyote (*Canis latrans*), mountain lion (*Puma concolor*), or wolf predation on mule (*Odocoileus hemionus*), white-tailed deer (*Odocoileus virginianus*), and black-tailed deer (*Odocoileus hemionus columbianus*). They concluded that deer populations at or near carrying capacity were food-limited and did not respond to predator removal experiments. Deer living in densities well below forage carrying capacity appeared limited by predation and their survival significantly increased when predator numbers were reduced (Ballard et al. 2001). Furthermore, other research has demonstrated different responses to wolf reduction among various species of North American ungulates, both locally (Kunkel and Pletscher 1999; Hayes et al. 2003) and along a latitudinal gradient (Crête and Manseau 1996).

Crête (1999) analyzed the pattern of deer biomass in North America and found that it increased from the High Arctic to the north of boreal forests and remained in the same range southward within wolf range. However, in the absence of wolves, deer biomass increased by a factor of five when latitude was held constant. South of wolf range, deer biomass was positively correlated with primary productivity. In a large-scale controlled experiment conducted in southwestern Yukon, Canada, Hayes et al. (2003) documented that 80% reduction in wolf numbers resulted in increased numbers of moose and woodland caribou (*Rangifer tarandus*), but not of Dall sheep (*Ovis dalli*). Wolf predation strongly limited recruitment of caribou and moose and survival of adult moose. The response of those two ungulate species to release from predation was rather rapid, having occurred during the five-year wolf reduction experiment (Hayes et al. 2003).

Other studies on the relative importance of top-down and bottom-up forces in structuring communities have provided evidence for the combined or synergistic effects of the two forces. In the Serengeti, Tanzania, migratory wildebeest (*Connochaetes taurinus*) were regulated by food supply (grasses), whereas resident herbivores were limited by top predators such as lions (*Panthera leo*) and hyenas (*Crocuta crocuta*) (Mduma et al. 1999; Sinclair et al. 2003). Furthermore, body size of ungulates appeared important in determining the impact of predators. Small species of herbivores experienced strong predation, whereas large ones had few natural predators and exhibited food limitation, the threshold occurring at about 150 kg. Sinclair et al. (2003) concluded that, due to great biodiversity of prey and predators, both top-down and bottom-up forces act simultaneously on herbivore populations, and proposed that this result applied generally in ecosystems with rich diversity of predators and prey. Indeed, this very conclusion came out of the long-term study on large carnivores and ungulates in the BPF. Wolves and lynx significantly suppressed red and roe deer below the forage carrying capacity, but larger-bodied bison and moose, as well as very fertile wild boar were resource limited.

Trophic Cascades: Indirect Effects of Large Carnivores on Forest Ecosystems

Can the propensity of large carnivores to limit their prey cause a trophic cascade by significantly releasing plants from grazing and browsing by ungulates? Following years of focused studies on predator–prey relations in the BPF, empirical research on cascading impacts of predation have only recently been initiated, with conclusive results not yet available. For the purposes of this chapter, we look to the literature from other temperate and boreal forests to aid us in understanding the possible role of top predators in the conservation of this last remaining intact forest ecosystem in temperate Europe.

Reports on heavy impact by deer on forest regeneration in predator-free environments abound in Europe and North America. High grazing and browsing levels change the composition and density of ground flora (e.g., Kirby 2001; Morecroft et al. 2001; Watkinson et al. 2001), recruitment rates of young trees (Rooney 2001), and density and structure of undergrowth (e.g., McShea, this volume). The modeling approach used by Jorritsma et al. (1999) suggests that, in the absence of top carnivores, the 100-year changes in forest structure and species composition would vary with the densities of ungulates and their species composition. Generally, however, a more detrimental effect can be expected from domestic species (cattle, horse) than from wild ungulates (red and roe deer). Grazing and browsing by the latter species may even indirectly promote the long-term presence of pine and birch in treestands (Jorritsma et al. 1999). Exclosure experiments or reduction of fauna have clearly shown that the removal of browsing and grazing mammals profoundly shapes the early stages of forest regeneration in both boreal and temperate forests (reviewed in Gill and Beardall 2001) and tropical forests (Dirzo and Miranda 1990).

Reintroduction of wolves to Yellowstone National Park (USA) has offered an opportunity to observe whether the return of a top carnivore would release woody vegetation from heavy browsing pressure (Ripple and Beshta 2003). Indeed, a significant increase in the height of cottonwood (*Populus* spp.) and the regeneration of willow (*Salix* spp.) was observed in sites of high predation risk, whereas riparian habitats with low predation risk continued to be heavily browsed. Thus Ripple and Beshta (2003) showed that, apart from reducing ungulate numbers, top carnivores also act on lower trophic levels through predation risk that alters habitat use by prey.

High densities of ungulates, released from predator pressure, can affect not only vegetation. Through browsing and grazing (cervids, bison) and soil rooting (boar), ungulates act as ecosystem engineers, altering the structure of the forest, its physical properties, and the amount of food and shelter for other organisms, such as small rodents, birds, and invertebrates (Flowerdew and Ellwood 2001; Fuller 2001; Stewart 2001). In England, deer occurring in excessively high numbers reduced the density of low foliage and ground cover of vegetation, which led or contributed to the decline of herbivorous bank voles (*Clethrionomys glareolus*) but not granivorous wood mice (*Apodemus sylvaticus*). An experiment in the Netherlands showed that exclusion of deer from heavily grazed areas reversed the process: vegetation quickly recovered, and field vole (*Microtus agrestis*) and wood mice numbers increased (Smit et al. 2001).

In the woods of England and the United States, changes in forest structure due to heavy browsing have caused significant shifts in the species composition of bird communities but less so in the overall abundance or diversity of birds (Casey and Hein 1983; McShea and Rappole 2000; Perrins and Overall 2001). Studies in Grand Teton National Park (Wyoming, USA) by Berger et al. (2001a) provided a convincing example of how the local extinction of large carnivores triggered a cascade of ecological events leading to reduction of avian species richness. Extermination of grizzly bears and wolves some 65 to 70 years earlier caused a fivefold increase in densities of moose, a riparian-dependent herbivore. The subsequent heavy browsing by moose on willow communities led to decline of neotropical migratory birds. Two species, the gray catbird (*Dumetella carolinensis*) and MacGillivray's warbler (*Oporornis tolmiei*), have even disappeared from regions with high densities of moose (Berger et al. 2001a).

Studies on invertebrates showed that areas with high densities of ungulates markedly differed from ungulate exclosures in respect to the structure and species composition of communities, but no consistent trend emerged regarding invertebrate abundance (Kozulko 1998; Suominen et al. 1999b; Feber et al. 2001; Stewart 2001). High ungulate densities can be detrimental for some species, and beneficial for others, such as thermophilous, gap-preferring butterflies or insects developing on feces.

Finally, large predators may promote increased biodiversity by subsidizing scavengers with unconsumed prey remains. In the BPF, carrion of ungulates has been an important food resource for about 30 species of small and medium-sized mammals and birds (Jędrzejewska and Jędrzejewski 1998). The presence of wolves and lynx ensures the predictable, year-round supply of carrion in the form of kill remains. In the absence of top carnivores, carrion would be available to scavengers more seasonally, mainly in late winter (Jędrzejewska and Jędrzejewski 1998). Similarly, Wilmers et al. (2003a) documented that wolves introduced to Yellowstone National Park (USA) prolonged the time period over which carrion was available and changed the availability of carcasses from a late winter pulse dependent on abiotic conditions to a relatively constant supply of kill remains throughout the year. Furthermore, wolves can facilitate scavenging by other species. Selva et al. (2003) showed that, in the case of European bison carcasses in the BPF, only wolves could open the thick skin of the carcass and make it available to smaller scavengers.

Conservation Implications

Extirpation of large carnivores from temperate and boreal forests often causes significant increases in ungulate numbers, though various species of ruminants may respond differently to release from predation. The resultant increased level of herbivory can cause changes in forest vegetation, especially woody plants, usually leading to a decline in ground cover and changes in species and height structure of forest regeneration. This results in changes in the numbers and/or occurrence of species that rely on forest vegetation as food resources or refuge: birds, small mammals, and invertebrates. The changes caused by increased levels of herbivory often lead to an erosion of biological diversity, particularly the decline of woodland species, and may promote influx of species typical of open areas.

Though many elements and mechanisms of the trophic cascade remain to be studied, the empirical support for its wide occurrence in predator–ungulate–forest systems of temperate and boreal zones is unquestionable. Therefore, conservation should consider the restoration of the role played by large carnivores in woodlands and/or restoration of the species themselves. The following options may be considered, depending on local conditions and public support: (1) reintroduction programs of wolves, lynx, and bear; (2) promotion of natural dispersal of large carnivores into previously occupied habitat; and (3) replacement of natural predation with (well planned) human harvest of ungulates by humans.

Summary

In this chapter, we discuss the role of top carnivores in the food chains of temperate and boreal forests. Studies in the Białowieża Primeval Forest, located on the Polish–Belarussian borderland, provide evidence that wolves and lynx limit the densities of ungulates below the carrying capacity of the habitat. Predation effect differed among species of ungulates. Large carnivores exerted strong pressure on red deer and roe deer and had little impact on European bison, moose, and wild boar. In addition, the analysis of long-term data (over 100 years) showed that predators' limiting effect on ungulate numbers was higher in periods of cooler climate and (presumably) lower ecosystem productivity. Finally, our data suggested that wolves and lynx affected the ecosystem biodiversity by promoting scavenger guilds (constant provision of kill remains).

A review of studies conducted in other temperate and boreal woodlands of Europe and North America corroborated the wide occurrence of top-down control of ungulates. Although conclusions of empirical studies investigating cascading impacts of predation on other trophic levels are not yet available in the BPF, research in similar habitats has demonstrated widely that extinction of large predators usually causes notable increases in densities or even eruption of ungulates. Increased levels of herbivory and alteration of forest vegetation often affect the numbers and/or occurrence of birds, small mammals, invertebrates, and nontarget plants. Though further studies are required to better understand the trophic cascades in the BPF and other terrestrial ecosystems, it is unquestionable that top carnivores play an important role in preservation of forest biodiversity.

ACKNOWLEDGMENTS

We are most grateful to Dr. Justina Ray for her encouragement, unfailing support during our work on this chapter, and comments on the chapter manuscript. Karol Zub prepared the figures. Dr. Kent Redford, Dr. Joel Berger, and three anonymous reviewers contributed comments to the earlier version.

Recovery of Carnivores, Trophic Cascades, and Diversity in Coral Reef Marine Parks

Tim R. McClanahan

Trophic cascades are commonly reported in aquatic ecosystems (Menge 1995; Brett and Goldman 1996; Vanni et al. 1997; Estes et al. 1998; Steneck 1998; Pace et al. 1999; Pinnegar et al. 2000). These cascades influence the abundance of species and are expected to influence species interactions and diversity (Duffy 2002). Predation, the force causing cascades, can have both positive and negative effects on numbers of species, and the relationship between predation and species diversity is often unimodal or hump shaped (McClanahan 1998; Worm et al. 2002). The recovery of carnivores is likely to increase predation rates but one cannot a priori expect predation to either decrease or increase the number of species unless one knows the position of a site and species assemblage on the predation-diversity continuum. The loss of carnivores is likely to lead to low levels of predation and losses of species through competitive exclusion of competing prey, but at high levels of predation the predator-susceptible species can be locally extirpated (McClanahan 1998). This complicates ecological predictions and conservation planning, particularly when species are assembled into larger and more complex food webs such as coral reefs.

Cascades may be weaker in species-rich ecosystems due to multiple parallel and intertwined pathways and trophic levels, omnivory, dominance of the detrital chain, symbiosis, parasitism, and control of production by many environmental factors (Polis and Strong 1996). Nonetheless, several reports addressing trophic cascades have been documented for marine and coral reef ecosystems (Pinnegar et al. 2000). The ability to detect a cascade effect may be influenced by the taxonomic or functional group resolution of the examined food web (Hall and Raffaelli 1991, 1993; Martinez 1991, 1993), and the inclusion or exclusion of species

(Goldwasser and Roughgarden 1997). Much remains to be examined concerning the factors that influence the detection and description of trophic cascades in aquatic ecosystems, particularly diverse food webs.

Trophic cascades have been reported in coral reefs largely when species abundance has been manipulated by fishing (Hay 1984; McClanahan and Shafir 1990; Hughes 1994; McClanahan et al. 1999b; Pinnegar et al. 2000; Halpern and Warner 2002; Halpern 2003), or when large-scale diseases have eliminated important predators (Lessios 1988; Carpenter 1990; Hughes 1994). A detailed study of coexistence among coral reef–inhabiting sea urchins showed the importance of predation in influencing abundance and maintaining the diversity of this species assemblage (McClanahan 1988, 1998). These studies indicate that coral reef trophic cascades occur and do influence species diversity, but are seldom characterized by simple linear cause and effect ripples that cascade predictably down the food chain determined largely by the number of levels in the food chains. The greater complexity of real food webs is likely to cause switches in the dominance of predator–prey energy pathways and compensation among species functions (Norberg 2000), such that it is difficult to make simple predictions about the effect on all species pooled into a single trophic level, a common simplifying assumption of food-chain models (i.e., Hairston et al. 1960).

The result of species deletions or additions may depend on the strength and species diversity associated with certain pathways in the intertwined web, with differences in interaction strengths among different trophic levels causing braided cascades rather than channeled or linear ones. Due to species redundancy and compensation it may also be common to have trophic-level effects that either trickle weakly down the food web or do not cascade to primary producers, such that top-level consumers do not have strong or predictable influences at the base of the food chain (Polis and Strong 1996). Finally, the level of resolution with which one observes the ecosystem—ranging from species to gross functional groups—can also influence the detection and types of reported cascades. To date, these possibilities have received only cursory exploration with regard to coral reefs, despite their relevance to management and conservation. Here I present findings on carnivore recovery and changes in community organization from two disparate coral reef ecosystems undergoing similar protective closed-area management, and discuss the effects of this management on carnivore recovery and effects on species diversity.

Statement of the Problem

I examined the changes in two recently created marine protected areas, one in the remote Glovers Reef atoll in the western Caribbean and the other in an East African fringing reef near Mombasa, Kenya. Both areas are part of marine protected area conservation programs that have been supported by local government bodies associated with marine protection since the early 1990s—the Fisheries Department in Belize and Kenya Wildlife Service in Kenya. I studied the benthic community—sea urchins and most of the visible fish species over the early stages of protection—and ask here whether the elimination of human fishers at the top of the food web resulted in population increases in top-level carnivorous fish and cascading effects on the rest of the food web. I also investigated whether these effects can be distinguished at the taxonomic (genus-species) or functional group level of resolution (Fig. 13.1) and for different levels in the food web, and whether there are clear changes in abundance of species for the various taxa. The study determines the role of these conservation programs in protecting key fisheries species and the ability of closed areas to enhance species diversity.

Study Sites and History of the Coral Reef Parks

Glovers Reef Atoll, Belize

Glovers Reef Atoll is an area of 260 km^2 that is dominated by coral and seagrass and located approximately 40 km off the Belizean (Central American) coastline. It is one of the largest and more remote atolls of the Caribbean, and was chosen for conservation programs of the Wildlife Conservation Society (WCS) as an example of one of the last marine wilderness sites in the Caribbean (Perkins and Carr 1985). Despite its remote location, the atoll is a popular fishing area and is used throughout the year by a small number of residents and a larger number of transient fishers who travel between the atoll and landing sites on the mainland. Nonetheless, fishers are transient and their densities are not high. Fishers largely use spear guns, but nets, traps, and lines are also used. The Belizean government designated the southern quarter of the atoll as a conservation zone in 1993, and the numbers of transient fishers were successfully reduced or excluded from this zone by 1995.

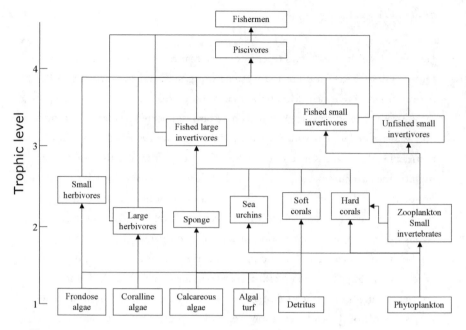

Figure 13.1

Food web diagram for a simplified coral reef food web based on analyses of diet and metabolisms (Opitz 1996). Trophic groups are plotted along the vertical axis based on estimates of their mean trophic level calculated from diet data (Pauly et al. 2001).

This ecological study was undertaken among the approximately 850 patch reefs located within the lagoon of the Glover's Reef Atoll. Moderate-sized replicate patch reefs of around 1000 m² were chosen in the conservation (no fishing) and general use zones (unrestricted fishing) for monitoring the benthos, mobile invertebrates, and fish populations. Patch reefs are largely formed by massive coral taxa in the genera *Montastrea, Porites,* and *Diploria,* with additional coral taxa consisting of *Agaricia, Acropora,* and *Millepora* in addition to a variety of seawhips, sponges, and algae growing on dead surfaces. Erect frondose algae are currently the dominant primary producers on these patch reefs. Similar to many Caribbean reefs (Carpenter 1990; Hughes 1994; Shulman and Robertson 1996; Ostrander et al. 2000), frondose algae has increased during the past 30 years (McClanahan and Muthiga 1998) since the patch reefs were first described (Wallace 1975).

Mombasa Marine National Park, Kenya

The Mombasa Marine National Park (Mombasa MNP) is a 6 km² coral reef and seagrass area located 1 km off shore on the Kenyan fringing reef and 10 km north of the city of Mombasa. Both tourists and fishers heavily use the reef, and the central section of the park was a popular "coral gardens" snorkeling area prior to the park's designation in 1987. Continued building of tourist facilities resulted in conflicts between tourists and fishers and, therefore, the most heavily utilized tourist area was designated and began to receive protection management by the Kenya Wildlife Service. This management was largely successful in excluding fishers by 1991. Fishing outside of the protected area is high with fisher densities around 13/km² with traps, spear guns, and various nets being used (McClanahan and Kaunda-Arara 1996).

The ecological studies described in this chapter were undertaken in two 30 m × 30 m sites in the center of the marine protected area in the fringing reef lagoon on the back reef in an area dominated by hard bottom and massive coral heads. Findings from these sites were compared with seven comparable sites at distances of 5 to 50 km from the park, on both the north and south side of the park. Corals in the genera *Porites, Pavona, Pocillopora, Galaxea, Favia,* and *Favites* dominate with a variety of algae and soft corals growing on the dead surfaces.

Field Sampling and Data Analysis

The benthos, invertebrates, and fish were sampled annually in both areas. In the case of Glovers, sampling was undertaken between 1998 and 2002, and in Mombasa from 1991 to 2003. The 1998 El Niño Southern Oscillation (ENSO) event in Kenya was one of the largest of the past 100 years (McPhaden 1999), and caused massive mortality of hard and soft corals in all sites (McClanahan et al. 2001a). It also triggered bleaching and mortality of corals in Belize (Aronson et al. 2002a), although coral mortality in the study site was minor (McClanahan et al. 2001b). Consequently, we analyzed the data in Kenya for the periods before and after 1998 to determine the possible influence of this rare event. Sampling methods are described in detail in McClanahan (1999) and McClanahan et al. (2001a,b). Briefly, benthic cover descriptions were based on three 10 m line intercept transects per patch reef in Belize and nine per site in Kenya. Corals were identified to species, frondose algae to genus, whereas filamentous turf, crustose coralline algae, sponges, soft coral, and gorgonia remained in these gross taxonomic groups and were not further classified. Benthic cover had to be greater than 0.6% of the sampled substratum to be included in the analyses. Sea urchins were identified to species, and counted in nine 10 m² plots per reef site.

The two study sites are in different biogeographic regions and they therefore differ in their total species richness, with the East African region having more than double the number of species as the Caribbean (Paulay 1997; McClanahan 2002). Sampling of the fish fauna differed between the two regions because of the effort required to describe the larger number of species in the Kenyan site. The Belizean sampling was based on a nearly complete list of species that were easily observed. Species such as those in the Holocentridae, or small Serranidae-like hamlets (*Hypoplectrus* spp.) that are cryptic and largely occupy small crevices were excluded from this study. In the case of Kenya, sampling was based on eight common, diverse, and noncryptic families; namely, the surgeonfish (Acanthuridae), triggerfish (Balistidae), butterflyfish (Chaetodonitidae), pufferfish (Diodontidae), wrasses (Labridae), angelfish (Pomacanthidae), damselfish (Pomacentridae), and parrotfish (Scaridae). Since these families do not include groups that are entirely piscivorous, data are lacking for the highest trophic levels in Kenya, such as groupers, jacks, sharks, and barracuda (*Sphyraena* spp.). Data on the Belizean fish were collected yearly whereas Kenyan fish data were collected only during the early and later stages. I only present data from a taxon if more than 10 individuals were sampled during data collection. The discrete group sampling (DGS) method was used (Greene and Alevizon 1989; McClanahan 1994; McClanahan et al. 2001b), where a limited number of species of a similar shape, position in the water column, or behavior were sampled during each data collection period.

Data Analysis

Response ratios were used to test for differences between the area closed to fishing and control areas where fishing continued (Hedges and Olkin 1985). Response ratios were defined as the difference between the means of park and nonpark estimates for each species or functional group, divided by the pooled standard deviations (Hedges and Olkin 1985). Response ratios of each species were plotted against their mean trophic level. Data on abundance were examined at two levels of resolution: the species or genus level and the functional group level. Functional groups were based on a combination of diet and taxonomic affinity as determined by the analyses of Opitz (1996). Algae were divided into four functional groups: coralline (red calcifying and encrusting algae); frondose (large brown, red, and green algae with leathery thalli); calcareous (green articulated calcifying algae); and turf (green, blue-green, red, and brown microscopic algae) (Steneck and Dethier 1994). Corals were divided into soft (internal skeleton) and hard (external skeleton) coral functional groups. Where only functional or gross taxonomic group data were collected (such as for filamentous turf algae), analyses are only presented at the functional level of resolution. The 77 and 131 species and the 13 functional groups used in the analysis for fishes are presented in the legends for Figure 13.2 (Glovers) and Figure 13.4 (Mombasa), respectively. Where taxa had mixed diets, their abundance was split between functional groups. All erect algae were identified to genus, coral to species in Belize and genus in Kenya. Fish were identified to species with the exception of *Stegastes* in Belize, which was identified to genus.

Decisions to pool species into functional groups were based on previous taxa aggregations based on diet and metabolism for the development of a Caribbean coral reef Ecopath model (see Fig. 13.1; Opitz 1996), and whether or not the taxa were harvested by humans. Sampled taxa were divided into groups based on their diet and whether or not they were fished, which largely reflected the maximum adult size of the species. The densities of a few species that had more than 25% of their diet mixed between two groups (based on Randall 1967) were split based on the percentage of the diet composed of each functional group. This was done for < 10% of the species. Figure 13.1 presents the proposed functional group model that describes the gross functional groups and their interactions.

The trophic level for most species was obtained from published studies (Christensen and Pauly 1993, 2000; Opitz 1996), with the exception of corals. Opitz's (1996) coral trophic level of 2.34 appears to be too high, the author having possibly only considered the animal host and not the symbiotic algae. Since most of the energy of corals is derived from the photosynthesis of symbiotic algae and not plankton (Edmunds and Davies 1986), I used a lower value of 1.5 with the understanding that this will vary with taxa and environment. Trophic levels for a few species not listed in these sources were either estimated from taxonomically or trophically similar species in these published works or calculated from the diet data of Randall (1967). To estimate the mean trophic level for functional groups, I multiplied the density of each species by the species' trophic level, summed across all species in each trophic level, and divided by the total density for that functional group (Pauly et al. 2001).

Data were tested for normality and homogeneity of variance by Shipiro-Wilk and Levene's test (Sall et al. 2001). The benthic cover data had a mix of distributions and variances. The most abundant fish were normally distributed with equal variances, but this was seldom the case for the less abundant fish. Benthic taxa were arcsine transformed and fish taxa log transformed, and tested again for normality and equal variance and statistical significance (Sokal and Rohlf 1981). Transformation did not alter the outcome of tests of statistical difference and, therefore, a mixture of parametric and nonparametric statistics were used to test for statistical significance based on the distribution of the data and the assumptions of the statistical tests. Conservative values for statistical significance were based on Bonferroni corrected p-values for multiple comparisons (Rice 1989). On the plots of the response ratios with trophic level I drew a line through the points to distinguish those that were and were not statistically significant. Plots of the taxonomic richness of the variously measured groups against time were examined to determine if the management and recovery of the heavily fished groups were associated with changes in taxonomic richness. Species density data were tested for significance with two-way analysis of variance (ANOVA) for time and the management treatment. If time was significant, a regression with time was performed (Sall et al. 2001).

Research Findings

Comparative analyses of field research results from the two marine protected areas provide necessary background for exploring the role of predator recovery in restoring biodiversity in coral reef ecosystems.

Glovers Reef, Belize

Plots of the response ratios of the 77 taxa comparing differences in the Conservation and general use zones indicate that trophic levels lie between 1 and 4.5 with a great deal of scatter but a generally increasing response with the trophic level of the taxa (Fig. 13.2a). Most of the statistically significant responses were found among the higher trophic groups, and this was most clear when the abundances of the taxa were pooled into functional groups (Fig. 13.2b). Piscivores and small and large invertivores have increased in the conservation zone relative to the general use zone, whereas sea urchins (largely *Echinometra viridis*) and small herbivores have decreased. Changes in the numbers of species between 1998 and 2002 indicate an overall increase in algal taxa in both zones, an increased number of fish species in the conservation zone, and no differences or changes in coral or sea urchins, which were the smallest components of species diversity (Fig. 13.3a–d).

Mombasa Marine Park, Kenya

Plots of response ratios are presented for the period before and after the 1998 coral mortality event (Fig. 13.4a–d). Differences in the initial conditions of these reefs influenced comparisons of the corals. The Mombasa park had a greater number of coral taxa than the control areas due to lower temperature variation (McClanahan and Maina 2003), and this resulted in positive response ratios for those corals (specifically those labeled with numbers 13 to 28 in Fig. 13.4a). The plots of the response ratios of the 131 taxa in the Mombasa MNP indicate high scatter with fewer statistically significant differences in the pre- than the post-1998 period (Fig. 13.4a,c). Once differences in the initial conditions of the coral were removed, however, there was also an increase in the response ratios with the increase in trophic level. Increases in herbivorous fishes were more evident here than in Belize, particularly during the first study period. When the abundances of the taxa were pooled into functional groups, there were more small and large herbivores and few differences in the higher trophic levels for the first period. These differences were more evident in the second period, with greater abundance of small and large invertivores and large herbivores in the unfished area and a reduction in the small herbivores (Fig. 13.4b,d). Coralline algae were consistently more abundant in the unfished reef for both periods, and turf algae less abundant in the first

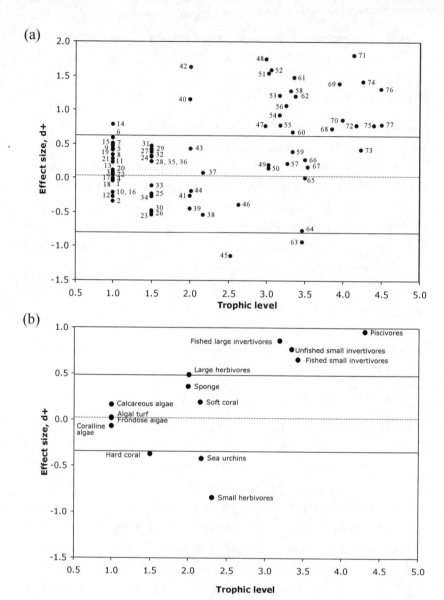

Figure 13.2

Response ratios or "effect size" for (a) taxa and (b) functional groups as a function of
the trophic level in Glovers reef. Methods for calculating the response ratios are described in the text. Horizontal lines depict the level of significance based on the
Bonferroni correction for multiple comparisons. Frondose algae: 1. *Lobophora*,
2. *Turbinaria*, 3. *Sargassum*, 4. *Dictyota*, 5. *Halimeda*, 6. *Dictyosphaeria*, 7. *Padina*,
8. *Laurencia*, 9. *Amphiroa*, 10. *Jania*, 11. *Centroceras*, 12. *Gelidiella*, 13. *Hypnea*,
14. *Galaxaura*, 15. *Caulerpa*, 16. *Coelothrix*, 17. *Acanthophora*, 18. *Avrainvillea*,
19. *Udotea*, 20. *Dasya*, 21. *Penicillus*. Scleractinian corals: 22. *Montastraea annularis*,

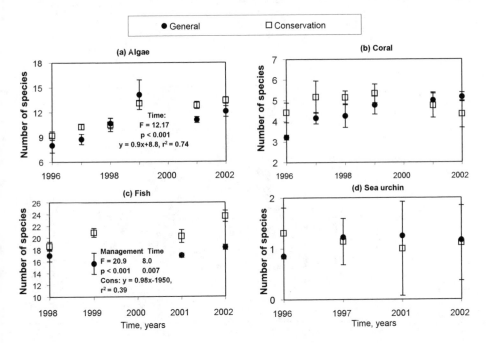

Figure 13.3

Plots of the numbers of species against time for four studied groups (a–d) in Glovers reef comparing the Conservation (no fishing) and General Use Zones (fishing) over the period following closure from fishing in 1995.

23. *Porites porites*, 24. *Agaricia agaricites*, 25. *Siderastrea siderea*, 26. *Porites asteroides*, 27. *Diploria aurolineatum*, 28. *Millepora alcicornis*, 29. *Millepora complanata*, 30. *Acropora cervicornis*, 31. *Diploria labyrinthiformes*, 32. *Diploria clivosa*, 33. *Montastraea carvernosa*, 34. *Porites colonensis*, 35. *Favia fragum*, 36. *Acropora palmata*. Sea urchins: 37. *Echinometra viridis*, 38. *Echinometra lucunter*. Bony fishes: 39. *Scarus croicensis*, 40. *Acanthurus coeruleus*, 41. *Acanthurus bahianus*, 42. *Sparisoma viride*, 43. *Acanthurus chirurgus*, 44. *Sparisoma aurofrenatum*, 45. *Stegastes* sp., 46. *Pomacanthus arcuatus*, 47. *Holocanthus tricolor*, 48. *Holocanthus ciliaris*, 49. *Chaetodon capistratus*, 50. *Chaetodon ocellatus*, 51. *Chetodon striatus*, 52. *Chromis cyanea*, 54. *Calamus bajonado*, 55. *Mulloidichthys martinicus*, 56. *Lachnolaimus maximus*, 57. *Haemulon flavolineatum*, 58. *Gerres cinereus*, 59. *Halichoeres bivittatus*, 60. *Thalassoma bifasciatum*, 61. *Haemulon sciurus*, 62. *Bodianus rufus*, 63. *Pseudupeneus maculates*, 64. *Haemulon plumieri*, 65. *Clepticus parrae*, 66. *Halichoeres pictus*, 67. *Halichoeres garnoti*, 68. *Lutjanus analis*, 69. *Lutjanus synagris*, 70. *Ocyurus chrysurus*, 71. *Epinephelus striatus*, 72. *Lutjanus apodus*, 73. *Epinephelus cruentatus*, 74. *Lutjanus griseus*, 75. *Caranx ruber*, 76. *Sphyraena barracuda*, 77. *Caranx bartholomaei*.

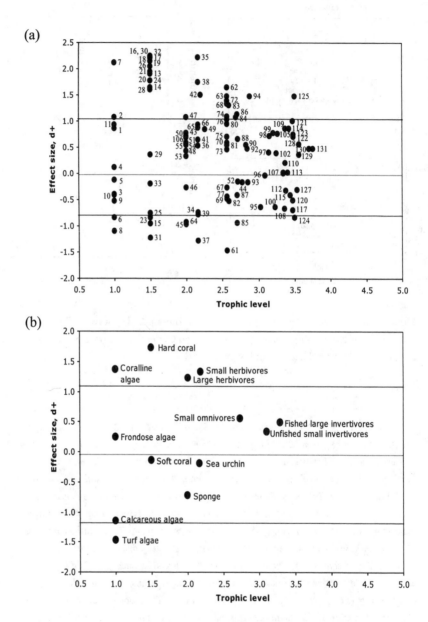

Figure 13.4

Response ratios for (a,c) taxa and (b,d) functional groups as a function of the trophic level for studies in southern Kenya. Methods for calculating the response ratios are described in the text. Data are presented for the period before and after the 1998

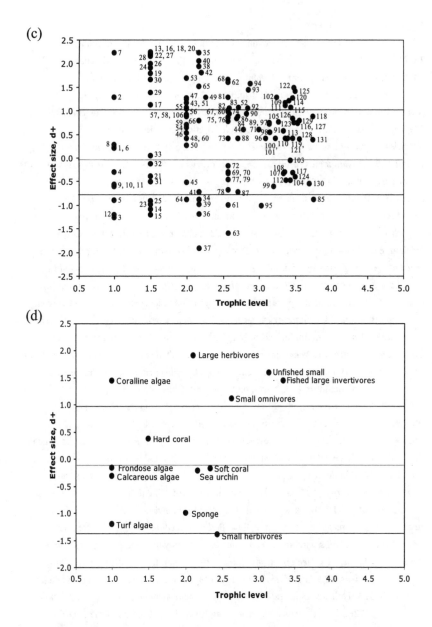

(c)

(d)

ENSO event (see text). Horizontal lines depict the level of significance based on the Bonferroni correction for multiple comparisons. Frondose Algae: 1. *Sargassum*, 2. *Turbinaria*, 3. *Padina*, 4. *Dictyota*, 5. *Hypnea*, 6. *Halimeda*, 7. *Amphiroa*, 8. *Jania*, 9. *Ulva*, 10. *Cystoseiria*, 11. *Dictyospheria*, 12. *Codium*. Scleractinian corals: 13. *Massive*

Figure 13.4 (continued)

Porites, 14. Branching *Porites*, 15. *Pavona*, 16. *Synarea*, 17. *Galaxea fascicularis*, 18. *Favites*, 19. *Favia*, 20. *Platygyra*, 21. *Pocillopora*, 22. *Leptoria*, 23. *Millepora*, 24. *Cyphastrea*, 25. *Stylophora*, 26. *Acropora*, 27. *Turbinaria*, 28. *Goniopora*, 29. *Hydnophora*, 30. *Echinopora*, 31. *Tubipora*, 32. *Montipora*, 33. *Astreopora*. Sea urchins: 34. *Echinometra mathaei*, 35. *Echinostrephus molaris*, 36. *Diadema savignyi*, 37. *Tripneustes gratilla*, 38. *Echinothrix diadema*, 39. *Diadema setosum*, 40. *Echinothrix calamaris*, 41. *Toxopneustes pileolus*. Bony fishes: 42. *Acanthurus nigrofuscus*, 43. *Calotomus carolinus*, 44. *Acanthurus triostegus*, 45. *Leptoscarus vaigiensis*, 46. *Scarus ghobban*, 47. *Ctenochaetus striatus*, 48. *Scarus rubroviolaceus*, 49. *Zebrasoma scopas*, 50. *Acanthurus dussumieri*, 51. *Scarus psittacus*, 52. *Naso annulatus*, 53. *Scarus sordidus*, 54. *Zebrasoma veliferum*, 55. *Acanthurus leucosternon*, 56. *Scarus niger*, 57. *Scarus frenatus*, 58. *Ctenochaetus strigosus*, 59. *Scarus falcipinnis*, 60. *Cetoscarus bicolour*, 61. *Chrysiptera biocellata*, 62. *Plectroglyphidodon dickii*, 63. *Chrysiptera unimaculata*, 64. *Stegastes nigricans*, 65. *Acanthurus nigricauda*, 66. *Stegastes fasciolatus*, 67. *Thalassoma herbraicum*, 68. *Chromis dimidiata*, 69. *Chromis viridis*, 70. *Amphiprion allardi*, 71. *Pomacanthus imperator*, 72. *Plectroglyphidodon lacrymatus*, 73. *Plectroglyphidodon johnstonian*, 74. *Pomacentrus caerueius*, 75. *Pomacentrus sulfureus*, 76. *Pomacentrus baenschi*, 77. *Chrysiptera annulata*, 78. *Neoglyphidodon melas*, 79. *Pomacentrus pavo*, 80. *Chromis nigrura*, 81. *Neopomacentrus azysron*, 82. *Abudefduf vaigiensis*, 83. *Abudefduf sexfasciatus*, 84. *Pomacanthus semicirculatus*, 85. *Dascyllus aruanus*, 86. *Chromis weberi*, 87. *Dascyllus trimaculatus*, 88. *Chaetodon kleinii*, 89. *Chaetodon melannotus*, 90. *Chaetodon guttatissmus*, 91. *Chaetodon trifasciatus*, 92. *Pomacanthus chrysurus*, 93. *Gomphosus coeruleus*, 94. *Centropyge multispinis*, 95. *Abudefduf sparoides*, 96. *Rhinecanthus aculeatus*, 97. *Chaetodon xanthocephalus*, 98. *Pseudochelinus hexataenia*, 99. *Stethojulis strigiventer*, 100. *Diodon holocanthus*, 101. *Diodon liturosus*, 102. *Balistapus undulatus*, 103. *Sufflamen fraenatus*, 104. *Sufflamen chrysopterus*, 105. *Chaetodon lunula*, 106. *Chaetodon auriga*, 107. *Labrichthys unilineatus*, 108. *Chelio inermis*, 109. *Anampses caerulopunctatus*, 110. *Anampses twistii*, 111. *Anampses meleagrides*, 112. *Halichoeres nebulosa*, 113. *Coris gaimardi africana*, 114. *Bodianus axillaries*, 115. *Thalasomma hardwicke*, 116. *Thalassoma amblycephalum*, 117. *Hemigymus melapterus*, 118. *Hologymnosus doliatus*, 119. *Hemigymus fasciatus*, 120. *Stethojulis albivittata*, 121. *Diprocanthus xanthurus*, 122. *Labroides dimidiatus*, 123. *Labroides bicolour*, 124. *Halichoeres scapularis*, 125. *Halichoeres hortulanus*, 126. *Macropharyngodon bipartitus*, 127. *Thalassoma lunare*, 128. *Coris Formosa*, 129. *Coris caudimacula*, 130. *Cheilinus trilobatus*, 131. *Coris aygula*.

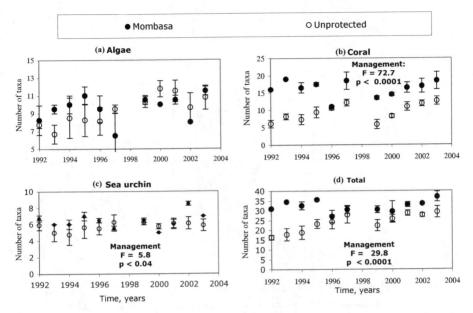

Figure 13.5

Plots of numbers of species against time comparing the Mombasa Marine National Park and unfished control areas.

period. The numbers of species between 1992 and 2003 indicated no differences with algae and sea urchin taxa for both management areas, although higher numbers of fish species in the conservation zone increased over time (Fig. 13.5a–d). This increase was solely attributable to increases in species that were fished elsewhere. Numbers of fished species decreased slightly over time in the fished reefs. Coral and the total numbers of all sampled species were higher in the unfished than fished areas, with this difference evident not only at the start of the study but also maintained across the mass mortality event in 1998 (Fig. 13.6).

Discussion

The high diversity of species and life histories of coral reef ecosystems has produced complex food webs, where only a few taxa exhibit discrete trophic positions such as primary producers, herbivores, and carnivores. Coral, the architect of this

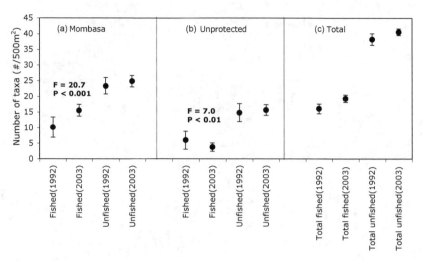

Figure 13.6

Plot of fish diversity separating species by those that are fished and unfished in the unfished Mombasa Marine National Park (a) and fished reefs (b) and for all sites combined (c). Fishing was largely excluded from the park in 1991.

ecosystem, is a good example of an animal–plant association that uses multiple sources of energy, nutrients, and food but is difficult to place in a discrete trophic position. Above the second trophic level, diets are mixed such that species lie along a trophic continuum rather than as discrete units. This makes it less likely that simple trophic linear cascades will dominate this system, particularly when viewing interactions at the species level of resolution. Nonetheless, species vary in their interaction strengths, and therefore cascades do occur, but they often do not affect the entire species assemblage. For example, sea urchins and small herbivorous fishes increased when their predators were removed by humans, but the effect on their algal food is not always clear due to other factors such as the abundance of large herbivorous fishes and coral. This is because when cascades occur they are likely to be braided and possibly attenuated or truncated due to species replacement and compensation (Norberg 2000), resulting in complex interactions and dynamics that influence only a subset of the diverse food web. This complexity is most clear when we view these data at the taxonomic level of genus or species. In the case of Glovers Reef, where all trophic levels were studied, including the piscivores, most fished species responded to the elimination of fishing. Cascading in this system seemed, however, to be largely restricted to decreases in two small herbivores, the sea urchin *Echinometra viridis* and the brown damselfish *Stegastes*. Both species are probably susceptible to predation. The brown damselfish is most likely to be reduced by piscivores such as barracuda and jacks, and the sea urchin by invertivores such as the jolthead porgy (*Calamus bajonado*), ocean and queen triggerfish (*Canthidermis sufflamen* and *Balistes vetula*, respectively), and hogfish (*Lachnolaimus maximus*) (McClanahan 1999). One strong interacting species for the Caribbean, the sea urchin *Diadema antillarum* (Hughes 1994), was rarely observed in these patch reefs and its absence may be one reason for these few strong interactions.

Some groups exhibited little or no response to closed-area management in Belize. The poor response of the large herbivores to no fishing could reflect a number of factors including lower fishing pressure on this group, poor recruitment during this time, susceptibility to piscivores in the no-fishing zone, or poor grazing conditions in the reef (McClanahan et al. 2001b). An experimental study that manipulated algae on these patch reefs found that the abundance of unpalatable late-successional algae was a likely factor in inhibiting herbivores (McClanahan et al. 2001b). Beneath the herbivore level there is no indication of a cascading effect. Two of the more likely and possibly synergistic explanations are the short time of the study and that abiotic factors are a greater influence than biotic factors on the benthic taxa. These reefs have experienced a loss in coral and an increase in algal cover over the past three decades (McClanahan and Muthiga 1998), with this being reflected in the continuing increase in algal cover and diversity and coincident decrease in coral cover and diversity. This change is most likely due to warmer temperatures resulting in coral bleaching and mortality, combined with diseases that have influenced corals and the sea urchin *Diadema antillarum* (Aronson et al. 2002b). An alternative hypothesis that high nutrients have caused this change has been unsupported by nutrient manipulation experiments (McClanahan et al. 2002). A bleaching event in 1998 appeared to have had a greater influence on coral cover in the conservation than in the general use zone, probably due to the poorer water circulation in this zone (McClanahan, unpubl. data). Increases in the number of species in this system are most easily explained by the recovery of fish in the no-fishing zone that were previously influenced by fishing. There was no clear indication of an increase in the number of unfished taxa.

The results from southern Kenya share some attributes with the Glovers study, but since there were no data collected on piscivores, no comparisons can be made for the highest trophic levels. Nevertheless, the biggest increases in numbers of individuals and species were among the fished functional groups occupying the highest sampled trophic positions. In Kenya, there was evidence for cascades at the lowest trophic positions, but changes in abundance were not associated with clear increases in numbers of species with time. There were more species in the no-fishing zone at the beginning of the study and this may be an attribute of the site and less an effect of management and recovery of carnivores. McClanahan and Maina (2003) found that this site is more connected to the open ocean, resulting in lower water temperature variation and the persistence of more coral taxa than the other studied sites. Consequently, the high diversity and selection of this site for closed-area management were appropriate, with evidence for the control of diversity by nonhuman carnivores not particularly strong.

The most notable cascading effect on the Mombasa benthos was the increase in coralline algae associated with decreased turf algae. This is likely due to the change in dominance of grazers from sea urchins to herbivorous fishes (McClanahan 1997). Although sea urchin numbers reported here are still relatively high, they have been reduced since the inception of the park (McClanahan 2000), and this has been associated with an increase in their fish predators and competitors.

Conservation Recommendations and Concluding Thoughts

The areas closed to fishing achieved one of the primary objectives of closed-area management—protecting species affected by fishing. This resulted in an increase in total number of species, largely through an increase in the numbers of fished species. These species are frequently the top-level carnivores and some of the larger herbivores, because fishing selectivity in these reefs is largely a function of the fish's adult body size. In some cases the return of these species reduced smaller herbivores that are common to heavily fished reefs. In the case of Mombasa this may have changed the abundance of coralline and turf algae. In Glovers this management did not, however, reduce frondose algae and increase corals and associated invertebrate species to levels reported from the early 1970s (Wallace 1975; Gardner et al. 2003). It has been the hope of many interested in the management of Caribbean coral reefs that the recovery of the higher trophic levels would have cascading consequences that will return reefs to their pre-Columbian ecology (Jackson 1997). This has not been the case at the time scale of this study and, if physicochemical factors such as water temperature and acidity have been the ultimate cause of this large-scale change, then closed-area management alone is unlikely to restore reefs. Clearly, longer studies are required to evaluate closed-area management for restoring reef ecology, but this study also indicates the need for management at the scale of climatic effects. Increased efforts are needed to reduce the rate of global climate change, and to select closed areas based on oceanographic conditions that favor the persistence of coral.

This study indicates the importance of site selection for coral reef closed areas. The potential to increase the number of species in the Glovers Reef conservation area may be undermined by physicochemical conditions that are both stressing this ecosystem and possibly arresting its full recovery. Higher temperature variation in the conservation area compared to the general use zone is associated with the position of the closed area in the southern end of the atoll where there is less water movement (McClanahan, unpubl. data). During calm periods this results in higher temperature variation and maximum temperatures across extreme events, such as warm ENSO years, thereby causing mortality of temperature-sensitive taxa. The opposite is the case for the Mombasa region where the park has the highest water flow and the lowest temperature variation of the studied reefs. This results in more coral taxa, which has been shown to maintain itself across warm ENSO events (see Fig. 13.4; McClanahan and Maina 2003). Clearly, the oceanographic conditions such as water flow and associated temperature patterns need to be an important part of the design of closed area management. Otherwise, reducing or eliminating fishing alone may not produce the intended effect of increasing the number of species apart from fishing-sensitive but stress-tolerant species.

The relationship between predation and numbers of species is not tight or univariate because a number of other factors, including colonization, extinction, productivity, and refuge and habitat complexity, can interact to either increase or decrease the effects of predation (McClanahan 1998; Worm et al. 2002). Time for species to be extirpated or to colonize is required and the time scales of these processes are only recently being understood (McClanahan 1998, 2000; Halpern and Warner 2002). Although target fisheries species may recover quickly from fishing pressure (Halpern and Warner 2002), some important species and nontarget species lower in the food web may not have had sufficient time in this decade-length study to change or recover (McClanahan 2000). Physicochemical factors are also highlighted by this study: Water flow may have been a particularly important influence through its effect on colonization rates and water temperatures and is likely to affect the number of species by changing colonization and extinction rates. Consequently, an increase in predation associated with a recovery of carnivores cannot by itself insure the recovery of species, which is an interaction between physicochemical and biological interactions. There is good evidence, however, that reduced predation by humans on coral reef carnivores can result in an increase in the density of harvested species (Halpern 2003). The question that cannot be answered by this study is the possible effects that the highest-level carnivores, such as sharks, have on these ecosystems. The size of closed areas in these and many other coral reef protected areas is not sufficient to support viable populations of the largest carnivores, and answering this question will require study of large undisturbed areas. Because these large carnivores have been an important component of coral reef ecosystems until recently their inclusion in the ecosystem is expected to play an important role in reef ecology.

Summary

This study compared the early recovery process for ~210 taxa in two marine protected areas—Glovers Reef, Belize, and Mombasa, Kenya—following initiation of closed-area management. I examined the response of the taxa and functional groups at different trophic levels and the numbers of species of major groups (algae, coral, sea urchins, and fish), and compared these reefs to those where fishing continues to take place. There was a great deal of scatter in the responses of the various taxa to closed-area management, suggesting complex trophic dynamics. The patterns were clearer and trophic cascades more easily identified when taxa were pooled into functional groups. Higher trophic levels generally increased in number, whereas lower trophic levels experienced fewer changes. Cascading effects were evident for some but not all groups, suggesting a mix of strong and weakly interacting species and functional groups. Such patterns also varied between the two protected areas. In both sites, numbers of species generally increased for fish, particularly those affected by fishing, but less so for other groups at the lower trophic levels, which may be more influenced by abiotic factors. Consequently, the control of species diversity in both studied sites results from an interaction between biotic and abiotic forces and not by predator-induced cascades alone.

ACKNOWLEDGMENTS

This study would not have been possible except for the numerous people that assisted with field studies and the long-term support of the Wildlife Conservation Society. Thanks to R. Arthur, B. Cokos, K. Bergman, Raphael Fremiot, M. Huitric, S. Jones, A. Kamukuru, R. Kiambo, B. Kaunda-Arara, H. Machano Ali, J. Maina, S. Mangi, J. Mariara, M. McField, N. Muthiga, I. Nordemar, M. Nystrom, D. Pietri, M. J. Rodrigues, E. Sala, and P. Stickels. This is publication number 20 of the Middle Key Research Station. I am grateful to two anonymous reviewers for their comments.

CHAPTER 14

Human-Induced Changes in the Effect of Top Carnivores on Biodiversity in the Patagonian Steppe

Andrés J. Novaro and R. Susan Walker

Because predator–prey systems can have more than one stable state (Holling 1973), the relative importance of top-down versus bottom-up control may depend on the density of the population being controlled (Bowyer et al., this volume). Predators can sometimes limit their prey at low density (Gasaway et al. 1992), but limitation at high density is more likely to occur as a result of environmental factors (Walker and Noy-Meir 1982; Sinclair 1989; Pech et al. 1992; Messier 1994), or a combination of environmental factors and predation (Krebs et al. 1995). On the other hand, endangered prey may be limited by exotic predators at both high and low densities (Kinnear et al. 1998). When external factors reduce prey density and simultaneously promote high predator density, limitation of prey by their predators is likely to occur.

Predators can affect biodiversity through limitation of prey numbers and other mechanisms. When prey occur at low densities, limitation by predators can increase the probability of local extinction of these prey, thereby reducing diversity. Several recent studies have documented cases of top predators becoming threats to the persistence of some of their prey (Gasaway et al. 1992; Estes et al. 1998; Roemer et al. 2002). When prey occur at high densities, predators can modify prey use of habitats and foraging patterns, reducing herbivory pressure on some habitats and releasing resources for other species (Berger et al. 2001a). Therefore, when prey are abundant, predation by carnivores may increase diversity by preventing some species from becoming overly abundant or reducing foraging pressure on certain habitats, thus promoting coexistence among competitors (Estes 1996).

Top carnivores usually occur at low densities and frequently come into conflict with humans, so they are often among the first species to be lost with human alteration of ecosystems (Redford, this volume). For this reason, most ecosystems that have suffered significant faunal alterations due to human activities have lost their top carnivores, even if prey are still relatively plentiful. As a result, ecosystems where top-down effects of large vertebrate predators currently limit prey densities and affect prey diversity may be relatively rare (Terborgh et al. 1999). On the other hand, drastic reductions of prey populations by human activities at specific sites can make prey more likely to be limited by large predators. This, in turn, represents a shift in the role of top carnivores in biodiversity conservation.

The Patagonia region of the Southern Cone of South America presents what is perhaps an uncommon opportunity to study the role of top carnivores in ecosystems, and particularly their place in biodiversity conservation. In Patagonia, populations of native herbivores and omnivores have suffered tremendous declines due to persecution by humans and competition from exotic species, while top native carnivores are thriving. The reasons for the abundance of native carnivores have only recently begun to be considered (Novaro et al. 2000) and are further discussed here. No study to date, however, has considered the potential implications of predation by the abundant carnivores on the greatly reduced native prey populations of Patagonia. In this chapter we ask three questions. First, have human activities in the Patagonian steppe indirectly induced widespread top-down control of native prey by large carnivores? Second, could such an alteration in the role of carnivores represent a new threat to the recovery of herbivores, reducing overall biodiversity of native species? And finally, is this change in the effect of top carnivores on their prey in Patagonia unusual? Or have the drastic, human-induced modifications of ecosystems in recent millennia repeatedly altered the role that top carnivores have on their prey and biodiversity?

Native Carnivore and Prey Communities of the Patagonian Steppe

The Patagonian steppe ecosystem, extending from the Andes Mountains to the Atlantic Ocean, consists of temperate grasslands and shrublands encompassing over 700,000 km^2 in southern Argentina and Chile (Fig. 14.1). The top carnivore

1 Auca Mahuida Provincial Reserve
2 Ranches of Southern Neuquen
3 Rio Negro site
4 Torres del Paine National Park

Figure 14.1
Areas in Patagonia referred to in the text, and location of Patagonia in South America.

of the Patagonian steppe is the puma (*Puma concolor*), with a mean body weight of 62 kg (Franklin et al. 1999). Where the puma is absent, the culpeo (*Pseudalopex culpaeus*), a 10-kg canid, is the top carnivore. Smaller carnivores include the chilla fox (*Pseudalopex chilla*), the pampas (*Lynchailurus colocolo*) and Geoffroy's cats (*Oncifelis geoffroyi*), two weasel-like mustelids, the grison (*Galictis cuja*) and the smaller huroncito (*Lyncodon patagonicus*), and two hog-nosed skunks (*Conepatus chinga* and *C. humboldti*).

From the time of the Pleistocene extinctions (Markgraf 1985), the dominant herbivores of the Patagonian steppe have been the guanaco (*Lama guanicoe*)—

a camelid of 100 to 120 kg—and the Darwin's rhea or choique (*Pterocnemia pennata*) —a large-bodied (20–25 kg), flightless bird that also consumes small animals. Early European explorers describe herds of guanacos that numbered in the thousands and large flocks of choiques (Musters 1964). Other native prey species include the mara (*Dolichotis patagonum*), a 7 to 9 kg rodent, the mountain vizcachas (*Lagidium* spp.), rodents of 2 to 3 kg, the fossorial tuco-tucos (*Ctenomys* spp.), the guinea pig–like cuises (Cavidae), two species of social geese, the cauquens (*Chloephaga* spp.), and two edentates, the hairy armadillo (*Chaetophractus villosus*) and the pichi (*Zaedyus pichiy*). Cricetine rodents are abundant in most habitats and highly diverse (Redford and Eisenberg 1992).

Human Impact on Patagonian Wildlife

Patagonia has few human inhabitants, but human activities over the past 100 years have forever altered the structure and composition of Patagonian wildlife communities. When the native inhabitants were decimated by introduced disease and defeated by the Argentine army in the late 1800s, Europeans and Argentines moved in with large herds of sheep, which reached a peak population of 22 million in the 1950s (INDEC 2002). Many exotic wildlife species were also introduced by the new settlers. The European hare (*Lepus europaeus*) and the wild swine (*Sus scrofa*) have subsequently colonized all of Argentine Patagonia (Bonino 1995), with the hare reaching densities of more than 2/ha (Novaro et al. 2000). The red deer (*Cervus elaphus*) first populated the Andean forests and ecotone in northwestern Patagonia but has recently begun to expand into the steppe, reaching high densities along river valleys (Funes et al., unpubl. data). The rabbit (*Oryctolagus cuniculus*) has spread throughout the northwestern and southernmost regions of Patagonia (Bonino 1995).

Impacts on Native Herbivores

The human-mediated processes that have most affected the native herbivores are habitat degradation, competition with livestock and exotic species, and hunting. Overgrazing by livestock and exotics has resulted in severe desertification of approximately 25% of the Patagonian rangelands (del Valle et al. 1998). In many parts of Patagonia the lands are so degraded that they can no longer support the large numbers of sheep they once did (Golluscio et al. 1998). Range degradation has likely also lowered the carrying capacity for native herbivores.

In addition to the direct effects of habitat degradation, competition with sheep and exotic wildlife may have negatively affected native species that use similar resources. Guanaco and sheep diets overlap significantly (Pelliza-Sbriller et al. 1997), and movement of sheep into an area quickly excludes guanacos, suggesting that sheep are strong competitors of guanacos for forage, although exclusion mechanisms are not yet clear (Baldi et al. 2001). In terms of dietary overlap and biomass, the foraging of one sheep is equivalent to that of five choiques or 20 cauquens (Bonino et al. 1986). Where European red deer and guanacos are sympatric, their diets overlap seasonally, with both species consuming mostly shrubs (Bahamonde et al. 1986). European hares have high dietary overlap with mountain vizcachas and maras (Bonino et al. 1997; Galende and Grigera 1998).

Throughout the 20th century, native Patagonian herbivores also suffered intensive hunting. Guanacos were hunted to reduce their competition with sheep. Commercial hunting of guanaco young for their skins and of choiques for their feathers was heavy and widespread. Between 1972 and 1979, 443,655 guanaco pelts were legally exported from Argentina (Ojeda and Mares 1982). Mountain vizcachas were heavily hunted during the 1950s for their hides (C. Menna, pers. comm. 1994). All three species are still hunted for food for subsistence purposes, and choique eggs are collected for human consumption (Funes and Novaro 1999).

For most native herbivores there are no good data on either past or present population sizes, so the exact extent of population reductions over the last century remains unknown. Based on explorer accounts of guanaco distribution, plant productivity, and forage consumption by guanacos, guanaco numbers in Patagonia prior to European colonization were estimated at 7 to 20 million (Raedeke 1979; Torres 1985; Lauenroth 1998). Guanaco numbers in recent times have been estimated at 400,000 to 600,000 individuals, or about 0.5 to 0.9 guanacos/km^2 (Raedeke 1979; Torres 1985; Amaya et al. 2001), representing 2 to 9% of the original population. Remaining populations are fragmented and largely relegated to the driest lands that are not suitable for livestock (Baldi et al. 2001). In the case of the choique, comparison of recent density estimates in northern and southern Patagonia with accounts of early explorers suggests a widespread collapse of the population (de Lucca 1996; Funes et al. 2000). Because threats for most other native herbivores and omnivores were similar, it is possible that other species have suffered declines of similar magnitude.

Impacts on Carnivores

With the introduction of sheep to Patagonia, pumas and culpeos were killed for their tendency to prey on these livestock. Bounty hunting of pumas was initiated in many places, and pumas were extirpated from most of their former range by the middle of the 20th century (Bellati and von Thüngen 1990). The small cats and skunks were also hunted heavily for their furs until the export of their skins was banned in the 1980s. Hunting of the two canid species for fur and to prevent livestock predation was intensive, and continues today. During the 1970s and 1980s, roughly 15,000 to 20,000 culpeo and over 100,000 chilla skins were traded in Patagonia annually (J. Rabinovich et al., unpubl. data). Regional densities of these species were reduced significantly, but apparently without widespread effects on the species' distributions (Novaro 1997). Indeed, culpeo distribution expanded to the east during the 20th century, perhaps in response to high availability of exotic prey (Crespo and de Carlo 1963), increased water availability due to artificial waterholes for livestock, or the extirpation of the puma.

Effects on Wildlife of Reduction of Sheep and Hunting

During the last 20 years some of the major threats to both native herbivores and carnivores have been reduced. Reduced wool prices since the early 1980s, in combination with the reduced carrying capacity of the range, resulted in a drastic reduction in sheep numbers. Overall, the Patagonian sheep herd has declined to about eight million, or 35% of the peak (INDEC 2002). Some ranchers in the more humid Andean foothills have replaced sheep with cattle (i.e., 90% of ranches in southern Neuquén), and in the southernmost province of Santa Cruz, many ranches have been abandoned.

Because fewer workers were needed on the ranches, the rural human population declined. This was accompanied in the 1980s and 1990s by a lower international demand for furs and other wildlife products, and increased regulation of

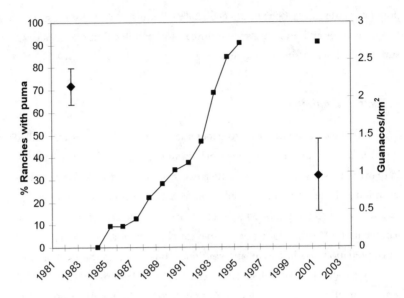

Figure 14.2

Trend in puma recolonization (◆: based on year of puma reappearance at 32 ranches) and mean guanaco density (■) in 1982 and 2001–2 (from transect estimates at 7 ranches) in southern Neuquén province.

wildlife harvest by the provinces. All of this resulted in decreased hunting pressure on both native herbivores and carnivores. In southern Neuquén, for example, the number of culpeos killed annually declined by 70% between 1989 and 2002 (Novaro et al. 2003).

Because of this reduction in competition with sheep and hunting pressure, an increase in wildlife populations could be expected throughout Patagonia, except in areas where the range had been too degraded or where other herbivores such as cattle or exotic wildlife became overabundant. Indeed, pumas have recolonized much of their former range throughout Patagonia during the last 20 years (Novaro et al. 1999; Fig. 14.2). Culpeos increased in number throughout Patagonia and continued their eastward expansion, their density doubling in southern Neuquén between 1989 and 2002 (Novaro et al. 2003; Fig. 14.3).

Guanaco populations have also recovered in some areas, recolonizing abandoned ranches in southern Santa Cruz province a few years following sheep

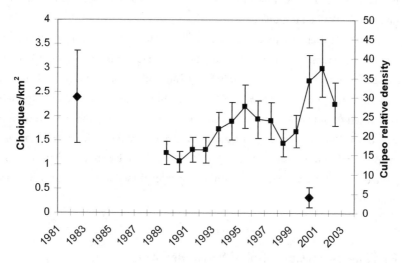

Figure 14.3

Trends in mean culpeo (◆: % stations with tracks at six ranches) and choique densities (■ : transect estimate in 1982 and 2000 at Corral de Piedra ranch) in southern Neuquén province.

removal. However, in many places where they were expected to recover, guanaco and other native herbivore populations have not rebounded, and in some cases have declined even further. Guanacos, choiques, and mountain vizcachas, for example, have continued to decline in southern Neuquén province. This has occurred even on large ranches where sheep were removed during the 1980s and early 1990s, and where subsistence hunting of native herbivores has been minimal. Regional guanaco density in southern Neuquén declined by 77% during the last 20 years (see Fig. 14.2; Gonzalez and Novaro, unpubl. data). Sixty-eight percent of ranchers throughout Neuquén report a decreasing trend of choiques over the last 10 years. At one large ranch with no hunting, an 86% decline was measured between 1982 and 2000 (see Fig. 14.3; Funes et al. 2000). Choiques have also declined at Laguna Blanca National Park in central Neuquén, in spite of being protected for 55 years (R. Pereira, pers. comm. 1998). Local people report a decline in mountain vizcachas in terms of both numbers of colonies and numbers of individuals within colonies (Walker et al. 2000). Since 1970 mountain vizcachas have disappeared from 15% of a sample of randomly chosen rock outcrops, with no new colonizations (Walker, unpubl. data).

Low densities of native prey have not prevented the recovery of puma and culpeo populations, even with the reduction in sheep numbers, because these carnivores have been capable of exploiting the huge prey base of introduced wildlife. In southern Neuquén, 94% of available prey biomass is of introduced animals, compared to only 6% of native species. European red deer and hares represent 90% of the biomass of the diet of pumas, and hares are the major prey for culpeos in southern Neuquén and eastern Río Negro. Selectivity analyses have demonstrated that pumas and culpeos strongly select for these introduced prey (Novaro et al. 2004).

Possible Top-Down Control of Native Herbivores by Pumas and Culpeos

We propose that predation by abundant pumas and culpeos limits populations of native prey at low densities and may prevent their recovery in parts of Patagonia. This impact of top carnivores on native herbivores may be exacerbated by the presence of exotic herbivores. European hares and red deer, in addition to competing with native species for resources, may help to maintain high carnivore numbers by supplementing their food base. This may result in high predation rates on native herbivores, further limiting these herbivores even in areas where sheep and other livestock are now scarce.

If a predator population is maintained at high levels by an abundant supply of a focal prey, the ability of the predator to limit the numbers of a less abundant prey may be increased, even if the latter prey is killed only occasionally (Holt and Lawton 1994; Bêty et al. 2002; Roemer et al. 2002). This process of indirect interaction between two prey species that share a predator has been termed apparent competition (Holt 1977). Additionally, increases in the consumption of less abundant prey, or prey switching, can occur when a more common, focal prey declines (Corbett and Newsome 1987; Sinclair et al. 1998). If predators switch back to native prey when exotic prey decline, they can have devastating effects on populations of rare native prey (i.e., Sweitzer et al. 1997; Roemer et al. 2002). Apparent competition and prey switching between native and exotic prey may also occur in portions of Patagonia where sheep and/or goats are still present and are an important prey of pumas and culpeos, and a decreased regional hunting pressure has allowed for recovery of these carnivores. In summary, we suggest that if the recent declines of native herbivores are even partially related to increased predation, the current effect of top carnivores on prey diversity in parts of Patagonia may be negative.

Alternative or complementary explanations exist, however, for the further decline or lack of recovery of native herbivores in Patagonia. First, habitat degradation may be so extensive in some areas that recovery is not possible even after sheep removal. Second, rapid colonization and population increase of exotics such as the red deer in northwestern Patagonia may have monopolized forage resources released by sheep removal before guanacos and other native herbivores could recover. Finally, habitat changes induced by overgrazing, such as increases in the shrub to grass ratio, may have favored colonization and increase of exotic herbivores such as the red deer. Unfortunately there are currently no data to assess these alternative hypotheses.

Evidence for Limitation of Guanaco Populations by Pumas

The recent regional decline documented for guanacos in southern Neuquén (see Fig. 14.2) has not occurred uniformly throughout this 7200 km² area where sheep were removed during the last 10 to 30 years. Guanaco populations have actually recovered to high densities, probably near carrying capacity for the steppe (Lauenroth 1998; Baldi et al. 2001), in specific ranches such as Achecó (23 guanacos/km²) and Alicura (17 guanacos/km²). At other nearby ranches, guanacos have not recovered, remaining at a density of ca. 0.1/km² at Quemquemtreu, and declining from 2.4 to 1.4 guanacos/km² at Los Remolinos (Gonzalez and Novaro, unpubl. data). Pumas have recolonized all of these ranches, often reaching high densities (Novaro et al. 1999; see Fig. 14.2).

Guanacos were historically the main prey of the puma and are killed by pumas at high rates where they are abundant (Iriarte et al. 1991; Franklin et al. 1999; Bank et al. 2002). Where guanacos are rare, they are consumed by pumas only occasionally (Novaro et al. 2000). Even sporadic killing by pumas, however, could have a significant effect on guanaco populations that occur at low densities. Although several studies have described puma predation on guanacos (Wilson 1984; Franklin et al. 1999; Bank et al. 2002), none has analyzed the potential for puma predation to limit guanaco populations or prevent their growth. Pumas have been reported to limit and even reduce the numbers of bighorn sheep (*Ovis canadensis*, Wehausen 1996), mule deer (*Odocoileus hemionus*, Hornocker 1970), and porcupines (Sweitzer et al. 1997) in North America when these occur at low densities.

A preliminary assessment of consumption of guanacos by pumas suggests that puma predation has the potential to prevent growth of low-density guanaco populations (Novaro et al. in prep.). The impact of predators on their prey depends on the predators' total response to changes in prey density, which is the product of their functional and numerical responses (Messier 1994). The numerical response of pumas to guanaco densities is unknown, but current puma numbers likely are high in southern Neuquén due to low hunting pressure and an abundant supply of introduced prey (Novaro et al. 2000), so their total response will depend mainly on their functional response to guanaco densities. The relationship between consumption of guanacos by pumas and guanaco density at a series of sites shows that the level of consumption may be density dependent and the curve resembles a Type III response (Holling 1959, 1965). This type of response has the greatest potential to contribute to prey regulation because predation rate accelerates as guanaco density increases (Fig. 14.4). As in most cases of regulatory effects of predation, the puma–guanaco response suggests that acceleration of the predation rate, and thus the potential to prevent population growth, only occurs at low guanaco density. A significant threshold for this acceleration occurs at about 8 guanacos/km², indicating that puma predation may regulate guanaco populations at densities below this threshold. Thus, in southern Neuquén, with a regional density in the range of 1 to 2 guanacos/km² during the last 20 years, pumas have had the potential to prevent growth in guanaco numbers throughout the area. Conversely, guanaco numbers may only increase in areas where density at the time of release from intense competition from sheep was above the threshold for regulation by predation.

This prediction was tested by comparing guanaco trends among ranches where sheep have been removed and guanacos were either above or below the 8/km² threshold in 1982 (Novaro et al. in prep.). Guanaco density was above this

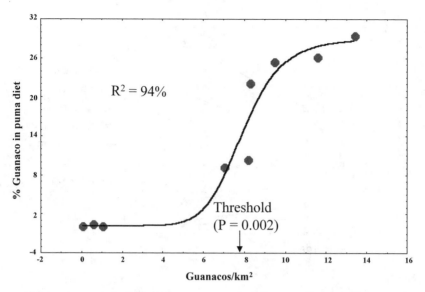

Figure 14.4

Response in puma consumption of guanacos as a function of density of guanacos at different sites in Patagonia (Novaro et al. in prep.). Data are from three ranches in southern Neuquén, Auca Mahuida Provincial Reserve, and five yearly means during guanaco density increase at Torres del Paine Park in Chile (Iriarte et al. 1991).

threshold only at Achecó Ranch and in the eastern portion of Alicura Ranch (Fig. 14.5), and as expected, increased up to 17 to 23 guanacos/km² at Achecó and Alicura but did not increase at the other five ranches. Thus the prediction appears to hold for this small sample of sites.

Densities below the potential threshold level for puma regulation occur throughout the guanaco's range in Patagonia, because average density is less than 1 guanaco/km² (Novaro et al. 2000; Amaya et al. 2001). Therefore, because pumas are widespread and increasing throughout Patagonia, we predict that puma predation may play a significant, increasing role in preventing guanaco recovery following the current trend of release of competition from sheep. This effect of pumas on guanacos may be more likely in areas where abundant, introduced wild prey or livestock support high puma densities.

If puma predation is responsible for the observed trends in guanaco populations, it is not necessarily density-dependent predation leading to population reg-

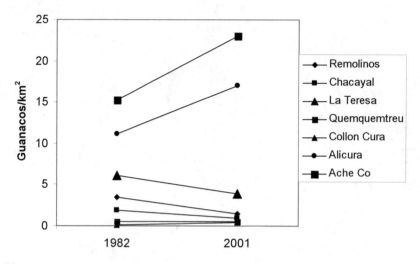

Figure 14.5

Trends in guanaco density at seven ranches where sheep were removed in southern Neuquén province (Novaro et al., in prep.).

ulation, but could be merely intense puma predation causing population limitation. Detecting limitation (processes that set populations at certain densities) may be simpler and more relevant to conservation than detecting regulation (processes that prevent growth or make populations return to certain levels through density-dependent mechanisms) (Krebs 2002). Regardless of whether the process is regulation or limitation, predation by abundant pumas in either case could be preventing the recovery of reduced guanaco populations. However, use of the theory of population regulation can help identify threshold densities that may provide useful targets for conservation (Sinclair et al. 1998).

Evidence for Limitation of Native Prey by Culpeos

Culpeos may also exert top-down control on some of their native prey, particularly where these have been greatly reduced in numbers by competition from exotics and hunting. The native prey of culpeos include small rodents, mountain vizcachas, edentates, cauquens, and eggs and young of choiques. All native prey combined now represent only 13 to 18% of the biomass in the culpeo diet at three

Patagonian sites (Johnson and Franklin 1994a; Novaro et al. 2004). Currently, culpeos prey mostly on European hares and sheep. However, as in the case of the puma, predation mortality by culpeos could be locally significant for native prey at low densities.

Among Patagonian wildlife species, choiques, like the closely related Pampas rheas (*Rhea americana*), may be particularly susceptible to predation because they nest in communal nests on the ground and produce large eggs (Fernández and Reboreda 1998; Funes et al. 2000). Also, reduced colony size of mountain vizcachas due to competition with exotics or hunting may make this rodent more susceptible to predation, as is the case for other social animals (Alexander 1974). We do not have information to document if the mountain vizcacha or the choique declines at the ranches in Neuquén or Laguna Blanca National Park were caused by increased predation. However, the available data suggest that an increase in predation pressure cannot be ruled out as one potential cause.

The relationship between consumption of native prey and prey densities has not been studied yet for culpeos. However, our preliminary data indicate that, as in the case of pumas, the density of certain native prey may affect culpeo predation on those species, leading to functional responses that may contribute to prey regulation. Consumption of mountain vizcachas is much greater where their density is higher across a range of sites in northern Patagonia and the southern puna (Walker et al. in prep.). Similarly, it appears that culpeos switch back to consume more native prey in areas where or at times when European hares and sheep are rare or decline temporarily (Novaro et al. 2004). The consumption of all native prey, including choiques and mountain vizcachas, increased in southern Neuquén after a decline in hare densities in the early 1990s (Novaro 1997). Overall, the functional response of culpeos and their capacity to switch between exotic and native prey may promote limitation and, in some cases, regulation, of certain native prey.

Increases in culpeo density in southern Neuquén and other parts of Patagonia during recent years also may have contributed to their role in limiting populations of some native prey. The twofold increase in culpeo density reported here occurred at six ranches monitored annually between 1989 and 2002, with a positive numerical response of culpeos to European hare densities across ranches (Novaro 1997; Novaro and Funes, unpubl. data). Sheep were removed from three of these ranches between 1992 and 1994, resulting in a reduction in hunting pressure, lower availability of sheep as prey and carrion, and increased consumption of native prey by culpeos (Novaro et al. 2000). Similarly, sheep were removed and culpeo hunting was banned by the ranch owner in the late 1980s at the Neuquén site of the documented choique collapse, which likely led to an increase in culpeo density as in the rest of the region. Ranchers throughout Patagonia report similar increases in culpeo and chilla numbers during recent years, in spite of the implementation of bounty systems (R. Cardón and A. Contreras, Direcciones de Fauna de Río Negro and Chubut, pers. comm. 2002). Increased culpeo density, combined with lower hare and sheep densities in some parts of Patagonia, may have provided a total predator response that could have limited populations of some native prey, especially if densities of these prey were already reduced by other causes.

Predation as a Potential Threat to Wildlife Conservation in Patagonia

The evidence presented here indicates that predation by two top native carnivores may prevent the recovery of and, in some cases, further reduce populations of, native herbivores and omnivores in parts of Patagonia. Therefore, under current conditions, predation may have a negative effect on Patagonian biodiversity. Current predation by pumas and culpeos may increase the likelihood of local extinction of native prey by further reducing their numbers and increasing their exposure to stochastic events such as droughts, severe winters, and disease. Although the main threats to native prey species are direct competition from introduced herbivores, habitat degradation, and hunting, predation by top native carnivores is an added threat that needs to be considered when addressing conservation of Patagonian wildlife.

Before the 20th-century colonization of the Patagonian steppe, top-down control by predators may have occurred locally, where populations of abundant native herbivores were reduced by environmental factors or hunting by indigenous people. When native herbivores were abundant elsewhere in the region, however, predation may have contributed to maintaining biodiversity by providing temporary releases on resource use that would have allowed for local increases of other species. Evidence from prevalence of plant defenses in the Patagonian steppe and historic accounts suggest that guanaco numbers and their foraging pressure were high throughout the region in the past (Lauenroth 1998). Therefore, predation that helped maintain low guanaco numbers locally would have contributed to local plant recovery and provided opportunities for other herbivores to increase in numbers, thereby increasing spatial heterogeneity in plants and herbivores.

The dramatic population decline of native herbivores and the supplementation of the carnivore food base by exotic prey are likely the most important factors that have determined the current role of predation as a threat to wildlife recovery in parts of Patagonia. In addition to reduction of population size, fragmentation of remaining prey populations can also augment the negative effect of predators (Macdonald et al. 1999; Schneider 2001). The extent of fragmentation of native wildlife populations in Patagonia, however, is even less known than that of population declines, so it is more difficult to speculate about its effects.

Biological characteristics of the top carnivores, in combination with the spatial pattern of hunting, have probably also contributed to the current role of carnivores in the ecosystem. First, pumas and culpeos are both generalist predators (Novaro et al. 2000) that can easily switch between native and exotic prey, which contributes to their negative impact on the more vulnerable prey (Holt and Lawton 1994). Additionally, high reproductive potential of pumas and culpeos and the existence in the vast Patagonian rangelands of refugia from hunting (Bellati and von Thüngen 1990; Novaro 1997) have contributed to a rapid population recovery of these carnivores. Carnivore population increases have likely exacerbated the effects of prey switching and provided a significant source of additional mortality to native herbivores.

The threat of increased effects of predation by top carnivores and apparent competition from introduced herbivores in Patagonia also exists in the temperate forest along the Andes Mountains. There the population of Andean deer or huemul (*Hippocamelus bisulcus*) has been drastically reduced to a few thousand individuals, European hares are ubiquitous, and European red deer and wild swine are locally abundant and expanding in distribution. Forest-dwelling puma prey on huemul (Serret 1995) and can also occur at high densities where exotic prey have become abundant. Thus, in areas where huemul are rare, puma predation could reduce population size of this endangered deer even further, increasing its risk of local extinction.

The population increase of top carnivores in Patagonia may also have ecological and conservation implications for the rest of the carnivore assemblage
through increased resource competition, interference, and even predation. Recent
reviews of interactions within carnivore guilds (Palomares and Caro 1999; Linnell
and Strand 2000) suggest that top carnivore increases may affect the composition
and structure of the entire carnivore community. Displacement of chillas by
culpeos from more productive habitats (Johnson and Franklin 1994b), reports of
small carnivores killed by larger ones (Johnson and Franklin 1994c; Novaro 1997),
and reports of culpeo declines in areas where puma densities are unusually high
(M. Anz, pers. comm. 2000) indicate that intraguild interactions may play an important role in the diversity of the Patagonian carnivore assemblage.

The effects of abundant top carnivores in Patagonia, however, may not all
be negative for biodiversity. Predation by abundant culpeos can maintain populations of European hares at low densities after these have declined due to other
factors (Novaro and Funes, unpubl. data). High risk of puma predation can alter
patterns of habitat use by European red deer that occur at high densities, lowering deer foraging pressure on meadows and riverine forests (Funes and Novaro,
pers. obs.). Hence, increased densities of top carnivores in Patagonia can help limit
negative effects of introduced herbivores on vegetation and competition for forage between introduced and native herbivores. The overall effect of top carnivores
on biodiversity in Patagonia will depend on the local balance of positive and negative effects, including those effects on direct competition and mediation of apparent competition between exotic and native prey.

Conservation Recommendations

Based on the evidence presented in this chapter, predation by abundant native top
carnivores should be considered an additional threat to native wildlife in parts of
Patagonia. Because of the interrelatedness of threats, to effectively promote the
recovery of native wildlife, predation needs to be considered and addressed simultaneously with hunting and competition from exotic species. For conservation
actions to be effective, however, conservation practitioners first need to determine
targets for restoration of wildlife communities.

Different targets need to be defined for both agricultural lands and protected areas in Patagonia. Livestock and exotic wildlife species are economically desirable and a significant portion of the regional economy depends upon them. However, historic stocking rates of livestock have been unsustainable, and the quality of valuable exotic red deer trophies has declined due to overpopulation. Reducing the numbers of livestock and exotic wildlife on economically productive range-lands could improve the condition of the range, livestock, and harvested exotics, while at the same time allowing more room for native species. Specific target communities on private lands must vary according to the history and type of land use, the will of the landowner, and the biogeographical characteristics of the site.

In many cases, along with a reduction in stocking rate of livestock, controlled, regulated hunting can be an important management tool to promote the coexistence of native herbivores with livestock and commercially desirable exotic species. The continued commercial hunting of European hares and increased sport and commercial hunting of European red deer could locally reduce densities of these species (Novaro and Funes, unpubl. data). Harvest of culpeos for fur can temporarily reduce their densities (Novaro 1997) and decrease predation pressure on choiques, mountain vizcachas, and edentates. Sport hunting of pumas was legalized in Neuquén in 2003 and will soon be implemented (M. Funes, pers. comm.). This may contribute to efforts to locally limit puma densities and could be used as a tool to temporarily reduce predation rates on populations of guanacos and choiques that need to be recovered (but see Evans 1983).

In protected areas, on the other hand, the desired state is as close to the pre-Hispanic condition as possible, with no livestock and few exotics, although complete eradication of exotics is probably impossible. Target densities for guanacos, choiques, maras, and perhaps huemuls in protected areas should be set above minimum thresholds that would allow them to escape threats like intense predation. Target densities for exotic wildlife should be below the levels where they are the main food source of native carnivores. This could help prevent excessive supplementation of carnivore numbers and decrease the risk of negative effects on native prey. The potential role of predation in limiting prey populations should be evaluated locally in protected areas (Bowyer et al., this volume). If the target population appears to be limited by predation, nonlethal measures to reduce predation could be implemented, such as guarding populations during birthing periods or at other times when they are most vulnerable. Lethal control of top carnivores should not be implemented inside protected areas until more information is available on their role in limiting native prey.

Conservation-oriented actions both on private lands and in protected areas should be coordinated within an adaptive management scheme (Walters 1986). This will require developing site-specific targets and hypotheses about species interactions, such as those presented in this chapter, and the effects of different threats on native wildlife. Predictions based on these hypotheses should be tested by implementing interventions and monitoring outcomes. This approach is particularly crucial because we know so little about the system. An adaptive management approach can ensure that strategies are changed quickly if undesired outcomes occur, which is especially important if the interventions involve control of top native carnivores. Target densities of native prey and predators and hypotheses about how the system functions can be adjusted as we learn more about each particular site and the system as a whole.

Our overall conclusion based on the Patagonian example is that humans can have a strong effect on the role that top carnivores play in the maintenance of biodiversity. Several studies have shown that major human-induced alterations can trigger the negative effects of top predators. These effects include not only limiting or reducing the abundance of certain prey but also causing trophic cascades in which the negative effects spread toward other components of food webs through complex pathways (Estes et al. 1998; Roemer et al. 2002). In some cases, the main human alteration was the reduction of the prey populations to densities so low that they became limited or even threatened by predation (Gasaway et al. 1992; Wehausen 1996; Sweitzer et al. 1997), while in another case it may have been the reduction of the predator's main prey (Estes et al. 1998). In the study by Roemer et al. (2002), the human alteration was the introduction of feral pigs that supplemented the food base of a top predator, the golden eagle (*Aquila chrysaetos*). The novel feature of the Patagonian case, unfortunately, is that the negative role of top predators on biodiversity was likely triggered by a combination of direct reductions in prey numbers by competition and hunting and supplementation of the predator food base by the introduction of exotic species. This perverse combination, though informative from the perspective of predator–prey interactions, creates unusual challenges for the conservation of Patagonian wildlife.

Concluding Thoughts: How Unique Is the Patagonian Example?

A final point should be borne in mind when designing research and conservation strategies for faunal assemblages of Patagonia and other areas where the role of top predators has been greatly altered. The dramatic effects that humans have had on native herbivores and on the role of carnivores in top-down control processes in Patagonia and in other extant ecosystems may not be unusual events in a historical perspective. The catastrophic extinctions of megaherbivores that occurred during the Pleistocene (Martin and Klein 1984) probably had effects on the functioning of ecosystems comparable to those produced by the recent collapse of native herbivores in Patagonia. The Pleistocene extinction of megaherbivores, which were mostly free from the limiting effects of predators due to their large size, may have increased the prevalence of top-down regulation of predator–prey interactions among large terrestrial vertebrates. Persistent hunting by humans on remaining herbivores through millennia likely increased the susceptibility of their populations to control by predators, as it appears to have done in Patagonia in the last century. In recent centuries, the extirpation or drastic decline of top predators from most of the earth's land masses has provided yet another stage in this dramatic play, removing predator controls and leading in some cases to overabundant herbivores that are severely depleting plant resources (McShea, this volume; Terborgh, this volume). Therefore, the relative roles of top predators may have shifted more than once since humans began to exert their influence on terrestrial ecosystems. The significance of top predators in limiting their prey and maintaining biodiversity may depend on the effects that humans have had on the particular system being observed.

Summary

In Patagonia, populations of native herbivores and omnivores declined dramatically during the last century, due to persecution by humans, competition from livestock and exotic species, and habitat degradation caused mostly by overgrazing of sheep. In spite of reduced levels of both hunting by humans and sheep numbers over the last 20 years, populations of native guanacos, choiques, and mountain vizcachas have not recovered concomitantly. Top native carnivores, the puma and the culpeo, on the other hand, have increased in abundance and expanded their ranges, probably due to the large prey base of exotic wildlife, mostly European hares and red deer, that have replaced the native prey. The results presented in this chapter strongly suggest that predation by two top native carnivores may prevent the recovery of native prey in parts of Patagonia. Comparisons of puma consumption of guanacos and guanaco population trends among sites, in particular, suggest puma predation is keeping guanaco populations at very low densities. We argue that human activities have induced top-down control of native prey by large carnivores in parts of Patagonia, likely as a result of apparent competition from nonnative herbivores. This alteration in the role of native carnivores may help prevent the recovery of the once abundant herbivores and locally reduce their diversity. Such a role change for carnivores may not be unusual because modifications of terrestrial ecosystems induced by humans throughout the world in the past millennia may have altered repeatedly the limiting role of top carnivores on their prey.

ACKNOWLEDGMENTS

We thank A. Gonzalez, P. Carmanchahi, O. Monsalvo, G. Sánchez, O. Pailacura, and M. Funes of CEAN for assistance in the field. Discussions with M. Anz and comments from W. McShea, A. Treves, M. Funes, and the book's editors helped us improve the manuscript. Financial support was provided by the Argentine Research Council (CONICET) and Science Agency (SETCIP) and the Wildlife Conservation Society.

Achieving Conservation and Management Goals through Focus on Large Carnivorous Animals

As some of the most charismatic members of a given community, large predators are symbols in the efforts to save species. As a result, they frequently receive inordinate conservation attention. Because of their direct competition with humans, their conservation may also involve extraordinary expense—social, political, and financial. This concentration of effort and resources may be justified in some cases if conserving large carnivores also conserves all elements of biodiversity. So far in this volume we have seen how, in some cases, removal of these top trophic levels can have profound consequences for the species composition of ecosystems as well as the way in which ecosystems function. In other cases, however, there have been no discernible impacts from the loss of large carnivores, and they appear to be relatively unimportant in the structure and function of biotic communities. In yet other instances, though top predators can be demonstrated to have an important functional role, the research provides no clear prescriptions for achieving conservation goals.

The five chapters in this section pay particular attention to the practical applications that may be derived from the science of understanding carnivory. The authors emphasize divergent themes and perspectives that relate to both conservation and management. David Maehr and coauthors (Chapter 15) launch the section by providing a clear example of a case where revisiting a long-term dataset collected for other purposes (i.e., species-specific biology) can be analyzed to provide valuable clues to broader questions regarding the relationship between Florida panthers, black bears, wolves, and broader biodiversity. Dr. Maehr and colleagues also use this opportunity to discuss how the role of top predators in shaping biotic communities in Florida, USA, can help the use of wide-ranging carnivores as conservation flagships. Joel Berger (Chapter 16) picks up the theme of functional redundancy examined in previous sections—this time focusing on the extent to which humans can substitute for large predators, when the latter have been removed. The relationship between hunting by humans and hunting by large carnivores is a complex problem with multiple facets and often becomes a central issue in conservation and management planning.

A prerequisite to answering the broader question of whether large carnivores are essential to the maintenance of biodiversity, is to assess to what degree food webs that enmesh them are regulated by top-down processes. If top-down control is strong, any changes in carnivore density, distribution, and behavior could cause profound disruption in an ecosystem. Terry Bowyer and coauthors (Chapter 17) make a compelling case for assessing the degree to which food webs are regulated by top-down or bottom-up processes in order to achieve both management and biodiversity conservation goals. They provide a prey-centric conceptual framework to aid in evaluating at what point in the top-down–bottom-up continuum a given predator–prey system is situated.

The final two chapters in this section provide insightful analyses of the rationale behind the use of top carnivores in conservation programs from two contrasting regions of the world. Stan Boutin (Chapter 18) explores the scientific evidence assessing the importance of top carnivores to boreal forest structure and function and, ultimately, to biodiversity maintenance. He then takes this an important step further to examine what this says about the utility of carnivores as central components of boreal biodiversity conservation programs. John Linnell and his colleagues from the Large Carnivore Initiative for Europe (Chapter 19) then turn to the heavily human-dominated European landscapes, where they declare that saving the continent's carnivores will not conserve biodiversity because the human influence is too substantial to allow functional population densities to be attained. From there, Dr. Linnell and coauthors focus their attention on the myriad conservation benefits that can be generated by working to restore top carnivores across the European continent.

Large Carnivores, Herbivores, and Omnivores in South Florida: An Evolutionary Approach to Conserving Landscapes and Biodiversity

David S. Maehr, Michael A. Orlando, and John J. Cox

Ecologists and wildlife managers are frequently challenged to justify their claims that large carnivores are valuable in ecosystems (Noss 2001). Equally frequently we resort to defenses such as the need to "keep every cog and wheel" (Leopold 1993: 146) and the value of wilderness experiences (Noss 2001), or we advocate ethical responsibility (Leopold 1949). Only some of the time do these arguments carry weight, if at all. However, there is increasing evidence that large carnivores are extremely important in shaping biotic communities, but the public, many hunters, politicians, and a surprising number of wildlife managers still find this difficult to accept. A recent debate in the *Bugle,* the official magazine of the Rocky Mountain Elk Foundation (RMEF), following a published interview with Michael Soulé, exemplified the disconnect among science, policy, and the public when it comes to restoring landscapes including large carnivores (Petersen 2002). Despite the fact that many of the objectives of RMEF pertain more to ecosystem and landscape restoration than to elk (*Cervus elaphus*) (and thus, very similar to the goals of many conservation organizations), the paying membership of this nonprofit organization is divided on the ecological and aesthetic benefits of large carnivores. The subsequent issue (November/December) contained a panoply of vitriolic editorials that condemned the magazine and its organization for embracing government landgrabs, siding with ecological radicals, being absurd, and being too idealistic. The unspoken message was clear: We like our elk just fine without the complication of big predators running around. Although the sport-hunting community has generally supported a wide variety of environmental issues over the decades, its anxiety over this one suggests that we have a long way to go before carnivore conservation becomes mainstream—especially if the scientific community has difficulty in communicating the ecological roles of large predators. The documentation and communication of the functions of carnivory will become increasingly important as efforts to restore large predators such as the cougar (*Puma concolor*) become more commonplace.

Recently, one of us observed a disturbing trend among wildlife management agencies in the United States, one in which the senior author was certainly a participant during a 14-year tenure as a state wildlife agency biologist. It is a phenomenon that can be termed the "get real syndrome"—a condition that accepts the status quo and that denies a suite of alternate futures involving landscape-scale ecological restoration, including the return of extirpated large mammals (Maehr 2001). This tendency is partly a function of wildlife agency focus on "real world" management questions that precludes the luxury of considering the evolutionary history and consequences of management actions, despite Leopold's (1949) repeated admonitions more than half a century ago. This "real world" focus we claim as the explanation for eschewing such basic yet critical-to-management science despite a wealth of single-species studies that span continents and centuries.

Several studies have indicated that large solitary cats have little impact on their prey populations (Hornocker 1970; Seidensticker et al. 1973; Sunquist 1981; Bailey 1993). Maehr et al. (2001a) suggested that the black bear (*Ursus americanus*) is even less influential than the Florida panther (*Puma concolor coryi*) or bobcat (*Lynx rufus*) with respect to landscape and evolutionary processes. Whereas both of these solitary felids base their use of space primarily on the availability of mobile prey, the black bear depends on the availability of sessile food resources and is a habitat generalist, using forested habitats in proportion to their availability (Maehr 1997a). Furthermore, the south Florida bear diet can be so diverse (at least 50 species) that no habitat specialization is needed for a local population to survive. A comparison between western and eastern North American black bear diets would reveal little similarity. In Florida, black bear foods vary considerably even between locations at similar latitudes (Maehr and Brady 1984). No special relations between the bear and its food have been discerned, and there are no foods that can universally explain the distribution of the species in North America. It might ap-

pear, therefore, that there is little compelling evidence to further examine potential relations between these species and the landscape.

In this chapter, we reexamine archives of data from southwest Florida studies on white-tailed deer (*Odocoileus virginianus*), the endangered Florida panther,
and the bobcat, and revisit field notes on North America's southeasternmost black
bear population. Specifically, how might south Florida predator–prey relations differ in the absence of the recently extirpated red wolf (*Canis rufus*), a potential deer
predator of relatively open terrain (Young 1946)? Does a remnant population of
panther affect the spatial patterns of white-tailed deer and its close relative, the
bobcat? And how might the naturalized coyote (*Canis latrans*), a relative newcomer
to this part of the world, influence panther, bobcat, and deer? How does the black
bear fit into this picture—is it too much of an ecological bumbler (Maehr 2001)
with food habits too general to exert any kind of top-down influence on this subtropical landscape? Might any of these interspecies relations translate into the hon-
ing of new or age-old evolutionary strategies? Finally, we examine the challenges
that such evolution-oriented thinking might present to land stewards and wildlife
managers in the only region in eastern North America where white-tailed deer,
black bear, bobcat, coyote, and cougar are known to coexist.

Study Area

South Florida is a flat, subtropical landscape that is experiencing rapid human population growth, particularly in coastal areas. Much of the area ($> 15,000 \text{ km}^2$) has
been set aside in some type of conservation designation (Maehr 1990, 1997b). The
Bear Island Unit of the Big Cypress National Preserve (BCNP) covers approximately 150 km² in Collier County, Florida, and is adjacent to other public lands
and private ranches (Fig. 15.1). Plant communities consist of a network of pine
flatwoods dominated by slash pine (*Pinus elliottii* var. *densa*) and saw palmetto
(*Serenoa repens*); hardwood hammock with live oak (*Quercus virginianus*) and sabal
palm (*Sabal palmetto*); open herbaceous habitats including inland marshes and
prairies dominated by sedges, grasses, and emergents; and shrub communities
with wax myrtle (*Myrica cerifera*) and willow (*Salix caroliniana*). Outside of this core

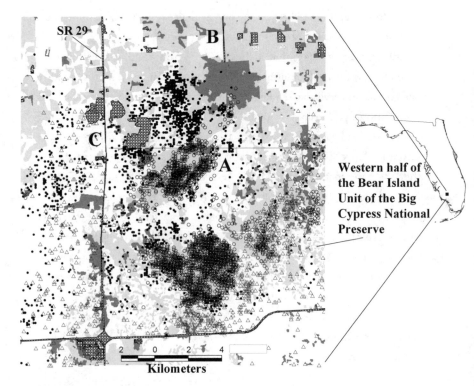

Figure 15.1

Location of the 400 km² study area in western Bear Island (A) in south Florida. To the north of Bear Island (dashed line) are private lands (B). Land to the west of State Road 29 is the Florida Panther National Wildlife Refuge (C). General habitat types include forest (light gray), open herbaceous (dark gray), human-dominated including open water (waffle texture), and shrub and brush (white). Symbols represent radiotelemetry locations of white-tailed deer (○), bobcats (△), and panthers (●) collected between 1986 and 1991.

of public reserves land uses range from citrus and row crops and low-density human residences to sprawling urban complexes with expanding networks of highways and other infrastructure. Most people in this region live near the coasts of the Atlantic Ocean and the Gulf of Mexico. Additional ecological and landscape attributes of the Big Cypress region can be found in Duever et al. (1986).

Anthropogenic influences that border on or are part of the study area include Interstate Highway 75 to the south, State Highway 29 to the west, a limestone quarry in the north of the study area, and limited cropland. As part of a national monument, limited hunting—administered by the Florida Fish and Wildlife Conservation Commission (FWC)—is permitted in Bear Island for a variety of species including white-tailed deer, waterfowl, wild turkey (*Melagris gallopavo*), wild hog (*Sus scrofa*), and small game. We chose a 400 km² area with the Bear Island Unit at its center because this was the site of a previous deer mortality study that was a part of earlier Florida panther investigations (Land 1991; Land et al. 1993). Although the wild hog is regularly consumed by the panther in higher-quality habitats (Maehr et al. 1990a) and by the bobcat in Bear Island (Land et al. 1993), virtually nothing is known about hog spatial or demographic characteristics in south Florida. The area of relevance for our examination of black bear ecology encompasses much of Florida because of the wider distributions of the species involved, but field observations were made in the Big Cypress region of south Florida.

The Bear and the Weevil

We examined black bear food habits data from south Florida (Maehr 1997a), field notes of the senior author, and life history information on the giant palm weevil (*Rhynchophorus cruentatus*) (Oliveria et al. 1989; Weissling and Giblin-Davis 1994; Giblin-Davis et al. 1996; Vanderbilt et al. 1998; Hunsberger et al. 2000) to investigate a potential mutualistic relation between the two species. We suspected that weevil populations inhabiting areas occupied by palms and black bear obtained feeding and breeding sites in damaged tissues created by bears consuming apical meristems (hearts) of sabal palm and saw palmetto. Black bears may benefit if return visits are made to damaged palms that were previously fed upon by a bear and subsequently colonized by the aggregating beetles. We hypothesized that bears are more important than stochastic disturbance forces in the creation of weevil breeding and feeding habitat.

The giant palm weevil is one of more than a dozen insect species consumed by the black bear in south Florida (Maehr 1997a). The larvae of the genus *Rhynchophorus* are among the richest sources of animal fat and are nutritionally important to aboriginal cultures in South America (Dufour 1987; Giblin-Davis 2001). *Rhynchophorus* seeks out damaged palms for mating aggregations, egg laying, pupal

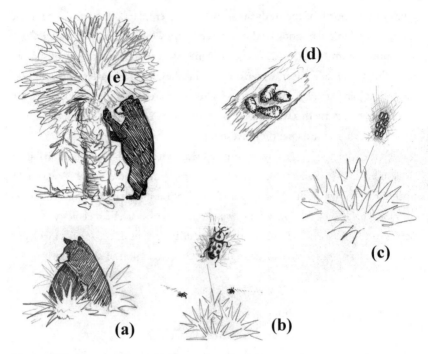

Figure 15.2

The giant palm weevil life cycle begins with the colonization of a palm damaged by a bear (a), then is followed by courtship (b), egg-laying (c), pupation (d), and metamorphosis into adults. These aggregations may be subsequently consumed by a black bear (e) that may have created the initial damage (a) by feeding on the apical meristem of a sabal palm or saw palmetto (either palm species can serve as the target for bear feeding and weevil reproduction). This tri-trophic, mutualistic relationship is likely a modern analog to ancient coevolutionary assemblages.

development, and maturation (Giblin-Davis 2001). Females can lay up to 400 eggs, with larvae growing to a length of about 4 cm (Fig. 15.2). Its high nutritive value and breeding aggregations make it similar to other colonial insects such as ants, termites, and wasps that are also important in the Florida black bear diet (Maehr 1997a). Although the giant palm weevil is widely distributed in the southeastern coastal plain, it has not been reported elsewhere in bear diets (Maehr and DeFazio 1985). It may be the tropical climate and abundance of palms in south Florida that help to explain the relative commonness of the weevil in this region.

We first became aware of the ability of male weevils to find damaged palm tissues during a south Florida field course that was instructed by the senior author in May of 1996. After extracting the heart of a sabal palm with an axe and machete at a Big Cypress Swamp hunting camp (to provide the prime ingredient in a traditional hunter's casserole), it was placed in a metal pan on a picnic table in a clearing surrounded by slash pine flatwoods and cypress (*Taxodium distichum*) swamp. Within minutes, a male giant palm weevil landed on the extracted heart after a loud flight of indeterminate distance. Our curiosity piqued by this unexpected visitation, we returned to the tree from which the heart had been extracted just hours earlier and discovered several adult weevils on the newly damaged, and exposed, internal palm tissues.

It is the need for *Rhynchophorus* to colonize damaged palms that sets it apart from other colonial insects that are eaten by the black bear. For instance, yellow jackets (*Vespula* spp.) will build nests in the ground, carpenter ants (*Campanotus floridanus*), bessie bugs (*Odontotaenius disjunctus*), and termites colonize rotting wood, while giant walking sticks (*Anisomorpha buprestoides*) aggregate on low-growing shrubs. This leads us to suspect that sabal palm and saw palmetto evolved with herbivores such as ground sloths (*Megalonyx* spp.), glyptodonts (*Glyptotherium* spp.), peccaries (*Mylohyus* spp., *Platygonus* spp.), llamas (*Hemiauchenia* spp.), and horses (*Equus* spp.) among others (see Kurten and Anderson 1980), and that these species were the driving force in creating the giant palm weevil niche. Such evolutionary history between weevil and herbivores may date back to the middle Eocene and browsing by extinct tapirs and horses (Taylor and Taylor 1993). The giant palm weevil and its Curculionid ancestors likely benefited from a wide range of potential palm eaters. The palms, generally noted for their elaborate defenses against herbivory (Uhl and Dransfield 1987), in turn developed anatomical strategies to reduce damage to the apical meristem—sabal palm outgrows its tormentors by rapidly growing to 30 m in height; whereas the low-growing saw palmetto exhibits razor-sharp petioles that surround the otherwise vulnerable terminal bud. Saw palmetto is also one of the few palms that can respond to a damaged apical meristem by sympodial branching from axillary buds. Prior to the wave of large mammal extinctions during the Pleistocene, herbivorized palms were likely commonplace and easily found by palm weevils via released volatile chemicals that indicated plant stress and damage (Giblin-Davis 2001). Today, other than tool-wielding humans, the black bear is likely the primary native agent of palm damage and creator of weevil breeding habitat. Although tropical storms and sea-level rise might also cause stress in palms, such influences do not occur predictably nor throughout the range of the weevil.

But the carnivore–insect relation does not end here. Aside from the benefits gained by a black bear from a carbohydrate-rich meal of palm heart that also facilitates weevil reproduction, bear scats often contain the remains of lipid-rich weevil larvae (Maehr 1997a). This suggests that bears revisit previously damaged palms and are rewarded with a high-fat meal. While we do not suggest that bears understand the cause-and-effect implications of their predilection for heart of palm, we do believe that, like gray squirrels (*Sciurus carolinensis*) forgetting the locations of buried acorns and hickory nuts, not all bear-damaged and weevil-colonized palms are revisited. Thus the loss of scattered weevil aggregations due to bear predation becomes offset by the creation of more weevil breeding habitat than bears can take advantage of. In the absence of damage caused by black bears, giant palm weevils must depend on stochastic events such as wind storms and lightning for the creation of stressed and damaged palms. In this sense, the black bear is the ecological replacement for perhaps dozens of palm heart predators from the Pleistocene and beyond.

The details of the relationship between *Rhynchophorus* and the black bear are the product of speculation based on anecdotes and systematic observations such as food habits analysis. For example, we have not seen weevils colonize a palm that was recently damaged by a black bear. We fully recognize the limitations of this approach. The existence of a symbiotic relation between bear and weevil could be tested experimentally, however, in forests with and without black bear. Weevils could be censused by using synthetic aggregation pheromones (cruentol) as a lure (Vanderbilt et al. 1998). If, indeed, the giant palm weevil does benefit from bear-induced palm damage, we would expect to see fewer weevils in forests that have lost the black bear through habitat fragmentation or other disturbances. Regardless, the interaction between bear, weevil, and palm succession (a pulled heart almost always results in the death of the tree) is one that deserves additional study. At the least, the black bear may be an important factor in maintaining abundance and distribution patterns of North America's largest weevil and its obligate palm breeding sites.

Felid Predators and Deer

We compared habitat characteristics of home ranges inhabited by radio-collared female deer, bobcats, and Florida panthers based on aerial telemetry locations collected from 1986 to 1991. Data were derived from studies conducted previously by the senior author and associates in southwest Florida (Land 1991; Foster 1992; Maehr 1997a). Details of capture and tracking can be found in Land et al. (1993), McCown et al. (1990), and Maehr et al. (1991). We constructed 95% minimum convex polygons for deer and bobcats that had at least 40 radio locations and for those portions of Florida panther home ranges that overlapped with the study area (individual panther home ranges can exceed the size of the study area). We also compared the distribution of deer home ranges relative to areas used by bobcats and panthers. We hypothesized that panthers influence the spatial arrangement of female white-tailed deer, and that this would be apparent in habitat use characteristics and geographic separation between species. We also hypothesized that panther use of forest would affect the distribution of the bobcat in Bear Island, and that habitat use between bobcat and deer would be more similar than between bobcat and panther. Home ranges, telemetry locations, and land cover were depicted and analyzed using ArcView (ESRI, Redlands, CA) and associated data management extensions. We combined land cover characteristics into five general categories: forest, nonforested herbaceous (open), shrub and brushland, open water, and human-dominated, including highways, surface mines, and agricultural land. These categories reflect general structural characteristics of the landscape and the panther's preference for forests (Belden et al. 1988; Maehr et al. 1991; Maehr and Cox 1995; Fuller and Kittredge 1996; Maehr 1997a; Maehr et al. 2001b).

We analyzed home range and habitat data based on 3628 radio locations from 41 female white-tailed deer in the Bear Island Unit of BCNP, collected from September 1986 through July 1991 (Land 1991). We also examined habitat use and distribution of five radio-collared panthers ($n = 2831$ radio locations) and 24 bobcats ($n = 977$ radio locations) that utilized this area during the study period (see Fig. 15.1). Small sample sizes (< 40 locations) reduced the bobcat home range analysis to six animals. Home ranges of deer were composed primarily of herbaceous vegetation (66%) and forest (23%) (Fig. 15.3a). Deer habitat use averaged 42% (sd = 22) in herbaceous, 38% (sd = 19) in forest, and 20% (sd = 17) in shrub habitats

(Fig. 15.3b). Those portions of panther home ranges in Bear Island were made up primarily of forest (58%) and herbaceous vegetation (30%) (see Fig. 15.3a). Panther habitat use, on the other hand, averaged 1% (sd = 1) in herbaceous habitats, 3% (sd = 1) in shrub habitat, and 96% (sd = 2) in forest (see Fig. 15.3b). With respect to available habitat in Bear Island (65% forest, 32% herbaceous), both deer and panther used habitat differently than what was available, with the former preferring open habitats and avoiding forests. Panthers exhibited an opposite tendency. Land cover in the six bobcat home ranges was dominated by forest (54%, sd = 21) and herbaceous cover (28%, sd = 11) (see Fig. 15.3a). Due to the small sample sizes for individual bobcats we combined habitat use data and presented tendencies as simple means of proportions. Like panthers, bobcats also tended to use forest cover (86%), but they were located primarily in smaller patches of woody vegetation and shrub habitat (13%) relative to large forested patches in which panthers were usually found (see Fig. 15.3b). Panthers tended to use shrub habitat less than it was available in Bear Island, whereas deer and bobcats exhibited a trend toward the selection of this cover type.

Panther, bobcat, and deer exhibited varying habitat use strategies in the Bear Island mosaic. Panther and bobcat locations were predominantly associated with forest cover—almost to the exclusion of herbaceous cover. At least for one of these species we doubt that this was an artifact of using only daytime telemetry data, because the crepuscular panther remains active well into daylight hours (Maehr et al. 1990b, Maehr et al. 2004). Interestingly, although the white-tailed deer is considered one of its most important prey (Maehr et al. 1990a), only four study animals died due to panther predation versus seven killed by bobcat (Land 1991). Although the deer-capture methodology (i.e., the capture of deer in open areas with netgun and helicopter; Land 1991) may have biased the sample toward a segment of the population that disdains forest cover, these deer appear to have reduced the likelihood of death by panther. Regardless, if panthers selectively hunt in forest, then deer that reside primarily in open habitats will enjoy a measure of immunity to panther predation. Given a statewide tendency to utilize small prey (Maehr and Brady 1986), we believe that bobcats in Bear Island were opportunistic predators of deer thereby avoiding interference competition (see Dalrymple and Bass 1996; Palomares and Caro 1999) from panthers by utilizing patchier forests and open habitats. As a corollary, forest deer may enjoy reduced predation by bobcats.

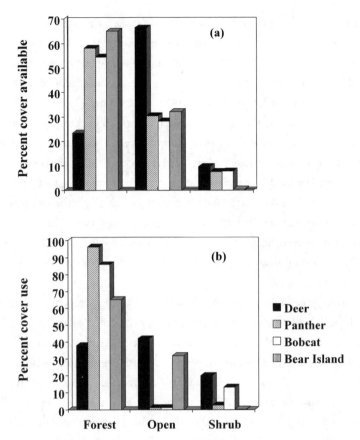

Figure 15.3

Mean home range composition (a) and mean habitat use (b) among white-tailed
deer, Florida panther, and bobcat in the Bear Island area. Habitats included are those
that accounted for the majority of the study area and cover types used by these
three species. "Open" refers to unforested herbaceous communities.

Furthermore, the tendency for bobcats and deer to use shrub habitat—a cover type
in which panthers were not commonly found—may be the result of the smaller
cat taking advantage of deer that seek out this cover type (Fig. 15.3a,b). Thus different kinds of selection may be operating on two demes within the Bear Island
deer herd—panther predation avoidance through habitat selection among open-
living deer, and panther predation avoidance through more direct behavioral
responses of forest-dwelling deer to the ambush tactics of a stealth hunter (the

panther). Prey species can reduce predation by stalkers through the avoidance of dense cover (Bailey 1993), a trait that may be a cultural artifact of learning by young from adults (Schaller 1972), and through nonfatal encounters with a predator (Maehr et al. 2001a). It is often the young, inexperienced individuals that succumb to ambush predation (Schaller 1967). Interestingly, because this system contains two predators that appear quite capable of killing the largest native ungulate in the area, selection may be operating in such a way that the evolution of antipredator responses in deer (at least with respect to habitat selection) may have been retarded by the differential habitat use of the bobcat and panther. In the sense of Bertram (1979: 232), Bear Island's native cat species may be benefiting one another by preventing the white-tailed deer from evolving escape behavior and adaptations that are completely effective against both cat species. Regardless, this may well be the most predator-conscious white-tailed deer population in the southeastern United States.

Although open-living deer may be actively avoiding potential panther predation through habitat selection, we did not have a comparable data set from forest-living deer in Bear Island to test this possibility. Regardless, deer in Osceola National Forest, an area without the panther in north Florida, used forested habitats virtually to the exclusion of open habitats—clearcuts and other open areas were avoided throughout the year (Kilgo et al. 1998). In this case, deer were more likely to inhabit forests to avoid human disturbance and hunting that occurs primarily during autumn. In the Everglades, an area inhabited by the panther in southeastern Florida, female white-tailed deer preferentially used open, wet prairie habitat despite the higher fawn mortality that may have been a function of periodic flooding of these herbaceous, lowland habitats (Smith et al. 1996). Thus, in the former example, antipredator vigilance may have been reduced due to the absence of a large predator and only seasonal human disturbance, whereas in the latter case, deer opted for suboptimal fawn-rearing habitat in the presence of the panther. The distribution of radiotelemetry data in Bear Island supports our hypothesis that the panther may have a shaping influence on deer habitat use. The effect of the panther on the bobcat was less clear, but the latter species tended to utilize areas that were characterized by natural patchworks of forest islands and herbaceous cover, rather than the more continuous forest habitat used by the panther. Whereas the bobcat may be more successful in avoiding direct encounters with panthers in this patchier habitat, it is also more likely to encounter the cotton rat (*Sigmodon hispidus*), one of its staple food items in Florida (Maehr and Brady 1986). Clearly, however, the unexpectedly high number of deer kills made by bobcats suggests that this smaller cat is an important ungulate predator in the system.

In south Florida landscapes, the presence of the panther and other predators is certainly exerting selective tension on prey species, forcing them to be alert, swift, and responsive to different kinds of predator attack. In Bear Island there are approximately 30 deer killed by hunters annually, versus up to 150 killed by panthers (based on an average of one deer killed per week per panther, and the equivalent of three fulltime panthers living in this area; J. Bozzo, and D. Land, FWC, pers. comm. 2004). The total take by the bobcat in Bear Island is speculative, but elsewhere in the Big Cypress region this species is notably adept at killing deer (Labisky and Boulay 1998) and may locally exceed the panther in total numbers killed. It is likely that all of south Florida's large carnivores prey on the wild hog to some degree. However, its role in shaping this predator–prey system awaits future study. Regardless, the hog's tendency to be a plodding, relatively noisy forager may make it equally vulnerable to ambush and coursing predators alike. Given the lower harvest and seasonal nature of human hunting in Bear Island, native predators are clearly the greatest constant threat to local deer. Elsewhere in the southeastern United States, deer inhabit areas containing a much narrower set of potential attackers. We believe that the panther is the primary modern influence on the spatial arrangement of white-tailed deer in south Florida, and that this felid also exerts an influence on local bobcat populations (which, in turn, influence deer as well). The involvement of the black bear in this multicarnivore assemblage is debatable at best, but we wonder if the unusual diurnal activities of this population (Maehr 1997a) may have been partly influenced by the nocturnal tendencies of the two cat species with which it shares the same landscape. The cost for a deer living in the Bear Island forest is to be constantly vigilant against predation by one of North America's most proficient hunters of large ungulate prey (Hornocker 1970). The cost of an open-living lifestyle in Bear Island is to withstand other forms of predation such as opportunistic bobcats that benefit from the avoidance of nonforested habitats by the panther, and threats associated with living in wetlands such as the alligator (*Alligator mississippiensis*) (Land 1991) as well as the physiological/pathological stresses of living in an aquatic habitat (Brokx 1984; Smith et al. 1996).

Although the top-down influences of large carnivores have mostly been presented as theoretical constructs with little empirical demonstration (Polis and Strong 1996), it has been suggested that the force of predation has long been underestimated (Terborgh 1990). It has been nearly three decades since Mech (1977) suggested that the spatial distribution of large carnivore populations directly influences the distribution of their prey. In Mech's Minnesota example, the buffer zones between wolf (*Canis lupus*) packs became refugia for white-tailed deer. The combination of mutual pack avoidance and the maintenance of a source population of deer facilitated the persistence of this predator–prey system. A similar relation between warring native American tribes also created virtual sanctuaries for deer (Hickerson 1965; Mech 1977). Mech (1977: 321) concluded that "such a possible evolutionary strategy of a prey species—taking advantage of the spatial organization of predators to provide greater security—should be sought in other predator–prey systems." There appears to be a degree of ecological separation among large mammals in Bear Island, and perhaps all of south Florida, that is even more complex than the boreal, wolf-dominated systems just mentioned.

Wolves and Big Cypress Deer

Bartram's (1996) 18th-century descriptions of the red wolf as a killer of large prey, as well as archaeological records (Nowak 2002), suggest that this species occurred throughout Florida, and that its ecological services and evolutionary tensions have been absent from the state for nearly a century. Habitat use patterns of white-tailed deer and their felid predators in Florida and other areas of the southeastern United States have certainly been altered by the regional extirpation of wolves.

Virtually nothing is known about the red wolf diet in Florida, although both white-tailed deer and the feral hog were taken to some extent elsewhere (Young and Goldman 1944). Generally, however, it is believed that the red wolf consumed smaller prey than was taken by the larger gray wolf (Paradiso and Nowak 1972). White-tailed deer rely upon individual and group tactics to escape predation (Mech 1984), with sociality and familiarity of escape terrain being important antipredation strategies (Nelson and Mech 1981). Where panther and red wolf once coexisted, deer coevolved to deal with both the ambush strategy of the panther and the cursorial team tactics of the red wolf (Kleiman and Eisnberg 1973). The simultaneous occurrence of these predators must have influenced habitat use, movements, social structure, and behavior of deer, although in different ways. Contemporary Florida panthers prefer forest habitat whereas wolves were likely associated, or at least active, in open areas of the landscape (Carley 1979), thus creating predatory tension that would have kept deer vigilant in both major habitat types. The exertion of this predatory "rebound effect" on deer may have altered herd movements that could have temporally limited use of preferred habitat, resulting in shifts to denser or swampier areas that may have provided better security against both predators. These may have become areas where the opportunistic bobcat now exerts an additional predatory influence on the deer population—primarily in the patchier networks of dense woody and herbaceous cover that the panther tends to avoid. Even if the bobcat and red wolf consumed similar prey, we suspect that the solitary bobcat would have been at a competitive disadvantage. The modern arrangement of bobcat and panther may be helpful in understanding how the red wolf fit into this system with little interference competition, especially if the panther uses available habitat in the same way that it did a century ago. Perhaps today, the bobcat has expanded its role as a predator in the absence of open-hunting wolves. It is also possible that panther and bobcat are more numerous in modern south Florida due to the absence of predation, kleptoparasitism, and interference competition that wolves would have caused (Fig. 15.4).

The establishment of the coyote in south Florida over the last 20 years (Maehr et al. 1996) has reentangled predator–prey relations in the Big Cypress Swamp. Arguably, the panther may have had more of an impact on south Florida deer than did the red wolf, but the ecologically similar coyote has a greater potential to overlap in diet with south Florida's native carnivores than they currently overlap with each other (Maehr 1997a). Regardless, whereas the white-tailed deer coevolved with the panther, wolves, due to their more recent colonization of North America, represented a relatively "novel" predatory threat. In addition, Florida panthers have large home ranges (> 400 km^2) (Maehr 1997a) as compared to red wolves in similar habitat (45–150 km^2) (Riley and McBride 1972; Shaw 1975). The difference in home range size, coupled with the fact that wolf packs have more individuals patrolling a given territory, suggest that the density of wolves in the southeastern

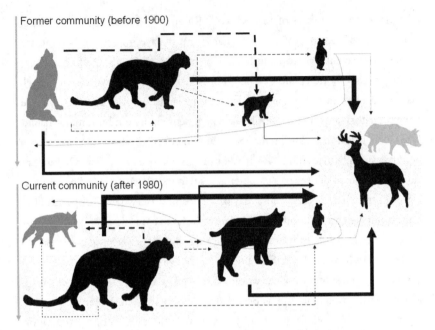

Figure 15.4

The former and current relations among south Florida's large mammals have consistently included predator–prey and interference competition among the white-tailed deer, Florida panther, bobcat, and black bear. Solid lines represent direct predation on deer, dashed lines represent interference competition. The thickness of each line suggests the relative strength of the interaction relative to other pathways. The dotted line associated with the black bear represents its omnivorous diet and occasional scavenging of deer. The half-tone profiles of the red wolf (above), coyote, and wild hog suggest the previous role of the former and equivocal roles of the latter two.

United States may have been higher than that of panthers. Thus their impact in a given locale certainly had the potential to exceed that of a resident panther. At the very least, the coyote has likely returned a predatory tension that has been missing for more than a century, and once again the white-tailed deer must be vigilant against predation from two felids and one canid.

The absence of the gray wolf and grizzly bear (*Ursus arctos*) from all or parts of the Greater Yellowstone Ecosystem has been implicated in increased herbivory on riparian vegetation and reduced songbird populations (Kay 1985; Berger et al. 2001a), despite the uninterrupted presence of a resident cougar population (Murphy 1998). The recent reintroduction of the wolf to Yellowstone has had profound implications on the park's ungulates, including increased vigilance (Berger et al. 2001b), and changes in habitat use patterns (Ripple et al. 2001). This has resulted in the recovery of vegetation that had previously been overbrowsed as the result of moose (*Alces alces*) populations that were unaffected by wolves (Berger et al. 2001a). Thus predator restoration and management can provide measurable benefits to landscapes. In large part, these functions have not been lost in south Florida.

Despite their modern top-down effects, the constraints imposed by bobcat, panther, and coyote predation on Big Cypress deer are only a subset of those exerted by extinct specialist carnivores and herbivore competitors that lived during the Pleistocene. It was only after the megafaunal extinctions that white-tailed deer became abundant and expanded their distribution in North America (Geist 1998). As with the black bear and its role in damaging palms and creating weevil habitat, the influence of today's Big Cypress carnivore community is likely only a fraction of what was done by the mammalian giants of the Pleistocene. Subsequently, the work that modern carnivores do in south Florida is of greater importance to ecosystem processes and evolutionary trajectories than at any time in the past, and at any time in these species' histories.

The Challenge to Managers

North America's large carnivores are survivors of the Pleistocene extinctions that coincided with human colonization and dramatic climate change (Marshall 1984). As a result, an array of evolutionary relations among species was lost in a geological bat of an eye. Regardless of the realities and complexities of deer, weevil, and predator relations in south Florida, the fact that this large mammal community exists at all is remarkable and unique in eastern North America. For this reason alone, it deserves special recognition and proactive stewardship. South Florida is also unique in North America because palms are at the ecological center of many upland and wetland communities (Simpson 1920; Craighead 1971; Brown et al. 1990). They are also at the center of large carnivore natural history—bears both den and feed in palms, and panthers seek out saw palmetto as preferred natal cover (Maehr et al. 1990b; Maehr 1997a). Panthers use it as stalking cover. The ubiquity of south Florida palms is certainly a product of a subtropical climate (Nelson 1994; Stevenson 1996), but also an accident of continental geography and plate tectonics, with every other land mass at the same latitude a desert (Ewel 1990). South Florida's position at the terminus of a peninsula in warm water and awash with humid air has allowed this part of the world to retain many ancient relics of long-vanished biota. Though one might argue that south Florida's Pleistocene throwbacks are merely trivial reminders of mostly extinct mammalian assemblages, it is the best that almost 10,000 km^2 of former *Smilodon* and *Glyptotherium* range can offer. In our view, the manager's challenge is in maintaining the evolutionary tensions that have so dramatically relaxed in the last 10 millennia and in using south Florida's large carnivores as conservation flagships to do so.

As Maehr et al. (2001b) suggested, the Florida panther embodies many of the attributes of the quintessential flagship species. It is wide ranging, it offers direct and indirect ecological services, it is a conservation priority among natural resource agencies, it is widely supported by the general public, and it has the potential for population growth and range expansion. This latter point is important. Like other large carnivores, however, the panther is also controversial (Fuller and Kittredge 1996). Regardless, recovery could mean that the evolutionary tensions that have been missing from so much of the region will be recovered as well. Further, when combined with the marginally broader forest requirements of the black bear (Maehr 1997a), an expanding distribution could spread a conservation umbrella over many other species and the landscapes they inhabit. The ultimate measure of panther recovery success will be the creation of large carnivore management needs throughout the southeastern coastal plain (Maehr et al. 2001b). Forest cover and prey appear ample in much of the species' former range. What remains is the difficult task of convincing the public and agencies that large carnivores belong there. Following the attainment of this understanding, managers will be able to consider more than just the historical role of large canids in the southeastern United States. For example, whereas the coyote may create a functional competitive tension among south Florida's three remaining native carnivores, the red wolf would presumably do this and at the same time return the product of thousands (or more) years of closer competitive ties, and thus, the coevolutionary relations that have now disappeared here. It is not too soon, therefore, to reconsider south Florida as a logical refuge for a future population of the red wolf.

Managers control many of south Florida's biotic communities with structures that impound regional water supplies, and drainage canals that divert water quickly to the ocean. These actions have resulted in hydroperiod extremes that exacerbate periods of flood and drought, and that threaten a variety of species and processes. They also serve as filters and barriers to the movements of species and communities (Harris et al. 2001). Managers also prescribe fires in virtually every habitat that supports above-ground vegetation, and most often during seasons that are convenient to people, rather than with the ecological history of the biota in mind (e.g., all three of south Florida's large carnivores tend to give birth to litters during winter when managers are most apt to burn the woods). Then there are the array of public lands and private sanctuaries that are universally promoted as biotic preserves of one sort or another, but that fall short as an integrated whole. Dealing with habitat loss and fragmentation is daunting enough without the additional complication of shortsighted management plans and the geocentric tendencies of natural resource agencies. Small, isolated, "postage stamp" preserves (Eisenberg and Harris 1989; Harris et al. 1996) will ultimately be as effective in exterminating species as were the mysterious forces that decimated the Rancholabrean fauna.

Schaller (1972) suggested that coursing and stalking predators affected prey differently where both coexisted, but that habitat conditions ultimately dictated the population dynamics of prey. Our case study of the Bear Island large mammal community suggests that both habitat and predation shape the spatial arrangement of prey. As Errington (1967: xi) noted nearly four decades ago, "The rules of order behind Life and Life's processes hold for both predators and their prey, and the fact that these rules are not always apparent is no argument against their validity." Despite his acknowledgment that the cougar can be a specialist on deer, Errington believed that a variety of smaller carnivores could do the predatory work in the absence of this felid. Following this logic we would conclude that a black bear is simply a large raccoon (Procyon lotor) in an ecological sense. On the contrary, the recent documentation of mesopredator release in the absence of large carnivores suggests that many small carnivores do not equate to a few big ones (Terborgh 1988; Crooks and Soulé 1999a). Whereas the bobcat and panther are both capable of killing deer, the latter appear to have a shaping influence on other sympatric large mammals. Similarly, the black bear performs work that can only be matched by tool-wielding humans, powerful abiotic forces, or long-extinct megafauna.

Large carnivore management in Florida has the potential to promote biodiversity, especially if panther and red wolf are permitted to rejoin the black bear, white-tailed deer, and bobcat in their ecological roles inside and outside of the Big Cypress region. Often, however, it is difficult or impossible to understand a large mammal community to the same extent as in south Florida. Decades of research, perpetual funding, and an expanding public land base facilitate work on species that can be very difficult to study. We were fortunate to have such a backlog of data and background from other published works. Similar resources on large carnivore ecology and behavior elsewhere might lie forgotten in agency and researcher files, or syntheses of previously published works may await a new set of questions and subsequent analyses. In our case, we were more than simply tantalized with the possibility that large carnivores actively shape terrestrial biotic communities in south Florida, and that their evolutionary work should be a part of landscape-scale conservation efforts here and elsewhere. We encourage other researchers, managers, and conservation planners to engage in similar data mining. The results could provide more compelling reasons to redouble current biodiversity conservation efforts that focus on carnivores and the work that they do.

Certainly, the restoration of pre-Columbian ecological processes and landscape connectivity is an important part of the future southeastern carnivore assemblage. Although it appears virtually intact in much of south Florida, it may need only a few gentle bureaucratic nudges to encourage its expansion to parts of the southeastern United States that have not felt the tension of large felid predation for centuries. The logic behind the capital and intellectual investments needed to reclaim the region's eroding biodiversity must include an appreciation for the forces that created the modern patterns of life in the southeastern United States. However, "to develop the public support for ecologically rational conservation options" (Berger et al. 2001a: 947) land managers must first understand the conservation of biodiversity from an evolutionary landscape perspective. Such an understanding will promote the effective employment of carnivores as conservation flagships and as ambassadors for community restoration. The resulting landscapes will reflect these organisms' evolutionary pedigrees and promote their long-term security.

Summary

We revisited archived radiotelemetry data and field notes on south Florida's largest terrestrial mammals to investigate the potential relationships among three carnivores, an omnivore, and an herbivore. Whereas these data were used previously to address single-species ecology, and the nutritional relation between the Florida panther and white-tailed deer, the site-specific complexities of this large-mammal community had yet to be fully examined. Further, although wide-ranging carnivores have often been suggested as conservation flagships in Florida, little attention has been given to their potential in shaping biotic communities. Whereas it has been suggested elsewhere that temporal separation facilitated similar spatial use among panther, bobcat, and black bear, the colonizing coyote adds a tension to competitive and evolutionary trajectories that were relaxed by the regional extermination of the red wolf. The total effects of these interactions, whether direct predation, interference competition, or more subtle influences on habitat use, movements, activity, and vigilance, can be found nowhere else in eastern North America.

Despite rampant development, the tremendous range reduction in the Florida panther, and the extirpation of the red wolf, the south Florida large mammal community remains remarkably intact. The black bear is primarily an opportunistic omnivore in this system but exhibits a mutualistic relation with the giant palm weevil, a gregarious beetle that likely evolved with extinct megafauna. Distribution of radio-collared study animals suggests that white-tailed deer may select habitat to avoid predation by bobcat and panther, but that the avoidance of one may increase susceptibility to the other. The naturalized coyote may be filling the niche vacated by the red wolf by exerting an additional influence on deer survival and bobcat spatial dynamics. In south Florida, a combination of geology, hydrology, plant community distribution, and behavior has shaped the predator–prey relations of the region's large mammals. The challenge to land managers is to maintain the evolutionary tensions among these species—whether they represent age-old coevolution (deer and panther) or are analogs to former interspecies relations (black bear and palm weevil). The absence of any one of the actors in this evolutionary play weakens the ties between the landscape and the processes driven by south Florida's unique carnivore assemblage. The conservation and expansion of this once widespread community phenomenon will require dedicated managers and support from a better educated public.

ACKNOWLEDGMENTS

We appreciate the efforts of the editors of this volume and the Wildlife Conservation Society for organizing the people and ideas that challenge the status quo of global carnivore conservation. J. Ray, J. Berger, and two anonymous reviewers made helpful suggestions on earlier drafts of this paper. This is contribution #03-09-139 of the journal series, University of Kentucky Agriculture Experiment Station and is published with permission of the director.

Hunting by Carnivores and Humans: Does Functional Redundancy Occur and Does It Matter?

Joel Berger

Throughout history some 50 billion people have inhabited the earth; 10,000 years ago most societies were involved in the harvest of meat. Five hundred years ago, 1% were; by 1900 less than 0.001% were (Lee and Devore 1968). A propensity to eat game meat has changed inversely with human population size and, predictably, so has the abundance of native large mammalian carnivores (Weber and Rabinowitz 1996; Woodroffe 2000). With the replacement of carnivores by human hunters it would be surprising if the ecological effects of native carnivores and humans did not differ in the modern world. One oft-cited ecological rationale for encouraging human hunting is that it substitutes for the role once played by native carnivores. To evaluate this issue, I ask a deceptively simple question: Are human hunters functionally equivalent to large mammalian carnivores?

Under ideal conditions, functional redundancy occurs when one species replaces another with no change in community structure. For the purposes of this chapter, I consider humans as functionally redundant to large carnivores when either individually or combined their presence (or absence) does not markedly change the bounds of a given system. In reality, any proof of functional redundancy is challenging since conditions across broad temporal and spatial scales are rarely constant. Time frames, densities of hunters (human or otherwise), dissimilarities in prey species and their abundance, variation in plant communities and productivity, historic conditions, environmental and demographic stochasticity, and other factors all contribute noise that muddies contrasts between systems with and without human exploitation (Ludwig 2001; Robinson 2001).

Nevertheless, it seems possible to make sense of how humans and carnivores affect biodiversity. In the following text, I (1) test predictions about functional redundancy, and (2) suggest criteria by which humans may replicate the roles of native carnivores. I also offer both a general framework for what is known about how hunting by both modern humans and native carnivores structures ecological communities and make a broader point about how little we know about the subsequent effects of predation by humans on biodiversity.

The issue of functional redundancy is relevant for conservation planning and policy for at least three reasons. First, in many areas where large carnivores have been lost they will never be restored even if sufficient prey is available. For instance, large carnivores will never be tolerated in and around most large cities. Second, where biological conservation is a goal, its potential achievement will be strengthened by knowing the extent to which humans are capable of fulfilling the role(s) once played by carnivores (Holsman 2000). Finally, given that hunting by humans is a pervasive global activity (in the United States, approximately one of every seven males between the ages of 16 and 65 hunts; Enck et al. 2000) and areas designated for hunting include some largely intact ecosystems with extant large carnivores, the possibility of competition for the same prey is high. By understanding the relative contribution of each of these two types of predators to ecological communities, it may be possible to disentangle expectations about how carnivore and human hunting contribute to the conservation of biological processes and attendant diversity.

Limitations of Approach

There are limitations to this analysis. I have not categorized the immense variety of methods used by human hunters nor their often differing harvest goals. Among the former will be cables, snares, leghold and pitfall traps, bows and arrows, spears, muzzle-loading rifles, pistols, and high-powered rifles. Among types of hunters will be those seeking trophies, high-quality meat, and any and all meat. Some will hunt year round (legally and illegally), and others only during a prescribed season. Given that human hunting styles also vary by culture as well as weaponry (Robinson and Redford 1992; Alvard 1998; Robinson and Bennett 2000), both factors must be accounted for when attempting to understand putative effects on the structure of food webs and biological diversity. Therefore, what I lay out here is focused on trying to understand whether and how generally "Westernized" hunters fulfill the role played by nonhuman carnivores. I refer to biodiversity in its broadest sense—populations, species, communities, and ecosystems as well as its attributes (processes, structure, and function; Redford and Richter 1999).

Current Overlap between Hunting Humans and Carnivores

For humans and carnivores to be ecologically equivalent as hunters, several prerequisites must be met, although not necessarily simultaneously. Both must (1) be strongly concordant in species selection, (2) overlap spatially in off-take, and (3) yield mortality that is compensatory rather than additive (Jorgenson and Redford 1993; Ginsberg and Milner-Gulland 1994; Murphy 1998). Although these conditions are in and of themselves insufficient to demonstrate functional redundancy, they represent a necessary first step. For example, it is not sufficient for carnivores and hunters to exhibit high dietary overlap. Rather, similarity in prey functional response is better evidence of functional redundancy. Where the strength of interaction(s) between prey and predator is tightly linked, prey responses may converge in antipredator behavior, activity patterns, or habitat use, irrespective of whether sources of mortality are carnivores or humans.

Evidence in support of the preceding three general prerequisites for ecological equivalency is diverse, spanning all continents where native carnivores and human hunters overlap (Table 16.1). Therefore, an a priori expectation of functional redundancy between carnivores and humans seems not unreasonable. The concept of functional redundancy however requires more than interactions at a single ecological level such as predator and prey. As elaborated in the following text, the nature of hunting styles, densities of humans and carnivores, variance in the strength and timing of predation, environmental heterogeneity, gender-specific harvest, and history are all salient to testing predictions about functional redundancy.

Table 16.1

Selective summary of prerequisite conditions (see text) to be met prior to reasonably assuming a functionally redundant relationship exists between carnivore and human hunting

	Prey Species in Common	Spatial Overlap	Mortality Compensatory	Reference
Europe	Red deer	Yes and no	Yes and no	1
	Roe deer	Yes and no	Yes and no	2
South America	Brocket deer	No[a]	Not likely	3
	Capuchin	No[a]	Unclear	4
	Paca	No[a]	Unclear	3
	Peccary	No[a]	Unclear	3
	White-tailed deer	No[a]	Not likely	3
North America	Bison	Yes	No	5
	Caribou	Yes	No	6
	Elk	Yes	No	7–9
	Moose	Yes	No	7, 8, 10
Asia	Sika deer	Yes	Not likely	11
	Chital	Yes	Not likely	12–14
	Muntjac	Yes	Not likely	14
	Red deer	Yes	Not likely	15, 16
Africa	Buffalo	Yes	Not likely	17
	Impala	Yes	Not likely	18
	Giraffe	Yes	Not likely	19
	Wildebeest	Yes	Not likely	19, 20
	Zebra	Yes	Not likely	19, 20
	Duiker	Yes	Not likely	21, 22

[a] Large carnivores may be locally extinct in these specific study regions.

1. Clutton-Brock et al. (1982); 2. Aanes and Anderson (1996); 3. Jorgenson and Redford (1993); 4. Hill et al. (1997); 5. Carbyn et al. (1998); 6. Mech et al. (1998); 7. Kunkel et al. (1999); 8. Kunkel and Pletscher (1999); 9. Boyd et al. (1994); 10. Berger et al. (2001a); 11. Makovkin (1999); 12. Karanth and Sunquist (1992); 13. Karanth and Sunquist (1995); 14. Madhusudan and Karanth (2002); 15. Miquelle (1998); 16. Miquelle et al. (1999); 17. Sinclair (1977a); 18. Creel and Creel (2002); 19. Caro (1999); 20. Arcese and Sinclair (1997); 21. Noss (1998); 22. Ray (2001).

Predictions: Concordance in Effects of Human and Carnivore Hunting

Central to testing predictions about functional redundancy is an understanding of the level of possible effect. "Indirect effects" refer to the strength and direction of interaction between two species that may change as a consequence of a third one (Strauss 1991; Wootton 1994). "Subtle effects," on the other hand, generally apply to systems dominated by humans and denote an array of changes, some of which trickle down because of our nondeliberate actions (McDonnell and Pickett 1993). A distinction between "indirect" and "subtle" is not always clean, but typically the latter involves two species, and the former only one other than humans.

In addition to specific effects, functional redundancy can occur at different tiers of biological organization. I focus here on three general tiers where humans have replaced carnivores: (1) predator–prey interactions, (2) prey–ecological dynamics, and (3) biological diversity (Table 16.2). What I shall not consider are effects of types of predation on prey populations per se, which are considered extensively elsewhere (e.g., age and gender selection [Ginsberg and Milner-Gulland 1994; Berger and Gommper 1999], as well as possible evolutionary change [Harris et al. 2002; Coltman et al. 2003]).

The following three sections and associated subheadings develop the framework to address the overarching question: Are humans functionally redundant to native carnivores? It is important to note that from the perspective of an optimal research design, four general conditions are possible: (1) humans and carnivores present, (2) humans and carnivores absent, (3) humans present–carnivores absent, and (4) humans absent–carnivores present (see Fig. 16.1). In reality, very few areas have empirical results available for each of these four conditions where the strength of predation is known.

Despite this, available data from a variety of sites allow evaluation of the possibility of functional redundancy. For example, both brown bears (*Ursus arctos*) and wolves (*Canis lupus*) exert strong effects on moose (*Alces alces*) population size, particularly through their influences on juvenile recruitment (Gasaway et al. 1983, 1992; NRC 1997). In Grand Teton National Park, at the core of the Greater Yellowstone Ecosystem, moose populations irrupted during a 60- to 75-year period after predation was removed as a selective force. High densities of moose and

Table 16.2

Summary of predictions and outcomes for functional redundancy with human replacing carnivore hunting given the assumption that other factors are equal[a]

Tier	Prediction	Outcome for Functional Redundancy			Comment
		Similar	Dissimilar	Uncertain	
I	Timing and biomass of mortality		xx		Temporal differences large
	Consequent mortality		xx		Wounding loss differs
	Prey densities	xx			Congruent effects possible
	Sexual segregation			xx	Overall role of predation unclear
II	Herbivore–vegetation				
	Behavior	xx			Congruent effects possible
	Foraging rates			xx	Perhaps species dependent
	Habitat shifts			xx	Data on strength of predation lacking
	Activity patterns			xx	Relevance to ecological events unclear
	Carrion–scavenger		xx		Temporal differences
	Intraguild predation		xx	xx	Exploitation and interference may vary
	Subtle				
	Roads, horses, ATVs, camps		xx		
	Dangerous prey and human safety			xx	Differences anticipated but data lacking
III	Biological Diversity	xx	xx	xx	All may apply but paths become increasingly muddled

[a] Generalized relationships (from text) can be similar, dissimilar, or uncertain.

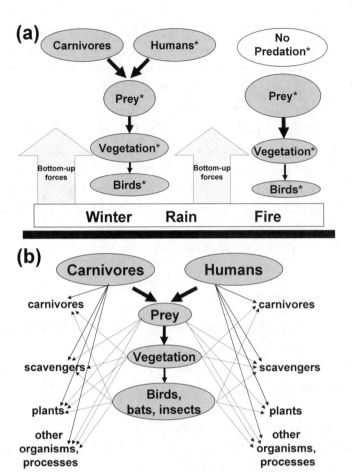

Figure 16.1

Summary of putative apex role of predators. (a) Scenarios with hunting by carnivores, humans, and no predation. Asterisks reflect empirical support derived from systems with moose, humans, riparian willows, and migratory Neotropical birds (after Berger et al. 2001a). Size of circles reflects changes in relative abundance. Possible bottom-up forces as shown. (b) Depiction of predicted outcomes if hunting by humans and by carnivores yield similar effects across various tiers of biological organization. Multiple simple and interactive pathways are illustrated by solid and dotted finer lines, respectively.

associated levels of herbivory decreased riparian willow vegetation and concomitantly avian species diversity (Berger et al. 2001a; Berger and Smith, this volume). In adjacent but nonprotected regions also lacking in brown bears and wolves for a similar period, human hunting reduced moose densities by a factor of 4½; coincident willow riparian biomass and avian species diversity were higher. In this particular case, human predation replaced that by carnivores and enhanced the abundance of neotropical migrants. Because moose in the arid intermountain West tend to be riparian dependent, it is also possible that other elements linked to riverine vegetation may also be affected. This example reinforces the idea that human hunters can affect biological diversity by suppressing prey abundance in the absence of native predators. Table 16.2 outlines predictions and tests for functional redundancy involving humans and carnivores.

Tier 1: Predator–Prey Interactions

An obvious truism is that, where predators are absent, interactions with prey cannot occur. However, from a biodiversity perspective, it is critical to know not only about the form but about the magnitude of prey responses to systems with and devoid of predators.

Timing of Predation and Distribution of Meat

If predation by carnivores and humans is functionally equivalent, then meat consumption schedules should be similar. This is not the case, as illustrated in three examples: (1) three species of nonhuman carnivore predation (Fig. 16.2a–c), (2) a total cessation of predation (Fig. 16.2d), and (3) the distribution of prey biomass as a function of predation by sympatric carnivores and humans (Fig. 16.2e). Although predation by wolves, brown bears, and cougars (*Puma concolor*) produces some similarities in the timing and inflection of curves (Fig. 16.2a–c), the total mortality by carnivores differs from reality in two important ways, one biological and one methodological. First, the respective carnivore–prey system represented in Figure 16.2 is that of a single predator only, whereas each of the study regions portrayed has multicarnivore assemblages. Second, the extent of predation is underestimated because: (a) neonates (except in the moose studies) were not specifically studied, and (b) summer mortality in adult deer is difficult to

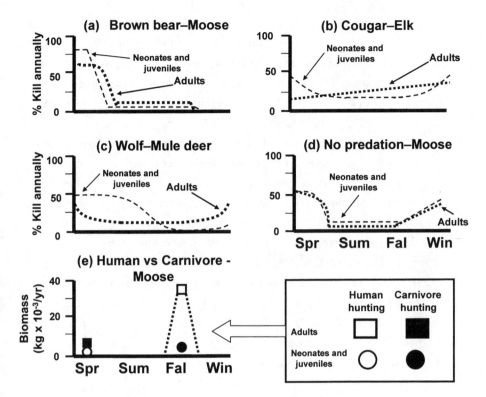

Figure 16.2

Summary of patterns in timing of predation, involving three scenarios: (1) carnivore-specific (brown bear, cougar, and wolf) predation on identified prey (a–c); (2) lack of human and carnivore predation (d); and (3) annual biomass off-take by both carnivores and humans in a moose population. (d) (assumes a population of 1500) and (e) are from the Grand Teton–Jackson Hole region of Wyoming. Depictions from the following: (a) composite of south- and east-central Alaska (Boerjte et al. 1996; Bowyer et al. 1999; Gasaway et al. 1983, 1992); (b) Yellowstone (Murphy 1998); (c) mule deer in northern Rocky Mountains, Montana (Boyd et al. 1994; Kunkel and Pletscher 1999; Kunkel et al. 1999); (d) mortality (N = 85) reflects starvation and other sources but not predation, which has been trivial; see (e) (Berger et al. 2001a, unpubl. data); and (e) is the mean for years 2000 and 2001; human off-take values based on hunter success from Wyoming Game and Fish (unpubl. data). The following values used: calf mass @ birth = 20 kg, at 6 months of age125 kg; adult female = 225 kg, adult male = 350 kg, adult female pregnancy rate = 74% (Berger et al. 1999).

detect because monitoring tends to be less intense (Boyd et al. 1994; Kunkel et al. 1999). The preceding caveats aside, the distribution of kills of juveniles or adults as a consequence of hunter type (human vs. carnivore) is striking. If the measure is not total proportion of kills but biomass removal by season, differences become more accentuated, with a fivefold difference in total yield between humans and carnivores (see Fig. 16.2).

From a study design perspective, the condition of "no predation" is portrayed (see Fig. 16.2d), but predator-free areas are unlikely to be realistic as controls since the historic condition across most systems was one that involved predation rather than one without. Predation-free anomalies can, however, characterize nonmanipulated wild ungulates. Native reindeer (*Rangifer tarandus platyrhynchus*) of the Svalbard Archipelago (Norway; Aanes et al. 2003) and caribou (*Rangifer tarandus*) of southwestern Greenland (Denmark) south of the permanent Melville Bulge (icefields) have apparently lived in the absence of native predators until recent human occupation (Melgaard 1988; Marquard-Petersen 1995). Despite such rare exceptions, the overall point is that functional redundancy does not appear to characterize many of the systems with carnivore and human hunting despite variation that inevitably depends on hunter densities.

Wounding Losses and Additional Mortality
Not all human hunters are good shots. As a result some animals are wounded and die later from hunter-related injuries. This proportion is referred to as wounding or crippling loss and varies by weapon type, but the losses can be as high as 21% for rifle hunters in the United States (Smith and Anderson 1998; Carpenter 2000). A similar rate of injury occurred for individuals snared by cable in central Africa (Noss 1998). Animals that die from wounds represent food that becomes available to scavengers and other organisms.

In contrast to humans, data on wounding loss by carnivores is even rarer, although values must be much lower than that for human hunters given the remarkable scarcity of injured prey in areas of Africa with large predators. At sites where large carnivores have been extirpated, such as areas occupied by red deer (*Cervus elaphus*) and bison (*Bison bison*), animals injured in intraspecific combat survive at comparably high rates (Clutton-Brock et al. 1982; Berger and Cunningham 1994). The variation in deaths due to wounding losses as a consequence of effectiveness in carnivore and human predation is suggestive of a lack of concordance in the distribution of death.

Prey Densities through Time

Since neither animal population size nor density remains constant in time, any expectation of congruent effects between carnivores and humans may not be realistic without accounting for temporal variability. If study of multiple systems with carnivores or with humans were possible over many generations, and if amplitudes of fluctuation in density were similar (with other factors equal), one might reasonably argue that human hunters fulfill an ecologically redundant role. But such broad contrasts are rarely possible due to the range of variation and change that characterizes systems (Sinclair and Arcese 1995b; Clark et al. 1999) as well as the degree of environmental perturbation and hunting pressures outside protected areas.

Much work has focused on the potential for predator-limitation of mammalian prey (reviews in Boutin 1992; Sinclair and Krebs 2001; Krebs 2002). That prey densities differ spatially is clear from a geographical snapshot of elk/red deer across 17 systems (Fig. 16.3) with variation from more than 30 animals/km^2 to as few as < 1 km^2. Although elk populations have been manipulated for generations on three continents and have not occurred in many regions with predation by wolves being unabated over time, there remains great interest in relationships between elk abundance and factors that influence population size. In contrast to elk, species that have interacted with predators for at least half a century, and where data on such interactions exist, include wildebeest (*Connochaetes* spp.), buffalo, caribou, and moose; where these have been monitored for decades, fivefold differences in density can occur (Messier 1994; Sinclair and Arcese 1995b; Peterson et al. 1998; Crête 1999). In essence, prey densities can vary widely in both the presence and the absence of predation by large carnivores.

That humans can limit prey density is clear because in the absence of this predation populations irrupt when carnivores are missing (McCullough 2001), although this can also occur when carnivores are present (Sinclair 1989, 1997). However, the question is not whether human predation plays a role in "controlling" runaway ungulate populations but whether it can partially or wholly substitute for carnivores. In systems with carnivores, multiple stable states with both high and low levels of predator and prey are achievable (Walters et al. 1975; Ludwig et al.

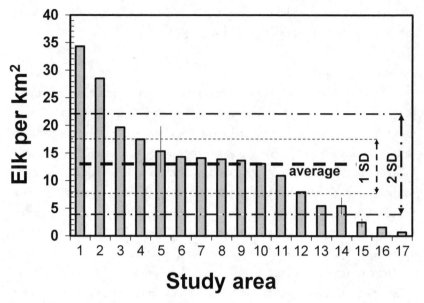

Figure 16.3

Mean densities of red deer–elk populations during winter or year-round. For populations where density estimates spanned multiple years, only the last of a given decade
was used. Bars indicate ranges spanning multiple decades. Sites and references as follows: 1–Scarba (United Kingdom), 2–Estes Park, CO (USA; Singer et al. 2002),
3–Rocky Mountain Park, CO (USA; Singer et al. 2002), 4–Yellowstone Park, WY
(USA; Houston 1982; Singer and Mack 1999; Smith et al. 2003), 5–Invermark (United
Kingdom), 6– Białowieża National Park (Jędrzejewska et al. 1994), 7–Jasper (Dekker
et al. 1995), 8–Rhum, 9–lands adjacent to number 4 (Houston 1982), 10–Glenfeshie
(United Kingdom), 11–Banff Park–central, AB (Canada; Hebblewhite et al. 2002),
12–Glen Fiddich (United Kingdom), 13–lands adjacent to Starkey in Oregon (Rowland et al. 2000), 14–Białowieża Primeval Forest (Jędrzejewska et al. 1994), 15–Wind
Cave Park, SD (USA; Lovaas 1973), 16–Banff–other, AB (Canada; Hebblewhite et al.
2002), 17–Glen Dye (United Kingdom). (UK sources for numbers 1, 5, 8, 10, 12, 16
from Clutton-Brock et al. 1982.)

1978). Agencies tasked with input from the hunting public understandably prefer
the higher state, which enables a higher yield. In these situations humans serves
as a de facto substitute for carnivores, and densities can be far lower than where
hunting is absent (Berger et al. 2001a). In areas with carnivores, the extent to which
consequent prey densities exceed, approximate, or fall below those with human
hunters will depend of course on the density of predators and their effectiveness.

An assumption held by some is that large mammal predators control their prey and that humans can functionally replace carnivores (NRC 1997). This assumption can be frail, especially for systems with deer both because weather (e.g., drought or winter) has strong effects on deer densities and because carnivore densities may be highly variable, particularly when alternate prey are available (Fuller and Keith 1980; Fuller and Sievert 2001; Krebs et al. 2001a). Hence, even, if human hunting pressure is more constant than that by carnivores, any effects on prey densities may arise due to a combination of other factors. Still, if prey density was used as a sole criterion by which to measure functional redundancy, then it would appear that human predation has the capacity to perform well as a surrogate for carnivores.

Tier 2: Prey Ecological Dynamics

Although studies have examined relationships between ungulate densities and community-level effects, only a few have extended the analysis to examine the role predators (human or nonhuman) might play in modifying these relationships (see Fig. 16.1). Nevertheless, since both humans and carnivores can modulate prey densities, direct, indirect, and subtle effects can be summarized with respect to the potential for functional redundancy (see Table 16.2).

Ungulate Densities and Herbivory

Predation can affect herbivore densities and therefore plant communities. Growth in balsam fir is modified by wolf abundance through moderating influences of moose (McLaren and Peterson 1994). Where wolves were extirpated and in the absence of heavy human hunting, elk densities were high and they exerted strong influences on plant canopies (Jędrzejewski and Jędrzejewska 1998; Singer et al. 2002). As with the Tier 1 predictions, when herbivore densities are the sole criterion, then humans may perform well as a surrogate for carnivores (Terborgh and Wright 1994; Berger et al. 2001a; Wright and Duber 2001).

Behavior

Prey species are not necessarily automatons. Even though many do not break the scale on intelligence or conscious decision making, some species learn about and react to the threat of predation. Both mongooses (*Helogale* spp.) and moose learn to key on birds that alert them to potential predators (Rasa 1983; Berger 1999; Berger et al. 2003). And, as noted by Georg Stellar some 250 years ago, sea otters (*Enhydra lutris*) apparently altered their habitat use in response to human hunting (Frost 1988). So how might an appreciation of behavior contribute to understanding the relative effects of human and carnivore predation on ecological dynamics? Some promising behaviors for investigation include: (1) foraging rates and grouping, (2) habitat shifts and refuge use, and (3) activity patterns.

Foraging Rates and Predator Detection. Individuals stop feeding to scan for predators (Lima 1998), a behavior that affects use of the landscape. This becomes relevant for assessing whether prey movements and habitat use are similar when prey are hunted by humans and by carnivores. The evidence is mixed.

Elk under pressure from reintroduced wolves apparently foraged less efficiently than when wolves were absent (Laundré et al. 2001) and have become less concentrated at local food patches (Ripple et al. 2001). Not surprisingly, elk also become highly alarmed when targeted by rifles (Bender et al. 1999). However, bison in Canada under predation threat from wolves and those from South Dakota living in the absence of all predation did not vary in their vigilance (Berger and Cunningham 1994). Similarly, Alaskan moose in areas with native carnivores did not vary in vigilance from Wyoming moose, the latter living in areas where either humans were the only predators or predation was lacking. Once predator cues were detected, however, differences in foraging rates and subsequent movements did occur (Berger et al. 2001b). Among African ungulates, behavioral differentiation by predator (carnivores vs. humans) may occur, although relative differences in predation intensity by type of predator remain unknown (Caro 1999). Finally, at several Neotropical sites where human predation occurs in the absence of large carnivores (see Table 16.1), densities of nine-banded armadillo (*Dasypus novemcinctus*), capuchin (*Cebus apella*), tapir (*Tapirus terrestris*), and brocket deer (*Mazama* spp.) were reduced; how much prey wariness contributed to lower detection rates by ecologists (and hence densities via redistribution across the landscape) was uncertain (Wright et al. 1994; Hill et al. 1997). Overall, a lack of evidence showing differing effects of carnivores and humans on foraging and predator detection can stem from many sources of variation. With respect to testing the prediction of functional redundancy, the appropriate series of experiments have not yet been conducted.

Habitat Shifts and Refuge Use. It is not clear what differences exist between humans and carnivores in inducing habitat shifts and changes in refuge use by prey. Although not strictly comparable, independent effects of both humans and carnivores can be inferred. In the absence of human predation, many prey species shift to or use as refuges areas with humans that are avoided by native carnivores. Such species include chital (*Axis axis*), wildebeest, moose, and vervet monkeys (*Cercopithecus* spp.) (Berger 1998). Perhaps the best example of broad-scale redistribution across a landscape in response to predation is the movement and migration of ungulates when hunted by humans to private lands or national parks where harvest is limited or prohibited.

For carnivores there is also evidence of refuge use when they may be prey. Coyotes (*Canis latrans*) and cheetahs (*Acinonyx jubatus*), each of which are killed by larger predators (Palomeres and Caro 1999), use refuges that may include roads or habitats low in prey density where encounters with larger carnivores are diminished (Thurber et al. 1992; Durant 2000). White-tailed deer are more frequent at territorial boundaries of wolves (Mech 1977). Parturient moose, after the expansion of brown bears into Grand Teton Park, increased their use of habitats within 500 m of paved roads (Berger, unpubl. data); such regions tend to be avoided by brown bears (Mattson et al. 1992). Overall, prey appear to use parallel approaches to decrease chances of being killed or eaten. The possibility clearly exists that predation both by carnivores and by humans may yield concordance in prey habitat use, but without well-designed comparative experiments that account for predation pressure current analyses are inconclusive.

Activity Patterns. A good deal of information indicates prey modification of activity in response to predation. Coyotes and elk both become more nocturnal in areas where hunted by humans (Kitchen et al. 2000; Rowland et al. 2000; McClennen et al. 2001). Whether a similar pattern characterizes prey when hunted by carnivores is not clear. However, carnivores themselves may partition their own activities to avoid competition with or predation by other carnivores. Whether such change is relevant to ecological or community dynamics is another issue. For coyotes (or elk), influences might be large; for example, consumption of ground squirrels (*Spermophilus* spp.) versus nocturnal mice or moles, but subsequent effects of variation in activity are mostly unknown. At this point, the relevance of activity patterns to functional redundancy is inconclusive.

Scavenging

Relationships involving scavengers, prey, and carnivores can be both direct and indirect. Ravens (*Corvus corax*) follow wolves and may locate potential prey before they are killed (Stahler et al. 2002). And both native folklore and naturalists report similarly; ravens follow and signal humans about the location of potential prey (Heinrich 1989). Just as the timing and distribution of meat that results from human hunters differs from that by carnivores (see Fig. 16.2), so too does the availability of carrion for scavengers (Selva et al. 2003; Wilmers et al. 2003a,b). However, differences in the timing of kills by humans and carnivores mean that magpies (*Pica* spp.), ravens, and other nesting scavengers can provision young with carrion from carnivores but not from humans since human predation typically occurs during the nonnesting season. Finally, high densities of ravens may be sustained in some systems by carrion from hunters. At the center of the Greater Yellowstone Ecosystem ungulate biomass may be the highest in North America (Berger, unpubl. data), and raven densities have increased over a 40-year span (Dunk et al. 1997).

Does carrion produced through carnivore and human predation yield similar effects among mammalian scavengers? An interesting case involves potential changes in the distribution of three sympatric large carnivores—wolves, brown bears, and cougars—each of which occurs inside Yellowstone National Park (where hunting by humans is prohibited) and outside (where hunting by humans is permitted). In response to the shooting and subsequent gut piles of elk, brown bears shifted to areas outside the park; the distribution of wolves was unchanged, and cougars moved away from areas with carrion-feeding bears (Ruth et al. 2003). In sum, there are both similarities and differences in the extent to which the pulse-phase availability of carrion resulting from hunting by humans and by carnivores may affect other levels of biological diversity.

Intraguild Predation and Mesocarnivore Release

Intraguild predation is relevant to understanding the potential for functional redundancy not because carnivores kill carnivores but because human removal of carnivores ultimately affects biodiversity. Here I briefly consider linkages among specific carnivores and how they relate to biodiversity.

Interference Competition. Among the best examples of mammalian intraguild predation (Polis et al. 1989; Polis and Holt 1992) are those involving canids where interactions and population level consequences are well documented (White and Garrott 1997, 1999; Crabtree and Sheldon 1999). Numerous cases are illustrative of the possibility of functional redundancy with respect to interference competition. Both humans and wolves kill coyotes, although all three species may coexist. Historically, coyotes were more restricted in distribution than wolves in North America, a relationship that today is the converse as coyotes have spread throughout the continent (Peterson 1995). Although habitat change may account for some expansion, the killing of wolves by humans released coyotes from one source of predation (e.g., carnivores). Still, the predation pressure exerted by wolves on coyotes has not been equivalent to that by humans, and the two forms of intraguild predation have not resulted in similar densities of coyotes. Such differences are relevant to an understanding of potential functional redundancy; variation in coyote abundance produces strong community effects that include an abundance and diversity of other carnivores, birds, and rodents (Crooks and Soulé 1999a; Henke and Bryant 1999), but just how humans and wolves modulate coyote densities has received little scrutiny. On the other hand, functional redundancy may be achievable at small scales. Both carnivores and humans have the capability of reducing coyote densities, although the adult harvests must regularly approximate 75% to be effective (Knowlton et al. 1999). Where coyotes tend to be abundant and are sympatric with pronghorn antelope (*Antilocapra americana*), fawn survival is generally low (Byers 1997). If a density reduction of coyotes (irrespective of source) yields similar levels of fawn survival in the absence of other mitigating factors, then functional equivalency is possible.

Exploitation Competition. Prey removal by human hunters will alter densities or biomass available to other carnivores. This scenario should more appropriately be considered an indirect food web interaction of human hunting and has real-world relevance, particularly where concern about competition between carnivores and humans for prey exists (Phillips and Smith 1996; Creel and Creel 2002). Exploitation competition is critically important for conservation because the densities and ranges of carnivores such as tigers (*Panthera tigris*) or bears are reduced where they cannot harvest prey at the same rates of hunting humans (Mattson et al. 1992; Karanth and Stith 1999).

Indirect and Key Subtle Effects

A consideration of functional redundancy involving humans and carnivores might be restricted to possible effects of human hunting alone, but hunting by humans (at least in Western society) usually relies on an infrastructure that is far different from that of native carnivores. This creates a range of indirect and subtle effects that are not usually acknowledged.

Roads and the Hunting Season in Western North America. Effects associated with roads are usually considered as large-scale disturbance (Trombulak and Frissell 2000; Havlick 2002). During the autumn hunting season these effects may include both direct and subtle ones, where hiking by tourists is replaced by all-terrain vehicles (ATVs), horses, and hunt camps. The situation is the opposite of summer when hordes of nonarmed hikers and mountain bikers visit montane regions. Although effects of recreation can be large (Knight and Landres 1998), I ignore these here, given this chapter's focus on potential for functional redundancy between human hunters and carnivores.

The pulse of activity on lands outside protected regions changes during the autumn. Roads, both gravel and paved (Fig. 16.4), are often extensive, but dirt two-tracks enable entry into regions that otherwise would be too far for many on-foot hunters. Access is thereby available to virtually all but legally designated wilderness (e.g., nonroaded) and some national park areas in the 60,000 km² Greater Yellowstone area (GYA). ATVs (not all associated with human hunters) have the potential to displace wildlife, and they compact soils, alter hydrological regimes, increase erosion, destroy meadows, crush plants, and increase noise, air, and water pollution (Havlich 2002). These effects are unique to human predators.

Firearms also produce outcomes that differ from those of carnivores beyond those of wounding loss discussed previously. A not infrequent occurrence in the GYA is the unintentional harvest of moose that were incorrectly identified as elk, a mix-up in species identity that also occurs between black (*Ursus americanus*) and brown bears (the former can be harvested legally, and the latter are an endangered species in the contiguous United States [Mattson and Craighead 1994]). Additionally, during the fall hunting season, competition for hunter-killed carcasses may occur, often resulting in bears being killed by hunters while attempting to appropriate the meat (Mattson et al. 1994).

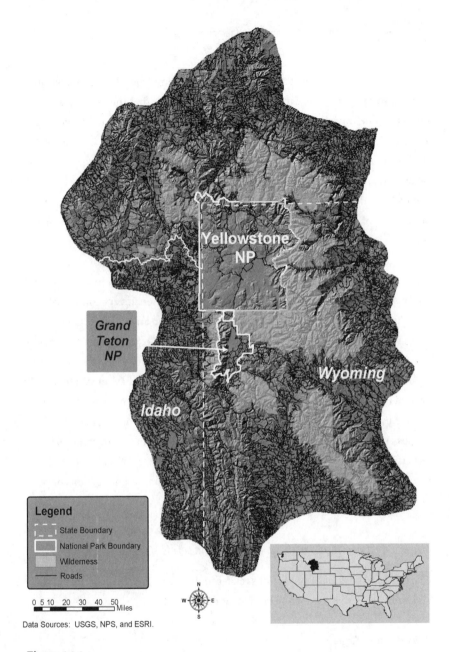

Figure 16.4
The road network in the southern Greater Yellowstone Ecosystem (GYE). Roads (paved roads, secondary, gravel, two-track) shown in black and the wilderness areas (that all lack roads) are highlighted in light gray. Map: sources are U.S. Geological Survey, National Park Service, and Environmental Systems Research Institute; map prepared by Jon Rieck, Wildlife Conservation Society.

A less direct but important broad-scale effect stems from the loss of available habitat. Both deer and elk move farther from roads during the hunting season, often become more nocturnal, and decrease use of open habitats (Kilgo et al., 1998; Rowland et al. 2000). Some carnivores are similarly displaced (Maehr 1997b; Ruth et al. 2003). Finally, numerous and uncounted species are killed as incidental off-take; porcupines (*Erethizon dorsatum*), skunks, snakes, coyotes, rabbits, and squirrels are designated as "varmints" in some western states. Effects of guns beyond those of targeted prey can therefore include either indirect or subtle effects on inadvertent prey, carnivores, and scavengers (Fig. 16.5).

A different effect derives from horses that offer access to remote wilderness areas. Some horse and mule trains contain up to 30 animals and may remain for 10 days or longer. When salt is placed for horses, it often attracts other wildlife including elk. Food for both horses as well as humans can attract bears. Conversely, some wildlife is displaced due to avoidance of humans. So, as indicated in the previous section on scavenging, the potential influx of human hunters may leave different-sized footprints. Most have yet to be measured but they must extend well beyond those of guns as they affect plants, other organisms, and processes (see Fig. 16.5). Finally, in seasonally cold environments camps involve trailers, mobile homes, and other forms of recreational vehicles. These may or may not have effects similar to those already described, but they will be far different from those incurred by carnivores hunting prey (see Fig. 16.5). To be fair, the millions of people that recreate on lands with wildlife also have very serious effects.

Tier 3: Effects on Biodiversity at Other Trophic Levels

The third and final set of predictions relates to the cascading effects on trophic levels beyond prey that emanate from predation.

A Few Clear Linkages
Because some prey are dependent on vegetation and other habitat components, animal and plant biodiversity may be modified through predation by carnivores (Terborgh et al. 1999, 2001). This is clear from studies in four protected areas of western North America (Yellowstone, Banff, and Rocky Mountain national parks, and the National Elk Refuge) where humans reduced large carnivores, and high

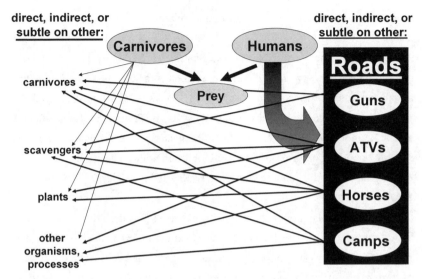

Figure 16.5
Interactive food web illustrating direct, indirect, and subtle effects of human hunting that differ in strength and direction from those of carnivore hunters.

elk densities compromised the productivity of aspen communities or coincident songbird diversity (White et al. 1998; Dieni et al. 2000; Soulé et al. 2003). Ungulate densities beyond the borders of these regions are regularly reduced by human hunting, and associated plant communities and avian diversity may be greater in such areas (see Berger et al. 2001a; Singer et al. 2002).

Although other work concentrates on effects of native ungulates at levels beyond those of herbivory alone (Frank et al. 1994; Frank and Groffman 1998; Singer et al. 2002), a general absence of direct contrasts between systems with predation by carnivores and by humans precludes strict assessments of functionality. On the other hand, before and after studies have been conducted in systems where carnivores were once extirpated and have now returned (e.g., wolves in Yellowstone Park) or in systems with great temporal variation in carnivore densities (Isle Royale Park). In these analyses both top-down and bottom-up forces have been shown to operate on biological diversity via carnivore–prey–vegetation dynamics (Pastor et al. 1993; Ripple et al. 2001). Cascading effects beyond these two trophic levels become less directly tied to carnivores per se, although density limitation of ungulates seems widely tied to enhanced biodiversity (Micheli et al. 2000).

Vegetation change and community processes linked to ecosystem function and biological diversity via herbivory by ungulates include nutrient cycling, fire, and decomposition (McNaughton et al. 1988; Hobbs 1996). Although high densities of grazers have the capacity to modify systems (Frank et al. 1994; Singer et al. 2002), understanding whether effects are positive or negative, are within reasonable bounds of natural variation, and/or attain points where predation dampens fluctuations is difficult and contentious (Kay 1994; Keigley 2000; Singer et al. 2000; Yochim 2001).

The difficulty lies not in understanding effects of prey densities on ecosystems but in interpreting relative roles of carnivores and humans in affecting those densities. Where wolves and moose have interacted for half a century in the absence of human predation (Peterson 1995; Jordan et al. 2000), high levels of herbivory change communities from deciduous to coniferous (Pastor et al. 1993). In Sweden, moose herbivory also affected forest leaf litter, which is associated with the diversity of ground-living invertebrates (Suominen et al. 1999a). However, making sense of change over time and understanding its relationships (if any) to the intensity of predation can be bewildering due to potentially confounding effects of fire regime, logging, and direct and indirect human effects. Linnaeus apparently never saw a moose in Scandinavia (how hard he tried is unclear) but more than 300,000 occurred there 10 years ago (Clutton-Brock and Albon 1992).

Fuzzier Pathways

How reasonably can effects of carnivore and human predation be extended to additional ecological components? In addition to indirect influences on vegetation, nutrient cycling, macroinvertebrates, and neotropical migrants (see preceding text and Figs. 16.1, 16.2), it is also possible that the loss of carnivores can change dynamics of predator–prey interactions to affect lower trophic levels. In the arid American West riparian zones contain as much as 90% of the species diversity (Stacey 1995; Fleishman et al. 1999, 2001). Because populations not limited by predation but by food alter vegetation structure (Schmitz et al. 2000; Berger et al. 2001a), and some insects, including midges and mosquitoes, show a strong preference for mesic vegetation (Darsie and Ward 1981), an interaction beyond that of mosquitoes feeding on moose may be involved. Insectivorous bats, for example, are more abundant and feed preferentially in and adjacent to mesic habitats where aquatic insects, midges, and mosquito abundance is greater than in less mesic habitats (Brigham et al. 1992; Seidman and Zabel 2001). If predation on ungulates by either large mammalian carnivores or humans alters insect-dependent vegetation, then the predators will also affect the strength and direction of bat–insect interactions. A similar speculation can be created for butterflies given the proclivity of some species toward riparian sites (Fleishman et al. 1999, 2001).

A particularly interesting case involves a link between elk browsing and the abundance of leaf-galling sawflies (*Phyllocolpa bozemanii*), which subsequently affect arthropod foraging and the diversity of insectivorous birds (Bailey and Whitham 2003). However, the extent to which carnivores have affected elk densities relative to that of humans has been less certain. One of the crucial challenges will be to unravel just how these likely pathways operate, and another challenge will be to develop rigorous tests to assay the role of factors alternative to predator and prey densities.

Conservation Recommendations: Functionality in Systems with Carnivores and Humans

Is knowledge of ecosystem structure and processes sufficient to assure that inferences about functional redundancy among types of predators are on target? The answer is likely to be yes in Westernized countries where there has been a history of exploitation, science, and conservation. For developing countries with far fewer resources, answers are generally less clear. Although predation by humans may or may not mimic that by carnivores (see summary in Table 16.2), it is possible to shape human hunting to simulate effects of carnivores. I focus here on biological attributes only, given the suite of social, political, cultural, and ethical challenges that otherwise will affect true implementation. It is important to note that, even where functional redundancy between humans and carnivores is a goal, some systems may never respond in the intended fashion simply because of vagaries associated with environmental variance and bottom-up forces that may have little to do with hunting per se.

How to Make Human and Carnivore Hunting Equivalent

To replicate predation by carnivores, humans should hunt year-round, kill a high proportion of neonates, remove young-of-the-year, and in some instances take a higher proportion of adult females. The kill by humans would have to be measured against the predicted off-take, site-specific densities, and nutritional needs of carnivores (where native predators are present) while to some extent balancing prey dynamics in a very general way. For modern humans to mimic predation by carnivores, hunts must be at close (rather than a distant) range since native carnivores ambush or chase prey to make their kills. Modern arsenals that include rifles, ATVs, bikes, and other "contemporary" products obviously did not evolve with prey. So, the best way to mirror carnivore hunting *might* be as an effective archer. The issue, to be sure, is not whether our human ancestral effects should be duplicated today (a controversial proposition; Redford 1992; Kay 1994; Redford and Feinsinger 2001; Yochim 2001) nor even a goal; instead, merely how to equilibrate effects of modern human hunters and carnivores on biological diversity at different levels.

Even if carefully designed experiments were conducted there are practical issues: how best will functional redundancy be evaluated with respect to biological diversity? Spatial considerations become an imposing issue. At least two areas of Alaska that exceed in size the state of Idaho are available for hunting by indigenous subsistence users, but what of smaller and more populated regions, whether in Europe, Asia, or the contiguous United States? Do we measure vegetation or behavior, butterflies or bats, soil microorganisms or macroinvertebrates? These are not trivial questions, because to truly understand whether hunting by humans and by carnivores can ever be semi-equal, follow-up monitoring will be necessary, yet the selection of appropriate measures is uncertain.

Why Care If Functional Redundancy Occurs?

Although large dangerous carnivores might have shaped human behavior and subsequent patterns of colonization when humans were unarmed thousands of years ago, today, the opposite is true—it is rare that the presence of carnivores controls human destiny. Moreover, hunting by humans is held up by many as a de facto replacement of predation by carnivores. That is, human hunting is regarded as necessary because carnivores have disappeared or become ecologically irrelevant in many places (Pyare and Berger 2003). If the top-down influence of predation is an important structural feature of ecosystems, then the putatively replacement function of humans has a great deal of relevance for the conservation of biodiversity. If, however, a great deal of scrutiny is not focused on how this replacement occurs, then either we risk further ecosystem erosion or we just do not care. An understanding of the manner in which human hunting can serve the same role as that served by carnivores will affect management decisions.

To understand whether human and carnivore effects can ever be equal, appropriate areas must be available to evaluate ecological change (Dassman 1972). Otherwise the relative roles played by so many factors, both within and beyond reserves, will remain conjectural (Arcese and Sinclair 1997). The restoration or elimination of carnivore populations represents experimental opportunities to test hypotheses related to the ecological role of carnivores. This currently is occurring in the Greater Yellowstone Ecosystem where wolves have been reintroduced and brown bears are expanding their geographical range (Smith et al. 2003; Pyare et al. 2004). In reality, however, the question for practical on-the-ground management has little to do with ecological dynamics of wolf recovery or even biodiversity. Instead, local state game boards and appointed wildlife commissioners are interested in the degree of resultant interference and exploitation competition between humans and carnivores. Although the quantity of prey acquired by carnivores and people has bearings on biodiversity, rarely at such local levels are concerns voiced beyond that of prey abundance.

Beyond the interest of local hunters, ecologists, wildlife biologists, environmental advocates, ranchers, business councils, and politicians, the public from afar may hold notions of existence value and healthy ecosystems. This is particularly true in places where large carnivore populations are in a state of recovery. But do these carnivores then exert some type of "normative" ecological role (Pyare and Berger 2003; Soulé et al. 2003)? The individual states of Montana, Idaho, and Wyoming have either introduced or discussed legislation to begin harvesting both bears and wolves once they are removed from federal protected status. Such action not only underscores competition between humans and carnivores for prey but for some highlights the expectation of functional redundancy between these two sorts of hunters.

In Yellowstone, Alaska, and most corners of the planet, human hunting does not and will not replicate effects of carnivores that resonate across tiers of biological organization (see Table 16.2). The potential to do so may exist. Indeed, in some carefully manicured areas human predation may already mirror the temporal variability in predation by native carnivores. But, where intact functioning systems with carnivores can never be achieved, perhaps the best we can do is recognize differences imposed by our own human culture and our hunting, and attempt to maintain places that are good for our souls.

Summary

This chapter asks whether hunting by humans is functionally redundant to that by carnivores. Systems with and without humans and carnivores are contrasted with respect to multiple tiers of biological organization, including prey dynamics (and behavior), vegetation, and species dependent upon plant mosaics. The evidence that humans can replace carnivores in an ecologically functional way is not strong, although few studies have been designed a priori to examine these issues. Despite much site-specific variability, primary similarities between hunting by carnivores and by humans include density-reduction in prey and consequent change in herbivory. As such, one way in which human hunting affects biodiversity is through a chain of events in which the reduction of prey results in reduced herbivory and consequent enhancement of native species diversity. Nevertheless, major differences between human and carnivore hunting include (1) alteration of the intensity and timing of predation, (2) removal of different prey age and sex classes, (3) off-take of species other than harvestable prey, (4) modulation of mesopredator densities, (5) infrastructure to support human hunting with consequent effects on vegetation and plant-dependent species, (6) manipulation of carrion–scavenger relationships, and (7) modification of patterns of intraguild predation. In an ideal world, if modern humans were to replicate predation by carnivores, humans must kill at close range, remove a disproportionate number of neonates, harvest year-round, approximate in a general way biomass removal by carnivores, and reduce reliance on "Westernized" hunting styles that rely upon elaborate modern infrastructure. Because the world is not ideal and neither carnivore repatriation nor hunting may be invoked in many regions, perhaps the best that can be achieved for biodiversity conservation is to recognize its local loss while reaffirming its existence value in areas where it remains.

ACKNOWLEDGMENTS

Many thanks to K. Redford, J. Ray, and R. Steneck for their critical insights. Reviews by J. Beckmann, L. Bennett, K. Berger, and J. Robinson greatly improved this chapter. Comments from S. Cain, E. Fleishman, T. Kerasote, S. Pyare, J. Reick, M. Reid, M. Soulé, A. Treves, and B. Weber, and support in manuscript preparation by J. Zigouris, were especially valuable. My work has been supported by the National Park Service, Biological Resource Division of USGS, the National Science Foundation, the Wyoming Game and Fish Department, and the Engelhard Foundation. I am especially grateful to the H. B. Gillman Foundation for sponsoring this workshop.

Detecting Top-Down versus Bottom-Up Regulation of Ungulates by Large Carnivores: Implications for Conservation of Biodiversity

R. Terry Bowyer, David K. Person, and Becky M. Pierce

Models of predator–prey dynamics have a long and rich scientific history (Taylor 1984; Berryman 1992; Boyce 2000). Indeed, such models have underpinned our understanding of predator–prey systems and helped define how we view and implement conservation strategies for predators, prey, and the environments they inhabit (Ballard et al. 2001). Although predator–prey models are of considerable heuristic value (Hutchinson 1980), they also have played a key role in the applied ecology of large mammals. A knowledge of predator–prey systems underlies decisions about whether predator control may be necessary to meet societal goals (Gasaway et al. 1992), is used to formulate tactics for conservation of endangered prey (Sinclair et al. 1998), holds implications for understanding competition among large carnivores (Creel 2001), and has relevance for inbreeding depression and thereby time to extinction for prey (Hartt and Haefner 1995). Moreover, predator–prey dynamics may interact with habitat fragmentation to determine predator–prey equilibria and subsequent persistence of populations (Swihart et al. 2001), an outcome that makes implementation of conservation schemes based on single species risky (Prakash and de Roos 2002). Large carnivores also influence community structure of their prey (Henke and Bryant 1999). Predator–prey disequilibria affect interspecific behavior among large carnivores, as well as antipredator responses of their ungulate prey (Berger 1999; Brown et al. 1999; Berger et al. 2001a). Such disequilibria may result in trophic cascades that affect ecosystem structure and function (Bowyer et al. 1997; Kie et al. 2003 for reviews). Hence, the failure to consider predator–prey dynamics, in particular whether regulation of prey is primarily top-down or bottom-up, has ramifications for the conservation of biodiversity.

From their inception, predator–prey models have emphasized the role of predation in regulating prey, and discounted or ignored the effect of environmental carrying capacity (K—the number of prey at or near equilibrium with their food supply—McCullough 1979; Kie et al. 2003) on population dynamics of large herbivores and, thereby, on predator prey–relationships. Kie et al. (2003) provide a detailed discussion of the role of a variable climate and successional changes on K. The classic Lotka-Volterra equation for growth of a single species incorporates K, but that parameter is conspicuously absent from original equations describing predator–prey dynamics. Nonetheless, May (1974) demonstrated that inclusion of a resource-limitation term could have a stabilizing influence on predator–prey dynamics. Numerous advances in predator–prey theory have been made (Vucetich et al. 2002), but models depicting how such systems work are still largely predator driven. Only May (1974), Eberhardt (1998), and Person et al. (2001) have placed emphasis on K in models of predator–prey dynamics. Likewise, initial attempts at understanding the biology of predator–prey dynamics of large herbivores, and the carnivores that rely on them, concluded that resources would seldom be limiting for herbivores in terrestrial environments and that predation was consequently the most important factor constraining population growth of prey (Slobodkin et al. 1967). This "world is green" approach has been reconstituted in most predator–prey models proposed for large mammals and illustrates how our view of an ecosystem is constrained by the models we use to emulate its processes. Indeed, several authors still persist in the view that food seldom will be limiting for populations of large herbivores (Bergerud et al. 1983; Boertje et al. 1996), despite considerable evidence to the contrary (McCullough 1979; Kie et al. 2003 for reviews). Furthermore, similar thinking concerning the role of predation in regulating prey permeates modern approaches to predator–prey systems and many models forwarded to explain their dynamics (Boutin 1992; Ballard et al. 2001; Vucetich et al. 2002).

We maintain that controversy over whether population regulation of large mammals is top-down or bottom-up has its origins in the manner in which we model predator–prey dynamics, and that a predator-centric perspective has hindered our understanding of such systems. Resolving factors responsible for the dynamics of predator–prey systems is crucial to the management of large mammals and may hinge on how well we understand those processes. Although effective wildlife conservation may help mitigate risk of extermination for some species (Linnell et al. 2001b), large mammals, in general, and carnivores, in particular, historically have been at risk of extinction (Van Valkenburgh 1999)—they remain so today (Maehr 1997b; Woodroffe 2001). Our purpose is to provide a new framework in which to examine top-down and bottom-up regulation of populations of large herbivores. We contend that the need to understand population dynamics of these unique large mammals and interactions with carnivores that prey upon them is paramount for the effective conservation of biodiversity.

Conceptual Models of Predator–Prey Dynamics

Large mammalian herbivores are useful for studying top-down versus bottom-up regulation of populations. These animals are relatively large bodied, have comparatively long life spans, delay reproduction, have small litters, and exhibit high maternal investment in young. Ungulates generally exhibit life-history characteristics that are related to density dependence and, therefore, have a strong potential for population regulation at K (McCullough 1979; Fowler 1987; McCullough 1999; Kie et al. 2003). Likewise, these large herbivores are preyed upon by an impressive array of large mammalian carnivores (Mills 1989; Gasaway et al. 1992; Prins 1996; Smith-Flueck and Flueck 2001). The need to incorporate life-history information to produce realistic predator–prey models recently has been recognized by those studying insects (Dostalkova et al. 2002). We concur, and similarly argue that studies of organisms such as insect parasitoids and other arthropods with markedly differing life histories are unlikely to provide sufficient insights into predator–prey dynamics for large mammals so as to resolve issues related to top-down and bottom-up population regulation. We acknowledge, however, that implementing an experimental approach for these vagile, and sometimes difficult-to-study animals provides a daunting impediment to understanding complex predator–prey systems (McCullough 1979; Boutin 1992; Stewart et al. 2002; Kie et al. 2003).

Predator–prey dynamics for large mammals often have been examined from the perspective of four conceptual models: recurrent fluctuations, low-density

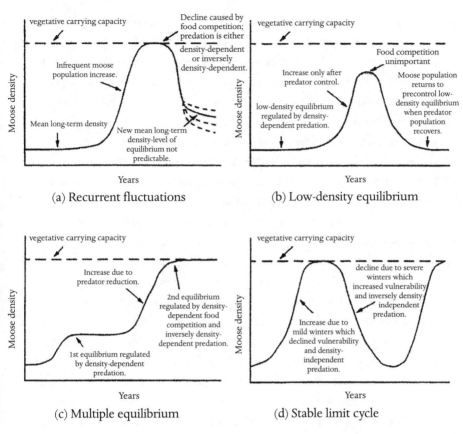

Figure 17.1

Four conceptional models for understanding population dynamics of ungulates and large mammalian carnivores (from Ballard et al. 2001, with permission—Copyright, The Wildlife Society).

equilibrium, multiple equilibria (predator pit), and the stable-limit cycle (Boutin 1992; Van Ballenberghe and Ballard 1994; Ballard et al. 2001; Fig. 17.1a–d). Although of considerable heuristic value, these conceptual models have had limited success in making empirically supported predictions concerning predator–prey dynamics (Boutin 1992 provides those predictions). This is a frustrating situation for those wishing to implement conservation and management initiatives for large mammals and their habitats based on these conceptual approaches.

A model of recurrent fluctuations (Fig. 17.1a) implies that an ungulate population fluctuates markedly in density but will not reach equilibrium. Although any perturbation can affect population numbers, such fluctuations are principally a result of severe weather, forage quantity and quality, and especially predation. Predation is inversely density dependent at high densities, and density dependent at low numbers of prey. Although prey may remain at low densities for extended periods of time, the long-term level of abundance cannot be predicted.

A low-density equilibrium (Fig. 17.1b) describes a system in which prey are held at low density by predation (i.e., density-dependent predation) for long periods. Should rates of predation lessen sufficiently (e.g., from predator control or a natural phenomenon), the prey population would rebound toward K but would never reach that level. Food limitation is unimportant under this model, and predation ultimately would reduce prey again to a low-density equilibrium.

A multiple-equilibria model (Fig. 17.1c) predicts regulation of prey by predators at low density, but allows for food limitation of the prey population at K. Prey populations are not thought to persist near $K;$ however, multiple equilibria at various densities of prey below K are possible. One result of this model is a Ricker-like (McCullough 1979) predator pit in which predation results in a strong point of equilibrium at low density of prey. When released from predation pressure, prey density will increase until it has reached a higher-density equilibrium with predators. This scenario is often an underlying assumption and justification for predator control (Gasaway et al. 1992).

Stable-limit cycles (Fig. 17.1d) are thought to be the result of interactions among density-independent processes (e.g., severe weather), population density of prey in relation to K, and predation. Here predation is density independent during periods of increasing prey abundance, and inversely density independent during declines in prey. Those processes are hypothesized to generate cycles with regular amplitudes and durations of 30 to 40 years.

Failure to Consider Effects of K

Few of the four conceptional models (Fig. 17.1a–d) adequately consider effects of K on the dynamics of predator–prey relationships. Low-density equilibrium ignores K, recurrent fluctuations and stable-limit cycles predict only short-term periods where prey are near or at K, and multiple equilibria allow for an equilibrium near K, although even that situation is thought to be transitory. All models incorporate the concept of predation rate relative to prey density as a driving force. None allows for an overshoot of K and subsequent decline in K from overexploitation of forage—a likely outcome from population irruptions that may occur under low rates of predation or from the lack of other important sources of mortality (Leopold 1943; Klein 1968; Caughley 1970; McCullough 1979; Andersen and Linnell 2000). In addition, no model satisfactorily addresses effects of the approach or decline of the prey population to and from a potentially changing K and the subsequent influence of those changes on recruitment of prey on dynamics of predator–prey systems, except via kill rate. The assessment of kill rate can be misleading because all models assume that K is constant and mortality of prey additive, an unlikely set of circumstances.

The failure to more fully incorporate K into models of predator–prey relationships has further ramifications. A small change in K may precipitate a large change in prey numbers—a conclusion also reached by McCullough (1979). Such an outcome stems from the nonlinear density-dependent relation between annual recruitment and population density of prey with respect to K. The area under the curve representing maximum sustained yield (MSY) declines in a negative-exponential fashion as K is reduced (Fig. 17.2). Consequently, net annual recruitment of prey, which represents the portion of a prey population that can be removed by predators (and other sources of mortality) without causing a decline in the population, is reduced disproportionately to the decline in K. Indeed, Sutherland (1996) noted that lowering K will disproportionally alter demographic rates along a declining spectrum of prey densities. Those varying densities of prey and their effects on availability and distribution of food, as well as their inputs of urine and feces, are the mechanism whereby large herbivores bring about key changes in ecosystem structure and function (Molvar et al. 1993; Wallis de Vries 1995; McShea, this volume). Accordingly, whether predators exert top-down influences on prey, or fail to do so (i.e., limitation is bottom-up), has ramifications for the biodiversity of ecosystems.

Relying on these four conceptional models to understand predator–prey dynamics has other shortcomings. The time necessary to recognize which model likely was correct is decadal or longer. Important conservation issues related to habitat or conservation of predators or prey likely would be resolved (for good

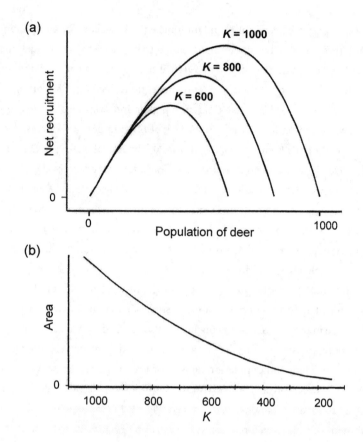

Figure 17.2
Hypothetical curves showing recruitment number of a large herbivore for varying levels of K (a), and relation between K and area under the recruitment curve as a function of K (b). As K is reduced, the area under the recruitment parabola declines in a nonlinear fashion.

or ill) long before an informed decision could be made. We argue that models that rely mostly on predation rate and fail to adequately consider K, or in some instances completely ignore this parameter, are ill suited to assess whether population regulation of prey is top-down or bottom-up. Indeed, the manner in which these models are conceptualized leads inexorably toward a conclusion of top-down regulation. For example, as K declines, stochastic events, such as severe winter weather (Sæther 1997; Solberg et al. 2001; Kie et al. 2003), or time lags in the numerical response of predators to changes in the density of their prey (O'Donoghue et al. 1998; Pierce et al. 1999; Keeling et al. 2000; Pierce et al. 2000a), may combine to limit number of prey to low levels without a concurrent reduction in predators. Under these circumstances, top-down control imposed by predation may appear to supersede effects of K on prey, an interpretation that could lead resource managers to undervalue the important role of habitat quality.

An example of how research might lead to potentially erroneous interpretations is provided by one study of wolves (*Canis lupus*) and black-tailed deer (*Odocoileus hemionus*) on Vancouver Island, Canada, which indicated that declines in populations of deer were the result of predation by wolves, and that changes in habitat because of logging had little effect on numbers of deer (Atkinson and Janz 1994; Hatter and Janz 1994). The authors suggested that numbers of deer declined in logged and unlogged landscapes and, thus, habitat change was not a factor influencing the decline of deer populations (Atkinson and Janz 1994; Hatter and Janz 1994). No information concerning relative densities of deer or K in logged and unlogged landscapes was provided. The authors further suggested that when numbers of deer were kept low by predation, deferring logging of winter habitat for deer was difficult to justify. In our view, studies such as these simply demonstrate that densities of ungulates are lower when exposed to predation by wolves than where wolves are absent. We hypothesize that the potential for populations of deer to rebound from low levels imposed by weather and predation is as dependent on K as it is on the reduction of predators. Indeed, using low densities of deer to justify reducing K for deer simply perpetuates a conceptual problem and risks a management catastrophe. We believe that failing to consider K of ungulate prey in dynamics of predators is an oversight that likely will result in misinterpretation of data and may hamper conservation efforts designed to assist predators and their prey or to maintain biodiversity.

Prey to Predator Ratios

Measures of the ratio of ungulate prey density (or their biomass) to predator density have been used widely to predict effects of predators on their ungulate prey (Keith 1983a; Fuller 1989; Gasaway et al. 1992; Person et al. 2001), thereby offering a potential mechanism to infer whether top-down or bottom-up processes were at work. One approach has been to use linear regression to predict density of predators from prey biomass (Keith 1983a; Fuller 1989; Gasaway et al. 1992; Messier 1995). The supposition is that density of predators is predicted by prey abundance, and this value represents an approximate carrying capacity or equilibrium density for predators (Gasaway et al. 1992). Accordingly, if predator densities are greater than predicted, or if prey–predator ratios are less than envisaged, then predators ostensibly would cause a decline in prey (i.e., regulation was top-down). An apparent time lag between numbers of mule deer (*Odocoileus hemionus*) and declining numbers of mountain lions (*Puma concolor*) indicates that interpretation of prey–predator ratios is difficult at best (Fig. 17.3). In that system, the mule deer population initially crashed during an extensive drought and only began recovering when the drought subsided—deer likely were tracking *K* (see Fig. 17.3). Mountain lion numbers initially remained high but ultimately declined with a substantial lag behind numbers of their principal prey (likely bottom-up forcing). Mule deer recovered much more slowly (well below their maximal intrinsic rate of increase) following the drought, even though the range and physical condition of deer had improved markedly (Pierce et al. 2000b; probably top-down limitation from mountain lion predation). The deer–mountain lion ratio in relation to number of mule deer during periods of deer recovery, however, is an exponentially increasing curve indicative of a prey population that had escaped effects of predation (bottom-up limitation). Differing conclusions concerning whether regulation is top-down or bottom-up are related to when predator–prey ratios are measured (Fig. 17.3). Even long-term data sets may not be sufficient to untangle potential biases in interpretation of prey–predator ratios.

Person et al. (2001) have cautioned that combining biomass from different species of prey is not advisable because this method obscures effects of variation in intrinsic rates of increase among prey species on predator–prey dynamics, including potential points of equilibria. Indeed, use of prey–predator ratios has been controversial (Theberge 1990; Messier 1994). Theberge (1990) further argued that changes in the functional response of predators to variation in prey density, prey-switching, and the nearness of the prey population to *K* would make interpretation of prey to predator ratios problematic. Moreover, predation rate per predator for a particular species of prey likely depends upon density of that prey, and the simultaneous density of alternative prey (Dale et al. 1994; Jędrzejewski et al. 2000).

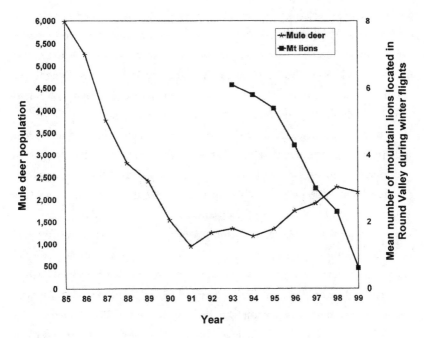

Figure 17.3
Numbers of mule deer and mountain lions in Round Valley, California, USA, during 1985–1999. A drought began in 1987 and ended in 1992. Note that numbers of prey to predators result in a ratio that increases exponentially from 1993 to 1999 (adapted from Pierce et al. 1999, 2000a).

Consequently, density-dependent changes in rate of killing by predators could require a reiterative interpretation of ungulate to predator ratios with changes in prey density for this method to offer meaningful insights into population regulation (Person et al. 2001)—a daunting task for those managing populations of either predators or their prey.

Kill Rates

Even the most sophisticated models (Vucetich et al. 2002) developed to explain predator–prey dynamics rely on the kill rate of predators in relation to either the abundance of prey (i.e., the functional response—Holling 1965) or the ratio of predator to prey. Although such modeling has become more elaborate over time, considerable debate exists over which model best describes predator–prey dynamics (Arditi and Ginzburg 1989; Ginzburg and Akçakaya 1992; Abrams 1994; Akçakaya et al. 1995; Abrams 1997). Many of these models are based on theoretical formulations, and much controversy has resulted from an absence of empirical data, especially for large mammals (Vucetich et al. 2002).

Vucetich et al. (2002) compared a variety of prey-dependent, ratio-dependent, and predator-dependent models against empirical data from wolves and moose (*Alces alces*) from 1971 to 2001 on Isle Royale, Michigan, USA. These authors concluded that, although both models may have value, they were overly simplistic —neither ratio-dependent nor prey-dependent models deserved primacy for understanding predator–prey relationships. We hypothesize that the relative poor fit ($R^2 \leq 0.36$) of the models examined by Vucetich et al. (2002) stems from the failure to include K in any model. Moreover, none of the models examined by these authors is tightly linked to the four conceptual models used to guide our understanding of predator–prey dynamics among large mammals (Fig. 17.1a–d).

Person et al. (2001) modeled density of wolves relative to moose by varying the population density of ungulates at which predation rate by wolves was halved (D), and the shape of the density-dependent growth curve for ungulates (θ). Those authors reported that, as the ratio of D to U (the prey population) became smaller, the influence of the functional response on the density of wolves decreased. Person et al. (2001) concluded that the functional response might have little effect on predator–prey systems of large mammals except at very low density with respect to K. Only at low density was there a discernable difference between simulations with D bounded by $[0, K/8]$ and simulations of $D = 0$ (which eliminated the functional response). Indeed, Marshal and Boutin (1999) cautioned that it was at such low densities, where reliable data were most difficult to obtain, that distinguishing between types of functional responses could be problematic because of low statistical power resulting from small sample size. Moreover, the effort and expense necessary to gather data for large mammals to estimate the type of functional response can be immense (Dale et al. 1994; Jędrzejewski et al. 2002). Studies by Marshal and Boutin (1999) and Person et al. (2001) draw into question the value of estimating the instantaneous kill rate of ungulates by large carnivores and, thereby, the worth of prey-dependent and predator-dependent models for un-

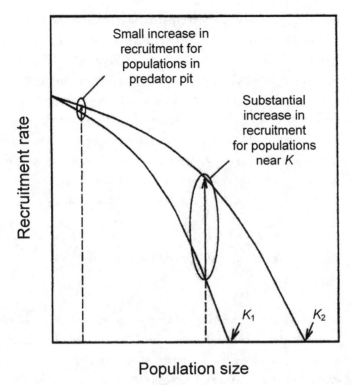

Figure 17.4
Variation in recruitment rate with increasing population size relative to long-term changes in carrying capacity (*K*) for an ungulate population. A substantial improvement in recruitment rate occurs only as the population increases from low to high density (from Kie et al. 2003, reprinted with the permission of Cambridge University Press).

derstanding predator–prey dynamics of large mammals and their subsequent effects on biodiversity.

Ratio-dependent models, likewise, have limitations for deciphering relationships between ungulates and the large carnivorous mammals that prey upon them. One prediction of these models is that an increase in *K* would result in an increase in both prey and predator. A simple population model for a large herbivore that includes *K* (McCullough 1979), however, indicates that an increase in *K* differentially affects recruitment rate relative to where the population is with respect to *K*. There is little increase in recruitment rate, therefore, for a population at low density because animals are not food limited and reproducing at near maximal rates (Fig. 17.4). The improvement in recruitment rate from enhancing K increases as the population moves toward the new carrying capacity (K_2) from its old one $(K_1$; see Fig. 17.4). This increase in recruitment rate from enhancing K results in differing prey availability to predators across a wide range of population densities with respect to K, which fits outcomes predicted by ratio-dependent models poorly. We note, however, it is at low density of prey where predator limitation (top-down forcing) has the greatest empirical support; these low densities of prey relative to K typically involve multiple-predator and multiple-prey systems (Gasaway et al. 1992; Bowyer et al. 1998; Hayes and Harestad 2000b).

A Prey-Based Approach for Understanding Top-Down and Bottom-Up Processes

If only bottom-up processes were involved in population regulation of ungulates, effects of large carnivores on prey numbers would be minimal, and conservation measures to benefit carnivores would have few consequences for biodiversity. Conversely, where these large predators alter the density of their prey relative to K (i.e., top-down forcing), the management of carnivores may have profound effects on biodiversity. We acknowledge that no system is regulated exclusively by either top-down or bottom-up processes and suggest that it is misleading to view such processes as a dichotomy. We also recognize that justifying the maintenance or restoration of large carnivores for the purpose of conserving biodiversity requires knowledge of their role in promoting ecosystem integrity. We caution, however, that effects of carnivores on their ungulate prey may change over time, and that predator-centric approaches, such as determining kill rates and prey to predator ratios, are poorly suited for determining whether forcing is primarily top-down or bottom-up.

We believe that an assessment of top-down versus bottom-up limitation of prey populations are most easily and accurately interpreted through simple models of prey population dynamics. Moreover, our approach does not require competing models that provide a yes-or-no answer to a process that is a continuum. Although questions concerning population regulation via predation or food often are framed as a dichotomy (populations of ungulates overshooting K, or being held at a low density by predation), a prey population might be regulated by top-down processes over one time period, and bottom-up effects during the next, a result suggested by May (1974). The most important consideration from a conservation perspective is to recognize what is regulating or limiting the population, and to take appropriate action relative to the conservation of predator, prey, or the biodiversity of their environment. Attempts to understand the intricate nature of predator–prey interactions are of considerable theoretical value but may hinder conservation efforts if they become the primary evaluative tool for making decisions concerning top-down and bottom-up processes and their effects on biodiversity.

We maintain that far too much reliance has been placed on the number of prey killed or the kill rate in interpreting predator–prey relationships. Although an adequately large kill of prey is necessary to invoke a predator-limited or regulated population, it is not sufficient to know that there is top-down regulation of prey. For instance, high mortality of young occurred in a mule deer population exposed to predation by coyotes (*Canis latrans*), in which both low reproductive rates of deer and poor range condition indicated the deer population was near K (Bowyer 1984, 1987, 1991). This outcome likely occurred because whether mortality of ungulate prey is additive or compensatory is related to proximity of the population to K. Mortality in prey populations becomes increasingly compensatory as the population grows from near MSY toward K, but it is largely additive at population densities below MSY (McCullough 1979; Kie et al. 2003; Fig. 17.5). Consequently, heavy losses of young in an ungulate population near K are not grounds for concern; those young would have died from other causes anyway (i.e., mortality was compensatory; Errington 1967). Simply documenting that predators are killing large numbers of prey is insufficient to infer top-down forcing and might lead to unnecessary control of predators.

We contend that life-history characteristics of ungulate prey (*sensu* Kie 1999; McCullough 1999; Keech et al. 2000; Kie et al. 2003) can be used to infer whether population limitation is top-down or bottom-up because of the strong density dependence exhibited by those large mammals (Table 17.1). Much of our knowledge concerning such processes comes from northern ecosystems. Nevertheless, our predictions are based on fundamental concepts of population

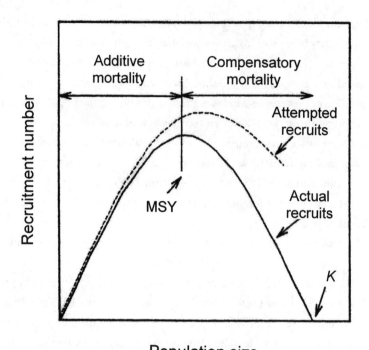

Population size

Figure 17.5
Changes in recruitment number and attempts to recruit young with increasing population size of an ungulate population. Females attempt to reproduce at a higher level than can be supported by the environment from densities ranging from maximum sustained yield (MSY) to carrying capacity (*K*), and that attempts to recruit young parallel the recruitment number below MSY because females are in good physical condition. Mortality tends to become increasingly compensatory from MSY to *K*, but is largely additive below MSY (from Kie et al. 2003, reprinted with the permission of Cambridge University Press; adapted from McCullough 1979).

ecology (Hutchinson 1980) and, consequently, should have wide applicability. Top-down processes would seldom result in ungulate populations near *K*, but rather in populations held at extremely low densities with respect to *K*. Consequently, measures of animal condition and reproduction in populations near *K* should be low, indicating bottom-up forcing. Conversely, top-down limitation implies that un-

Table 17.1

Life-history characteristics of ungulates that reflect the relative differences in a population regulated by top-down versus bottom-up processes

Life-History Characteristic	Population Top-Down Regulated	Population Bottom-Up Regulated
Physical condition of adult females	Better	Poorer
Pregnancy rate of adult females	Higher	Lower
Pause in annual production by adult females	Less likely	More likely
Yearlings pregnant[a]	Usually	Seldom
Corpora lutea counts of adult females[a]	Higher	Lower
Litter size[a]	Higher	Lower
Age at first reproduction for females	Younger	Older
Weight of neonates	Heavier	Lighter
Mortality of young	Additive	Compensatory
Age at extensive tooth wear	Older	Younger
Diet quality	Higher	Lower

[a] Some species of ungulates may show limited variability in particular characteristics.

gulates would be held at a low density relative to K, and the physical condition and reproductive performance of individuals in such populations would be high (see Table 17.1). Likewise, dietary quality should vary with population density of ungulate prey relative to K, with intensified intraspecific competition near K resulting in a lower-quality diet than would be expected for populations held far below K by predation.

This approach has limitations but may offer the only data readily available to help determine if populations of ungulates are predator-limited, and allow biologists to respond with appropriate management in a timely manner (Kie et al. 2003). Obviously, factors other than predation can drive populations to low levels or cause them to oscillate near K. Difficulties in sorting among other potential causes of population change, however, are minimal compared with trying to determine which conceptual model of predator–prey dynamics is appropriate (Fig. 17.1a–d), or in trying to determine kill rate, especially at low densities where it is most likely to result in an equilibrium. Moreover, either indices of overgrazing and hedging of trees and shrubs (Caughley 1977; Riney 1982; Kie et al. 2003) or other forage-based estimates of K (Hobbs et al. 1982; Stewart et al. 2000) may be used to help calibrate where the prey population is with respect to K.

Future Directions for Predator–Prey Modeling

There is an obvious need to incorporate values of K in future models of predator–prey dynamics for large mammals. We can still engage in conservation efforts that require knowledge of whether limitation is top-down or bottom-up using the approach we have recommended (Table 17.1), but having realistic and predictive models ultimately would be of theoretical and applied value. There is also a clear necessity to manipulate populations of predators and prey to fully understand these systems (Boutin 1992). Such manipulations will be difficult to perform with populations of large mammals, but opportunities for adaptive management should be sought out with an eye to resolving existing issues concerning how these systems work, and specifically how predator–prey dynamics are linked to biodiversity.

Including more information related to the life-history characteristics of predators and prey is also likely to provide new insights into their dynamics (Gittleman 1993; McCullough 1999). For instance, populations of large polygynous ungulates sexually segregate for much of the year (Bowyer 1984; Bowyer et al. 1996; Bleich et al. 1997; Kie and Bowyer 1999; Barboza and Bowyer 2000). Consequently, the population density of adult females, rather than adult males, relative to K has the greatest effect on recruitment of young and thereby the dynamics of the population (McCullough 1979). Accordingly, a male ungulate killed by a predator will have a proportionally lower effect on recruitment of young into the ungulate population than would the death of a female. Because males of dimorphic ungulates are considerably larger than females (Weckerly 1998), the food they provide is likely to affect reproduction of predators more than that of smaller-bodied females or young. Both outcomes have potential to affect predator–prey dynamics, including top-down and bottom-up processes, in ways that are not considered in existing models. The manner in which the sexes of ungulates are distributed spatially upon the landscape and the effects of this pattern on predator–prey dynamics is a topic in dire need of additional research. Perhaps an initial approach would be to modify the classic Lotka-Volterra equations for resource competition (Tilman

1982) to represent different sexes rather than different species. In addition, Pierce et al. (2000b) documented that female mountain lions with young killed a disproportional number of young mule deer compared with other sex, age, and reproductive classes of lions. Such selectivity could also have effects on productivity of prey populations and, in consequence, predator–prey dynamics. Despite such potential improvements in theoretical modeling, we concur with Person et al. (2001) that limited resources for conducting research on predator–prey dynamics of large mammals should be concentrated on understanding the growth of prey populations with respect to habitat quality in relation to the predation behavior of carnivores. Indeed, few studies concerning the conservation of carnivores consider the habitat necessary to support adequate densities of associated prey, a point also raised by the National Research Council and its Committee on Management of Wolf and Bear Populations in Alaska (1997).

Linking Predator–Prey Dynamics to Ecosystem Processes and Biodiversity

Large carnivores affect prey other than via population regulation, including influencing degree of sociality, habitat use, foraging dynamics, and distribution of ungulates across the landscape (Berger 1991; Molvar and Bowyer 1994; Berger 1999; Kie 1999; Berger et al. 2001b; Bowyer et al. 2001; Mills, this volume). Likewise, species of available prey and their dispersion hold import for the social organization of predators (Mills 1989; Pierce et al. 2000a). Predators make for a rich environment that embodies a full array of natural behaviors in ungulates that are absent from depauperate ecosystems lacking these unique mammals—an element of biodiversity that is seldom considered.

Systems without large carnivores often experience trophic cascades in which ungulates have deleterious effects on vegetation and other animals (Hobbs 1996; Bowyer et al. 1997; Kie et al. 2003 for reviews). Changes in densities of large herbivores have the potential to drive nutrient cycling in terrestrial and aquatic systems (McNaughton 1984; Ruess and McNaughton 1987; Irons et al. 1991; Frank and McNaughton 1993; Molvar et al. 1993) and affect successional pathways of vegetation communities (Pastor et al. 1993; Wallis de Vries 1995), resulting in "ecological meltdowns" of some systems that markedly lower their biodiversity (Terborgh et al. 2001). Indeed, ecological cascades from foraging by ungulates on rodents (Keesing and Crawford 2001), birds (deCalesta 1994; McShea and Rappole 2000; Berger et al. 2001a), and insects (Suominen et al. 1999a; Souminen et al. 1999b) are well documented. It is axiomatic that top-down forcing of ungulate prey by carnivores, or regulation of ungulates by bottom-up processes, holds potential to affect the relation of the prey population to K and thereby ecosystem structure and function, and ultimately biodiversity. Consequently, predator–prey dynamics hold important consequences for the well-being and richness of ecological systems. Many challenges exist in conserving large carnivores (Miquelle et al. 1999; Ginsberg 2001) and, thus, the biodiversity of landscapes they inhabit. We believe that understanding the role of top-down and bottom-up regulation of prey is an essential step in this critical process.

Summary

We set forth predictions for determining whether populations of large herbivores are regulated primarily via top-down or bottom-up processes. We contend that existing models of predator–prey dynamics based on kill rates—including prey-dependent, ratio-dependent, and predator-dependent approaches—are not well suited for understanding top-down and bottom-up regulation of ungulates by their predators. These models make predictions that are not realistic, do not cope with carrying capacity (K) of ungulate prey, fail to consider that some mortality of prey may be compensatory, or do not explicitly deal with multiple-prey–multiple-predator systems. Similarly, the four conceptual models—recurrent fluctuations, low-density equilibrium, multiple equilibria (predator pit), and stable-limit cycle—are predator-centric and offer limited promise to explain population dynamics of large mammals. We have demonstrated that, except at very low density of prey relative to K, where kill rates are most difficult to measure, population density of prey with respect to K is most important in determining potential points of equilibria, and thereby whether regulation is strongest from above or below. Moreover, funding necessary to collect data sufficient to fit models that predict kill rates across seasons for a sufficient number of years are seldom available; conservation issues would be long resolved before the best model could be selected. We have constructed a conceptual framework to make predictions about whether populations of large herbivores are regulated by top-down or bottom-up processes, and propose criteria to assess whether predator control would be effective in releasing ungulate populations from low-density equilibria. Because of the critical role of large carnivores in influencing biodiversity, primarily through their effects on dynamics of ungulate populations and their subsequent influences on ecosystem processes, understanding the role of top-down and bottom-up regulation of prey is an essential step to conserving large carnivores and the biodiversity of landscapes they inhabit.

ACKNOWLEDGMENTS

This paper benefited from helpful discussions of predator–prey dynamics with V. Van Ballenberghe, K. M. Stewart, J. G. Kie, and other authors of this book, as well as comments from the editors and two anonymous reviewers. Our work was funded in part by the Institute of Arctic Biology at the University of Alaska Fairbanks, the Alaska Department of Fish and Game, the California Department of Fish and Game, and the Department of Biological Sciences at Idaho State University.

CHAPTER 18

Top Carnivores and Biodiversity Conservation in the Boreal Forest

Stan Boutin

Large carnivores have long been the focus of many conservation programs because of their vulnerability to extinction, the public interest they generate, and their need for large protected areas (Noss et al. 1996; Ray, this volume). Recent attention to the global loss of species has spawned a biodiversity crisis that has resulted in the broader mandate to conserve biodiversity at multiple levels, including genes, species, habitats, and ecosystems (Noss 1990). This is a formidable challenge that has scientists, government, industry, and conservation organizations struggling to find effective approaches that achieve the overall objective without getting mired in the complexity.

One strategy might be to focus on those species and/or processes that play important or keystone roles in the structure and function of ecosystems in the hopes of carrying other elements of biodiversity with them (Simberloff 1998; Bergeron et al. 1999). Is there a scientific case to be made for top carnivores serving in this capacity and thus justifying continued emphasis on these relatively well-studied species? Estes (1996) summarized the growing evidence that mammalian carnivores have measurable effects on ecosystem processes, while at the same time pointing out that the actual scientific evidence for the functional importance of large carnivores in terrestrial ecosystems is weak (see also Linnell et al. 2000). With a growing number of alternative approaches to biodiversity conservation currently being developed, it is an opportune time to critically evaluate the scientific basis for continuing with a large carnivore focus to conserve biodiversity.

The boreal forest biome harbors some of the last great tracts of intact forest in the world (Nellemann et al. 2001). Unfortunately such areas are fast being changed by industrial activity, including forestry, and oil and gas, and agriculture

(Schneider et al. 2003). Biodiversity conservation concerns are therefore timely, with the typical range of approaches being considered. These include establishment of protected areas, ecosystem management, conservation of rare and endangered species, and conservation of top carnivores such as wolves (*Canis lupus*) and grizzly bears (*Ursus arctos*). The boreal forest is known for large-scale disturbances such as fire (Johnson 1992) and for the striking population cycles in many of its herbivores, including voles (*Clethrionomys* and *Microtus* spp.), snowshoe hares (*Lepus americanus*), forest grouse, and insect defoliators (Körpimaki and Krebs 1996). Midsized to small predators such as lynx (*Lynx canadensis*), fox (*Vulpes vulpes*), coyotes (*Canis latrans*), and weasels (*Mustela* spp.) figure prominently in vertebrate herbivore cycles (Körpimaki and Krebs 1996). The boreal forest is also the important summer home to a diverse array of neotropical migrant birds (Schmiegelow and Mönkkönen 2002). If the objective is to conserve the boreal forest's biodiversity (species, structure, and function), can we do so by focusing on its top carnivores or are there alternative species or groups that represent better keystones or umbrellas? How might conservation programs based on processes like natural disturbances such as fire and insect outbreak serve as an alternative strategy? All of these approaches have merit but does current scientific information provide some clear direction for setting priorities?

My objective in this chapter is to examine the scientific evidence for the importance of carnivores to boreal forest structure and function and, ultimately, to maintenance of species diversity. Based on this, I will assess how a conservation program focused on conserving large carnivores might achieve the broader objective of biodiversity conservation. I define "biodiversity" as the total species complement in a natural boreal forest with the assumption that this will encompass associated habitats and landscapes. Most carnivore conservation programs tend to focus on large or apex species (wolves or grizzly bears) as opposed to midsized predators such as lynx or fox. To assess whether this priority is justified, I will, where possible, compare the ecological effects of different carnivores—namely, wolves, brown bears, lynx, and coyotes. Specifically, I will review the literature to answer the following questions: (1) What is the relative dominance (relative biomass) of various carnivores and their prey in boreal ecosystems? (2) Do boreal forest carnivores exert strong limiting/regulatory effects on their prey? (3) Is there evidence that loss of carnivores can lead to trophic cascades? and (4) Do carnivores influence species diversity in other ways? Although the focus of this chapter will be on the North American boreal forest I will also draw on work from Fennoscandia where possible. Answers to the above questions will help to determine the relative influence of various carnivores in the boreal forest, but, where possible, I will also try to assess how the importance of carnivores compares to other species groups or to other ecological processes in affecting biodiversity. It is this comparison that is crucial when trying to decide on the broad merits of a conservation program focused on carnivores versus alternative approaches.

Scientifically, the best way to determine the importance of a phenomenon is to remove it and observe the subsequent effects. This "experiment" has been all too common for species like wolves and grizzly bears but unfortunately the loss of these species from much of their range has also been associated with other human-caused changes, so it is impossible to rule out confounding explanations for the ecosystem changes that have occurred. Actual controlled carnivore removal experiments are rare. However, there has been one experimental case study in the boreal forest at Kluane Lake, Yukon, where my colleagues and I tried to understand how the boreal forest vertebrate community was organized by quantifying the food web and performing a series of experiments designed to "kick" each trophic level to see how the community responded (Krebs et al. 2001a). I will draw heavily on the results of this study in addressing the questions posed in this chapter.

The Boreal Forest Context

The boreal forest cuts a wide swath across North America and Eurasia (Rowe 1972; Esseen et al. 1997). It is a treed biome dominated by a mix of coniferous and deciduous species. Wolves and wolverine (*Gulo gulo*) are the apex carnivores (i.e., they largely escape predation) across this vast area and their prey base consists of moose (*Alces alces*), caribou (*Rangifer tarandus*), and a variety of deer species. Black bears (*Ursus americanus*) are widely distributed whereas grizzly bears are largely limited to mountainous terrain. Although wolverine have not been well-studied, it is safe to say that densities are low in the boreal forest. Wolves have experienced some reduction in distribution in boreal North America but their populations remain robust (Hayes and Gunson 1995). Wolves, bears, and wolverine have been reduced to very low levels in Fennoscandia, but numbers appear to be recovering (Swenson et al. 1998; Linnell et al. 2000).

The midsized to small carnivore guild is more diverse and variable across the boreal forest. In North America, the Canada lynx is widespread whereas the red fox and coyote show more restricted distributions with the tendency toward coyotes excluding foxes and wolves excluding coyotes (Peterson 1995). In Scandinavia, foxes are prominent (Lindström et al. 1994), coyotes are absent, and the European lynx (*Lynx lynx*) is an ungulate predator feeding on roe deer (*Capreolus capreolus*). Weasels and stoats (*Mustela* sp.) play important roles in regulating populations of small mammals (Hanski et al. 1993). In North America, marten (*Martes americana*), fisher (*Martes pennanti*), and various species of weasel are widespread but none of these species has been identified as having important influences on their prey base. However, there have been no studies that have addressed this question directly.

The boreal forest is notable for the widespread occurrence of population cycles in the small to midsized herbivorous mammals and in a number of phytophagous insect species. These cycles could be considered "signature" phenomena in the boreal biome. In Canada, vertebrate cycles are dominated by the 10-year snowshoe hare cycle, whereas in Scandinavia various vole species are the dominant players, with cyclic periods of 4 to 6 years (Körpimaki and Krebs 1996). Weasels and stoats play a key role in the vole cycles (Hanski et al. 1993; Körpimaki et al. 2002) whereas the lynx and coyote are important for the hare cycle (O'Donoghue et al. 2001a; Keith et al. 1984). The boreal forests of Russia are less well known in the scientific literature but it appears that both 3- to 4- and 10-year cycles exist in the small herbivore guild (Danell et al. 1998).

The Kluane Study took place in the southwestern Yukon in forest dominated by white spruce (*Picea glauca*). The forest remains relatively pristine. There has been no commercial forestry and much of the area has not burned for over 100 years (Dale et al. 2001). The study area was bisected by the Alaska Highway and there was some hunting and fur trapping around but not within the study area. The research focused on the vertebrate food web and we tried to measure both standing biomass and major trophic flows during a 10-year snowshoe hare cycle on a 350 km^2 region. Our experiments (conducted on 65–100 ha plots) were relatively straightforward: we added fertilizer to pulse the plant trophic level, supplemented snowshoe hare and ground squirrel (*Spermophilus* spp.) diets with rabbit chow to increase the herbivore trophic level, and excluded terrestrial carnivores by electric fencing to alter the predator trophic level. In addition we excluded snowshoe hares from small 4 ha exclosures and ran a series of interaction experiments (food addition plus predator exclusion, hare exclosure plus fertilization) at a variety of scales. Details of the experimental design can be found in Krebs et al. (2001a).

How Do Carnivores Affect the Boreal Forest?

How carnivores impact boreal forest ecosystem properties can be evaluated at various levels. In this section, I take a stepwise approach in examining this issue using results from the Kluane study and other published boreal forest research. First, I undertake a comparative analysis of predators and prey biomass across various boreal forest sites as a way of assessing the potential ecosystem influence of these animals. Next, I review the evidence for top-down control of the herbivore trophic level and, in turn, the potential for loss of the top trophic level to precipitate trophic cascades or to influence biodiversity in other ways.

Numerical Dominance of Carnivores and Their Prey

I begin by asking how the numerical dominance (biomass) of various mammalian carnivores and their prey compare in the boreal forest, the rationale being that numerically dominant species would be more likely to exert greater influences on the ecosystem in general. It is actually extremely rare for any study to measure the abundance of more than a handful of vertebrate species within the same community. At Kluane, we estimated biomass of 10 species and derived density indices for another 33 species over an entire 10-year cycle (Boutin et al. 1995; Smith and Folkard 2001). Lynx and coyotes dominated the predator trophic level and their average biomass was four to seven times that of wolves (Fig. 18.1; Krebs et al. 2001a). Wolves were infrequent visitors to our study area based on winter tracks (O'Donoghue et al. 2001b), and density estimates, as determined by government surveys in the region, confirmed that their numbers were relatively low (Gasaway et al. 1992). Peak lynx biomass was 17 times higher than that of wolves and there was no indication that wolf densities changed in relation to the hare cycle. Win-

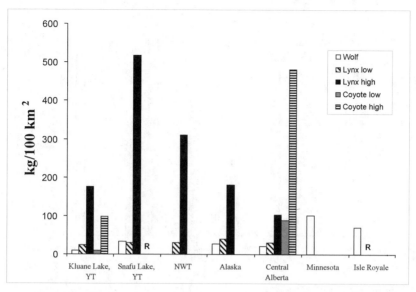

Figure 18.1
Biomass estimates of mammalian carnivores in the boreal forest. Values are kg/
100 km². Low and high estimates refer to the low and peak phase of the 10-year
cycle; R = rare. Data sources: Kluane Lake, Yukon (Gasaway et al. 1992;
O'Donoghue et al. 2001a), Snafu, Yukon (Slough and Mowat 1996; Gasaway et al.
1992), Fort Smith, NWT (Poole 1994), Alaska (Gasaway et al. 1992; Aubry et al.
2000), Rochester, Alberta (Brand et al. 1976; Keith et al. 1977; Todd et al. 1981), Minnesota (Fuller 1989), Isle Royale, Minnesota (Peterson and Page 1988; Post et al. 2002).

ter track surveys suggested that foxes were rare on the study site except during the
year of the hare peak, and the same could be said for wolverine (O'Donoghue et
al. 2001b). Fisher and marten were absent from the study area.

Snowshoe hares dominated the herbivore trophic level, particularly during
the hare peak (Fig. 18.2; Krebs et al. 2001a). Moose biomass (estimated by government surveys in the region; see Gasaway et al. 1992) was roughly one-third that
of average hare biomass over the 10-year cycle and only 15% that of hares during
the hare peak. Red squirrels (*Tamiasciurus hudsonicus*) and ground squirrels also
constituted more herbivore biomass than moose (Boonstra et al. 2001a; Krebs et
al. 2001a). Deer were extremely rare as were woodland caribou. Thus, for the Kluane system, snowshoe hares dominated the herbivore trophic level whereas lynx

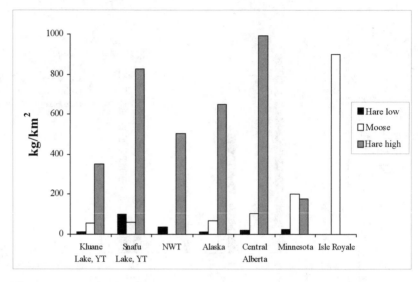

Figure 18.2

Biomass estimates of mammalian herbivores in the boreal forest. Values are kg/km². Low and high estimates refer to the low and peak phase of the 10-year hare cycle. Data sources: Kluane Lake, Yukon: Gasaway et al. 1992; Hodges et al. 2001; Snafu,Yukon: Slough and Mowat 1996; Gasaway et al. 1992; Fort Smith, NWT: Poole 1994; Alaska: Gasaway et al. 1992; Wolff 1980; Rochester, Alberta: Keith and Windberg 1978; Keith 1983b; Rolley and Keith 1980; Minnesota: Fuller 1989; Green and Evans 1940; Isle Royale, Minnesota: Peterson and Page 1988; Post et al. 2002.

and coyote were the principal mammalian predators. The top carnivores, wolves and wolverine, along with their prey, moose, were relatively less abundant.

Figures 18.1 and 18.2 provide biomass estimates for predators and selected mammalian herbivores from a number of study sites in boreal North America for comparison to the Kluane findings. As suggested, wolf densities at Kluane were relatively low with more typical densities for Alaska and the Yukon being two to three times higher (30 kg/100 km²; Gasaway et al. 1992). Fuller (1989, Appendix B) reported wolf densities in Minnesota as high as 100 kg/100 km² and the Isle Royale studies reported values as high as 300 kg/100 km² but 70 kg/100 km² was more common (Peterson and Page 1988; Post et al. 2002). Fuller et al. (2003) conclude that wolf densities rarely exceed 140 kg/100 km² on mainland North America. Studies at Snafu Lake, Yukon, and near Fort Simpson, Northwest Territories

(NWT) recorded lynx densities as high as 517 kg/100 km^2 and 310 kg/100 km^2, respectively (Poole 1994; Slough and Mowat 1996). In central Alberta, peak lynx densities were half of those at Kluane whereas coyote densities were five times higher (Brand et al. 1976; Keith et al. 1977; Todd et al. 1981). Coyotes were virtually absent from the Snafu and NWT studies. Bears are not included in these studies because it is difficult to estimate bear densities, they are omnivores, and they are active predators for only half of the year. The latter two reasons suggest that the relative dominance of bears in the predator trophic level of the boreal forest would be considerably less than that of wolves, lynx, or coyotes.

Peak snowshoe hare density estimates at Kluane were actually low relative to other studies in northwestern North America, where hares achieved maximum biomasses of near 1000 kg/km^2 (see Fig. 18.2). Moose densities at Kluane were slightly below average for Alaska and the Yukon where wolves are lightly harvested (69 kg/km^2; Gasaway et al. 1992). Moose can reach biomasses of 100 kg/km^2 in the eastern boreal forest (Fuller et al. 2003), but values are still well below those of hares at peak densities. The eastern boreal forest also supports high densities of deer (e.g., 330 kg/km^2 in Minnesota; Fuller 1989). Isle Royale represents a special case where moose regularly achieve biomasses similar to those of peak hare densities (see Fig. 18.2; Peterson and Page 1988), with maximum densities reaching as high as 2800 kg/km^2 (Post et al. 2002). This is as much as 50 times higher than levels observed in the Yukon and Alaska. I should point out that all of these estimates for moose were taken from studies where viable wolf and/or bear populations existed. Moose densities can be much higher in the absence of these predators (see following).

To summarize, although wolves and moose densities were slightly below average at Kluane, the detailed estimates of other mammalian carnivores and their prey from the Kluane study fall within the range found in other boreal forest studies and could be considered to be fairly typical. Based on current information then, it appears that midsized carnivores and their prey are more dominant in their respective trophic levels than are wolves and moose throughout large portions of the boreal forest. I should also point out that avian predators such as great horned owls (*Bubo virginianus*) can form a significant component of the predator trophic level (Rohner et al. 2001; Ruesink and Hodges 2001). Similarly, red squirrels can equal biomasses of moose (Krebs et al. 2001a). To put these vertebrate biomasses in a broader context we can compare them to that of forest tent caterpillars (*Malacosoma disstria*), a defoliating insect that shows 10-year population cycles. It can reach biomasses of 82,000 kg/km^2 and its parasitoids can reach 6,800 kg/km^2 (Batzer et al. 1995). This would be equivalent to 175 moose or 68,000 snowshoe hares per km^2 and some 2000 wolves per 100 km^2. In the broader context then, it is clear that mammalian predators and their prey do not necessarily dominate the predator and herbivore trophic levels of the boreal forest.

Do Boreal Forest Carnivores Limit Their Prey?

If carnivores play keystone roles in the ecosystems they inhabit, one would expect them to exert strong top-down control of the herbivore trophic level. In other words, they should be important limiting or regulatory factors for their primary prey. The case for wolves being an important limiting factor for their prey has grown over the years, with three main arguments having emerged. First, wolves are a significant source of ungulate mortality (Linnell et al. 1995; Berger et al. 2001a; Mech and Peterson 2003). Second, these losses do not appear to be compensatory, because wolf removal programs have usually resulted in increased calf survival rates and population increase (Orians et al. 1997). Third, moose densities are double on average in northern boreal areas where predators are heavily hunted (Gasaway et al. 1992) and densities can be 5- to 10-fold higher in areas where wolves have been extirpated (1000 kg/km^2; Crête 1987). Moose have reached very high biomasses (2000–5000 kg/km^2) on parts of the island of Newfoundland following their introduction in the early 1900s and the extirpation of wolves in the 1930s (Connor et al. 2000; Mercer and McLaren 2002). The same can be said for Sweden (Angelstam et al. 2000).

Both grizzly and black bears are known to prey on ungulates and there is strong evidence that they are a significant proximate cause of calf mortality (Ballard and Larsen 1987; Van Ballenberghe 1987). However, given that bears hibernate and much of their diet is vegetation, it seems unlikely that their degree of limitation on ungulate populations will be as substantial as that of wolves. As for wolverines, there have been no studies that have examined their ability to limit ungulate densities but this too seems unlikely, given their low density, small body size compared to bears and wolves, and the fact that studies of boreal ungulate population dynamics never list them as an important factor.

In the Kluane study, predation was the proximate cause of the majority of mortalities (> 90%) of adult snowshoe hares (Boutin et al. 1986; Krebs et al. 1995). Similar results were obtained by Keith et al. (1977, 1984) in central Alberta, Canada. At Kluane, lynx and coyotes accounted for 65 to 75% of all depredated hares over the 10-year study (O'Donoghue et al. 2001a). Survival of hares increased 10-fold on the predator exclusion site, although this did not translate into equivalent increases in densities of hares. During the initial hare decline, densities were as much as 2.5 times greater than on control areas but peak densities were unchanged. This can be attributed to the fact that great horned owls partially compensated for the lack of terrestrial predators on the predator exclusion area and hares also moved outside of the fence established to keep lynx and coyotes out. Once the hares were outside of the fence they were quickly killed (Hodges et al. 2001).

In Fennoscandia, specialist predators have been implicated as having a key role in the three- to six-year vole cycles (Hanski et al. 1993). Experimental removal of predators (weasels, stoats, and generalist avian predators) led to increased abundances of voles and disruption of the cycle (Körpimaki et al. 2002). Furthermore, studies have also demonstrated that as generalist predators increased in relative abundance (in more southern regions) the cycles disappeared (Erlinge et al. 1983; Hanski et al. 1993). To summarize, there is compelling evidence that wolves can have strong limiting effects on moose populations and there is also good evidence that small to midsized carnivores play a significant role in the population dynamics of cyclic boreal vertebrate populations.

Would the Loss of Boreal Forest Carnivores Trigger a Trophic Cascade?

One manner in which predators can affect more than their immediate prey is through trophic cascades, where the loss of a top predator leads to a release of herbivores that in turn triggers a reduction in the vegetation they feed on. Such effects appear to be widespread in aquatic environments (Carpenter and Kitchell 1993), and the sea otter (*Enhydra lutris*)–sea urchin–kelp cascade provides an example of how a mammalian carnivore can have striking and far-reaching effects on an ecosystem (Estes, this volume). Are there similar examples of trophic cascades in terrestrial systems? Polis (1999) suggested that trophic cascades will be less pronounced in terrestrial systems. A review by Schmitz et al. (2000) of the effects of carnivore removal on plants in terrestrial systems concluded that top-down effects were more common than originally thought, but these effects were most often expressed as plant damage rather than complete plant loss. None of the 61 studies included in the analyses involved mammalian carnivores because experimental manipulation of carnivore density was a requirement, and "natural" or uncontrolled manipulations were not permitted.

Terborgh et al. (2001; this volume) described a trophic cascade on predator-free islands in the tropics created by hydroelectric development where they found herbivore increases of one to two orders of magnitude and major reductions in recruitment of a variety of plant species. Evidence for trophic cascades involving top predators in North America comes from mountain protected areas where, in the absence of wolves, high density ungulate populations have significantly impacted tree regeneration and shrub growth (White et al. 1998). Similarly, the deer explosion in the eastern United States is at least partially related to predator release, and the effects of high deer density clearly affect vegetation type, structure, and succession (McShea et al. 1997; this volume). Berger et al. (2001a) showed that riparian areas within the Greater Yellowstone Ecosystem had extremely high moose densities (2400 kg/km^2; see Fig 18.2 for comparison), which led to measurable effects on willow (*Salix* spp.) biomass that, in turn, affected bird species abundance and diversity. This cascade effect was relatively moderate, however. Willow height was decreased by 20% and volume was reduced by 10%. Bird species diversity was reduced from 23 to 18 species, and abundances were significantly affected in 6 of 12 species measured.

In the eastern boreal forest, McLaren and Peterson (1994) provided evidence of top-down effects on Isle Royale where chronically high moose densities have led to suppressed balsam fir (*Abies balsamea*) regeneration and conversion to white spruce. McInnes et al. (1992) measured tree and shrub biomass inside and out of 30-year-old moose exclosures on the island. Balsam fir and aspen (*Populus tremuloides*) densities were lower outside the exclosures whereas spruce density was higher. Shrub biomass was actually higher outside of the exclosures and only the most preferred species showed decreased biomass. They found no differences in plant species diversity inside and outside of the exclosures. Extreme moose densities (1800–2700 kg/km^2) for 20 to 30 years on Newfoundland have resulted in suppressed recruitment and actual elimination of some deciduous species. Balsam fir was also heavily browsed (Connor et al. 2000; Mercer and McLaren 2002). Angelstam et al. (2000) describe how decades of heavy browsing by high moose densities has suppressed deciduous tree recruitment in Swedish forests. This browsing pressure may prevent the reestablishment of these species and the associated biodiversity.

It appears that high moose densities over long time periods (20+ years) can lead to changes in forest succession whereby palatable species are reduced relative to unpalatable species. However, browsing pressure must be very high for this to occur. Crête (1989) measured the production of browse and consumption by moose in a park in eastern Québec. Moose densities were high (two moose/km^2; 900 kg/km^2), but consumption of deciduous twigs was less than annual production and browsing on balsam fir was very low (< 1%). Crête (1989) also suggested that productivity in southwestern Québec could support moose biomass as high as 16,000 kg/km^2.

There is much less evidence for trophic cascades in the western boreal forest. In the Kluane study, moose were found to browse only trace levels of willow (*Salix* sp.) and birch (*Betula glandulosa*), the dominant shrubs in the system (Krebs et al. 2001b). Hares, in contrast, browsed up to 90% of birch twigs and 19% of willow twigs during the hare peak. Since willow constituted 90% of the shrub biomass, hares had little overall effect on shrubs in the system. In fact, shrub biomass actually increased during the hare peak to reach a maximum some two to three years postpeak. Hares have been known to suppress growth of small conifers at the peak, but browsing pressure generally resulted in delayed growth rather than mortality (Sinclair et al. 1993). In the Kluane predator exclusion experiment, snowshoe hare densities were only slightly higher relative to controls (Hodges et al. 2001). It was only when food was supplemented and predators excluded that numbers reached densities four times that of controls at the population peak and an order of magnitude higher than on control areas over the entire hare cycle (Krebs et al. 1995; 2001a). Hare densities on this grid were maintained at or above levels equivalent to peak control densities and this chronically high hare density led to increased browsing of willows and stimulated shrub growth rates such that there was no net decrease in biomass (Krebs et al. 2001b). Overall, the Kluane results suggested that midsized carnivore removal would not trigger any sort of trophic cascade. The only other experimental attempt to examine trophic cascades in boreal systems was done by Körpimaki et al. (2002) in Finland. Here, all predators were excluded from small exclosures, which led to increased vole densities, but there was no effect on vegetation growth and composition.

To summarize, there is some evidence of a potential for some degree of trophic cascades in boreal systems. This can occur, for example, when moose reach densities that are 5 to 10 times greater than those observed when wolves and bears are present. These very high densities however, do not lead to the major transformations seen in marine systems (Estes, this volume). Rather, the main effect appears to be increased browsing on deciduous shrubs and some biomass reduction but not outright mortality. There is no evidence for trophic cascades in the cyclic predator–herbivore systems. All in all, these findings tend to support the conclusions made by Polis (1999) and Schmitz et al. (2000) that trophic cascades are less pronounced in terrestrial systems.

Do Top Carnivores Affect Biodiversity in the Other Ways?

Are there other ways in which carnivores may affect biodiversity aside from generating trophic cascades? Crooks and Soulé (1999a) described the "release" of mesocarnivores following the reduction of coyotes in a highly fragmented and human disturbed system in California. Increased mesocarnivore densities affected bird nesting success and species diversity. Based on the suggestion that wolves exclude coyotes and foxes, it is believed that the loss of wolves from large parts of Scandinavia has led to red foxes becoming the dominant predator in the system (Lindström et al. 1994). Ballard et al. (2003) summarize the available data for interactions between wolves and other carnivores. There are reports of wolves killing coyotes, foxes, wolverine, and black bears and good evidence that wolves can have strong negative effects on the distribution and abundance of coyotes. Crabtree and Sheldon (1999) reported that coyotes reached densities of 45/100 km^2 in the absence of wolves in the Greater Yellowstone Ecosystem and suggested that this has led to reduced densities of small herbivores. Scavengers such as ravens (*Corvus corax*) and gray jays (*Perisoreus canadensis*) are known to utilize wolf kills extensively (Mech and Peterson 2003) but it is not known if wolves actually affect the distribution and abundance of these species.

The Kluane study revealed that the predator–herbivore trophic levels in the boreal forest are linked by a number of complex interactions, but despite this, only a few interactions dominate the system. The principal one is the suite of midsized carnivores and raptors feeding on snowshoe hares—the dominant herbivore in the system. This generates strong cycles in these predators, and they in turn impart 10-year cycles in grouse and ground squirrels through predator switching. The study revealed that hares also appeared to affect vole densities but the linkage was through nutrient cycling as opposed to predation (Boonstra et al. 2001b). The "hare web" was not absolute however. Red squirrels, the second most abundant herbivore, the moose–wolf system, and the diverse forest passerine guild (28 species; Smith and Folkard 2001) showed no relation to the cycle.

In Scandinavia, the three- to four-year vole cycle entrained four-year cycles in hares, grouse, and muskrats (*Ondatra zibethicus*) (Danell 1985; Hansson and Henttonen 1985; Linden 1988). The linkage appeared to be through red fox predation. Small to midsized carnivores therefore, provide important linkages in the boreal forest vertebrate food web. The really fundamental species, however, appeared to be the small cyclic herbivore: snowshoe hares in North America and voles in Fennoscandia.

I have argued that the population cycles are a signature process in the boreal forest, so it is relevant to ask what would happen to this process if various carnivores were to be lost from the system. It appears that the loss of wolves and bears would have little effect. Although moose densities would be likely to increase, they would have to do so substantively to reach levels that would lead to competition with hares. Even if this were to happen, the likely outcome would be the lowering of hare density but no disruption of the cycle. From results of the Kluane study we concluded that the boreal forest was more top-down than bottom-up controlled. That is, the experiments that manipulated the predator and herbivore trophic levels had much stronger effects than the ones where nutrients and plants were altered (Krebs et al. 2001a). There were strong reciprocal effects between hares and vegetation, and it was clear that hare densities drove predator population dynamics. Predators in turn, affected hare numbers but hare food supply was also involved.

It follows that the loss of both lynx and coyotes would clearly have a substantive effect on the hare cycle. Lynx are certainly vulnerable to heavy human harvest and they have been extirpated from southern portions of their range (Ruggiero et al. 2000), begging the question whether the extirpation of lynx alone would alter the hare cycle. For many parts of the boreal forest the answer appears to be no. There are some data to suggest that as coyotes expand their distribution into the boreal forest, they are playing a role very similar role to that of the lynx. Indeed, at Kluane, these two species had virtually identical feeding and habitat niches (O'Donoghue et al. 1998, 2001a). There also appears to be a reciprocal relation between coyote and lynx abundances. In areas where the effects of humans have been relatively extensive such as central Alberta, coyotes are more abundant than lynx. In less impacted areas like Kluane, lynx are slightly more abundant than coyotes, whereas at Snafu Lake and NWT, coyotes are rare and lynx attained their highest densities (see Fig. 18.1). In all cases, the hare cycle seemed to be intact. Consequently, the current information suggests that there is considerable redundancy between lynx and coyote such that the loss of lynx alone may not have strong effects on the hare cycle. That being said, it is possible that the increasing presence of coyotes in the boreal forest is directly linked to recent changes brought about by human activity. The actual mechanism may be one of mesocarnivore release associated with fewer wolves near industrial development or food supplementation through agricultural and ranching activities. Hares would likely continue to cycle following the disappearance of lynx but their declines would be caused by food shortage, and the cycle period would depend on vegetation recovery rates. Avian predators, like great-horned owls, could compensate for these changes to some degree although not completely (Hodges et al. 2001). Unfortunately, the Kluane experiments could not be conducted at a large enough scale to address this important issue.

To summarize, current research suggests that mammalian carnivores in the boreal forest play a significant role in limiting the density of their prey. There is some evidence for cascading effects when wolves disappear from the system and moose reach high densities, but these effects are limited to only moderate changes in vegetation structure and composition. There is limited evidence that these cascades could reduce species diversity but the time required for this to occur seems on the order of decades. In general then, the scientific case for mammalian carnivores serving a keystone role in the boreal forest is not strong at this time.

Top Carnivores as Umbrellas for Biodiversity

Most conservation programs focused on top carnivores are linked closely with a protected areas network, and because such areas must be large to be effective, there is the potential for top carnivores to function as umbrellas for biodiversity. Does a protected area strategy designed around top carnivores protect a broader array of biodiversity? Recent studies report that selecting priority conservation areas based on top carnivores does not necessarily ensure that other species are included (Andelman and Fagen 2000). Carroll et al. (2003) developed a reserve-selection algorithm using a suite of mammalian carnivores inhabiting the Rocky Mountains. The location and size of reserves selected to conserve viable populations of carnivores covered 76% of the ecosystem types present but less than 50% of targeted vascular and nonvascular plants, birds, amphibians, and butterflies.

The immense size of protected areas required to house viable populations of carnivores makes it unlikely that finding and protecting adequate areas will be likely, given other societal demands. This means that the surrounding matrix must also be managed for top carnivores. In contrast to many other species, industrial activities like forest harvesting are unlikely to destroy wolf or lynx habitat directly. In fact many practices create habitats that would favor increases in prey such as moose and hares (Linnell et al. 2000). That being said, forestry and the associated human access network bring a host of other threats. These come in the form of direct competition for prey (high human harvest of moose can affect wolves, Gasaway et al. 1992) and direct mortality due to hunting, poaching, and removal to protect livestock. It has proven very difficult to control losses of carnivores to humans when human access to a region is high, prompting initiatives to develop roadless areas as a conservation practice.

One could imagine then, a biodiversity conservation program designed around top carnivores, with protected areas made as large as possible and placed within a matrix of land use that carefully controlled the loss of carnivores to humans. Such a program would not provide any direct management to maintain a wide range of vegetation structures and habitat types so necessary for the maintenance of species diversity. A carnivore-oriented conservation program potentially achieves this by default rather than by design. By this I mean that structure and habitat diversity is achieved simply through the need for large protected areas. These protected areas would still require some form of natural disturbance management to maintain important vegetation structure and stand diversity. Further, as outlined, such a program has no plan for how to maintain the natural range of habitat types in the industrial matrix that will make up the majority of the landscape. In Scandinavia, where industrial forestry has a long history in the boreal forest and there are hundreds of red-listed forest insect species, the primary causes of species loss are forest cutting practices and loss of natural fire cycles (Linnell et al. 2000), not the loss of mammalian carnivores. Timber harvesting practices designed for carnivore conservation will do little to conserve species requiring dead wood, big deciduous trees, or recent burns with high densities of snags. That being said, a biodiversity conservation program planned around dead wood specialists might have little relevance to mammalian carnivore conservation. It seems likely that any conservation program organized around any single "umbrella" species or group is likely to have many other species "left out in the rain."

A Biodiversity Conservation Approach Focused on Maintaining the Range of Natural Variability

The carnivore-centric approach to biodiversity conservation is an example of a focal species or fine-filter approach. The scientific rationale for such an approach must focus on either the keystone or the umbrella capacity of these species. In addressing the former, it is clear that carnivores play important roles in limiting the population size of their prey, there is the potential for some degree of trophic cascades, and predators play important roles in boreal forest cyclic populations. However, in the boreal forest at least, there appears to be no strong scientific rationale for a conservation program focused on apex carnivores like wolves, bears, and wolverine versus one focused on lynx. In fact, if the intention is to focus on keystone species, then the snowshoe hare may be a much better candidate (Krebs et al. 2001a).

The need to conserve biodiversity in the broad sense has prompted interested parties to consider a completely different approach to conservation. It begins with the simple premise that habitat loss represents the greatest threat to biodiversity and the challenge is to maintain a wide range of habitats at relevant spatial scales. The concept is that the frequency, size, and intensity of disturbances like wildfire and insect outbreaks act to create the variability in stand type and stand structure that appears to be fundamental in creating and maintaining the species diversity in the boreal forest (Attiwill 1994; Angelstam and Pettersson 1997; Bergeron et al. 1999). This coarse-filter approach focuses on the processes that create habitat variability, and management is then designed to find creative ways to exploit natural resources while maintaining the range of natural variability. Lindenmayer and Franklin (2002) provide a broad summary of an approach to manage the industrial forest matrix in a way that maintains the range of habitats created by natural disturbance and succession. This puts emphasis on fundamental ecosystem processes rather than on single species and on managing the industrial landscape for conservation purposes along with protected areas. The removal of fire, the loss of dead wood, or the truncation of forest age classes in the boreal forest is likely to have much broader biodiversity ramifications than would the loss of a top carnivore.

The coarse-filter approach holds considerable promise as a conceptual basis for planning human activities and biodiversity on large spatial scales. As usual though, the devil is in the details and it remains unclear as to whether social and economic considerations related to human activities provide limitations so strong that the range of natural variability in habitats and landscapes is still greatly reduced. Ironically, this coarse-filter approach by itself may do little to conserve top carnivores given that hunting and poaching constitute the major threats to their survival. By the very nature of biodiversity it is clear that we need a diversity of approaches to its conservation.

Summary

My objective in this chapter was to examine the scientific evidence for the importance of carnivores to boreal forest structure and function and, ultimately, to maintenance of species diversity. The loss of top carnivores such as wolves from the boreal system will produce measurable effects on the ecosystem and it is likely that the loss of both lynx and coyote may incur even more dramatic impacts. However, the significance of these measurable effects to overall biodiversity is still largely in the eye of the beholder. For the most part, much of the boreal forest biodiversity as we know it would remain intact. It appears that top carnivores in terrestrial environments do not have the capacity to affect ecosystems to the degree shown by their aquatic counterparts. Thus we should be cautious about making bold statements as to the scientific rationale for the keystone role of top carnivores in terrestrial biodiversity conservation programs. It seems likely that, when it comes to conservation of boreal forest biodiversity as a whole, a focus on a single species or group will not prove successful in the long run. Top carnivores hold significant promise as useful components of a conservation program because of their immense appeal to the public, but conservation of boreal biodiversity will require a far broader approach that has a clear plan for how diversity of structure and habitat will be maintained over meaningful ecological space and time.

ACKNOWLEDGMENTS

I wish to thank Ainsley Sykes for editorial help throughout. The manuscript benefited from the comments of two anonymous reviewers, Justina Ray, Kent Redford, and Joel Berger. I also want to thank all of the participants in the carnivore conservation workshop for a very stimulating meeting. Finally, thanks to all of the people who toiled through good and bad field conditions to collect the Kluane data.

CHAPTER 19

The Linkage between Conservation Strategies for Large Carnivores and Biodiversity: The View From the "Half-Full" Forests of Europe

John D. C. Linnell, Christoph Promberger, Luigi Boitani, Jon E. Swenson, Urs Breitenmoser, and Reidar Andersen

Although the role of top predators may vary (Strong 1992; Chase 2000), there can be little doubt that such animals have been instrumental in shaping the evolution, behavior, and ecology of many species (e.g., Byers 1997), and the community structure of many natural ecosystems (chapters in this volume). However, there are very few systems left on Earth that can be called "natural" (in the sense that human influence is absent), given that the human footprint is detectable in virtually all ecosystems (Nellemann et al. 2001; Western 2001; Matson et al. 2002; Sanderson et al. 2002a). The areas where ecosystems still function without major human influence are sadly very few. For a variety of economic, practical, ethical, scientific, and aesthetic reasons such areas should be conserved, and treated like the crown jewels of our planet's natural heritage. There is mounting evidence that for an ecosystem to be considered functionally intact, the full complement of top predators is required. Top predators may also be useful from a strategic point of view to promote the conservation of these systems (umbrellas and flagships) (Simberloff 1998; Leader-Williams and Dublin 2000; Ray, this volume).

Anthropogenic changes to the planet's ecosystems span a wide gradient from seminatural habitats through various forestry and agricultural systems to suburban and urban habitats (Sanderson et al. 2002c). Biodiversity, including top predators, can be found throughout this gradient of modification. It is in the context of these modified habitats that most conservation will have to take place in the future (Rosenzweig 2003), and with this background that we aim to examine the potential linkage between conserving top predators and biodiversity (Linnell et al. 2000). Our focus is on all of Europe with the exception of Russia, and we mainly consider wolves (*Canis lupus*), bears (*Ursus arctos*), and Eurasian lynx (*Lynx lynx*), which have the most widespread distributions on the continent.

This is not an analytical work where we can justify all our conclusions with reference to statistical analyses of experimental data. Rather, it represents an attempt to combine the experience of the authors to identify elements of current European environmental philosophies ("visions" or "strategies" might be more modest words), and place the linkage between large carnivores and biodiversity into this context. At present we have no unifying philosophy to guide us in Europe. Given the diversity of social, cultural, and ecological conditions that categorize Europe, it is highly unlikely that any single philosophy could be accepted, and it may not even be desirable. In fact, this diversity, and especially mutual tolerance for this diversity, must lie at the heart of the European approach.

The Nature of Biodiversity

So much confusion exists around the term "biodiversity" that it may be useful to separate it out into its constituent elements, before going further. At its most simple, biodiversity is often perceived as a list of species, some of which may be endangered (species diversity). However, a far broader definition of biodiversity now exists such that all levels of *interactions* between species, ecological and behavioral *processes,* and *landscapes* can all be viewed as biodiversity (Redford and Richter 1999; Pyare and Berger 2003).

Europe: A Continent Shaped by Humans

Humans in their various forms have occurred in Europe during the many expansions and contractions of the Pleistocene ice age since at least 40,000 BP (Cunliffe 1994), and rapidly recolonized the land as the glaciers melted. These early humans were effective hunters and must be regarded as intrinsic members of the carnivore guild (Smith 1992; West 1997). Already from around 8000 BP humans

began farming in southeastern Europe and had begun modifying habitats throughout western and northern Europe by 3000 BP (Cunliffe 1994). Therefore, from the
first moments that the European landmass was released from the ice age's embrace, humans have been influencing the structure and functioning of the ecosystem in various ways to a far greater extent than in other continents, such as North
America (Kay 1994). Through their predation on prey species (Breitenmoser 1998),
intraguild predation on other carnivores (Boitani 1995), and habitat modification
(Berglund 1991), it is clear that from an ecological point of view humans have long
asserted both top-down and bottom-up effects on the ecosystem and have clearly
had a dominant influence over ecosystem processes on the continent.

    This human influence has been complex, dynamic, and far from linear. Al-
though habitats have been extensively modified, the changes in forest cover, the
manner of forest exploitation, and patterns of human distribution and density have
fluctuated widely under the influence of developing technology, climate change,
disease (e.g., the black death), warfare, and social upheaval (e.g., Björse and Bradshaw 1998; Verheyen et al. 1999; Farrell et al. 2000). Despite these changes, most
postglacial species have persisted through to modern times, albeit in greatly reduced numbers and in reduced ranges. This includes all four species of large carnivores: brown bear, wolf, Eurasian lynx, and wolverine (*Gulo gulo*) plus the smaller
Iberian lynx (*Lynx pardinus*); four cervid prey: wild reindeer (*Rangifer tarandus*),
moose (*Alces alces*), red deer (*Cervus elaphus*), and roe deer (*Capreolus capreolus*);
bison (*Bison bonasus*); wild boar (*Sus scrofa*); and three mountain ungulates: chamois (*Rupicapra rupicapra*), isard (*Rupicapra pyrenaica*), and Alpine ibex (*Capra ibex*).
Only two ungulate species, wild horses (*Equus ferus*) and auroch (*Bos primigenius*),
have gone extinct. For all these species and most forested habitats, the 19th and
early 20th centuries were the bottlenecks when a variety of factors coincided such
that human pressure on land and resources was at its maximum.

    At the start of the 21st century the situation has changed dramatically. Forest
cover is generally higher than it has been for at least several centuries (average is 34%
in continental Europe and 56% in Fennoscandia), and wild ungulates are now so
widespread that they have not been so abundant for centuries (if ever), and in many
areas are "overabundant" (*sensu* McShea et al. 1997; McShea, this volume). Despite
high human population densities (121 km^{-2} in continental Europe and 16 km^{-2}
in Fennoscandia), people are increasingly becoming concentrated in urban areas

(MacDonald et al. 2000). The fact remains, however, that Europe is home to ~580 million people, all trying to make a living, and these people are not going anywhere. Clearly, there are no wilderness areas left in Europe, although a few such fragments remain (Jędrzejewska and Jędrzejewski 1998). It is therefore important to set conservation goals that take into account the reality of the situation. Accordingly, European nature conservation is focused on integrating as much biodiversity into a human-dominated landscape as possible. Conservation efforts focus on all habitats, including urban, agricultural, and seminatural habitats, and the tiny fragments of relatively intact nature that remain (Redford et al. 2003).

Even in intensive agricultural areas, there have been attempts to integrate many species through subtle changes in land management, such as leaving slightly wider field margins and reducing the use of pesticides and herbicides (Sutherland 2002). Meanwhile, forestry is attempting to learn from the mistakes of the past and develop methods that better replicate natural disturbance processes (Bengtsson et al. 2000) and provide for "multiple uses," including biodiversity (Farrell et al. 2000). Because pure "naturalness" is not a goal, it is not an intrinsic problem for conservation if land is used, and thereby has had its original biodiversity and functionality changed (because it is inevitable that all use has an impact on biodiversity at some level; Redford and Richter 1999). Indeed many of the human-modified landscapes are preserved because of their visual or aesthetic appeal (e.g., Hunziker 1995). The integration of human structures into these landscapes is also regarded as positive in many contexts. Because of the very long period of human modification of Europe, the biotic landscape is as much a form of cultural heritage as any castle, cathedral, or monument. The landscapes that result from the combined effects of biotic and cultural processes are also associated with cultural identity (Sörlin 1999), and, interestingly, there appears to be a positive geographic relationship between species diversity and cultural diversity (Sutherland 2003), with areas of high biodiversity being linked to high human population density (Araújo 2003).

The interconnection of natural and cultural heritage is so extreme that in many cases some species and landscapes are totally dependent on constant human activity. This is most obvious for the biodiversity associated with grazing and hay production (Warren 1998). Since the Iron Age, pastoralists have been dependent on using an infield–outfield system, where livestock were grazed and hay was collected on outfields, while manure from the livestock was used to fertilize the arable infields (Bruteig et al. 2003). Many species of plant, fungi, and insects, for example, depend on the grazing pressure to keep the landscape open and suppress tree and shrub growth, or on the hay meadows that have a net nutrient loss (Smallidge and Leopold 1997; Moen et al. 1999). For example, in Norway, up to 30% of red list species across all taxa are associated with agricultural landscapes (i.e., dependent on a certain type of land-use). A further extreme example lies with the importance attached to the conservation of livestock breeds (Hall and Bradley 1995) and local crop types (Wood and Lenne 1997). This appreciation of grazing-dependent species and landscapes goes to the extent that grazing is allowed, and even encouraged, in many national parks and nature reserves.

As a result, European conservation goals are complex in that nature and culture heritage are regarded as being closely linked in our landscapes. This recognition of the desirability of continued human activity, and the importance of aesthetics as well as biodiversity, is apparent in the European Landscape Convention (2000). In the preamble to the Convention it is noted that "landscape has an important public interest role in the cultural, ecological, environmental and social fields" and that "landscape contributes to the formation of local cultures and that it is a basic component of the European natural and cultural heritage, contributing to human well being and consolidation of the European identity." Finally, according to the Convention's definition, "Landscape means an area, as perceived by people, whose character is the result of the action and interaction of natural and/or human factors." The goals are therefore very focused on the conservation of landscapes aesthetics, species diversity, and rare species.

Thus far there has been relatively little in the way of movement toward conserving ecological processes. Recently, Europe has been focused on preventing the extinction of species and populations in the face of 19th- and 20th-century pollution, agricultural intensification, and urbanization. As success is achieved at this stage of conservation, there is little doubt that there will be an increased focus on conserving and restoring at least some processes, and moving toward an ecosystem approach, although human presence will be regarded as an integral part of the ecosystem. In focusing our attention on processes rather than species it is likely that the large carnivores may represent a powerful driving role.

Table 19.1

The current size of large carnivore populations in Europe[a]

	Eurasian Lynx	Wolf	Bear	Wolverine
Fennoscandia	2800	170	2000	700
Alps	300	Occasional	30	
Baltic/Poland	1000	1800	400	
Carpathians	2000	3000	6000	
Southeast Europe	300	2000	3500	
Iberia and Italy		3000	120	

[a] The numbers are very approximate because estimates are uncertain for many populations. Data are mainly taken from the LCIE action plans (Swenson et al. 2000; Landa et al. 2000; Boitani 2000; Breitenmoser et al. 2000) with more recent updates when available.

European Large Carnivore Populations

Direct persecution of large carnivores was once widespread throughout Europe. The earliest bounties stretch back over 2000 years, and enormous efforts were made to exterminate all predators. Direct persecution combined with widespread extermination of their ungulate prey and forest clearance ensured that many countries succeeded in driving their large carnivore populations to local extinction (Boitani 1995; Breitenmoser 1998; Linnell et al. 2001b). This was especially evident in the densely populated British Isles (even here wolves held on until the 18th century) and countries of western Europe. However, even in the sparsely populated Fennoscandian countries, large carnivore populations were severely reduced and even exterminated. In the south and east of Europe somewhat larger populations survived.

By the late 1960s the tide had turned, and today most populations are increasing or stable (Boitani 2000; Breitenmoser et al. 2000; Landa et al. 2000; Swenson et al. 2000; Table 19.1). This stems in part from the introduction of favorable legislation in most countries, and great improvements in habitat quality (forest cover and ungulate density). In addition, a number of reintroductions and population supplementations have been conducted, many of which have been successful (Table 19.2). Most notable has been the return of lynx to several west and central European mountain ranges, and the return of bears to Austria (Breiten-

Table 19.2
Large carnivore reintroductions / translocations in Europe

Species	Country	Area	Years	Number of animals	Result
Eurasian lynx	Russia	Rominter Heide	1941	5	Failure
	Germany	Bavaria	1970–75	5–7	Failure
	Switzerland	Jura	1871–80	10	Success
		Alps	1971–82	14–18	Success
		Jorat	1989	3	Uncertain
		E. Switzerland	2002–		Just started
	Italy	Gran Paradiso	1975	2	Failure
	Slovenia	Kocevje	1976	6	Success
	Austria	Alps	1977–79	9	Failure
	Czech Rep.	Sumava	1982–89	17	Success
	France	Vosges	1982–92	16–18	Uncertain
	Poland	Kampinoski	1993–95	5	Uncertain
Brown Bear	Poland	Białowieża	1938–44	10	Failure
	France	Pyrennes	1996	2	Running
	Italy	Trentino	1999–	7	Running
	Austria	Eastern Alps	1989–93	3	Success
Wolverine	Finland	Central Finland	?	?	Success?

From Breitenmoser et al. (2001).

moser et al. 2001). Natural expansion is also occurring (e.g., the return of wolves to the Alps, Germany, and Norway in recent years) (Wabakken et al. 2001; Lucchini et al. 2002). There are still some critically small and isolated populations (especially of bears, e.g., in northern Spain and the Pyrenees; Swenson et al. 2000), and some populations are suffering from overharvest and high rates of poaching, but in general the recovery picture is positive. The exceptions are the Iberian lynx which is suffering from the combined effects of habitat fragmentation and the loss of prey (Delibes et al. 2000), and some local wolverine populations (Landa et al. 2000).

The landscape within which large carnivores are recovering is heavily modified (albeit highly diverse), with relatively high human densities that have been associated with extinction under different management regimes (Woodroffe 2000; Linnell et al. 2001b). European wolves, lynx (Eurasian), and bears appear to be very tolerant of human disturbance, and all three species have shown an ability to live close to people, even within suburban and urban environments (J. D. C. Linnell, pers. obs.).

The protected areas of Europe are generally too small to support more than a handful of individual large carnivores (Table 19.3), thus requiring them to live in the multiuse landscapes where conflicts are most likely to occur (Linnell et al. 2001a). These can be divided into material conflicts that have physical and/or economic components, and psycho-social conflicts that occur in the minds of individuals or between groups of people within society (Linnell et al. in press). The major material conflict is with domestic livestock. In areas where large carnivores have always been present, intensive husbandry using the traditional shepherd, guarding dog, or night-time enclosure system minimizes the conflicts (Linnell et al. 1996; Kaczensky 1996). However, in areas where large carnivores have returned following an absence, or where wild prey is otherwise scarce, depredation rates can be very high. Locally, depredation on horses, cattle, beehives, domestic dogs, and semidomestic reindeer, and competition with hunters for wild ungulates can be significant causes of material conflict.

Social conflicts associated with large carnivores include those between different knowledge systems (experience-based and hegemonic systems), economic and cultural modernization of rural communities, and urban–rural tensions (Skogen 2001; Ericsson and Heberlein 2003; Skogen et al. 2003). In most of these social conflicts, carnivores take on highly symbolic roles as the most important proximate factor that threatens rural lifestyles under perceived attack by national and international (e.g., globalization) forces. Fear of injury and death is also a factor in areas where wolves and bears have returned after an absence. There have been cases of both wolves and bears injuring and killing people in Europe (Swenson et al. 1996; Linnell et al. 2002), although most of the wolf cases belong to past centuries, and bears in Europe cause far fewer problems than elsewhere in Eurasia or North America.

The political landscape is also highly complex. In the last 14 years, Europe has seen the fall of the Iron Curtain and the increasing expansion of the European Union (EU). This is bringing enormous socioeconomic changes to the entire continent, which will greatly affect patterns of land use, human distribution, socioeconomics, and infrastructure. These in turn will influence large carnivores both

Table 19.3

The current number of protected areas (IUCN classes I to IV) of various sizes in selected regions of the world[a]

Region	100–999 km²	1000–9999 km²	> 10,000 km²
Fennoscandia	63	17	0
Continental Europe	92	3	0
Canada and Alaska	137	118	37
South and East Africa	146	60	15
West Africa	33	34	5
Central Africa	11	30	4
Indian subcontinent	240	37	0
West USA	231	56	0
Northeast USA	35	2	0
East USA	34	5	0
Midwest USA	19	0	0

[a] Data taken from IUCN website, http://www.unep-wcmc.org/index.html?http://www. unep-wcmc.org/protected_areas/~main

positively and negatively. Large carnivore conservation is very active in Europe, with two major international pieces of legislation: the Directive on the Conservation of Natural Habitats and of Wild Flora and Fauna (1992) within the European Union (25 countries), and the Bern Convention (1979) within the Council of Europe (45 countries), both of which provide clear mandates for their conservation. Although the quality of research and management/conservation institutions varies widely, the overall trend is positive, and there is a high degree of transboundary cooperation. For example, the Large Carnivore Initiative for Europe (LCIE) is an expert group supporting both the EU and the Council of Europe and a wide spectrum of nongovernmental organizations (NGOs) that are involved in the process of carnivore conservation (LCIE 2004).

Goals for Large Carnivore Conservation in Europe

Given the historic bottlenecks that almost all European large carnivore populations have been through, conservation goals for the species have been fairly modest (see Box 19.1 for a conceptual overview). The short-term goals have been to

Box 19.1

Seven Levels of Conservation Ambition for Large Carnivores

In transferring the modern view of biodiversity into achievable conservation objectives (either for maintaining the status quo or for guiding restoration), there are at least seven possible levels of ambition and ecological complexity at which goals for large carnivore conservation can be set.

1. Species *presence*—e.g., lynx (*Lynx lynx*) persist in an area following recolonization or reintroduction.

2. Some ecosystem *processes* occur—e.g., the lynx begin to eat roe deer (*Capreolus capreolus*) (predation), kill red foxes (*Vulpes vulpes*) (intraguild predation), and leave carcasses for scavengers (secondary effects).

3. Species *demographic viability* is achieved—e.g., this lynx population increases to a level of demographic viability.

4. The *evolutionary potential* of the species to adapt to future conditions is maintained—e.g., the population increases to a level (of size or connectivity with other populations) where genetic viability (evolutionary potential) is ensured.

5. The full *community* of carnivores (and their prey) is restored—e.g., lynx, wolves (*Canis lupus*), and bears (*Ursus arctos*) occur in the same area, together with roe deer, red deer (*Cervus elaphus*), and moose (*Alces alces*).

6. The *limitation* and/or *regulation* of numbers of predators and prey are primarily determined by trophic interactions—e.g., prey density and intraguild interactions, rather than by human intervention, will limit the density of lynx and wolf populations.

7. The system is able to exist in a *dynamic* state, fluctuating under the influence of climate, disease, and other external factors.

The first three levels are relatively easy to define and document, and conservation can usually be achieved given the application of enough knowledge and resources. Level 4 concerns the issue of genetic viability and has received far less technical focus than the issue of demographic viability, and relatively little attention within the conservation movement (Myers and Knoll 2001). If we are to think long term, then far greater attention needs to be placed on this level (Bowen 1999). Although it may be desirable to aim for level 6 or 7 conservation in wilderness areas, it may only be possible to achieve more modest goals in areas that have already been heavily modified by humans. In fact, even obtaining level 1 for some conflict-causing species (such as the large carnivores) has often proven to be problematic (Linnell et al. in press). Each step involves an increase in population size and the number of species considered. The choice of goal will clearly influence the extent to which conserving the large carnivores will also conserve biodiversity in its wider sense.

prevent the extinction of existing populations. We are now moving into a phase where attempts are being made to restore top predators to suitable areas and achieve population (demographic) viability. This requires their integration into human-dominated landscapes such that coexistence with human activities and increased acceptance for their presence are required. The mission statement of the LCIE reflects this approach: "To maintain and restore, in coexistence with people, viable populations of large carnivores as an integral part of the ecosystems and landscapes across Europe." Due to aspects of scale it is understood that this viability can be achieved only as a common effort through international cooperation. The focus at this early stage has been very much on the conservation of carnivores as species and much less on the ecological processes of which they are a part.

It is obvious that, when carnivores return to an area, some of the ecological processes also resume. Wolves and lynx (and bears to some extent) kill ungulates, the scavengers will probably benefit, and intraguild predation will resume (e.g., Linnell et al. 1998). We can also expect some changes in behavior and habitat selection of their ungulate prey and some resumption of selection effects on prey. We are just at the start of estimating the real effect restored populations of carnivores are having on ungulates, which is likely to vary dramatically between different areas. For example, the effect of returning carnivores will be greatest in areas with low prey densities, especially if hunting pressure on the ungulates is also heavy. However, in areas where ungulates occur at high densities (a common feature of highly modified landscapes where human land use enhances productivity), it will require very high densities of carnivores to have any significant effect on prey demographics. In many situations human tolerance for carnivore presence may well be lower than the ecological carrying capacity. As a result legal harvest and/or poaching may limit their numerical response, and thereby their potential impact on prey numbers and their top-down influence on ecosystem processes (Andersen et al. in press). At present, even the return of these species and the resumption of these processes on a limited scale are controversial. Given the constraints on space, habitat, and human tolerance, it seems unlikely that Europe will ever get to a stage where carnivore and ungulates numbers are determined primarily by trophic interactions and nonhuman factors.

In terms of the seven levels of conservation (see Box 19.1) it is clear that in Europe we have come to stages 1 (presence) and 2 (process resumption) in many areas, and that stage 3 (demographic viability) has been achieved or maintained in at least some populations. Of these maybe a few have reached stage 4 (genetic viability or maintenance of evolutionary potential), although the lack of precise estimates of required numbers makes it difficult to assess. In some few areas (Sweden, the Baltic States, the Carpathians) we have intact guilds of both predators and their prey (stage 5). There are probably no areas where the numbers of predators and prey are determined by trophic interactions (stage 6) because of the extent to which humans directly influence predator and prey populations. Likewise, stage 7 (dynamic state) remains elusive and controversial. It seems unlikely that stages 6 and 7 will ever be achieved any place in Europe, and in many ways they may not even be desired because they exclude human influence from the system. As we have argued earlier, European nature conservation philosophy is somewhat uniquely built on the integration of people and nature.

A frequently asked question is, Why try and conserve carnivores at all in such a landscape? Although there are many motivations, it appears to us that aesthetic and ethical reasons dominate. In other words, the carnivores are being conserved largely for themselves. The same goes for the processes associated with the carnivores. The desire to see predation and scavenging is mainly for the abstract aesthetics of knowing they occur, rather than out of an expectation that they will dramatically affect the ecosystem services provided by the European landscape. If manipulation of wild ungulate density (the main potential link between large carnivores and habitat) is desired, it would be far more effective to act through hunter pressure than carnivore restoration. Certainly, Europeans are under few illusions that the presence of large carnivores will produce a net economic benefit, although some of the costs may be mediated through ecotourism and trophy hunting. However, just because it is not possible to go all the way along the conservation ambition scale, it does not follow that it is futile to achieve as much presence, viability, and process as possible.

All this implies a clear focus on the intrinsic value of carnivores, which is a potentially legitimate and powerful conservation argument in and of itself (Lawton 1991; Redford and Richter 1999; Ghilarov 2000; Collar 2003; Jepson and Canney 2003) even if scientific or moral justifications remain elusive (Oksanen 1997; Rosenfeld 2002). Given the uncertainty of the relationship between biodiversity and ecosystem function (Ghilarov 2000; Schwartz et al. 2000; Hector et al. 2001; Loreau et al. 2001), it may well turn out that the preservation or restoration of fully functional "natural" ecosystems is also only built on similar aesthetic/ethical arguments. In a continent where the distinction between nature-dominated and human-dominated environments is often lost, we believe that large carnivores are present to remind people of the needs of nature, of the existence of some limits to the extent that humans can encroach on the environment if we wish to have an entity that we call "wild nature." Large carnivores embody an idea of nature that is otherwise lost to Europeans. The presence of large carnivores is what makes a difference between a "wild" area and an extended city park. Many view the return of carnivores as highly symbolic, almost as the ultimate test of human ability to coexist with biodiversity. In other words, although we cannot achieve wilderness as many conservationists hope for (e.g., in North America, Soulé and Terborgh 1999; Pyare and Berger 2003), we can at least restore some of the wildness to the landscape.

The danger of using carnivores as symbols is that they can symbolize very different things to different people. To the conservationist a wolf might represent beauty and wildness (in a positive sense), whereas to many others it may symbolize wastefulness, evil, and, in a more modern context, the intervention of powerful social forces in conservative rural lifestyles. In extreme cases, the return of the wolf, for example, may actually serve as the focus for an increased unity among rural people against central powers—a situation that will hardly benefit conservation of biodiversity (Ericsson and Heberlein 2003; Skogen et al. 2003).

Because of the dominant effect of humans in the European landscapes there are clear limits to how far we can restore species and processes. We probably can never approach the level of ambition that many conservation biologists hope for in the Rocky Mountains of North America, for example (Soulé and Terborgh 1999; Pyare and Berger 2003). In effect, in much of Europe the large carnivores may be ecologically extinct (in the sense of Redford and Feinsinger 2001)—but they are very much alive. In terms of strict definition, large carnivore recovery in Europe is more of an exercise in reconciliation ecology (*sensu* Rosenzweig 2003) than in strict restoration ecology.

How Does Conserving Carnivores Conserve Biodiversity in Europe?

To best explore how the conservation of large carnivores and biodiversity interact we can begin by looking at some of the major threats to species diversity in Europe. The greatest long-term threats are in the field of large-scale processes and global change. These include climate change, pollution (acid rain, heavy metals, persistent toxins), and long distance transport of nitrogen (Matson et al. 2002). The second group of threats concern patterns of land use and habitat conversion, which not only remove habitat but fragment areas of unmodified habitat (Andrén 1994). Wetlands are very susceptible to drainage, forests to cutting and fire regimes, and grasslands to fertilization, grazing intensity, and potential reforestation. The European strategy to conserve biodiversity is therefore heavily focused on establishing a network of relatively small reserves under the Natura 2000 network in the EU, and on regulating patterns of land use and other human activity through incentives, subsidy, and legislation. Most of this conservation is occurring on private lands (typically even national parks are predominantly made up of private land in Europe), reflecting the greater extent to which landowner activities are regulated in Europe, compared to the United States.

Against this background, the threats that face large carnivores in Europe vary from country to country and species to species but include the following (Boitani 2000; Breitenmoser et al. 2000; Landa et al. 2000; Okarma et al. 2000; Swenson et al. 2000):

(1) lack of human tolerance due to depredation on livestock, competition for game, and fear; (2) inappropriate quotas for large carnivore hunting; (3) illegal killing; and (4) infrastructure development and human disturbance that can lead to both direct mortality and population fragmentation. Solutions therefore, need

to be focused on reducing livestock depredation and improving both the regulation of harvest and the enforcement of legislation. It is important to note that there is relatively little overlap between the first three of these threats and those facing species diversity at large.

So, does the conservation of large carnivores relate to the conservation of biodiversity in general? Large carnivore populations have three basic requirements for persistence: (1) careful regulation of human persecution, (2) large areas of connected habitat, and (3) adequate prey. Providing these things can assist other biodiversity up to a point. First, reducing persecution requires effective legislation and enforcement of legislation on the ground. Whether this takes the form of protection or regulated harvest is irrelevant; the process of setting up management structures that embrace carnivores will go a long way to serving the needs of other species.

Second, providing habitat for carnivores will provide habitat for at least many other species. Because of their large area requirements, large carnivores have been particularly useful for focusing attention on the importance of continuous habitat. For example, in Croatia reducing bear mortality in traffic was the main motivation behind costly investment in "green bridges" over new highways that benefit not only bears but many other species (Huber et al. 2002). In other cases the presence of carnivores; for example, bears in Austria (Norber Gerstl, pers. comm.) and Spain (Juan Carlos Blanco, pers. comm. 2003), have been used to justify the creation and expansion of relatively large protected areas. However, large carnivores are very tolerant of habitat quality, and patterns of land use (both inside and outside protected areas) that are compatible with large carnivores will not be compatible with the needs of many other threatened species. Conservation of species diversity in Europe will always require very detailed design and control of land use and the creation of reserves and protected areas, irrespective of the presence of large carnivores.

Thirdly, large carnivores need adequate prey populations. In an effort to reduce depredation on livestock by wolves in Portugal (Vos 2000) there is an active reintroduction program to reestablish roe deer in the area. Although such examples are few, they do indicate the ability of large carnivores to motivate the restoration of severely degraded habitats. Even though the predation process may not be restored to the level of allowing trophic regulation of carnivore and ungulate densities, just the fact that some predation, scavenging, and avoidance processes resume is a form of conservation of processes. Finally, large carnivores are very successful at focusing the public's attention on conservation in general, although we have no quantitative data on the extent to which this benefits other species.

Overall, carnivores, together with their ungulate prey (Bruinderink et al. 2003), may be highly suitable umbrellas for a coarse-filter approach to conservation, but the fine-filter approach will also be necessary for many other species (*sensu* Noss 1996; Redford et al. 2003). This can be envisaged on a landscape scale where the needs of large carnivores and their prey are used to ensure the maintenance of large areas of connected habitat that is at least seminatural, but where a series of specially managed habitats and reserves are embedded in this favorable matrix to satisfy the needs of other species with more specific habitat requirements.

It is important to mention that the focus on large carnivores may also conflict with other conservation objectives. For example, the presence of large carnivores makes unsupervised extensive grazing of livestock almost impossible. In high-cost countries this may be the only way of farming livestock to maintain the grazing pressure needed to conserve grazing-dependent biodiversity. Furthermore, as discussed previously, large carnivores do create very real conflicts, and it seems that people with the most direct experience with carnivores (at least wolves) have the most negative views of them (Williams et al. 2002; Ericsson and Heberlien 2003). Although it is not clear if such negative views will result in a backlash in the form of decreasing the public's willingness to conserve biodiversity at large, there are reasons to believe that it may be more effective to increase the public's ecocentric values through a focus on less controversial species (Brainerd and Bjerke 2003).

Is Europe Unique?

Given the generally improving global attitudes toward conservation (at least in the industrialized world) there can be little doubt that there is sufficient incentive and motivation to preserve (or rewild) the last great wilderness areas and biodiversity hotspots on Earth in a manner that minimizes human influence on their systems (Soulé and Terborgh 1999; Myers et al. 2000). Clearly, in these situations the European approach has very little transfer value. However, those wilderness areas that do remain worldwide are unfortunately few, relatively small, and surrounded by increasingly human-dominated landscapes. It is in heavily modified landscapes that the European approach of integrating biodiversity conservation and humans may work best, either alone in the absence of protected areas or to soften the contrast between large protected areas and the landscape matrices in which they occur. The role of these multiuse landscapes in biodiversity conservation has been receiving increased attention even in areas where protected areas have long been the main focus for conservation activity (North America, Rosenzweig 2003; Africa, Western 2001). Areas such as the eastern United States come to mind as discussions about the potential for wolf recovery (Elder 2000) occur in a landscape where conservation of the cultural landscape is also an issue (Foster 2002). What remains unclear is whether the European approach can work in areas of extremely high human densities, or in areas where socioeconomic conditions and poor institutional development render law enforcement ineffective, and where pressure on even seminatural habitat is intense (Woodroffe 2000; Linnell et al. 2001b). Obviously there is no single approach to conservation, and no single conservation goal that applies to all areas. Recognition of this technical and philosophical diversity is going to be vital in the unification of conservation effort to save biodiversity in all its facets and glory (Redford et al. 2003).

Conservation Recommendations

There is no doubt that large carnivores are highly suitable species for grabbing the public's attention, and that the umbrella and keystone concepts (Simberloff 1998) are generally easy to understand, elegant, intuitive, and sellable. Many aspects of conservation, however, are highly context-dependant. Although the role of large carnivores in conserving biodiversity may be highly relevant in wilderness areas, it becomes much less so in human-modified ecosystems. It is therefore necessary to tailor the message to the individual situation. In other words, there is no "one-size-fits-all" in conservation. Conservation is complex, and we must communicate this complexity to the public. The fact that in many areas the conservation of large charismatic carnivores may have very little impact on the rest of the biodiversity implies that we cannot use "scientific" arguments for their conservation. In fact, there is a need for more honesty about the roles of science in conservation. Science is a provider of information and a tool for conservation, and scientists as individuals can be powerful and eloquent advocates for conservation. However, science cannot substitute for choices among values (Lackey 1998). This implies that the role of values in motivating the conservation of biodiversity in all its forms needs to be made much clearer (Collar 2003; Jepson and Canney 2003). A further consequence is that we need to tolerate a wide range of conservation philosophies, especially in how these place people with respect to nature.

Summary

Conserving large carnivores alone will not be enough to save the species diversity of Europe. Reciprocally, strategies aimed at conserving other endangered species through the creation of a system of reserves will not always be adequate for conserving large carnivores. However, an effective synergy could be produced by combining these two approaches into an overall concept where large carnivores are used as umbrellas to conserve a connected matrix of seminatural habitat within which a series of specially managed habitats and reserves can be embedded. Although large carnivore populations are recovering throughout Europe, and processes such as predation, predation avoidance, and scavenging have resumed, it is unlikely that they will ever recover to the stage where trophic interactions are the major determinant of abundance for either carnivores or their prey. The influence of humans is just too strong in Europe to allow a return to a "natural" system. However, this should not necessarily be viewed as a problem, because European visions of nature do not divorce humans from nature. Rather, the continentwide conservation ethic that is emerging is to integrate as much biodiversity into our lives as possible. Rather than creating high contrasts between reserves and intensive use, Europe is trying to minimize the gradients. The future path is not a return to the nostalgia of the Garden of Eden at noon on the sixth day of creation (after the creation of the fish, birds, and beasts, but before humans arrived on the scene), but a test of our ability to share a crowded continent with other species. It is clear that human activity is, and has been for millennia, the major selective force in all European ecosystems, and that this is unlikely to change. What we can hope is that we can find ways to integrate humans into ongoing evolutionary processes (Barry and Oelschlaeger 1996). In this context, large carnivores have two main functions. First, they can put some of the wild back into our lives, and second they remind us that biodiversity consists of dynamic processes and interactions as well as species diversity.

ACKNOWLEDGMENTS

The ideas presented here are the product of many discussions via e-mail and over glasses of wine with our colleagues in the LCIE and its sister organization the Large Herbivore Initiative, and in our various institutes and work places. We are also grateful to the many hunters, shepherds, farmers, foresters, journalists, bureaucrats, and ordinary people with whom our work has brought us into contact. These interactions have often bought us down to earth and shown us the real arena in which conservation occurs, far removed from the ivory tower of academia, and taught us that people are important. Finally, each of us has been privileged to come into close contact with large carnivores as part of our work. While the actual number of minutes of contact have been few, the inspiration that these moments has provided lasts. The senior author's work on large carnivores has been funded by the Research Council of Norway, the Norwegian Directorate for Nature Management, and the Norwegian Institute for Nature Research. Two anonymous reviewers and the editors provided very helpful comments on the manuscript and managed to decode our thoughts to such an extent that they might be understandable. Finally, the discussions with the other workshop participants provided inspiration and much food for thought.

CHAPTER 20

Conclusion: Is Large Carnivore Conservation Equivalent to Biodiversity Conservation and How Can We Achieve Both?

Justina C. Ray, Kent H. Redford, Joel Berger, and Robert Steneck*

Over the past half-century, slowly evolving attitudes toward large carnivorous animals have resulted in dramatic shifts in how they are integrated into conservation strategies. Once viewed predominantly as impediments to the achievement of conservation goals due to their predatory habits, top predators are increasingly regarded as essential players in the efforts to save biodiversity. This transition has been enabled by a progressive enlargement of the scale and scope of conservation activity to the ecosystem level, with the accompanying recognition of predation as a key structuring process. Although viewed as vital actors by front-line conservationists large carnivores are most commonly seen as responsible for loss of property and life, and their targeted killing continues to be the most important threat to their survival on land and sea alike. Therefore, at the same time large carnivores are facilitating the development of conservation tools for biodiversity, conservationists are engaged in a fight to ensure their survival. A critical but largely unexplored question that arises is, How compatible are large carnivores and biodiversity conservation? How strong is the link between the two, and will a focus on the top predator element in a system result in the realization of the conservation of broader biodiversity?

---

* This chapter reflects the views of the authors and not necessarily the views of the other contributors to this volume.

400

Framing the Question

There are two fundamental components to the overall question of whether saving large carnivorous animals is equivalent to the conservation of biological diversity: First, is the functioning of top predators important for conserving biodiversity—in other words, can critical ecosystem roles be demonstrated for these species at the highest rung in the food chain? Second, assuming the above is confirmed, how can knowledge of the importance of top predators as structuring agents for ecosystems be used to develop biodiversity conservation tools?

The fact that these questions remain largely unanswered is an indication that conservation efforts focusing on large carnivorous animals and mixed with an emphasis on the achievement of biodiversity conservation are based on many assumptions that have yet to be tested. Indeed, the more general question is whether these assumptions can be tested (Ray, this volume). If such use is not based on correct assumptions, how might the tools that have been developed be improved? How and when can an increase in our understanding of when, where, and whether the failure to keep large carnivorous animals in a system be used to improve the prospects for biodiversity conservation?

This concluding chapter concentrates on three more general issues. First, we will examine the evidence for an ecosystem role of top predators in marine and terrestrial realms—addressing the adequacy of the knowledge available to link carnivore conservation with the conservation of biodiversity. Next, we explore whether carnivore and biodiversity conservation are compatible, examining the usefulness of current understanding to informing the actions of conservation practitioners. Finally, we analyze when large carnivores can be used appropriately as tools to achieve biodiversity conservation and what we can do better to achieve both large carnivore and biodiversity conservation.

The Link between Large Carnivorous Animals and Biodiversity

Biodiversity has become both a major objective and the currency of conservation over the past 15 years (Redford and Richter 1999). In most definitions biodiversity has three components: genetic, population / species, and community / ecosystem.

By far the most commonly used of these components is the species one—in fact, to many people biodiversity is seen as equivalent to species diversity. This is in part because the most straightforward way to measure biodiversity, and hence conservation loss or gain, is through monitoring the abundance, disappearance, or recovery of individual species (Meffe and Carroll 1997). However, species presence or absence is only part of the conservation story because it ignores what Soulé et al. (2003) refer to as second-order consequences of species loss, or the disappearance of species interactions. This concept is vital in tying the species component of biodiversity to the community and ecosystem components of biodiversity. Considering species as important due to the functional roles they play implies that some species are less expendable than others, depending on their interaction strength within the community (Harley 2003; Soulé et al. 2003). Successful conservation would therefore require protection of both species and the processes that maintain ecosystems.

Theory suggests that ecosystem impacts caused by the loss of top carnivores should be stronger than those caused by the loss of lower trophic levels because this group is characterized by less diversity (and hence redundancy) and stronger interaction strengths through predation (Duffy 2003). Large carnivorous animals are of course elements of biodiversity, but as consumers, they have the potential to be determinants of biodiversity structure and function (Steneck, this volume). At the same time, the top consumers in an ecosystem are some of the most challenging components of biodiversity to preserve due to their tendency to be lost prior to other members of the food web. Hence, many conservation projects focus on top carnivores as both the target and the means to achieve biodiversity conservation (Redford; Ray, this volume).

The direct effects of a consumer on its prey are obvious to any observer of nature. Extending such effects from the individual to populations takes no great leap of faith. In contrast, the existence of indirect effects of predators cascading over several trophic levels is a matter of considerable debate, due to the difficulty of establishing such causal links through observation. The chapters in this book have revealed the enormous complexity of this issue, adding to a body of work examining the role of large carnivorous animals in the maintenance of biodiversity. These studies do not provide one simple answer, but some general patterns do emerge.

Where There Is Strong Evidence for Biodiversity Impacts through Predation by Large Carnivorous Animals

The scientific literature is replete with works that provide both theoretical and empirical perspectives on the ecological functions played by different trophic levels in structuring ecosystems (top-down vs. bottom-up control; Steneck, this volume). Most experimentation that has offered evidence of predator-induced impacts on lower trophic levels has been drawn only from systems dominated by smaller, usually invertebrate, predators (see metaanalyses by Schmitz et al. 2000; Halaj and Wise 2001; Shurin et al. 2002). The theoretical groundwork explaining similar impacts generated by larger-bodied species has been established for some time (Hairston et al. 1960; reviewed in Steneck, this volume) but is much less understood in either more complex systems or those in which large-bodied carnivores constitute the upper echelons of the food web.

A substantial body of work in the area of game management has focused on the direct effects top predators have on prey (see NRC 1997; Hayes et al. 2003; Bowyer et al., this volume). Although the details will be subject to vigorous debate, most would agree that there is ample demonstration of the limiting and even regulating roles played by top predators through their ability to prevent prey populations from reaching the carrying capacity set by food resources (Jędrzejewska and Jędrzejewski; Bowyer et al., this volume). Because the goal of game management generally pertains to increasing available meat for human hunters to meet societal goals (Berger, this volume), the effects of predation that extend beyond the prey level are seldom explored or discussed in this context (Maehr et al.; Berger, this volume).

The practical challenges and costs inherent in the study of large carnivorous animals have generally limited careful scrutiny of their roles in structuring communities. It is only relatively recently that empirical research in a broad array of environments has extended beyond the prey level to demonstrate indirect or cascading ecosystem effects directly attributable to the loss of large predators. The most famous examples derive from the documentation of impacts on vegetation and lower trophic levels caused by hyperabundant herbivores whose numbers increased dramatically following the disappearance of top predators from coastal Pacific kelp forests (Estes, this volume), the North American Yellowstone ecosystem (Berger and Smith, this volume), or island systems in the Neotropics (Terborgh, this volume).

The fact that these relatively few examples are widely cited in the conservation literature as substantiation for cascading effects due to removal of large mammalian carnivores is indication of both the unambiguous support for reported effects and the lack of a larger body of evidence. Additional authoritative support, although less widely cited, comes from a large body of research in the marine realm (Steneck and Sala, this volume). Other work hints at top-down forcing by large predators, though definitive conclusions are confounded by abiotic factors, anthropogenic disturbance, and oscillating productivity (e.g., Peterson 1999; Jędrzejewska and Jędrzejewski; McClanahan; Miquelle et al., this volume). These factors appear to drive shifts in the relative influence of top-down and bottom-up factors over time and space (Meserve et al. 2003; Sinclair et al. 2003). Like the top-down/bottom-up literature in general, the evidence for indirect effects of predation is heavily weighted toward apex predators that are resident in aquatic environments (Steneck, this volume). Such realm differences could result from greater productivity in benthic aquatic and marine ecosystems. On the other hand, the relatively small sizes of intertidal predators and prey, along with their limited mobility and range, are characteristics that lend themselves to predator manipulations and experimental studies, which may make any top-down impacts more apparent (Shurin et al. 2002; Steneck, this volume).

Even when predation has little or no discernible effect on population densities of adjacent trophic levels, the so-called nonlethal attributes of predation (*sensu* Lima 1998) do influence the evolution of prey, shaping their behavior and ecology (Berger and Smith; Bowyer et al.; Maehr et al.; Mills; Woodroffe and Ginsberg, this volume). Indirect effects resulting from behavioral changes in response to the relative risk of predation can be as profound as those that emanate from direct predation (Schmitz et al. 1997). By affecting factors such as habitat selection of prey, predators can indirectly influence plant composition and structure through increased usage of certain areas with low relative predation risk. Such effects can and do appear to trickle down to users of such habitats other than the prey itself (Berger et al. 2001a,b; Ripple and Beschta 2003; Berger and Smith, this volume). Those large carnivores that at first glance do not have particularly important roles through direct predation may exert subtle but critical effects that influence the state of biodiversity (Woodroffe and Ginsberg, this volume).

Where the Evidence Is Less Compelling or Absent

If the evidence for impacts of predation beyond the adjacent trophic level were unequivocal everywhere, there would be no need for a book of this nature. Frequently, however, documented top-down effects do not appear to be as strong as expected or to produce the predicted cascading effects. In seeking a synthetic conclusion as to whether predation plays an important role when there is no direct evidence from the system in question, it is important to make the distinction between effects of predation not occurring because there are indeed no impacts and effects that are difficult or impossible to discern due to difficulties in measurement. Both merit careful exploration and are discussed in turn.

Large Carnivores Do Not Always Have Important Impacts on Biodiversity

Predation is only one of many factors hypothesized to limit the capacity of herbivores to regulate plant biomass (Polis 1999), and carnivores are not the only force contributing to the structure and function of biological communities. In fact, populations at lower trophic levels may limit the coexistence of predators themselves due to their availability and dispersion (Powell 2001). Polis (1999) has argued that consumers are held in check by the same diverse factors that limit herbivores— including resource defenses, abiotic factors, spatial and temporal heterogeneity, self-regulation, and even other predators. Therefore, predators may be able to exert influence on prey only in situations where such constraints are relaxed.

Conditions under which predation is not likely to exert an important ecosystem structuring force will come about either when (1) prey are not limited by predation at all or (2) the impacts of predation do not extend beyond the prey level. The first condition can occur for a variety of reasons that relate to characteristics of the predator population, the prey population, or environmental conditions. Intrinsic characteristics of predator species are such that certain consumers will be more powerful top-down agents than others (Polis 1999; Miquelle et al.; Woodroffe and Ginsberg, this volume). In other situations, predators may be exerting the appropriate forcing, but at levels not high enough to be ecologically effective (e.g., if their population numbers have been depressed) (Pyare and Berger 2003; Soulé et al. 2003). In some cases prey may be able to escape control by predators by virtue of their migratory habits or large body size (Krebs et al. 2003; Sinclair et al. 2003; Terborgh; Mills, this volume). Finally, features of the environment itself may play a significant role in determining the extent of top-down forcing by large predators in a system, even if they are abundant. For example, productivity or prey densities relative to the carrying capacity of the environment appear to have heavy influences on the extent to which predators will be able to limit their prey (Oksanen and Oksanen 2000; Bowyer; Mills, this volume). Whether predators exert a top-down influence, therefore, will vary in time and in space (e.g., Meserve et al. 2003).

Even if predation does exert important impacts on prey populations, not all of these effects propagate to lower trophic levels or have significant impacts on ecosystem processes (Pace et al. 1999). The attenuation of top-down effects may occur because direct effects of predators on herbivores appear to be stronger than indirect effects of carnivores on plants (Schmitz et al. 2000). However, this varies among ecosystems, with attenuation less pronounced in most benthic marine and aquatic ecosystems than in terrestrial systems (Shurin et al. 2002). Although trophic-level perturbations in an ambitious 10-year experiment in Kluane, Yukon, caused direct top-down effects such as increase in biomass (Boutin, this volume), such effects tended to attenuate rapidly (Sinclair et al. 2000). There was little evidence produced by this study of indirect cascading impacts on other trophic levels (Sinclair et al. 2000).

The Ecological Importance of Large Carnivorous Animals May Be Difficult or Impossible to Discern

The various challenges and mandates facing many researchers studying large carnivores have meant that they were not able to examine ways in which such animals affect elements of biodiversity other than prey. As a result, the lack of evidence for impact of top predators on lower trophic levels has resulted in an interpretation that predation is not an important structuring force. This conclusion may be unwarranted because of the myriad realities that have constrained our ability to discern the impacts, if any, of predation on biodiversity (Steneck; Estes, this volume).

Measuring the Right Thing

The failure to discern whether large carnivores are significant interactors is often rooted in the lack of appropriate experimental conditions brought about by the great difficulty of designing suitable experiments with appropriate temporal and spatial dimensions (Duffy 2002; Terborgh, this volume). It is tremendously challenging to properly detect impacts unless the system is perturbed (Boutin; Estes, this volume). Even then, change might only be noted in systems that have been under continuous monitoring (Mills, this volume). The only other experimental option is to compare perturbed with unaltered systems, though such matched comparisons are difficult and costly, or impossible. Demonstrating the true functional role of some top predators may never be possible because baselines are elusive or gone (Steneck, this volume), and there may be no truly "natural" benchmarks against which to measure impacts. This latter is particularly the case for many marine systems (Dayton et al. 1998).

It is easy to draw incorrect conclusions about the true role of predation. For example, the role of sheephead (*Semicossyphus pulcher*) in regulating red sea urchins (*Strongylocentrotus franciscanus*) was at one time not evident because urchins had a relatively low importance rank in the diet of the fish. It was only following removal of this predatory fish that Cowen (1983) was able to recognize the regulatory role of predation in the maintenance of sea urchin densities. Experimental design difficulties such as these are of even more concern in investigations of the predatory role of larger carnivores. This kind of research has generally confined inquiry to the next trophic level and not examined the broader question of the role of predation on other components of biodiversity (Maehr et al., this volume). Impacts of the extinction of or large reductions in terrestrial species such as the passenger pigeon (*Ectopistes migratorius*), American chestnut (*Castanea dentata*), and bison (*Bison bison*), escaped the notice of the scientific community because no assessments were made at the time (Simberloff 2003). Extraordinary as it is, we may never know the impact, if any, of the disappearance of such visible members of their ecosystems.

Confounding Factors
It is not possible to predict what consequences will derive from species loss without understanding how the nature of species interactions themselves is affected by environmental context, including physical, environmental, and biotic conditions (Menge 2003). Such predictions have been difficult even in a system with as few components as Isle Royale—an essentially closed system with relatively few species and a simple food web—that has been studied in detail and monitored over 40 years. The significant yet seemingly random influence of external drivers such as weather seems to override relationships with other species in determining the

outcome of species loss (Smith et al. 2003). The interplay of confounding factors with ecosystem properties, therefore, will always hamper our ability to evaluate the link between top predators and the maintenance of biodiversity.

Human actions also confound the ability to discern the functional role of top predators in contemporary settings (Ray, this volume). With the largest predators completely or nearly extirpated from most of the world's land- and seascapes, it is often not possible to determine whether negative ecosystem changes are the direct result of the loss of predation as a structuring force or are due to other anthropogenic changes (e.g., through habitat change) (Dayton et al. 1998). In northeastern North America, for example, predators were extirpated by humans long before deer populations became overabundant, indicating that enhanced vegetation productivity due to seed crops, disturbance, and agriculture, rather than decreases in predator abundance, may have been the primary catalyst for the hyperabundance of ungulates apparent today (McShea, this volume). A similar conclusion can be drawn from the marine realm, where, in coral reefs, climate change was found to be more important to ecosystem function than the return of predators following fishing closure (McClanahan, this volume). Indeed, Jameson et al. (2002) point out that external atmospheric, terrestrial, and oceanic stressors that degrade the environment will often persist in marine protected areas in spite of cessation of fishing, overriding the potential recovery role of predation. It is clear that even if predation continues to be a force in such systems, other agents of change may be more important drivers of biodiversity shifts, making it impossible to understand the role of predation in the final analysis.

In the Eye of the Investigator

Researchers are influenced by numerous factors, including their education, their cultural background, and the systems on which they work. Yet they rarely realize, or acknowledge, these influences when phrasing conclusions from their work. One result of this is that ecologists who specialize in some biological systems often favor particular hypotheses and discount others developed to inform work on other systems (Chase 2000). This can be illustrated by comparing the views of ecologists working in terrestrial versus marine systems, where the proof for predator-induced trophic cascades in the former has been less consistently demonstrated than in the latter.

Such contrasts also exist when comparing differences in definitions of "true" impact of loss of carnivory. Impact can be measured in different ways as illustrated by the measurements made by plant ecologists. They have measured the effects of herbivory in a variety of ways: as plant damage, or changes in growth rate, species composition, or plant biomass (Schmitz et al. 2000). Hence, the various ways to measure impact influence the perception of the importance of the role of top predators in causing changes to plants. Although changes in vegetation composition may be equally important as changes in total plant biomass to ecosystem functioning (Duffy 2003), the latter are generally more obvious to the observer. Effects of loss of predators on plant community biomass have been shown much less often than such loss on plant damage or changes in species composition. This raises the question, How much evidence is enough to declare the existence of predator-mediated impacts? Will it simply be increased browse levels, significant reduction in biomass, or wholesale landscape transformation? Different interpretations of impacts of carnivore loss may even result in different interpretations of the same study regarding the overall strength of top-down control by predators (e.g., Kluane, Boutin; Terborgh, this volume). For some, it is enough to demonstrate potential impacts; for others, the results must demonstrate immediate and obvious impacts.

Unknowns and Unknowable

It is important to acknowledge the factors that remain unknown. It is one thing to document the loss of large carnivorous animals from systems around the globe (Gittleman et al. 2001; Myers and Worm 2003), but quite another to determine what this loss means for other parts of their ecosystems. Most of the research attention has focused on large terrestrial predators such as wolves (*Canis lupus*) and lions (*Panthera leo*). Impacts of lesser-known large carnivores are alternately assumed or ignored. The danger is in supposing that there is a lack of important ecological role where information is lacking.

In marine systems, comprehensive studies of the impacts of predation are often only feasible for benthic species having populations with modest geographic ranges (McClanahan, this volume). Expanding the area of inquiry to include broad-ranging pelagic species like bluefin tuna (*Thunnus thynnus*) and sharks has been difficult, if not impossible. Although the largest group of large marine carnivores, the sharks, have declined worldwide through overfishing (Baum et al. 2003), there has been no reported evidence of the impacts of their loss on other elements of marine biodiversity.

In the terrestrial realm, most research has focused on mammalian predators. Other taxa that occupy the top predator role, such as raptors and reptiles, have most often been ignored in ecosystem-level studies. It is a fairly novel approach even to include such species as members of the same top predator guild as mammals (but see Jaksic et al. 1993) The availability of basic ecological information (the first priority for research) on many raptor and reptile species already lags far behind that collected for many mammals and is generally collected in isolation from assessing their ecological roles.

Another intriguing group worthy of consideration is parasites and pathogens that occupy predator roles in virtually all ecosystems (Huxham et al. 1995). The relative invisibility of this class of organisms to the human investigator as well as the complexity of the interaction pathways in which they are imbedded account for the fact that they are seldom incorporated in food web models (Marcogliese and Cone 1997). Yet, because parasites and pathogens are active players in many species interactions, ignoring them risks profoundly misleading conclusions (Sousa 1991). In their recent review of trophic cascades in benthic marine systems, Pinnegar et al. (2000) speculated that including pathogenic interactions would substantially increase the number of examples of trophic cascades. In the few cases that pathogens and parasites have been considered as interactors, they are often classified as top predators (Huxham et al. 1995; Marcogliese and Cone 1997). Like predators, parasites and pathogens bring about both direct and indirect impacts on other organisms through changes in abundance and behavior (Mouritsen and Poulin 2002). This occurs in spite of the fact that energy flow from hosts to parasites constitutes a mere fraction of that from prey to predators.

As mediators of other trophic or competitive interactions in a given system, parasites and pathogens may well qualify as keystone species whose potentially strong effects are disproportionate to their biomass (Power et al. 1996). In such cases, their omission from a community web is "no more defensible than the omission of the principal vertebrate predators" (Huxham et al. 1995: 169). Yet, because we are talking about much smaller organisms feeding on larger ones (reversal of the size hierarchy; Marcogliese and Cone 1997), focus on the largest-bodied top predators naturally leads us to overlook the role played by these small organisms.

Parasites or pathogens may exert an über-apex predator role that could cascade down to release herbivore populations through elimination of predators. Although there are multiple examples of carnivore-specific diseases that cause high mortality levels (Funk et al. 2001), the cascading impacts of such outbreaks are not often documented. Glimpses, however, have been provided by disease events that have swept through systems that are under ecological study. Isle Royale, where wolves were reduced in number by a canine parvovirus, resulted in a marked increase in moose (*Alces alces*) on the island in spite of heavy wolf predation (Peterson 1999). In another example, lion and spotted hyena (*Crocuta crocuta*) populations were reduced significantly by a canine distemper outbreak in the early 1980s, with demonstrated effects on a subset of prey species (Sinclair et al. 2003). Parasites and pathogens can also work the other way, involving themselves in feedback loops following the demise of predators. In this context, the absence of predation as a regulatory force renders ecosystems more vulnerable by making elevated herbivore populations more susceptible to disease (Smith 2001; Ostfeld and Holt 2004).

The extent to which parasites and pathogens can be considered ultimate ecosystem regulators needs to be treated in the same manner as other confounding influences, such as humans who may similarly override impacts of predators. Although parasites and pathogens are present virtually everywhere, the fact that disease outbreaks are a relatively rare phenomenon suggests that the regulatory power of such organisms is not consistently strong (Polis 1999). Nevertheless, because the strength of their occurrence can be so unpredictable, ignoring them altogether would be ill advised for conservation practitioners.

Is Carnivore Conservation Compatible with Biodiversity Conservation?

Some in the conservation community endorse carnivore conservation based on their intrinsic value whereas others base their support on the structuring roles played by these members of the highest trophic level (Ray, this volume). The distinction between these positions is not always clean because the weighting of values and science is often different for different people, with some using science (e.g., the pivotal ecosystem role of carnivores) as a justification for a position primarily rooted in aesthetic or ethical values (Ray, this volume). Determining whether broader biodiversity interests require focus on carnivores demands a solid grasp of both the degree to which top predators influence biodiversity patterns and the extent to which the conservation needs for top carnivores and biodiversity can be described as compatible. Regarding the former, it is useful to pause and examine the state of knowledge on this issue, and specifically, whether there is enough information available to effect conservation. Next, to determine the relationship between conservation of carnivores and conservation of biodiversity, it is important to know the extent to which the goals of each overlap (Linnell et al.; Miquelle et al., this volume).

How Much Do We Need to Know about the Ecological Role of Top Carnivores to Do Conservation?

There has been enough demonstration of the trophic cascades resulting from the top-down roles played by large carnivores to justify the conclusion that they can and do occur (Schmitz et al. 2000; Estes, this volume). Still lacking, however, is the ability to predict exactly where and when such cascades take place and to predict the roles that both indirect and cascading effects will have on trophic levels other than prey.

Exploring the ecological role of large top predators has been primarily limited to an intellectual exercise with few practical considerations. The one exception has been cases where managers have needed to know the extent to which carnivores exert a top-down influence on their prey (Bowyer et al., this volume). Increasing research on the tangled complexity of food webs and trophic interactions resulting from predation has not led to a better understanding of how to conserve predators, let alone other components of biodiversity. It is high time that we forge an explicit link between the science of large carnivorous animals and the science and conservation of biological diversity to answer the many urgent conservation questions. For example, is it important to disprove whether factors other than predation occur as a regulatory or controlling force? Or is it enough to simply know that there has been ample demonstration of predators as a fairly universal influential limiting factor? Once there are answers available to such questions, how can we translate these into conservation action?

Are Conservation Needs for Top Carnivores and Biodiversity Equivalent?

Large carnivorous animals in both terrestrial and marine environments are commonly evoked as symbols of wilderness. Accordingly, wilderness is thought to be a precondition for conservation of these wide-ranging animals. Widespread human-induced erosion of biodiversity has meant that such remote areas are also counted on to preserve the other components of biodiversity. On the face of it, therefore, preserving large carnivores and preserving the remainder of biodiversity appear to be overlapping goals.

This supposition is true only when the root causes of decline are the same for both large carnivores and other components of biodiversity in all ecological settings. But is this true in both marine and terrestrial systems? Generally speaking, a much stronger case can be made for the compatibility between the conservation needs of top predators and general biodiversity in marine than in terrestrial environments. In the marine realm, where overexploitation has been the prime cause of ecosystem decay (Jackson et al. 2001; Pauly and Maclean 2003), even the most remote areas of the sea have been impacted by humans (Myers and Worm 2003). This impact has occurred in a very predictable sequence, with the largest predators the primary targets of fishing activities, with most species affected either directly or indirectly. Hence, one would be justified in invoking similar conservation remedies (reducing current exploitation levels), whether the target was the largest predators or other elements of biodiversity that share the same water column.

Unlike in marine settings, in terrestrial environments the predominant driver of most species decline is habitat change (Wilcove et al. 1998) with the areas that are most remote from human occupation least subjected to such change (Sanderson et al. 2002a). Although it is true that many terrestrial large carnivores as well as many components of biodiversity today find refuge in these remote wilderness areas, the two groups do not share identical root causes for endangerment. This means that conserving them may require different strategies. Large carnivores are generally not specialized animals, and pristine conditions are not needed for their continued survival. Rather, the principal factor explaining their distribution and abundance is security from human conflict (Linnell et al. 2000; Musiani and Paquet 2004; Treves and Karanth 2003; Miquelle et al.; Linnell, this volume). Many large carnivores have intrinsic ecological and life history characteristics that allow a resilient response in the face of many human perturbations, provided that baseline conditions of adequate prey and limited direct killing exist. In other words, there can be flexibility in selecting habitat for conserving terrestrial carnivores, whereas the same cannot be said for other highly specialized fauna and flora that are confined to areas with limited human impact.

In systems where top predators structure ecosystems, designing conservation plans around their needs can achieve conservation for the rest of biodiversity. Where most components of biodiversity, including top predators, are threatened by overexploitation, cessation of this activity would promote recovery of all trophic levels (Estes; Steneck and Sala, this volume). On the other hand, where threat reduction actions necessary to conserve the two categories of targets are different, focusing actions on large carnivores for their putative structuring role may allow species vulnerable to other threats to be overlooked by conservation actions (Kunkel 2003; Novaro and Walker, this volume).

It is vital that all biomes of the earth be included in the search for robust conservation conclusions, and time that marine and terrestrial systems be dealt with together instead of perpetuating the "separate but equal" doctrine. More to the point, perhaps these systems are not that equal. Pelagic marine populations are enormous and since isolated populations with small populations are rare, so too are extinctions in marine systems, except among some shore-dwelling vertebrates. The large areas, linked by connectivity, differ significantly from terrestrial ecosystems. But no matter what the system, it is clear that numerous pitfalls make it dangerous to assume that the mere presence of large carnivores will ensure persistence of biodiversity, such that working to ensure their conservation will inevitably result in achievement of other conservation goals. This does not, however, rule out the active employment of large carnivores as conservation tools, even in terrestrial regions. Areas that are protected from human influence form the basis of the great majority of our efforts to simultaneously conserve both large carnivores and biodiversity. However, concluding that protecting such areas will conserve both large carnivores and other components of biodiversity requires careful attention to the assumptions that form the basis of their use (Ray, this volume).

A Reality Check

Our knowledge of the roles of large carnivores in affecting biodiversity comes from only a few types of areas, many of which are far from dense human populations. It seems prudent to ask if results from these few studied regions are applicable to most other areas. But perhaps it is more relevant to ask whether carnivores can be used as markers of biodiversity in the many areas where the top predators have disappeared. Is there a difference in appreciation for biodiversity in areas where carnivores have disappeared as compared with areas where they still persist?

In landscapes dominated by humans, appreciation for overall biodiversity may be greater than in areas where large carnivores still exist. No one would claim that the biodiversity of smaller organisms is more highly valued in Alaska or Wyoming than in Europe, Toronto, or Tokyo. Similarly, where the primary clientele for the natural world of a region consist of hunters who value game they can hunt, there is likely to be less interest in pollinators and flowering plants than on the interaction between large carnivores and game species. Our point is simply that, where carnivores have been extirpated, the general populace becomes more enamored with biodiversity in general than in regions where they persist. Indeed, we might expect an inverse relationship between carnivore presence and human attitudes that believe biodiversity is relevant to environmental health (Rapport and Whitford 1999). Sadly, as top predators become less numerous, the overall appreciation of biodiversity may very well improve.

Large Carnivores as Conservation Tools

In spite of the concerns raised by those advocating the employment of carnivores as biodiversity surrogates, top predators have great potential as mechanisms to achieve broader biodiversity conservation goals. By virtue of their relatively large area requirements, their symbolic value, and their structuring roles, large carnivores have the potential to help conservation programs to (1) achieve higher-scale conservation ambitions (Miquelle et al., Maehr et al.; Boutin; Linnell et al., this volume) and (2) restore degraded ecosystems (Estes, this volume). The extent to which carnivores can be deployed to achieve such conservation goals depends on the following factors.

Large Carnivores, Large Dreams

For members of the public, there is a significant difference between a conservation program that includes large carnivores and one in which they are absent. The former suggests lofty aspirations, whereas the latter implies settling for somewhat less than the best. If large carnivores still persist in a conservation area, they will invariably figure prominently in conservation planning, even though this will inevitably result in a more expensive and complicated project. Conversely, when large carnivores are not included, the result is usually a project that is considerably more modest in scope.

Despite the limitations of using large carnivores as the umbrella or conservation symbol (discussed in Boutin; Linnell et al.; Miquelle et al.; Ray; Woodroffe and Ginsberg, this volume), they can be effectively used to help bring about big dreams (Linnell, this volume). This means, for example, that, although carnivores might not be useful in delineating the best habitat for conserving biodiversity, they are of great potential utility in keeping the public's focus on planning and protection at large scales. This important role that carnivores can play in helping create large spatial scales for conservation should not be confused with their ecological role and hence their importance to the persistence of biodiversity writ large.

Healing a Broken System: Restoring Carnivory

Carnivore restoration is often promoted as a mechanism to restore function to ecosystems that have been degraded. With ecosystem decay increasingly being attributed to the demise of top predator components, the restoration of these top predators as a first step has been promoted as a priority action (e.g., Soulé and Noss 1998; Terborgh et al. 1999; DeBoer 2000). Clearly, predators are returning in numerous places following prey recovery and security from human conflict (Berger, this volume), thereby providing some excellent opportunities to test this notion. In some cases, the return of top predators has directly resulted in the healing of disrupted systems. The most famous example has been illustrated by cycles of recovery and collapse of sea otters (*Enhydra lutris*) that have unequivocally demonstrated not only the consequences of heightened herbivory following the decline of sea otter populations but the relatively rapid recovery of denuded seascapes subsequent to the recovery of sea otter populations (Estes, this volume). By and large, however, the variety of ecosystem responses exhibited demonstrates that the return of top predators to areas from which they had been eliminated will not predictably result in restoration of biodiversity.

This dissimilarity is not species-dependent, as shown by the contrasting examples of wolf recovery in the United States. Recorded changes over the eight years since wolves were reintroduced into the Greater Yellowstone Ecosystem have included reduction in population size of the dominant herbivores, control of mesocarnivores, such as coyotes (*Canis latrans*), and recovery of woody riparian vegetation and associated biodiversity (Berger and Smith, this volume). This contrasts with the less publicized and gradual recolonization of wolves in the Great Lakes region of the United States (Wisconsin, Minnesota, and Michigan) over the past 30 years (Treves et al. 2002). Wisconsin is known for its exceptionally high-density deer populations (up to five times those of presettlement times), which have resulted in numerous adverse direct and indirect effects on the structure and composition of forest plant and animal communities (Rooney and Waller 2003). Although high deer populations have certainly facilitated the ongoing recovery and range expansion of wolves in the Great Lakes region (Mladenoff et al. 1997), there is little evidence that wolves have yet had any real impact on these prey populations or forest structure, except on a localized basis (Mech and Nelson 2000; Anderson et al. 2003; Anderson et al. in prep.).

Of course, part of the presumed lack of effect in the Great Lakes region could be due to the relative lack of research attention or temporal lags in responses. The most likely explanation, however, lies in the extent of human-induced habitat change and accompanying human conflict that has occurred over the last century in Wisconsin and Minnesota. The former arguably did more to promote changes in biodiversity than the disappearance of top predators. This may translate into a more significant time lag being required in the Great Lakes region for predation to exert an important influence—if it occurs at all. It is possible, however, that the ability of wolf populations to reach the densities necessary to play a restorative role will be limited by population control measures that have arisen in response to increases in livestock depredation accompanying wolf recolonization (Treves et al. 2002). In Yellowstone National Park, by contrast, the most significant change to an otherwise intact system has been the demise of its top predator (Smith et al. 2003; Berger and Smith, this volume), which may account for the relatively rapid time frame in which changes have ensued following the wolf's return. However, wolf impacts on ecosystem processes and biological diversity in areas outside the park, where habitat change and conflict are more prominent, are likely to be less substantial than inside park boundaries (Berger and Smith, this volume).

Two study areas where restorative effects of top predator recovery on overall biodiversity have not been readily apparent are in the Patagonian Steppe and some coral reef protected areas. It appears that, rather than contributing to the restoration of biodiversity, recovering puma (*Puma concolor*) and culpeo fox (*Pseudalopex culpaeus*) populations in southern Argentina and Chile (Novaro and Walker, this volume) have responded to high densities of nonnative prey and may themselves be responsible for keeping native prey at very low densities. This example —although counterintuitive regarding the potential for top predators to incur dysfunction rather than restoration—is not unique. In western desert ecosystems, for example, Berger and Wehausen (1991) and Sweitzer et al. (1997) subsequently documented ecosystem change as a consequence of colonization by mountain lions, including the possible elimination of small populations of native vertebrates. In coral reefs off Kenya and Belize, although recovery of top predators did occur following the initiation of closed-area management, the cascading impacts on other organisms were modest (McClanahan, this volume). Explanations for this lack of response are hypothesized to involve the complexity of the system, as well as confounding abiotic factors, such as warming ocean temperatures, that may have overridden predator-induced cascades.

Interpretation of these cases is confounded by the complicating effects of ecosystem changes other than predator removal. In the Yellowstone wolf and sea otter examples, physical modification of habitat by humans was relatively minimal, and few factors other than the disappearance of the top predator element seem to have been responsible for biodiversity changes. In marine systems, the larger issue is that humans continue to hunt large predators, so by the time we began to study ecosystem structure, most had already been extirpated (Steneck and Sala, this volume). In the few reefs where large predators are known to persist, they feed on the most important herbivores (e.g., the parrotfish) in the ecosystem (Sudekum et al. 1991). Yet even larger carnivores, such as tiger sharks that exceed 5 m in length, were known to eat large groupers on reefs and thereby may have created an additional trophic level. Thus it is not only possible that trophic cascades were more common in the past prior to systemic extirpations but that they may have involved an additional trophic level (Steneck and Sala, this volume).

The best chances for using predator restoration to reverse the negative condition of biodiversity may be in systems where the demise of predators has been clearly shown to result in adverse ecosystem impacts and where the system has not been importantly degraded by other factors. As previously discussed, the finest examples may emanate from marine systems, where overexploitation is the primary driving factor in biodiversity decline (Jackson et al. 2001). In the terrestrial realm, on the other hand, where habitat conversion has brought on so many changes to biodiversity, the return of predators to many places may require lengthy periods of time, if recovery is achieved at all. However, in all such systems, restoring top predators may still be one important component of a restoration plan with many other elements, and by itself may still nudge along the healing process in some fashion. This may come about simply by introducing an additional mortality for overabundant ungulate populations (McShea, this volume), or by spiritual restoration of some semblance of the wild past (Linnell et al., this volume).

Conservation Recommendations: Achieving Carnivore and Biodiversity Conservation

Evaluation of the cases presented in this book and of the supporting literature suggests a set of important issues that should inform conservation of both large carnivorous animals and biodiversity.

Clarifying the Targets of Conservation Action

Conservation must specify a target or set of targets, and the condition in which these are to be conserved (Redford et al. 2003). Although prevention of species loss is arguably the overarching goal of most conservation and management activities, the target of action will differ according to the mandate of the practitioner. It is, therefore, necessary for conservation practitioners and managers to specify in a proactive and transparent fashion the targets of their action. There are at least four possible targets of conservation action that may involve large carnivores, each with different prospects for carrying biodiversity along with it.

The first target is conservation of large carnivores themselves. Although the presumed functional importance of that species may be used as a way to justify specific attention the target is the species itself. The second target of action is biodiversity. In this case, large carnivores may be employed as tools to achieve this as discussed previously in this chapter. Third, the targets of many wildlife management agencies are species of importance as game (Gasaway et al. 1992; Van Ballenberghe and Ballard 1994; Boertje et al. 1996). In this situation, large carnivores may be seen more as an impediment to achieving management goals, than as tools (Berger, this volume). The fourth and last target is a multiple-use landscape where human values and aesthetic considerations play important roles in specifying conditions, as in Europe (Linnell et al., this volume).

Humans as Predators

When discussing the list of top carnivores in any given system humans are rarely listed. This, despite the fact that human predation has been a part of almost all continental biological systems for at least tens of thousands of years (Kay 1998; Berger, this volume). In this volume many authors have discussed the impact of large carnivorous animals, but few have specifically included humans in this list of actors (but see Jorgenson and Redford 1993; McShea; Berger, this volume). Instead, the analyses have treated humans as external to the list of large carnivores, with the impacts of hunting and fishing by humans assumed to be different from those of nonhuman carnivores. But is this true? Berger (this volume) takes on this question, asking whether hunting by humans is functionally redundant to that by carnivores.

Though this question has rarely been explicitly asked, available evidence does not indicate that humans can replace carnivores in an ecologically functional way. Major differences between human and carnivore hunting include (1) alteration of the intensity and timing of predation, (2) removal of different prey age and sex classes, (3) off-take of species other than harvestable prey, (4) modulation of meso-predator densities, (5) infrastructure to support human hunting with consequent effects on vegetation and plant-dependent species, (6) manipulation of carrion–scavenger relationships, and (7) modification of patterns of intraguild predation (Berger, this volume).

Given these differences, the impacts of human hunting on biodiversity remain largely undocumented and little appreciated. Humans are killing a great many animals throughout the world, and in most cases this hunting is unsustainable (Robinson and Bennett 2000). Hunting by humans often becomes incorporated in contentious debates from whether sport hunting is morally acceptable to whether subsistence hunting by indigenous peoples should be considered "natural" (Redford 2000). There is no shortage of rhetoric on these issues, but there is a marked shortage of data that would allow managers and the public to make informed decisions.

Is One Predator More Important Than Another?

Not all top carnivores are similar in their impacts on lower trophic levels, which likely means that they are not functionally redundant (Polis 1999; Kunkel 2003; Miquelle et al.; Woodroffe and Ginsberg, this volume). If it is possible to choose between conserving one species of carnivore or another, the intuitive choice might be to spend the most effort on those predators that have the strongest ecosystem roles (and hence are the most important determinants of biodiversity). However, it may be that in some cases this approach may hinder rather than help conservation goals. In the Russian Far East, the very fact that wolves have a higher potential than sympatric tigers (*Panthera tigris*) to exert strong impacts on biodiversity may in fact bolster the argument for promoting tiger over wolf conservation. Because large carnivore conservation would then be likely to incur less conflict with local hunters, better prospects for biodiversity conservation in the region would result (Miquelle et al., this volume). On the other hand, the diversity of functional roles exhibited among species in top predator guilds may bolster arguments that all should be conserved. In African savannas, where the megapredator guild includes as many as five species, Woodroffe and Ginsberg (this volume) make passionate arguments for the retention of intact carnivore guilds on the basis that there is little evidence for functional redundancy, so less important predators may in fact have roles that in the context of the complete guild serve to conserve ecological processes.

Not all carnivores are equal in their ecological roles so the replacement of one by another does not necessarily result in the same impacts. In boreal forest regions, for example, loss of lynx has been demonstrated to cause few disruptions to prevailing predator–prey cycles, with coyotes readily able to take over their role (Boutin, this volume). The problem is that both decreases in lynx populations and increases in coyote populations appear to be related to the degree of human influence in a given landscape (Boutin, this volume). One can therefore imagine a scenario where conservation goals regarding the integrity of the system are still very much compromised following smooth transferral of the top predator role from one species to another, even with the process of predation having remained unaffected.

Restoring Carnivory through Attaining Functional Population Sizes and Conserving Sufficiently Large Areas

Carnivore restoration or conservation designed to restore the ecological functions of the species must consider the appropriate densities that such functions require (Pyare and Berger 2003; Soulé et al. 2003). This way of thinking extends conservation action beyond the concept of minimum viable population sizes, which form the basis of the goals of most endangered species recovery plans (Estes, this volume). Extending action from focusing on presence or absence to critical ecological densities challenges practitioners to embrace more ambitious recovery goals. Such a new way of thinking is challenged by the difficulties both in determining how to measure functional population densities and in balancing ecologically functional population goals with socially acceptable predator densities. However, restoring carnivore function may not be enough to conserve nature, for, as Terborgh (this volume) reminds us, humans have also eliminated the largest of the herbivores from over 90% of the globe and, immune from regulation from predators, the ecological functions of these species have been extinguished. Similar forward thinking will be necessary in decision making surrounding the size of conservation areas. Almost all authors in this volume stress the need to think big when ensuring both the persistence of the species and the functional roles they play. This need is particularly relevant in light of the aggression that often characterizes relationships between predators (Woodroffe and Ginsberg, this volume). If indeed conservation of the intact carnivore guilds, and not just one or two species, is necessary for predation processes to fulfill their potential in influencing biodiversity patterns, conservation will be most effective in large and heterogeneous areas.

Adopt a Precautionary Approach and Use Adaptive Management

It may prove impossible to ever fully comprehend the roles that species play in ecosystems (Kareiva and Levin 2003). Despite the difficulties of showing the subtle ecosystem changes caused by the changing role of predation, it is becoming increasingly apparent that there are thresholds of change. Evidence from phase shifts that have been documented in a variety of ecosystems suggest that gradual changes in certain factors might have little effect until a threshold is reached, at which point a large shift occurs that might be difficult to reverse (Scheffer and Carpenter 2003). There is ample evidence that the disappearance of strong interactors, such as top predators, can play a role in such phase shifts (Soulé et al. 2003). Available evidence clearly suggests that the precautionary principle should be front and center in conservation action dealing with large carnivorous animals.

There is enough compelling evidence of the repercussions of removal of large carnivores across a diversity of systems to shift the burden of proof to those who discount the role of predation (Estes; Terborgh; this volume). Assuming that predators are not important could lead to dangerous assumptions, such as the idea that perturbations can be reversed easily (Scheffer and Carpenter 2003). This means that managers and practitioners should be compelled to prove a lack of predators' importance before assuming that they do not play important ecological roles.

In the absence of unequivocal answers to the set of questions addressed in this book, we strongly urge adoption of an adaptive management approach that incorporates research into conservation action. Specifically, it is the integration of design, management, and monitoring to systematically test assumptions in order to adapt and learn (Salafsky et al. 2001). Several conditions warrant an adaptive management approach: (1) conservation takes place in complex systems, (2) the world is constantly and unpredictably changing, (3) immediate action is required, (4) there is no such thing as complete information, and (5) learning is possible and essential (modified from Salafsky et al. 2001). Using an adaptive management approach would help us learn more about the large carnivores that share our world by working to ensure their conservation.

Understanding Values and Communicating to the Public

Large carnivores have a strong emotional valence. Virtually everyone has a strong opinion about their worth—negative, or positive. In this climate, there has been a woefully inadequate articulation of science and management/conservation and an even greater disconnect between advocacy and science. Advocates both for and against the continued existence of carnivores have often either skipped the use of data entirely or selectively used data to justify their positions. Left largely ignored is that often the positions held by advocates are based on value propositions—the perceived value of carnivores.

We suggest that it is important to distinguish between value-based and science-based reasons for carnivore conservation—understanding that the two can be integrated. Too often scientifically grounded principles to justify carnivore conservation have obscured the more fundamental aesthetic and ethical values that lie at the root of many who argue for their conservation (Jepson and Canney 2003). Warren et al. (1990: 587) document an important example of the need to clearly articulate for the public the real reasons for carnivore restoration. Having told the public that bobcats (*Lynx rufus*) were to be reintroduced to control herbivore populations on Cumberland Island, Georgia, they then had to face evidence that bobcats were doing well but that deer numbers had not decreased. As they concluded: "We erred by not emphasizing the original objective of restoring biodiversity." We feel that it is vital to be honest in enumerating the reasons for carnivore conservation or restoration based on an explicit statement of the underpinning scientific understanding. If we make arguments based only on scientific claims, and data are subsequently presented that refute our claims, our position is substantially weakened. Value-based arguments speak for themselves. We recognize that this recommendation is complicated by the way that scientific investigation is structured through testing hypotheses and not through providing unified "truths." As such, scientists may often not know the answer to a question posed by managers or the public. But in cases such as these, and the Cumberland Island bobcat example, what is vital is that the best available scientific understanding not be used to cloak uncertainty from the eyes of other stakeholders.

Concluding Thoughts

Large carnivores must be conserved, whether for their ecological function, their intrinsic value, or their value to humans. In raising the issues discussed in this volume we have strived to contribute to the conservation not only of carnivores but of all biodiversity. The question posed by this volume, whether conservation of carnivores is equivalent to the conservation of biodiversity, cannot be answered with a simple yes or no. This can come as no surprise because it stems from the conclusion that it is impossible to say that all ecological systems are either regulated exclusively by top-down or by bottom-up processes (Bowyer et al., this volume). Simple answers elude us.

It is clear, however, that large carnivores *can* impact biodiversity through direct impacts on prey and through indirect effects on other trophic levels. Most questions that would help in conservation practice remain unanswered. We would advocate that those interested in studying predators seek to more clearly elucidate (1) when and where predators play ecologically significant roles, (2) where and when restoration of large carnivores would result in restoration of biodiversity, (3) what densities of large carnivores are necessary to produce these effects, and (4) what the interactions are between hunting by carnivores and hunting by humans.

Keeping large carnivores is a measure of how we are doing in the battle to save the planet's biodiversity. We must redouble our efforts to ensure their conservation in all possible settings, in both the terrestrial and the marine realms. Top predators can be powerful symbols and useful tools for conservation. But they cannot, and should not be asked to, themselves carry the weight of biodiversity conservation.

Summary

There has been an increasing tendency for conservation programs to place large carnivores as centerpieces of conservation strategies. This, despite the lack of understanding about whether or not large carnivore conservation is compatible with biodiversity conservation. Through synthesis of the material presented in this volume, as well as other published literature, this concluding chapter examines two parts of the central question of this volume. The first addresses whether top predators play an important role in structuring ecosystems, thereby affecting biodiversity at multiple levels. The second discusses practical considerations for using such knowledge in developing effective use of large carnivorous animals as biodiversity conservation tools. Experimental evidence of both direct and cascading effects of large predators on biodiversity is scarce and confined to a few well-known examples, mainly from marine systems. Although this can be partly explained by the practical difficulties associated with such study, the challenge lies in making the distinction between effects of predation not occurring because there are indeed no impacts, and effects that are difficult or impossible to discern due to the many constraints placed on the ability of scientists to perceive true impact. These constraints include difficulty in measurements, confounding factors, and investigator biases. There also exist significant knowledge gaps regarding the role of many lesser-known predators, including sharks and other apex predators of the sea, reptiles and raptors, and parasites and pathogens.

Although there is enough knowledge of the potential impact of predation, scientists will likely never be able to reliably predict cascading impacts on elements of biodiversity other than prey. The path to resolving how the science of understanding predation can be translated into practical considerations is rooted in understanding the degree to which biodiversity and large carnivore conservation are equivalent. This will be the case only when the root causes of species decline in a given area are similar for both large carnivores and other biodiversity elements— a condition that is more likely to be the case in marine than in terrestrial environments. Even in places where large carnivore and biodiversity conservation are not compatible, the former can still be used as a tool to help achieve broader biodiversity goals, as long as careful attention is paid to the underlying assumptions. Specifically, integration of large carnivorous animals into a conservation strategy can help conservation programs "think big" and restore degraded ecosystems. The best chances for predator restoration may be in areas where the demise of predators has been clearly shown to result in adverse ecosystem impacts and where the system has not been importantly degraded by other factors, such as human-induced habitat change. We conclude by laying out a number of important issues for conservation practitioners to address when working to conserve both large carnivorous animals and biodiversity.

ACKNOWLEDGMENTS

We thank the participants in the original White Oak workshop (and the authors of these chapters) for their insightful discussions, which catalyzed many of the ideas contained in this chapter. The conclusions we have presented, however, are not those of the participants, but of the authors themselves. Adrian Treves, Adrian Wydeven, and Joanna Zigouris provided important information or support in the preparation of this manuscript.

References

Aanes, R., and R. Anderson. 1996. The effects of sex, time of birth, and habitat on the vulnerability of roe deer fawns to red fox predation. *Canadian Journal of Zoology* 74: 1857–1865.

Aanes, R., B-E Sæther, E. J. Solberg, S. Aanes, O. Strand, and N. A. Oritsland. 2003. Synchrony in Svalbard reindeer population dynamics. *Canadian Journal of Zoology* 81: 103–110.

Abramov, K. G. 1940. Wolves in Sikhote-Alin Zapovednik and control over them. *Nauchno-metodicheskie Zapiski Glavnogo Upravleniya Po Zapovednikam, Zooparkam i Zoosadam* 6: 133–135.

Abrams, P. A. 1994. The fallacies of "ratio-dependent" predation. *Ecology* 75: 1842–1850.

Abrams, P. A. 1997. Anomalous prediction of ratio-dependent models of predation. *Oikos* 80: 163–171.

Aizen, M. A., and P. Feinsinger. 1994. Forest fragmentation, pollination, and plant reproduction in a Chaco dry forest, Argentina. *Ecology* 75: 330–351.

Akçakaya, H. R., R. Arditi, and L. R. Ginzburg. 1995. Ratio-dependent predation: an abstraction that works. *Ecology* 76: 995–1004.

Alexander, R. D. 1974. The evolution of social behavior. *Annual Review of Ecology and Systematics* 5: 325–383.

Alroy, J. 2001. A multispecies overkill simulation of the end-Pleistocene megafaunal mass extinction. *Science* 292: 1893–1896.

Alvard, M. 1998. Indigenous hunters in the Neotropics: conservation or optimal foraging? Pages 474–500 *in* T. M. Caro, editor. *Behavioral ecology and conservation biology.* Oxford University Press, New York, New York, USA.

Alverson, W. S., D. M. Waller, and S. L. Solheim. 1988. Forests too deer: edge effects in northern Wisconsin. *Conservation Biology* 2: 348–358.

Amaya, J. N., J. von Thüngen, and D. A. Delamo. 2001. *Relevamiento y distribución de guanacos en la Patagonia.* Comunicación Técnica 111, INTA, Bariloche, Argentina.

Ames, E. P. 2003. Atlantic cod stock structure in the Gulf of Maine. *Fisheries Research* 291: 10–19.

Andelman, S. J., and W. F. Fagan. 2000. Umbrellas and flagships: efficient conservation surrogates or expensive mistakes? *Proceedings of the National Academy of Sciences* 97: 5954–5959.

Andersen, R., and J. D. C. Linnell. 2000. Irruptive potential in roe deer: density-dependent effects on body mass and fertility. *Journal of Wildlife Management* 64: 698–706.

Andersen, R., J. D. C. Linnell, and E. J. Solberg. In press. The future role of large carnivores on terrestrial trophic interactions: the northern temperate view. *In* K. Danell, R. Bergström, P. Duncan, and J. Pastor, editors. *The impact of large Mammalian herbivores on biodiversity, ecosystem structure and function.* Cambridge University Press, Cambridge, UK.

Anderson, D. P., T. P. Rooney, M. G. Turner, D. M. Waller, A. P. Wydeven, J. E. Wiedenhoeft, D. E. Beyer, W. S. Alverson, and J. D. Forester. 2003. Tri-trophic interactions among wolves, deer, and plant communities in Northern Wisconsin and Michigan. The 88th Annual Meeting of the Ecological Society of America. Savannah, Georgia. (Oral presentation).

Anderson, J. E. 1991. A conceptual framework for evaluating and quantifying naturalness. *Conservation Biology* 5: 347–352.

Anderson, R. C. 1994. Height of white-flowered trillium (*Trillium grandiflorum*) as an index of deer browsing intensity. *Ecological Applications* 4: 104–109.

Anderson, R. C., and O. L. Loucks. 1979. White-tailed deer (*Odocoileus virginianus*) influence on the structure and composition of *Tsuga canadensis* forests. *Journal of Applied Ecology* 16: 855–861.

Anderson, R. C., S. S. Dhillion, and T. M. Kelley. 1996. Aspects of the ecology of an invasive plant, garlic mustard (*Alliaria petiolata*) in Central Illinois. *Restoration Ecology* 4: 281–291.

Anderson, R. J., P. Carrick, G. J. Levitt, and A. Share. 1997. Holdfasts of adult kelp *Ecklonia maxima* provide refuges from grazing for recruitment of juvenile kelps. *Marine Ecology Progress Series* 159: 265–273.

Andrén, H. 1994. Effects of habitat fragmentation on birds and mammals in landscapes with different proportions of suitable habitat: a review. *Oikos* 71: 355–366.

Andrew, N. L. 1993. Spatial heterogeneity, sea urchin grazing and habitat structure on reefs in temperate Australia. *Ecology* 74: 292–302.

Andrew, N. L. 1994. Survival of kelp adjacent to areas grazed by sea urchins in New South Wales, Australia. *Australian Journal of Ecology* 19: 466–472.

Andrew, N. L., and J. H. Choat. 1982. The influence of predation and nonspecific adults on the abundance of juvenile *Evechinus chloroticus* (Echinodea: Echinodermata). *Oecologia* 54: 80–87.

Andrew, N. L., and A. L. O'Neill. 2000. Large-scale patterns in habitat structure on subtidal rocky reefs in New South Wales. *Marine and Freshwater Research* 51: 255–263.

Andrew, N. L., and A. J. Underwood. 1993. Density-dependent foraging in the sea urchin *Centrostephanus rodgersii* on shallow subtidal reefs in New South Wales, Australia. *Marine Ecology Progress Series* 99: 89–98.

Andrew, N. L., Y. Agatsuma, E. Ballesteros, A. G. Bazhin, E. P. Creaser, D. K. A. Barnes, L. W. Botsford, et al. 2002. Status and management of world sea urchin fisheries. *Oceanography and Marine Biology: An Annual Review* 40: 343–425.

Angelstam, P., and B. Pettersson. 1997. Principles of present Swedish forestry biodiversity management. *Ecological Bulletins* 46: 191–203

Angelstam, P., P. Wikberg, P. Danilov, W. E. Faber, and K. Nygrén. 2000. Effects of moose density on timber quality and biodiversity restoration in Sweden, Finland, and Russian Korelia. *Alces* 36: 133–145.

Araújo, M. B. 2003. The coincidence of people and biodiversity in Europe. *Global Ecology and Biogeography* 12: 5–12.

Arcese, P., and A. R. E. Sinclair. 1997. The role of protected areas as ecological baselines. *Journal of Wildlife Management* 61: 587–602.

Arditi, R., and L. R. Ginzburg. 1989. Coupling in predator–prey dynamics: ratio dependence. *Journal of Theoretical Biology* 139: 311–326.

Aronson, R. B., W. F. Precht, M. A. Toscano, and K. H. Koltes. 2002a. The 1998 bleaching event and its aftermath on a coral reef in Belize. *Marine Biology* 144: 435–447.

Aronson, R. B., I. G. Macintyre, W. F. Precht, T. J. T. Murdoch, and C. M. Wapnick. 2002b. The expanding scale of species turnover on coral reefs in Belize. *Ecological Monographs* 72: 233–249.

Atkinson, K. T., and D. W. Janz. 1994. *Effect of wolf control on black-tailed deer in the Nimpkish Valley on Vancouver Island.* British Columbia Ministry of Environment, Lands, and Parks, Wildlife Bulletin B-73, British Columbia, Canada.

Attiwill, P. M. 1994. The disturbance of forest ecosystems: the ecological basis for conservative management. *Forest Ecology and Management* 63: 247–300.

Aubry, K. B., G. M. Koehelr, and J. R. Squires. 2000. Ecology of Canada lynx in southern boreal forests. Pages 373–396 *in* Ruggiero, L. F., K. B. Aubry, S. W. Buskirk, G. M. Koehler, C. J. Krebs, K. S. McKelvey, and J. R. Squires, editors. *Ecology and conservation of lynx in the United States.* University Press of Colorado, Niwot, Colorado, USA, and USDA, Rocky Mountain Research Station, Missoula, Montana, USA.

Augustine, D. J. 2002. Large herbivores and process dynamics in a managed savanna ecosystem. Ph.D. thesis. Syracuse University, Syracuse, New York, USA.

Augustine, D. J., and L. E. Frelich. 1998. Effects of white-tailed deer on populations of an understory forb in fragmented deciduous forest. *Conservation Biology* 12: 995–1004.

Augustine, D. J., and S. J. McNaughton. 1998. Ungulate effects on the functional species composition of plant communities: herbivore selectivity and plant tolerance. *Journal of Wildlife Management* 62: 1165–1183.

Babcock, R., S. Kelly, N. T. Shears, J. W. Walker, and T. J. Willis. 1999. Changes in community structure in temperate marine reserves. *Marine Ecology Progress Series* 189: 125–134.

Bahamonde, N., S. Martin, and A. P. Sbriller. 1986. Diet of guanaco and red deer in Neuquén Province, Argentina. *Journal of Range Management* 39: 22–24.

Bailey, J. K., and T. G. Whitham. 2003. Interactions among elk, aspen, galling sawflies, and insectivorous birds. *Oikos* 101: 127–134.

Bailey, T. N. 1993. *The African leopard: ecology and behavior of a solitary felid.* Columbia University Press, New York, New York, USA.

Baird R. W., and L. M. Dill. 1996. Ecological and social determinants of group size in transient killer whales. *Behavioral Ecology* 7: 408–416.

Baker, B. W., and E. P. Hill. 2003. Beaver (*Castor canadensis*). Pages 288–310 *in* G. A. Feldhamer, B. C. Thompson, and J. A. Chapman, editors. *Wild mammals of North America: biology, management, and conservation.* Johns Hopkins University Press, Baltimore, Maryland, USA.

Baldi R., S. D. Albon, and D. A. Elston. 2001. Guanacos and sheep: evidence for continuing competition in arid Patagonia. *Oecologia* 129: 561–570.

Balgooyen, C. P., and D. M. Waller. 1995. The use of *Clintonia borealis* and other indicators to gauge impacts of white-tailed deer on plant communities in northern Wisconsin, USA. *Natural Areas Journal* 15: 308–318.

Ballard, W. B., and D. G. Larsen. 1987. Implications of predator–prey relationships to moose management. *Swedish Wildlife Research* 1 (Supplement): 581–602.

Ballard, W. B., D. Lutz, T. W. Keegan, L. H. Carpenter, and J. C. deVos Jr. 2001. Deer–predator relationships: a review of recent North American studies with emphasis on mule and black-tailed deer. *Wildlife Society Bulletin* 29: 99–115.

Ballard, W. B., L. N. Carbyn, and D. W. Smith. 2003. Wolf interactions with non-prey. Pages 259–271 *in* L. D. Mech and L. Boitani, editors. *Wolves: behavior, ecology, and conservation.* The University of Chicago Press, Chicago, USA.

Bangs, E. E., and S. H. Fritts. 1995. Reintroducing the gray wolf to central Idaho and Yellowstone National Park. *Wildlife Society Bulletin* 24: 402–413.

Bank, M. S., R. J. Sarno, N. K. Campbell, and W. L. Franklin. 2002. Predation of guanacos (*Lama guanicoe*) by southernmost mountain lions (*Puma concolor*) during a historically severe winter in Torres del Paine National Park, Chile. *Journal of Zoology,* London 258: 215–222.

Barboza, P. S., and R. T. Bowyer. 2000. Sexual segregation in dimorphic deer: a new gastrocentric hypothesis. *Journal of Mammalogy* 81: 473–489.

Barrales, H. L., and C. Lobban. 1975. The comparative ecology of *Macrocystis pyrifera* with emphasis on the forests of Chubut Argentina. *Ecology* 63: 657 – 677.

Barry, D., and M. Oelschlaeger. 1996. A science for survival: values and conservation biology. *Conservation Biology* 10: 905–911.

Bartram, W. 1996. *Travels and other writings (with travels through North and South Carolina, Georgia, East and West Florida).* The Library of America. New York, New York, USA.

Batzer, H. O., M. P. Martin, W. J. Mattson, and W. E. Miller. 1995. The forest tent caterpillar in aspen stands: distribution and density estimation of four life stages in four vegetation strata. *Forest Science* 41: 99–121.

Baum, J. K., and R. A. Myers. 2003. Shifting baselines and the decline of pelagic sharks in the Gulf of Mexico. *Ecology Letters* 7: 135–145.

Baum, J. K., R. A. Myers, D. C. Kehler, B. Worm, S. J. Harley, and P. A. Doherty. 2003. Collapse and conservation of shark populations in the northwest Atlantic. *Science* 299: 389–392.

Bekoff, M. 2003. Cunning coyotes: tireless tricksters and protean predators. Available from http://www.nosnare.org/articles/bek.html (accessed October 2004).

Belden, R. C., W. B. Frankenberger, R. T. McBride, and S. T. Schwikert. 1988. Panther habitat use in southern Florida. *Journal of Wildlife Management* 52: 660–663.

Bellati, J., and J. von Thüngen. 1990. Lamb predation in Patagonian ranches. Pages 263–268 *in* L. R. Davis and R. E. Marsh, editors. *Proceedings of the 14th Vertebrate Pest Conference.* University of California, Davis, California, USA.

Bellwood, D. R., A. S. Hoey, and J. H. Choat. 2003. Limited functional redundancy in high diversity systems: resilience and ecosystem function on coral reefs. *Ecology Letters* 6: 281–285.

Bellwood, D. R., T. P. Hughes, C. Folke, and M. Nyström. 2004. Confronting the coral reef crisis. *Nature* 429: 827–833.

Bender, L. C., D. E. Beyer Jr., and J. B. Haufler. 1999. Effects of short-duration, high intensity hunting on elk wariness in Michigan. *Wildlife Society Bulletin* 27: 441–445.

Bengtsson, J., S. G. Nilsson, A. Franc, and P. Menozzi. 2000. Biodiversity, disturbances, ecosystem function and management of European forests. *Forest Ecology and Management* 132: 39–50.

Benson, E. P. 1997. *Birds and beasts in ancient Latin America.* The University Presses of Florida, Gainesville, Florida, USA.

Bentley, A. 1998. *An introduction to the deer of Australia.* Australian Deer Research Foundation, Ltd., Melbourne, Australia.

Berger, J. 1991. Pregnancy incentives, predation constraints and habitat shifts: experimental and field evidence for wild bighorn sheep. *Animal Behavior* 41: 61–77.

Berger, J. 1998. Future prey: some consequences of the loss and restoration of large carnivores. Pages 80–100 *in* T. M. Caro, editor. *Behavioral ecology and conservation biology.* Oxford University Press, New York, Oxford, USA.

Berger, J. 1999. Anthropogenic extinction of top carnivores and interspecific animal behavior: implications of the rapid decoupling of a web involving wolves, bears, moose and ravens. *Proceedings of the Royal Society of London B* 266: 2261–2267.

Berger, J. 2002. Wolves, landscapes and the ecological recovery of Yellowstone. *Wild Earth* 12: 32–37.

Berger, J., and C. Cunningham. 1994. *Bison: mating and conservation in small populations.* Columbia University Press, New York, New York, USA.

Berger, J., and M. E. Gompper. 1999. Sex ratios in extant ungulates: products of contemporary predation or past life histories? *Journal of Mammalogy* 80: 1084–1113.

Berger J., and J. D. Wehausen. 1991. Consequences of a mammalian predator–prey disequilibrium in the Great Basin Desert. *Conservation Biology* 5: 244–248.

Berger, J., J. W. Testa, T. Roffe, and S. L. Montfort. 1999. Conservation endocrinology: a noninvasive tool to understand relationships between carnivore colonization and ecological carrying capacity. *Conservation Biology* 13: 980–989.

Berger, J., P. B. Stacy, L. Bellis, and M. P. Johnson. 2001a. A mammalian predator–prey imbalance: grizzly bear and wolf extinction affect avian neotropical migrants. *Ecological Applications* 11: 947–960.

Berger, J., J. E. Swenson, and I. Persson. 2001b. Recolonizing carnivores and naïve prey: conservation lessons from Pleistocene extinctions. *Science* 291: 1036–1039.

Berger, J., S. L. Monfort, T. Roffe, P. B. Stacey, and J. W. Testa. 2003. Through the eyes of prey: how the extinction and conservation of North America's large carnivores alter prey systems and biodiversity. Pages 133–156 *in* M. Appolilio and M. Festa-Bianchet, editors. *Animal behavior and wildlife conservation*. Island Press, Covello, California, USA.

Bergeron, Y., B. Harvey, A. Leduc, and S. Gauthier. 1999. Forest management guidelines based on natural disturbance dynamics: stand- and forest-level considerations. *Forestry Chronicle* 75: 49–54.

Bergerud, A. T. 1988. Caribou, wolves and man. *Trends in Ecology and Evolution* 3: 68–72.

Bergerud, A. T., W. Wyette, and B. Snider. 1983. The role of wolf predation in limiting a moose population. *Journal of Wildlife Management* 47: 977–988.

Berglund, B. E. editor. 1991. The cultural landscape during 6000 years in southern Sweden: the Ystad project. *Ecological Bulletins* 41: 1–495.

Berryman, A. A. 1992. The origin and evolution of a predator–prey theory. *Ecology* 73: 1530–1535.

Bertram, B. C. R. 1979. Serengeti predators and their social systems. Pages 221–248 *in* A. R. E. Sinclair and M. Norton-Griffiths, editors. *Serengeti: dynamics of an ecosystem*. University of Chicago Press, Chicago, Illinois, USA.

Bêty, J., G. Gauthier, E. Korpimäki, and J. Giroux. 2002. Shared predators and indirect trophic interactions: lemming cycles and arctic-nesting geese. *Journal of Animal Ecology* 71: 88–98.

Biggs, H. C., and K. M. Rogers. 2003. An adaptive system to link science, monitoring and management in practice. Pages 59–80 *in* J. Du Toit, K. H. Rogers, and H. C. Biggs, editors. *The Kruger experience*. Island Press, Washington, District of Columbia, USA.

Bjorge, R. R., and J. R. Gunson. 1989. Wolf, *Canis lupus*, population characteristics and prey relationships near Simonette River, Alberta. *Canadian Field Naturalist* 103: 327–334.

Björse, G., and R. Bradshaw. 1998. 2000 years of forest dynamics in southern Sweden: suggestions for forest management. *Forest Ecology and Management* 104: 15–26.

Bleich, V. C., R. T. Bowyer, and J. D. Wehausen. 1997. Sexual segregation in mountain sheep: resources or predation? *Wildlife Monographs* 134: 1–50.

Bodkin, J. L., A. M. Burdin, and D. A. Ryazanov. 2000. Age- and sex-specific mortality and population structure in sea otters. *Marine Mammal Science* 16: 201–219.

Boertje, R. D., P. Valkenburg, and M. E. McNay. 1996. Increases in moose, caribou, and wolves following wolf control in Alaska. *Journal of Wildlife Management* 60: 474–489.

Boitani, L. 1992. Wolf research and conservation in Italy. *Biological Conservation* 61: 125–132.

Boitani, L. 1995. Ecological and cultural diversities in the evolution of wolf human relationships. Pages 3–12 *in* L. N. Carbyn, S. H. Fritts, and D. R. Seip, editors. *Ecology and conservation of wolves in a changing world.* Canadian Circumpolar Institute, Alberta, Canada.

Boitani, L. 2000. Action plan for the conservation of the wolves (*Canis lupus*) in Europe. Council of Europe, *Nature and Environment*, Council of Europe Publishing. 113: 1–86.

Bonino, N. A. 1995. Introduced mammals in Patagonia, southern Argentina: consequences, problems, and management considerations. Pages 406–408 *in* J. A. Bissonette and P. R. Krausman, editors. *Integrating people and wildlife for a sustainable future. Proceedings of the First International Wildlife Management Congress.* The Wildlife Society, Bethesda, Maryland, USA.

Bonino, N., G. Bonvissuto, and A. Sbriller. 1986. *Composición botánica de la dieta de herbívoros silvestres y domésticos en el área de Piclaniyeu (Río Negro) 2: Calculo de los índice de diversidad trófica y similitud.* Comunicación Técnica 10, INTA, Bariloche, Argentina.

Bonino, N., A. Sbriller, M. M. Manacorda, and F. Larosa. 1997. Food partitioning between the mara (*Dolichotis patagonum*) and the introduced hare (*Lepus europaeus*) in the Monte desert, Argentina. *Studies on Neotropical Fauna and Environment* 32: 129–134.

Boomgaard, P. 2001. *Frontiers of fear: tigers and people in the Malay world 1600–1950.* Yale University Press, New Haven, USA.

Boonstra, R., S. Boutin, A. Byrom, T. Karels, A. Hubbs, K. Stuart-Smith, M. Blower, S. Antpoehler. 2001a. The role of red squirrels and arctic ground squirrels. Pages 179–214 *in* C. J. Krebs, S. Boutin, and R. Boonstra, editors. *Ecosystem dynamics of the boreal forest: The Kluane Project.* Oxford University Press, New York, New York, USA.

Boonstra, R., C. J. Krebs, S. Gilbert, and S. Schweiger. 2001b. Voles and mice. Pages 215–239 *in* C. J. Krebs, S. Boutin, and R. Boonstra, editors. *Ecosystem dynamics of the boreal forest: The Kluane Project.* Oxford University Press, New York, New York, USA.

Boutin S. 1992. Predation and moose population dynamics: a critique. *Journal of Wildlife Management* 56: 116–127.

Boutin, S., C. J. Krebs, A. R. E. Sinclair, and J. N. M. Smith. 1986. Proximate causes of losses in a snowshoe hare population. *Canadian Journal of Zoology* 64: 606–610.

Boutin, S., C. J. Krebs, R. Boonstra, M. R. T. Dale, J. Hannon, K. Martin, A. R. E. Sinclair, et al. 1995. Population changes of the vertebrate community during a snowshoe hare cycle in Canada's boreal forest. *Oikos* 74: 69–80.

Bowen, B. W. 1999. Preserving genes, species, or ecosystems? Healing the fractured foundations of conservation policy. *Molecular Ecology* 8: S5–S10.

Bowen, L., and D. van Vuren. 1997. Insular endemic plants lack defenses against herbivores. *Conservation Biology* 11: 1249–1254.

Bowers, M. A. 1997. Influence of deer and other factors on an old-field plant community. Pages 310–326 *in* W. J. McShea, H. B. Underwood, and J. H. Rappole, editors. *The science of overabundance: deer ecology and management.* Smithsonian Institution Press, Washington, District of Columbia, USA.

Bowyer, R. T. 1984. Sexual segregation in southern mule deer. *Journal of Mammalogy* 65: 410–417.

Bowyer R. T. 1987. Coyote group size relative to predation on mule deer. *Mammalia* 51: 515–526.

Bowyer, R .T. 1991. Timing of parturition and lactation in southern mule deer. *Journal of Mammalogy* 72: 138–145.

Bowyer, R. T., V. Van Ballenberghe, and K. R. Rock. 1994. Scent marking by Alaskan moose: characteristics and spatial distribution of rubbed trees. *Canadian Journal of Zoology* 72: 2186–2191.

Bowyer, R. T., J. G. Kie, and V. Van Ballenberghe. 1996. Sexual segregation in black-tailed deer: effects of scale. *Journal of Wildlife Management* 60: 10–17.

Bowyer, R. T., V. Van Ballenberghe, and J. G. Kie. 1997. The role of moose in landscape processes: effects of biogeography, population dynamics, and predation. Pages 265–287 *in* J. A. Bissonette, editor. *Wildlife and landscape ecology: effects of pattern and scale.* Springer-Verlag, New York, New York, USA.

Bowyer, R. T., V. Van Ballenberghe, and J. G. Kie. 1998. Timing and synchrony of parturition in Alaskan moose: long-term versus proximal effects of climate. *Journal of Mammalogy* 79: 1332–1344.

Bowyer, R. T., V. Van Ballenberghe, J. G. Kie, and J. A. K. Maier. 1999. Birth site selection by Alaskan moose: strategies for coping with a risky environment. *Journal of Mammalogy* 80: 1070–1083.

Bowyer, R. T., D. R. McCullough, and G. E. Belovsky. 2001. Causes and consequences of sociality in mule deer. *Alces* 37: 371–402.

Boyce, M. S. 2000. Modeling predator–prey dynamics. Pages 253–287 *in* L. Boitani and T. Fuller, editors. *Research techniques in animal ecology.* Columbia University Press, New York, New York, USA.

Boyd, D. K., R. R. Ream, D. H. Pletscher, and M. W. Fairchild. 1994. Prey taken by colonizing wolves and hunters in the Glacier National Park area. *Journal of Wildlife Management* 58: 289–295.

Brainerd, S. M., and T. Bjerke. 2003. Informasjonstiltak om store rovdyr i Norge. *NINA fagrapport* 069: 1–71.

Bramwell, D., editor. 1979. *Plants and islands.* Academic Press, London, UK.

Brand, C. J., L. B. Keith, and C. A. Fischer. 1976. Lynx responses to changing snowshoe hare densities in Alberta. *Journal of Wildlife Management* 43: 416–428.

Brandner, T. A., R. O. Peterson, and K. L. Risenhoover. 1990. Balsam fir on Isle Royale: effects of moose herbivory and population density. *Ecology* 71: 155–164.

Bratton, S. P. 1979. Impacts of white-tailed deer on the vegetation of Cades coves, Great Smoky Mountains National Park. *Proceedings of the Annual Conference of the Southeastern Association of Fish and Wildlife Agencies* 33: 31–35.

Brawley, S. H., and W. H. Adey. 1977. Territorial behavior of threespot damselfish (*Eupomacentrus planifrons*) increases reef algal biomass and productivity. *Environmental Biology of Fish* 21: 45–51.

Breitenmoser, U. 1998. Large predators in the Alps: the fall and rise of man's competitors. *Biological Conservation* 83: 279–289.

Breitenmoser, U., C. Breitenmoser-Würsten, H. Okarma, T. Kaphegyi, U. Kaphegyi-Wallmann, and U. M. Müller. 2000. Action plan for the conservation of the Eurasian lynx in Europe (*Lynx lynx*). Council of Europe, *Nature and Environment*, Council of Europe Publishing. 112: 1–69.

Breitenmoser, U., C. Breitenmoser-Würsten, L. N. Carbyn, and S. M. Funk. 2001. Assessment of carnivore reintroductions. Pages 241–281 *in* J. L. Gittleman, S. M. Funk, D. W. Macdonald, and R. K. Wayne, editors. *Carnivore conservation.* Cambridge University Press, Cambridge, UK.

Brett, M. T., and C. R. Goldman. 1996. A meta-analysis of the freshwater trophic cascade. *Proceedings of the National Academy of Science USA* 93: 7723–7726.

Brigham, R. M., H. D. J. N. Aldridge, and R. L. Mackey. 1992. Variation in habitat use and prey selection by Yuma bats, *Myotis yumanensis. Journal of Mammalogy* 73: 640–645.

Brokx, P. A. 1984. South America. Pages 525–546 *in* L. K. Halls, editor. *White-tailed deer ecology and management.* Stackpole Books, Harrisburg, Pennsylvania, USA.

Bromley, G. F. 1953. Experience of elimination of wolves in the zapovedniks of Primorski Krai. Pages 140–144 *in Preobrazovanie fauny pozvonochnykh nashey strany (biotekhnicheskie meropriyatia)* (in Russian). Izd. Mosk. Obschestva Ispytateley Prirody, Moscow, Russia.

Bromley, G. F., and S. P. Kucherenko. 1983. *Ungulates of the southern Far East of Russia.* Nauka Publishing, Moscow, Russia (in Russian).

Brooks, J. L., and S. I. Dodson. 1965. Predation, body size, and composition of plankton. *Science* 150: 28:35.

Brooks, R. T. 1999. Residual effects of thinning and high white-tailed deer densities on northern redbacked salamanders in southern New England oak forests. *Journal of Wildlife Management* 63: 1172–1180.

Broomhall, L. S. 2001. Cheetah *Acinonyx jubatus* ecology in the Kruger National Park: a comparison with other studies across the grassland–woodland gradient in African savannas. Unpublished master's thesis. University of Pretoria, Pretoria, South Africa.

Brown, J. S., J. W. Laundré, and M. Gurung. 1999. The ecology of fear: optimal foraging, game theory, and trophic interactions. *Journal of Mammalogy* 80: 385–399.

Brown, R. B., E. L. Stone, and V. W. Carlisle. 1990. Soils. Pages 35–69 *in* R. L. Myers and J. J. Ewel, editors. *Ecosystems of Florida.* University of Central Florida Press, Orlando, Florida, USA.

Bruinderink, G. G., T. Van der Sluis, D. Lammertsma, P. Opdam, and R. Pouwels. 2003. Designing a coherent ecological network for large mammals in northwestern Europe. *Conservation Biology* 17: 549–557.

Bruno, J. F., and M. D. Bertness. 2001. Habitat modification and facilitation in Benthic marine communities. Pages 201–218 in M. D. Bertness, S. D. Gaines, and M. E. Hay, editors. *Marine community ecology*. Sinauer Associactions, Inc. Sutherland, Massachusetts, USA.

Bruno, J. F., J. J. Stachowicz, and M. D. Bertness. 2003. Inclusion of facilitation into ecological theory. *Trends in Ecology and Evolution* 18: 119–125.

Bruteig, I. E., G. Austrheim, and A. Norderhaug. 2003. Beiting, biologisk mangfald og rovviltforvalting. *NINA fagrapport* 71: 1–65.

Bryant, D., D. Nielsen, and L. Tangley. 1997. *The last frontier forests: ecosystems and economies on the edge*. World Resources Institute, Washington, District of Columbia, USA.

Burrows, R. 1995. Demographic changes and social consequences in wild dogs 1964–1992. Pages 400–420 in P. Arcese, editor. *Serengeti II: research, management, and conservation of an ecosystem*. Chicago University Press, Chicago, Illinois, USA.

Bustnes, J. O., O. J. Lonne, H. R. Skjoldal, C. Hopkins, K. E. Erikstad, and H. P. Leinass. 1995. Sea ducks as predators on sea urchins in a northern kelp forest. Pages 599–608 in H. R. Skjoldal, C. Hopkins, and K. E. Erikstad, editors. *Ecology of Fjords and Coastal Waters*. Elsevier Science, Amsterdam, Netherlands.

Butfiloski, J. W., D. I. Hall, D. M. Hoffman, and D. L. Forster. 1997. White-tailed deer management in a coastal Georgia residential community. *Wildlife Society Bulletin* 25: 491–495.

Byers, J. A. 1997. *American pronghorn: social adaptations and the ghosts of predators past*. University of Chicago Press, Chicago, Illinois, USA.

Cabin, R. J., S. G. Weller, D. H. Lorence, T. W. Flynn, A. K. Sakai, D. Sandquist, and L. J. Hadway. 2000. Effects of long-term ungulate exclusion and recent alien species control on the preservation and restoration of a Hawaiian tropical dry forest. *Conservation Biology* 14: 439–453.

Calenge, C., D. Maillard, J-M. Gaillard, L. Merlot, and R. Peltier. 2002. Elephant damage to trees of a wooded savanna in Zakouma National Park, Chad. *Journal of Tropical Ecology* 18: 599–614.

Carbone, C., and J. L. Gittleman. 2002. A common rule for the scaling of carnivore density. *Science* 295: 2273–2276.

Carbone, C., S. Christie, K. Conforti, T. Coulson, N. Franklin, J. R. Ginsberg, M. Griffiths, et al. 2001. The use of photographic rates to estimate densities of tigers and other cryptic mammals. *Animal Conservation* 4: 75–79.

Carbyn, L. N., N. J. Lunn, and K. Timoney. 1998. Trends in the distribution and abundance of bison in Wood Buffalo National Park. *Wildlife Society Bulletin* 26: 463–470.

Carley, C. J. 1979. *Status summary: the red wolf.* U.S. Fish and Wildlife Service, Albuquerque, New Mexico, USA.

Caro, T. M. 1994. *Cheetahs of the Serengeti Plains*. University of Chicago Press, Chicago, Illinois, USA.

Caro, T. M. 1999. Demography and behaviour of African mammals subject to exploitation. *Biological Conservation* 91: 91–97.

Caro, T. M., and G. O'Doherty. 1999. On the use of surrogate species in conservation biology. *Conservation Biology* 13: 805–814.

Caro, T. M., and C. Stoner. 2003. The potential for interspecific competition among African carnivores. *Biological Conservation* 110: 67–75.

Carpenter, L. H. 2000. Harvest management goals. Pages 192–213 *in* S. Demarais and P. R. Krausman, editors. *Ecology and management of large mammals in North America*. Prentice Hall, Upper Saddle River, New Jersey, USA.

Carpenter, R. C. 1984. Predator and population density control of homing behavior in the Caribbean echinoid *Diadema antillarum*. *Marine Biology* 82: 101–108.

Carpenter, R. C. 1990. Mass mortality of *Diadema antillarum*, I: Long-term effects on sea urchin population-dynamics and coral reef algal communities. *Marine Biology* 104: 67–77.

Carpenter, S. R., and J. F. Kitchell, editors. 1993. *The trophic cascade in lakes*. Cambridge University Press, Cambridge, UK, and New York, New York, USA.

Carr, M. H., J. E. Neigel, J. A. Estes, S. Andelman, R. R. Warner, and J. L. Largier. 2003. Comparing marine and terrestrial ecosystems: implications for the design of coastal marine reserves. *Ecological Applications* 13: S90–S107.

Carroll, C., R. F. Noss, and P. C. Paquet. 2001. Carnivores as focal species for conservation planning in the Rocky Mountain region. *Ecological Applications* 11: 961–980.

Carroll, C., R. E. Noss, P. C. Paquet, N. H. Schumaker. 2003. Use of population viability analysis and reserve selection algorithms in regional conservation plans. *Ecological Applications* 13: 1773–1789.

Casey, D., and D. Hein. 1983. Effects of heavy browsing on a bird community in deciduous forest. *Journal of Wildlife Management* 47: 829–836.

Castilla, J. C., and C. A. Moreno. 1982. Sea urchins and *Macrocystis pyrifera*: experimental test of their ecological relations in southern Chile. Pages 257–263 *in* J. M. Lawrence, editor. *Proceedings of the International Echinoderm Conference*, A. A. Balkema, Rotterdam, the Netherlands.

Castilla, J., and R. Paine. 1987. Predation and community organization on Eastern Pacific, temperate zone, rocky intertidal shores. *Revisita Chilena de Historia Natural* 60: 131–151.

Caughley, G. 1970. Eruption of ungulate populations, with emphasis on Himalayan thar in New Zealand. *Ecology* 51:53–72.

Caughley, G. C. 1977. *Analysis of vertebrate populations*. John Wiley and Sons, New York, New York, USA.

Caughley, G. 1981. Overpopulation. Pages 7–19 *in* P. A. Jewell and S. Holt, editors. *Problems in Management of Locally Abundant Wild Mammals*. Academic Press, New York, New York, USA.

Chase, A. 1987. *Playing God in Yellowstone*. Atlantic Monthly Press, New York, New York, USA.

Chase, J. M. 2000. Are there real differences among aquatic and terrestrial food webs? *Trends in Ecology and Evolution* 15: 408–412.

Chesson, J. 1978. Measuring preference in selective predation. *Ecology* 59: 211–215.

Choat, J. H., and A. M. Ayling. 1987. The relationship between habitat and fish faunas of New Zealand reefs. *Journal of Experimental Marine Biology and Ecology* 110: 228–284.

Choat, J. H., and D. R. Schiel. 1982. Patterns of distribution and abundance of large brown algae and invertebrate herbivores in subtidal regions of northern New Zealand. *Journal of Experimental Marine Biology and Ecology* 60: 129–162.

Christensen, V., and D. Pauly. 1993. *Trophic models of aquatic ecosystems*. ICLARM, Manila, Philippines.

Christensen, V., and D. Pauly. 2000. *Fishbase*. ICLARM, Manila, Philippines.

Clapham, P. J., and C. S. Baker. 2002. Modern whaling. Pages 1328–1332 *in* W. Perrin, B. Wursig, and J. G. M. Thewissen, editors. *Encyclopedia of Marine Mammals*. Academic Press, San Diego, California, USA.

Clark, T. W., P. C. Paquet, and A. P. Curlee. 1996. Introduction: Special Section: Large carnivore conservation in the Rocky Mountains of the United States and Canada. *Conservation Biology* 10: 936–939.

Clark, T. W., A. P. Curlee, S. C. Minta, and P. M. Kareiva, editors. 1999. *Carnivores in ecosystems: the Yellowstone experience*. Yale University Press, New Haven, Connecticut, USA.

Clementz, M. 2002. The evolution of herbivorous marine mammals: ecological and physiological transitions during the evolution of the order Sirenia Desmostylia. Ph.D. dissertation. University of California, Santa Cruz, California, USA.

Cliff, G., and S. F. J. Dudley. 1992. Protection against shark attack in South Africa, 1952–1990. *Australian Journal of Marine and Freshwater Research* 43: 263–272.

Clutton-Brock, T. H., and S. H. Albon. 1992. Trial and error in the highlands. *Nature* 358: 11–12.

Clutton-Brock, T. H., F. E. Guinness, and S. D. Albon. 1982. *Red deer: behavior and ecology of two sexes*. University of Chicago Press, Chicago, Illinois, USA.

Coalition to Restore the Eastern Wolf (CREW). 2004. Eastern wolf recovery. Available from http://www.restore.org/Wildlife/wolf.html (accessed October 2004).

Coates, A., and A. Carr III. 2001. Preface. Pages xi–xiv *in* A. G. Coates, editor. *Central America: a natural and cultural history*. Yale University Press, New Haven, Connecticut, USA and London, UK.

Coblentz, B. E. 1978. Effects of feral goats (*Capra hircus*) on island ecosystems. *Biological Conservation* 13: 279–286.

Coblentz, B. E. 1990. Exotic organisms: a dilemma for conservation biology. *Conservation Biology* 4: 261–265.

Cody, M. 1981. Habitat selection in birds; the roles of vegetation structure, competitors, and productivity. *BioScience* 31: 107–113.

Cole, R.G., and R. C. Babcock. 1996. Mass mortality of a dominant kelp (*Laminariales*) at Goat Island, Northeastern New Zealand. *Marine Freshwater Resource* 47: 907–911.

Cole, R.G., and C. Syms. 1999. Using spatial pattern analysis to distinguish causes of mortality: an example from kelp in north-eastern New Zealand. *Journal of Ecology* 87: 963–972.

Coley, P. D. 1980. Effects of leaf age and plant life history patterns on herbivory. *Nature* 284: 545–546.

Coley, P. D., and J. A. Barone. 1996. Herbivory and plant defenses in tropical forests. *Annual Review of Ecology and Systematics* 27: 305–335.

Coley, P. D., J. P. Bryant, and F. S. Chapin III. 1985. Resource availability and plant antiherbivore defense. *Science* 230: 895–899.

Colinvaux, P. 1979. *Why big fierce animals are rare: an ecologist's perspective.* Princeton University Press, Princeton, New Jersey, USA.

Collar, N. J. 2003. Beyond value: biodiversity and the freedom of the mind. *Global Ecology and Biogeography* 12: 265–269.

Collette, B. B., and G. Klein-MacPhee, editors. 2002. *Bigelow and Schroeder's fishes of the Gulf of Maine.* 3rd edition. Smithsonian Institution Press, Washington, District of Columbia, USA.

Coltman, D. W., P. O'Donoghue, J. T. Jorgenson, J. T. Hogg, C. Strobeck, and M. Festa-Bianchet. 2003. Undesirable evolutionary consequences of trophy hunting. *Nature* 426: 655–658.

Connell, J. H. 1961a. Effects of competition, predation by *Thais lapillus,* and other factors on natural populations of the barnacle, *Balanus balanoides. Ecological Monographs* 31: 61–104.

Connell, J. H. 1961b. The influence of interspecific competition and other factors on the distribution of the barnacle *Chthamalus stellatus. Ecology* 42: 710–723.

Connell, J. H. 1978. Diversity in tropical rain forests and coral reefs. *Science* 199: 1302–1310.

Connell, J. H. 1983. On the prevalence and relative importance of interspecific competition: evidence from field experiments. *American Naturalist* 122: 661– 696.

Connell, S. 1998. Spatial, temporal and habitat-related variation in the abundance of large predatory fish at One Tree Reef, Australia. *Coral Reefs* 17: 49–57.

Connor, K. J., W. B. Ballard, T. Dilworth, S. Mahoney, and D. Anions. 2000. Changes in structure of a boreal forest community following intense herbivory by moose. *Alces* 36: 11–132.

Conover, M. R. 1997. Monetary and intangible valuation of deer in the United States. *Wildlife Society Bulletin* 25: 298–305.

Cooke, A. S., and Farrell, L. 2001. Impact of Muntjac deer (*Muntiacus reevesi*) at Monks Wood National Nature Reserve, Cambridgeshire, eastern England. *Forestry* 74: 241–250.

Coomes, D. A., R. B. Allen, D. M. Forsyth, and W. G. Lee. 2003. Factors preventing the recovery of New Zealand forests following control of invasive deer. *Conservation Biology* 17: 450–459.

Corbett, L. K., and A. E. Newsome. 1987. The feeding ecology of the dingo, III: Dietary relationships with widely fluctuating prey populations in arid Australia: an hypothesis of alternation of predation. *Oecologia* 74: 215–227.

Couturier, S., J. Brunelle, D. Vandal, and G. St. Martin. 1990. Changes in the population dynamics of the George River caribou herd, 1976–87. *Arctic* 43: 9–20.

Cowen, R. K. 1983. The effect of sheephead (*Semicossyphus pulcher*) predation on red sea urchin (*Strongylocentrotus franciscanus*) populations: an experimental analysis. *Oecologia* 58: 249–255.

Crabtree, R. L., and J. W. Sheldon. 1999. Coyotes and canid coexistence in Yellowstone. Pages 127–164 *in* T. W. Clark, A. P. Curlee, S. C. Minta, and P. M. Kareiva, editors. *Carnivores in ecosystems: the Yellowstone experience.* Yale University Press, New Haven, Connecticut, USA.

Craighead, F. C., Sr. 1971. *The trees of south Florida.* University of Miami Press, Coral Gables, Florida, USA.

Crawley, M. J. 1983. *Herbivory; the dynamics of animal–plant interactions.* University of California Press, Berkeley, Los Angeles, California, USA.

Creel, S. 2001. Four factors modifying the effect of competition on carnivore population dynamics as illustrated by African wild dogs. *Conservation Biology* 15: 271–274.

Creel, S., and N. M. Creel. 1996. Limitation of African wild dogs by competition with larger carnivores. *Conservation Biology* 10: 526–538.

Creel, S., and N. M. Creel. 2002. *The African wild dog.* Princeton University Press, Princeton, New Jersey, USA.

Creel, S., G. Spong, and N. M. Creel. 2001. Interspecific competition and the population biology of extinction-prone carnivores. Pages 35–60 *in* J. L. Gittleman, S. M. Funk, D. W. Macdonald, and R. K. Wayne, editors. *Carnivore conservation.* Cambridge University Press, Cambridge, UK.

Crespo, J. A., and J. M. de Carlo 1963. Estudio ecológico de una población de zorros colorados, *Dusicyon culpaeus culpaeus* (Molina) en el oeste de la provincia de Neuquén. *Revista Museo Argentino de Ciencias Naturales "Bernardino Rivadavia," Ecología* 1: 1–55.

Crête, M. 1987. The impact of sport hunting on North American moose. *Swedish Wildlife Research* 1 (Supplement): 553–563.

Crête, M. 1989. Approximation of *K* carrying capacity for moose in eastern Quebec. *Canadian Journal of Zoology* 67: 373–380.

Crête, M. 1999. The distribution of deer biomass in North America supports the hypothesis of exploitation ecosystems. *Ecology Letters* 2: 223–227.

Crête, M., and M. Manseau. 1996. Natural regulation of cervidae along a 1000 km latitudinal gradient: change in trophic dominance. *Evolutionary Ecology* 10: 51–62.

Cronk, Q. C. B. 1980. Extinction and survival in the endemic vascular flora of Ascension Island. *Biological Conservation* 13: 207–219.

Crooks, K. R., and M. E. Soulé. 1999a. Mesopredator release and avifaunal extinctions in a fragmented system. *Nature* 400: 563–566.

Crooks, J. A., and M. E. Soulé 1999b. Lag times in population explosions of invasive species: causes and implications. Pages 103–125 *in* O. T. Sandlund, P. Johan, and Å. Viken, editors. *Invasive Species and Biodiversity Management*. Kluwer Academic Publishers, Dordrecht.

Cunliffe, B. 1994. *The Oxford illustrated prehistory of Europe*. Oxford University Press, Oxford.

Cyr, H., and M. L. Pace. 1993. Magnitude and patterns of herbivory in aquatic and terrestrial ecosystems. *Nature* 361: 148–150.

Dale, B. W., L. G. Adams, and R. T. Bowyer. 1994. Functional response of wolves preying on barren-ground caribou in a multiple-prey ecosystem. *Journal of Animal Ecology* 63: 644–652.

Dale, M. R. T., S. Francis, C. J. Krebs, and V. O. Nams. 2001. Trees. Pages 116–140 *in* C. J. Krebs, S. Boutin, and R. Boonstra, editors. *Ecosystem dynamics of the boreal forest: the Kluane project*. Oxford University Press, New York, New York, USA.

Dalrymple, G. H., and O. L. Bass Jr. 1996. The diet of the Florida panther in Everglades National Park, Florida. *Bulletin of the Florida Museum of Natural History* 39: 173–193.

Danell, K. 1985. Population fluctuations of the muskrat in coastal northern Sweden. *Acta Theriologica* 30: 219–227.

Danell, K., T. Willebrand, and L. Baskin. 1998. Mammalian herbivores in the boreal forests: their numerical fluctuations and use by man. *Conservation Ecology* 2: 9.

Danilkin, A. A. 1999. *Deer (Cervidae)*. GEOS, Moscow, Russia.

Darsie, R. E., and R. A. Ward. 1981. Identification and geographical distribution of the mosquitoes of North America. *Mosquito Systematics* Supplement 1: 1–313.

Daskalov, G. M. 2002. Overfishing drives a trophic cascade in the Black Sea. *Marine Ecology Progress Series* 225: 53–63.

Dassman, R. F. 1972. Toward a system for classifying natural regions of the world and their representation by national parks and reserves. *Biological Conservation* 4: 247–255.

Davis, W. 1996. *One river*. Simon and Schuster, New York, New York, USA.

Dayton, P. K. 1975. Experimental evaluation of ecological dominance in a rocky intertidal algal community. *Ecological Monographs* 45: 137–159.

Dayton, P. K. 1985. Ecology of kelp communities. *Annual Review of Ecology and Systematics* 16: 215–245.

Dayton, P. K. 2003. The importance of the natural sciences to conservation. *American Naturalist* 162: 1–13.

Dayton, P. K., M. I. Tegner, P. B. Edwards, and K. L. Riser. 1998. Sliding baselines, ghosts, and reduced expectations in kelp forest communities. *Ecological Applications* 8: 309–322.

DeBoer, K. 2000. Dreams of wolves. Pages 64–107 *in* B. McKibben, J. B. Theberge, K.

DeBoer, R. Bass, and J. Elder, editors. *The return of the wolf: reflections on the future of wolves in the Northeast.* Middlebury College Press, Hanover, New Hampshire, USA.

deCalesta, D. S. 1994. Effect of white-tailed deer on songbirds within managed forests of Pennsylvania. *Journal of Wildlife Management* 58: 711–718.

deCalesta, D. S. 1997. Deer and ecosystem management. Pages 267–279 *in* W. J. McShea, H. B. Underwood, and J. H. Rappole, editors. *The science of overabundance; deer ecology and management.* Smithsonian Institution Press, Washington, District of Columbia, USA.

Defenders of Wildlife 2004. Restoring wolves to the Northeast. Available from http://www.defenders.org/wildlife/new/facts/background.html (accessed October 2004).

Dekker, D., W. Bradford, and J. R. Gunson. 1995. Elk and wolves in Jasper National Park, Alberta from historic times to 1992. Pages 85–94 *in* L. N. Carbyn, S. H. Fritts, and D. R. Seip, editors. *Ecology and conservation of wolves in a changing world.* Canadian Circumpolar Press, Edmonton, Alberta, Canada.

Delibes, M., A. Rodriguez, and P. Ferreras. 2000. Action plan for the conservation of the Iberian lynx in Europe (*Lynx pardinus*). *Council of Europe Nature and Environment* 111: 1–44.

de Lucca, E. 1996. Censos de choiques (*Pterocnemia p. pennata*) en el sur patagónico. *El Hornero* 14: 74–77.

de Mazencourt, C., and M. Loreau. 2000. Effect of herbivory and plant species replacement on primary production. *American Naturalist* 155: 735–754.

del Valle, H., N. Elissalde, D. Gagliardini, and J. Milovich. 1998. Status of desertification in the Patagonian region: Assessment and mapping from satellite imagery. *Arid Soil Research and Rehabilitation* 12: 95–122.

deMaynadier, P. G., and M. L. Hunter Jr. 1995. The relationship between forest management and amphibian ecology: a review of North American literature. *Environmental Review* 3: 230–261.

Dey, D. 2002. The ecological basis for oak silviculture in eastern North America. Pages 60–79 *in* W. J. McShea and W. M. Healy, editors. *Oak Forest Ecosystem: ecology and management for wildlife.* Johns Hopkins University Press, Baltimore, Maryland, USA.

Diefenbach, D. R., W. L. Palmer, and W. K. Shope. 1997. Attitudes of Pennsylvania sportsmen towards managing white-tailed deer to protect the ecological integrity of forests. *Wildlife Society Bulletin* 25: 244–251.

Dieni, J.S., B. L. Smith, R. L. Rogers, and S. H. Anderson. 2000. Effects of ungulate browsing on aspen regeneration in northwestern Wyoming. *Intermountain Journal of Sciences* 6: 49–55.

Dirzo, R., and A. Miranda. 1990. Contemporary Neotropical defaunation and forest structure, function and diversity: a sequel to John Terborgh. *Conservation Biology* 4: 444–447.

Domning, D. P. 1972. Steller's sea cow and the origin of North Pacific aboriginal whaling. *Syesis* 5: 187–189.

Done, T. 1995. Ecological criteria for evaluating coral reefs and their implications for managers and researchers. *Coral Reefs* 14: 183–192.

Doroff, A. M., J. A. Estes, M. T. Tinker, D. M. Burn, and T. J. Evans. 2003. Sea otter population declines in the Aleutian archipelago. *Journal of Mammalogy* 84: 55–64.

Dostalkova, I., P. Kindlmann, and A. F. G. Dixon. 2002. Are classical predator–prey models relevant to the real world? *Journal of Theoretical Biology* 218: 323–330.

Du Toit, J., K. H. Rogers, and H. C. Biggs, editors. 2003. *The Kruger experience*. Island Press, Washington, District of Columbia, USA.

Dublin, H. T., A. R. E. Sinclair, and J. McGlade. 1990. Elephants and fire as causes of multiple stable states in the Serengeti-Mara woodlands. *Journal of Animal Ecology* 59: 1147–1164.

Duever, M. J., J. E. Carlson, J. F. Meeder, L. C. Duever, L. H. Gunderson, L. A. Riopelle, T. R. Alexander, R. L. Myers, and D. P. Spangler. 1986. *The Big Cypress National Preserve*. Research Report No. 8. National Audubon Society, New York, New York, USA.

Duffy, J. E. 2002. Biodiversity and ecosystem function: the consumer connection. *Oikos* 99: 201–219.

Duffy, J. E. 2003. Biodiversity loss, trophic skew and ecosystem functioning. *Ecology Letters* 6: 680–687.

Dufour, D. L. 1987. Insects as food: a case study from the northwest Amazon. *American Anthropologist* 89: 383–397.

Duggins, D. O., C. A. Simenstad, and J. A. Estes. 1989. Magnification of secondary production by kelp detritus in coastal marine ecosystems. *Science* 245: 170–173.

Dulvy, N. K., R. E. Mitchell, D. Watson, C. J. Sweeting, and N. V. C. Polunin. 2002. Scale-dependant control of motile epifaunal community structure along a coral reef fishing gradient. *Journal of Experimental Marine Biology and Ecology* 278: 1–29.

Dunham, K. M. 1992. Response of a lion (*Panthera leo*) population to changing prey availability. *Journal of Zoology* 227: 330–333.

Dunk, J. R., R. N. Smith, and S. L. Cain. 1997. Nest-site selection and reproduction success in common ravens. *Auk* 114: 116–120.

Dunishenko, Y. M. 1987. *The Amur tiger*. Khabaro. kn. izd, Khabarovsk (in Russian).

Dunlap, T. R. 1983. "The coyote itself": ecologists and the value of predators, 1900–1972. *Environmental Review* 7: 54–70.

Durant, S. M. 1998. Competition refuges and coexistence: an example from Serengeti carnivores. *Journal of Animal Ecology* 67: 370–386.

Durant, S. M. 2000. Predator avoidance, breeding experience, and reproductive success in endangered cheetahs, *Acinonyx jubatus*. *Animal Behavior* 60: 121–130.

Eberhardt, L. L. 1998. Applying difference equations to wolf predation. *Canadian Journal of Zoology* 76: 380–386.

Eberhardt, L. L., and R. O. Peterson. 1999. Predicting the wolf–prey equilibrium point. *Canadian Journal of Zoology* 77: 494–498.

Eberhardt, L. L., R. A. Garrott, D. W. Smith, P. J. White, and R. O. Peterson. 2003. Assessing the impact of wolves on ungulate prey. *Ecological Applications* 13: 776–783.

Ebling, F. S., A. D. Hawkins, J. A. Kitching, P. Muntz, and V. M Pratt. 1966. The ecology of Lough Ine, XVI: Predation and diurnal migrations in the *Paracentrotus* community. *Journal of Animal Ecology* 35: 559–566.

Edelstein-Keshet, L., and M. D. Rausher. 1989. The effects of inducible plant defenses on herbivore populations, I: Mobile herbivores in continuous time. *American Naturalist* 133: 787–810.

Edgar, G. J., and N. S. Barrett. 1999. Effects of the declaration of marine reserves on Tasmanian reef fishes, invertebrates and plants. *Journal of Experimental Marine Biology and Ecology* 242: 107–144.

Edmunds, P. J., and P. S. Davies. 1986. An energy budget for *Porites porites* (Scleractinia). *Marine Biology* 92: 323–347.

Eisenberg, J. F. 1984. Life history strategies of Felidae: variations on a common theme. Pages 293–303 *in* S. D. Miller and D. D. Everett, editors. *Cats of the world: biology, conservation and management.* National Wildlife Federation, Washington, District of Columbia, USA.

Eisenberg, J. F., and L. D. Harris. 1989. Conservation: a consideration of evolution, population, and life history. Pages 99–108 *in* D. Western and M. Pearl, editors. *Conservation for the twenty-first century.* Oxford University Press, New York, New York, USA.

Eisenberg, J. F., and M. Lockhart. 1972. An ecological reconaissance of Wilpattu National Park, Ceylon. *Smithsonian Contributions to Zoology* 101: 1–118.

Elder, J. 2000. *The return of the wolf: reflections on the future of wolves in the northeast.* Middlebury College Press, London, UK.

Elliott, J. P., and I. M. Cowan. 1978. Territoriality, density and prey of the lion in Ngorongoro Crater, Tanzania. *Canadian Journal of Zoology* 56: 1726–1734.

Eloff, F. C. 1964. On the predatory habits of lions and hyaenas. *Koedoe* 7: 105–113.

Elton, C. 1927. *Animal ecology.* Sidgwick and Jackson. Ltd., London, UK.

Emmons, L. H. 1987. Comparative feeding ecology of felids in a neotropical rainforest. *Behavioural Ecology and Sociobiology* 20: 271–283.

Enck, J. W., D. J. Decker, and T. L. Brown. 2000. Status of hunter recruitment and retention in the United States. *Wildlife Society Bulletin* 28: 817–824.

Englund G. 1997. Importance of spatial scale and prey movements in predator caging experiments. *Ecology* 78: 2316–2325.

Entwistle, A. C., and N. Dunstone. 2000. Future priorities for mammalian conservation. Pages 369–387 *in* A. Entwistle and N. Dunstone, editors. *Priorities for the conservation of mammalian diversity: has the panda had its day?* Cambridge University Press, Cambridge, UK.

Entwistle, A. C., S. Mickleburgh, and N. Dunstone. 2000. Mammal conservation: current contexts and opportunities. Pages 1–7 *in* A. Entwistle and N. Dunstone, editors. *Priorities for the conservation of mammalian diversity: has the panda had its day?* Cambridge University Press, Cambridge, UK.

Ericsson, G., and T. A. Heberlein. 2003. Attitudes of hunters, locals, and the general public in Sweden now that the wolves are back. *Biological Conservation* 111: 149–159.

Erlandson, J. M. 2001. Anatomically modern humans, maritime voyaging, and the Pleistocene colonization of the Americas. Pages 1–9 *in* N. G. Jablonski, editor. *The First Americans: The Pleistocene Colonization of the New World.* California Academy of Sciences, San Francisco, California, USA.

Erlandson, J. M., and T. C. Rick. 2002. A 9700-year-old shell midden on San Miguel Island, California. *Antiquity* 76: 315–316.

Erlinge, S., G. Goransson, L. Hansson, G. Hogstedt, O. Liberg, I. N. Nilsson, T. Nilsson, T. von Schantz, and M. Sylven. 1983. Predation as a regulating factor in small rodent populations in southern Sweden. *Oikos* 40: 36–52.

Errington, P. L. 1967. *Of predation and life.* Iowa State University Press, Ames, Iowa, USA.

Esseen, P-A., B. Ehnstrom, L. Ericson, and K. Sjoberg. 1997. Boreal Forests. *Ecological Bulletins* 46: 16–47.

Estes, J. A. 1990. Growth and equilibrium in sea otter populations. *Journal of Animal Ecology* 59: 385–400.

Estes, J. A. 1995. Top-level carnivores and ecosystem effects. Pages 151–158 *in* C. G. Jones and J. H. Lawton, editors. *Linking species and ecosystems.* Chapman and Hall, New York, New York, USA.

Estes, J. A. 1996. Predators and ecosystem management. *Wildlife Society Bulletin* 24: 390–396.

Estes, J. A., and D. O. Duggins. 1995. Sea otters and kelp forests in Alaska: generality and variation in a community ecological paradigm. *Ecological Monographs* 65: 75–100.

Estes, J.A., and J. F. Palmisano. 1974. Sea otters: their role in structuring nearshore communities. *Science* 185: 1058–1060.

Estes, J. A., N. S. Smith, and J. F. Palmisano. 1978. Sea otter predation and community organization in the western Aleutian Islands, Alaska. *Ecology* 59: 822–833.

Estes, J. A., D. O. Duggins, and G. B. Rathbun. 1989. The ecology of extinctions in kelp forest communities. *Conservation Biology* 3: 252–264.

Estes, J. A., M. T. Tinker, T. M. Williams, and D. F. Doak. 1998. Killer whale predation on sea otters linking oceanic and nearshore ecosystems. *Science* 282: 473–476.

Estes, J. A., K. Crooks, and R. Holt. 2001. Predators, ecological role of. Pages 857–878 *in* S. A. Levin and J. Lubchenco, editors. *Encyclopedia of Biodiversity*, Volume 4. Academic Press, San Diego, California, USA.

Estes, J. A., E. M. Danner, D. F. Doak, B. Konar, A. M. Springer, P. D. Steinberg, M. T. Tinker, and T. M. Williams. 2004. Complex trophic interactions in kelp forest ecosystems. *Bulletin of Marine Science* 74:621–638.

Evans, W. 1983. *The cougar in New Mexico: biology, status, depredation of livestock and management recommendations.* New Mexico Department of Game and Fish, Santa Fe, New Mexico, USA.

Ewel, J. J. 1990. Introduction. Pages 3–10 in R. L. Myers and J. J. Ewel, editors. *Ecosystems of Florida.* University of Central Florida Press, Orlando, Florida, USA.

Faliński, J. B. 1986. *Vegetation dynamics in temperate lowland primeval forest.* Dr W. Junk Publishers, Dordrecht, Netherlands.

Fanshawe, J. H., and C. D. Fitzgibbon. 1993. Factors influencing the hunting success of an African wild dog pack. *Animal Behavior* 45: 479–490.

Farrell, E. P., E. Führer, D. Ruyan, F. Andersson, R. Hüttl, and P. Piussi. 2000. European forest ecosystems: building the future on the legacy of the past. *Forest Ecology and Management* 132: 5–20.

Farrow, E. P. 1916. On the ecology of the vegetation of Breckland. *Journal of Ecology* 4: 57–604.

Feber, R. E., T. M. Brereton, M. S. Warren, and M. Oates. 2001. The impacts of deer on woodland butterflies: the good, the bad, and the complex. *Forestry* 74: 271–276.

Fernández, G. J., and J. C. Reboreda. 1998. Effects of clutch size and timing of breeding on reproductive success of greater rheas. *Auk* 115: 340–348.

Ferraz, G., G. J. Russell, P. C. Stouffer, R. O. Bierregaards Jr., S. L. Pimm, and T. E. Lovejoy. 2003. Rates of species loss from Amazonian forest fragments. *Proceedings of the National Academy of Sciences* (USA) 100: 14069–14073.

Filonov, K. P., and M. L. Kaletskaya. 1985. Influence of wolf predation on wild ungulates. Pages 336–354 in Bibikov, D. E., editors. *Wolf: history, systematics, morphology, ecology.* Nauka Publishers, Moscow.

Finch, D. M. 1989. Habitat use and habitat overlap of riparian birds in three elevational zones. *Ecology* 70: 866–880.

Fitzgibbon, C. D. 1990. Why do hunting cheetahs prefer male gazelles? *Animal Behavior* 40: 837–845.

Flannery, T. 1994. *The future eaters.* Grove Press, New York, New York, USA.

Flannery, T. 2001. *The eternal frontier.* Atlantic Monthly Press, New York, New York, USA.

Fleishman, E., G. T. Austin, P. F. Brussard, and D. D. Murphy. 1999. A comparison of butterfly communities in native and agricultural riparian habitats in the Great Basin. *Biological Conservation* 89: 209–218.

Fleishman, E., R. Mac Nally, J. P. Fay, and D. D. Murphy. 2001. Modeling and predicting species occurrence using broad-scale environmental variables: an example with butterflies of the Great Basin. *Conservation Biology* 15: 1674–1685.

Fletcher, J. D., W. J. McShea, L. A. Shipley, and D. Shumway. 2001a. Use of common forest forbs to measure browsing pressure by white-tailed deer (*Odocoileus virginianus* Zimmerman) in Virginia, USA. *Natural Areas Journal* 21: 172–176.

Fletcher, J. D., L. A. Shipley, W. J. McShea, and D. L. Shumway. 2001b. Wildlife herbivory and rare plants: the effects of white-tailed deer, rodents and insects on growth and survival of Turk's Cap Lily. *Biological Conservation* 101: 229–238.

Flowerdew, J. R. and S. A. Ellwood. 2001. Impacts of woodland deer on small mammal ecology. *Forestry* 74: 277–287.

Flueck, W. T. 2000. Population regulation in large northern herbivores: evolution, thermodynamics, and large predators. *Zeitschrift für Jagdwissenschaft* 46: 139–166.

Fosbrooke, H. 1963. The *Stomoxys* plague in Ngorongoro, 1962. *East African Wildlife Journal* 1: 124–126.

Foster, D. R. 2002. Thoreau's country: a historical–ecological perspective on conservation in the New England landscape. *Journal of Biogeography* 29: 1537–1555.

Foster, M. L. 1992. Effectiveness of wildlife crossings in reducing animal/auto collisions on Interstate 75, Big Cypress Swamp, Florida. Master's thesis. University of Florida, Gainesville, Florida, USA.

Foster, M. S., and D. R. Schiel. 1985. *The ecology of giant kelp forests in California: a community profile*. U.S. Fish and Wildlife Service Biological Report 85(7.2.), Slidell, Louisiana, USA.

Fowler, C. W. 1987. A review of density dependence in populations of large mammals. Pages 410–441 *in* H. H. Genoways, editor. *Current mammalogy 1*. Plenum Press, New York, New York, USA.

Fraenkel, G. 1959. The raison d'etre of secondary plant substances. *Science* 129: 1466–1470.

Frank, D. A., and P. M. Groffman. 1998. Ungulate vs. landscape control of soil C and N processes in grasslands of Yellowstone National Park. *Ecology* 79: 2229–2241.

Frank, D. A., and S. J. McNaughton. 1992. The ecology of plants, large mammalian herbivores, and drought in Yellowstone National Park. *Ecology* 73: 2229–2241.

Frank, D. A., and S. J. McNaughton. 1993. Evidence for the promotion of aboveground grassland production by native large herbivores in Yellowstone National Park. *Oecologia* 96: 157–161.

Frank, D. A., S. J. McNaughton, R. S. Inouye, N. Huntley, G. W. Minsall, and J. E. Anderson. 1994. The biogeochemistry of a north-temperate grassland with native ungulates: nitrogen dynamics in Yellowstone National Park. *Biogeochemistry* 10: 163–180.

Franklin, J. F. 1993. Preserving biodiversity: species, ecosystems, or landscapes? *Ecological Applications* 3: 202–205.

Franklin, W. L., W. E. Johnson, R. J. Sarno, and J. A. Iriarte. 1999. Ecology of the Patagonia puma *Felis concolor patagonica* in southern Chile. *Biological Conservation* 90: 33–40.

Freitag-Ronaldson, S., and L. C. Foxcroft. 2003. Anthropogenic influences at the ecosystem level. Pages 391–421 *in* J. Du Toit, K. H. Rogers, and H. C. Biggs, editors. *The Kruger experience*. Island Press, Washington, District of Columbia. USA.

Fretwell, S. D. 1977. The regulation of plant communities by food chains exploiting them. *Perspectives in Biology and Medicine* 20: 169–185.

Fretwell, S. D. 1987. Food chain dynamics: the central theory of ecology? *Oikos* 50: 291–301.

Freyfogle, E. T., and J. L. Newton. 2002. Putting science in its place. *Conservation Biology* 16: 863–873.

Friedel, M. H. 1991. Range condition assessment and the concept of thresholds: a viewpoint. *Journal of Range Management* 44: 422–426.

Friedlander A. M., and DeMartini E. E. 2002. Contrasts in density, size, and biomass of reef fishes between the northwestern and the main Hawaiian islands: the effects of fishing down apex predators. *Marine Ecology-Progress Series* 230: 253–264.

Fritts, S. H., and L. D. Mech. 1981. Dynamics, movements, and feeding ecology of a newly protected wolf population in Northwestern Minnesota. *Wildlife Monographs* 80: 6–79.

Fritts, S. H., E. E. Bangs, J. A. Fontaine, M. R. Johnson, M. K. Phillips, E. D. Koch, and J. R. Gunson. 1997. Planning and implementing a reintroduction of wolves to Yellowstone National Park and central Idaho. *Restoration Ecology* 5: 7–27.

Froese, R., and D. Pauly, editors. 2004. FishBase. World Wide Web electronic publication. Available from http://www.fishbase.org, version (accessed October 2004).

Frost, H. C., G. L. Storm, M. J. Batcheller, and M. L. Lovallo. 1997. White-tailed deer management at Gettysburg National Military Park and Eisenhower National Historic Site. *Wildlife Society Bulletin* 25: 462–469.

Frost, O. W. 1988. *Georg Wilhelm Stellar: journal of a voyage with Bering, 1741–42.* Stanford University Press, Stanford, California, USA.

Fryxell, J. M., J. Greever, and A. R. E. Sinclair. 1988. Why are migratory ungulates so abundant? *American Naturalist* 131: 781–798.

Fujita, D. 1998. Strongylocentrotid sea urchin-dominated barren grounds on the Sea of Japan coast of northern Japan. Pages 659–664 *in* R. Mooi and M. Telford, editors. *Echinoderms.* A. A. Balkema Publishers, Rotterdam, Netherlands.

Fuller, T. K. 1989. Population dynamics of wolves in north-central Minnesota. *Wildlife Monographs* 105: 1–41.

Fuller, R. J. 2001. Responses of woodland birds to increasing numbers of deer: a review of evidence and mechanisms. *Forestry* 74: 289–298.

Fuller, R.J., and R. M. A. Gill. 2001. Ecological impacts of increasing numbers of deer in British woodland. *Forestry* 74: 193–199.

Fuller, T. K., and L. B. Keith. 1980. Wolf population dynamics and prey relationships in Northeastern Alberta. *Journal of Wildlife Management* 44: 583–602.

Fuller, T. K., and L. B. Keith. 1981. Nonoverlapping ranges of coyotes and wolves in Northeastern Alberta. *Journal of Mammalogy* 62: 403–405.

Fuller. T. K., and D. B. Kittredge Jr. 1996. Conservation of large forest carnivores. Pages 137–164 *in* R. M. DeGraaf and R. I. Miller, editors. *Conservation of faunal diversity in forested landscapes.* Chapman and Hall, New York, New York, USA.

Fuller, T. K., and P. R. Sievert. 2001. Carnivore demography and the consequences of changes in prey availability. Pages 163–178 *in* J. L. Gittleman, S. M. Funk, D. MacDonald, and R. K. Wayne, editors. *Carnivore conservation.* Cambridge University Press, Cambridge, UK.

Fuller, T. K., L. D. Mech, and J. F. Cochrane. 2003. Wolf population dynamics. Pages 161–191 *in* L. D. Mech and L. Boitani, editors. *Wolves: behavior, ecology, and conservation*. University of Chicago Press, Chicago, Illinois, USA.

Funes, M. C., and A. J. Novaro. 1999. Rol de la fauna silvestre en la economía del poblador rural, provincia del Neuquén, Argentina. *Revista Argentina de Producción Animal* 19: 265–271.

Funes, M. C., M. M. Rosauer, G. Sanchez Aldao, O. B. Monsalvo, and A. J. Novaro. 2000. *Proyecto: manejo y conservación del choique en la Patagonia*. Informe segunda etapa: Análisis de los relevamientos poblacionales. Report to Centro de Ecología Aplicada del Neuquén, Neuquén, Argentina.

Funk, S. M., C. V. Fiorello, S. Cleaveland, and M. E. Gompper. 2001. Pages 443–466 *in* J. L. Gittleman, S. M. Funk, D. Macdonald, and R. K.Wayne, editors. *Carnivore conservation*. Cambridge University Press, Cambridge, UK.

Funston, P. J. 1999. Predator–prey relationships between lions and large ungulates in the Kruger National Park. Unpublished Ph.D. thesis. University of Pretoria, Pretoria, South Africa.

Funston, P. J., M. G. L. Mills, H. C. Biggs, and P. R. K. Richardson. 1998. Hunting by male lions: ecological influences and socioecological implications. *Animal Behavior* 56: 1333–1345.

Galende, G., and D. Grigera. 1998. Relaciones alimentarias de Lagidium viscacia (Rodentia, Chinchillidae) con herbívoros introducidos en el Parque Nacional Nahuel Huapi, Argentina. *Iheringia, Séria Zoologica, Porto Alegre* 84: 3–10.

García-Rubies, A., and M. Zabala. 1990. Effects of total prohibition of fishing on the rocky fish assemblages of Medes Islands marine reserve. *Scientia Marina* 54: 317–328.

Gardner, T. A., I. M. Cote, J. A. Gill, A. Grant, and A. R. Watkinson. 2003. Long-term region-wide declines in Caribbean corals. *Science* 301: 958–960.

Garrott, R. A., P. J. White, and C. A. Vanderbilt White. 1993. Overabundance: an issue for conservation biologists? *Conservation Biology* 7: 946–949.

Gasaway, W.C., R. O. Stephenson, J. L. Davis, P. E. Shepherd, and O. E. Burris. 1983. Interrelationships of wolves, prey, and man in interior Alaska. *Wildlife Monographs* 84: 1–50.

Gasaway, W. G., R. D. Boertje, D. V. Grangaard, D. G. Kelleyhouse, R. O. Stephenson, and D. G. Larsen. 1992. The role of predation in limiting moose at low densities in Alaska and Yukon and implications for conservation. *Wildlife Monographs* 120: 1–59.

Geist, V. 1998. *Deer of the world: their evolution, behavior, and ecology*. Stackpole Books, Mechanicsburg, Pennsylvania, USA.

Gell, F. R., and C.M. Roberts. 2003. Benefits beyond boundaries: the fishery effects of marine reserves. *Trends in Ecology and Evolution* 18: 448–455.

Gertenbach, W. P. D. 1983. Landscapes of the Kruger National Park. *Koedoe* 26: 9–121.

Ghilarov, A. M. 2000. Ecosystem functioning and intrinsic value of biodiversity. *Oikos* 90: 408–412.

Giblin-Davis, R. M. 2001. Borers of palms. Pages 267–304 *in* F.W. Howard, D. Moore, R.M.

Giblin-Davis, and R.G. Abad, editors. *Insects on palms*. CABI, New York, New York, USA.

Giblin-Davis, R. M., A. C. Oehlschlager, A. Perez, G. Gries, R. Gries, T. J. Weissling, C. M. Chinchilla, et al. 1996. Chemical and behavioral ecology of palm weevils Curculionidae: Rhynchophorinae). *Florida Entomologist* 79: 153–167.

Gill, R. M. A., and V. Beardall. 2001. The impact of deer on woodlands: the effects of browsing and seed dispersal on vegetation structure and composition. *Forestry* 74: 209–218.

Ginsberg, J. R. 2001. Setting priorities for carnivore conservation: what makes carnivores different? Pages 498–523 *in* J. L. Gittleman, S. M. Funk, D. W. Macdonald, and R. K. Wayne, editors. *Carnivore conservation*. Cambridge University Press, Cambridge, UK.

Ginsberg, J. R., and D. W. Macdonald. 1990. *Foxes, wolves, jackals and dogs: an action plan for the conservation of canids*. IUCN, Gland, Switzerland.

Ginsberg, J. R., and E. J. Milner-Gulland. 1994. Sex-biased harvesting and population dynamics in ungulates: implications for sustainable use. *Conservation Biology* 8: 157–166.

Ginsberg, J. R., G. M. Mace, and S. D. Albon. 1995. Local extinction in a small and declining population: wild dogs in the Serengeti. *Proceedings of the Royal Society of London B* 262: 221–228.

Ginzburg, L. R., and H. R. Akçakaya. 1992. Consequences of ratio-dependent predation for steady-state properties of ecosystems. *Ecology* 73: 1536–1543.

Gittleman, J. L. 1989. Carnivore group living: comparative trends. Pages 183–207 *in* J. L. Gittleman, editor. *Carnivore behavior, ecology, and evolution*. Cornell University Press, New York, New York, USA.

Gittleman, J. L. 1993. Carnivore life histories: a reanalysis in the light of new models. *Symposia of the Zoological Society of London* 65: 65–86.

Gittleman, J. L., and P. H. Harvey. 1982. Carnivore home range size, metabolic needs and ecology. *Behavioural Ecology and Sociobiology* 10: 57–63.

Gittleman, J. L., S. Funk, D. W. Macdonald, and R. K. Wayne. 2001a. *Carnivore conservation*. Cambridge University Press, Cambridge, UK.

Gittleman, J. L., S. M. Funk, D. W. Macdonald, and R. K. Wayne. 2001b. Why "carnivore conservation"? Pages 1–7 *in* J. L. Gittleman, S. M. Funk, D. W. Macdonald, and R. K. Wayne, editors. *Carnivore conservation*. Cambridge University Press, Cambridge, UK.

Glander, K. E. 1981. Feeding patterns in mantled howler monkeys. Pages 231–257 *in* C. Kamil and T. D. Sargent, editors. *Foraging behavior: ecological, ethological, and psychological approaches*. Garland Press, New York, New York, USA.

Glowacinski, Z., and P. Profus. 1997. Potential impact of wolves *Canis lupus* on prey populations in eastern Poland. *Biological Conservation* 80: 99–106.

Goldwasser, L., and J. Roughgarden. 1997. Sampling effects and the estimation of food-web properties. *Ecology* 78: 41–54.

Golluscio, R., J. Paruelo, J. Mercau, and V. Deregibus. 1998. Urea supplementation effects on low-quality forage utilisation and lamb production in Patagonian rangelands. *Grass and Forage Science* 53: 47–56.

Gompper, M. E., and J. L. Gittleman. 1991. Home range scaling: intraspecific and comparative trends. *Oecologia* 87: 343–348.

Goodman, P. S., and J. Hearn. 2003. Managing commercial wildlife rangelands in southern Africa: principles and practice. *African Journal of Range and Forage Science* 20: 131.

Goodrich, J. M., L. L. Kerley, B. O. Schleyer, D. G. Miquelle, K. S. Quigley, Y. N. Smirnov, I. G. Nikolaev, H. B. Quigley, and M. G. Hornocker. 2001. Capture and chemical anesthesia of Amur (Siberian) tigers. *Wildlife Society Bulletin* 29: 533–542.

Gorman, M. L., M. G. Mills, J. P. Raath, and J. R. Speakman. 1998. High hunting costs make African wild dogs vulnerable to kleptoparasitism by hyaenas. *Nature* 391: 479–481.

Graham, N. A. J., R. D. Evans, and G. R. Russ. 2003. The effects of marine reserve protection on the trophic relationships of reef fishes on the Great Barrier Reef. *Environmental Conservation* 30: 200–208.

Grant, C. C., and J. L. van der Walt. 2000. Towards an adaptive management approach for the conservation of rare antelope in the Kruger National Park: outcome of a workshop held in May 2000. *Koedoe* 43: 103–111.

Grant, C. C., T. Davidson, P. J. Funston, and D. J.Pienaar. 2002. Challenges faced in the conservation of rare antelope: a case study on the northern basalt plains of the Kruger National Park. *Koedoe* 45: 45–62.

Grayson, D. K., and D. J. Meltzer. 2003. A requiem for North American overkill. *Journal of Archaeological Science* 30: 585–593.

Green, R.G., and C. A. Evans. 1940. Studies on a population cycle of snowshoe hares on the Lake Alexander area, I: Gross annual censuses, 1932–1939. *Journal of Wildlife Management* 4: 220–238.

Greene, L. E., and W. S. Alevizon. 1989. Comparative accuracies of visual assessment methods for coral reef fishes. *Bulletin of Marine Science* 44: 899–912.

Grigione, M. M., P. Beier, R. A. Hopkins, D. Neal, W. D. Padley, C. M. Schonewald, and M. L. Johnson. 2002. Ecological and allometric determinants of home-range size for mountain lions (*Puma concolor*). *Animal Conservation* 5: 317–324.

Gromov, E. I., and E. N. Matyushkin. 1974. Analysis of the competitive relations between tigers and wolves in the Sikhote-Alin. *Nauchnye doklady vysshey shkoly, Biologicheskiye Nauki* 17: 20–25.

Groves, C. R. 2003. *Drafting a conservation blueprint: a practitioner's guide to planning for biodiversity.* Island Press, Washington, District of Columbia, USA.

Hacker, S. D., and R. S. Steneck. 1990. Habitat architecture and the abundance and body-size-dependent habitat selection of a phytal amphipod. *Ecology* 71: 2269–2285.

Hagen, N. T. 1983. Destructive grazing of kelp beds by sea urchins in Vestfjorden, northern Norway. *Sarsia* 68: 177–190.

Haines, A. L. 1977. *The Yellowstone story.* Volumes 1 and 2. Yellowstone Library and Museum Association. Yellowstone National Park, Wyoming.

Hairston, N. G., F. E. Smith, and L. B. Slobodkin. 1960. Community structure, population control, and competition. *American Naturalist* 94: 421–425.

Halaj, J., and D. H. Wise. 2001. Terrestrial trophic cascades: how much do they trickle? *American Naturalist* 157: 262–281.

Hall, S. J. G., and D. G. Bradley. 1995. Conserving livestock breed biodiversity. *Trends in Ecology and Evolution* 10: 267–270.

Hall, S. J., and D. Raffaelli. 1991. Food web patterns: lessons from a species-rich web. *Journal of Animal Ecology* 60: 823–842.

Hall, S. J., and D. G. Raffaelli. 1993. Food webs: theory and reality. *Advances in Ecological Research* 24: 187–239.

Halpern, B. 2003. The impact of marine reserves: do reserves work and does reserve size matter? *Ecological Applications* 13 (Supplement): S117–137.

Halpern, B. S., and R. R. Warner. 2002. Marine reserves have rapid and lasting effects. *Ecology Letters* 5: 361–366.

Hanby, J. P., and J. D. Bygott. 1979. Population changes in lions and other predators. Pages 249–262 *in* A. R. E. Sinclair and M. Norton-Griffiths, editors. *Serengeti: dynamics of an ecosystem.* University of Chicago Press, Chicago, Illinois, USA.

Hansen, A. J., and J. J. Rotella. 2002. Biophysical factors, land use, and species viability in and around nature reserves. *Conservation Biology* 16: 1112–1122.

Hansen, L. P., C. M. Nixon, and J. Berringer. 1997. Role of refuges in the dynamics of outlying deer populations. Pages 327–345 *in* W. J. McShea, H. B. Underwood, and J. H. Rappole, editors. *The science of overabundance; deer ecology and management.* Smithsonian Institution Press, Washington, District of Columbia, USA.

Hanski, I., P. Turchin, E. Körpimaki, and H. Henttonen. 1993. Population oscillations of boreal rodents: regulation by mustelid predators leads to chaos. *Nature* 364: 232–235.

Hansson, L., and H. Henttonen. 1985. Rodent dynamics as community processes. *Trends in Ecology and Evolution* 3: 195–200.

Harcourt, A. H. 1998. Ecological indicators of risk for primates, as judged by species' susceptibility to logging. Pages 56–79 *in* T. M. Caro, editor. *Behavioral ecology and conservation biology.* Oxford University Press, New York, New York, USA.

Hardin, G. 1968. The tragedy of the commons. *Science* 162: 1243–1248.

Harley, C. D. G. 2003. Species importance and context: spatial and temporal variation in species interactions. Pages 44–68 *in* P. Kareiva and S. A. Levin, editors. *The importance of species.* Princeton University Press, Princeton, New Jersey, USA.

Harrington, R., N. Owen-Smith, P. C. Viljoen, H. C., Biggs, D. R. Mason, and P. J. Funston. 1999. Establishing the causes of the roan antelope decline in the Kruger National Park, South Africa. *Biological Conservation* 90: 69–78.

Harris, L. D., T. Hoctor, D. Maehr, and J. Sanderson. 1996. The role of networks and corridors in enhancing the value and protection of parks and equivalent areas. Pages 173–197 in R. G.Wright, editor. *National parks and protected areas: their role in environmental protection.* Blackwell Science, Cambridge, Massachusetts, USA.

Harris, L. D., L. C. Duever, R. P. Meegan, T. S. Hoctor, J. L. Schortemeyer, and D. S. Maehr. 2001. The biotic province: minimum unit for conserving biodiversity. Pages 321–343 in D. S. Maehr, R. F. Noss, and J. L. Larkin, editors. *Large mammal restoration: ecological and sociological challenges in the 21st century.* Island Press, Washington, District of Columbia, USA.

Harris, R. B., W. A. Wall, and F. W. Allendorf. 2002. Genetic consequences of hunting: what do we know and what should we know? *Wildlife Society Bulletin* 30: 634–643.

Hartt, L., and J. W. Haefner. 1995. Inbreeding depression effects on extinction time in a predator–prey system. *Evolutionary Ecology* 9: 1–9.

Hatcher, B. G, H. Kirkman, and W. F. Wood. 1987. Growth of the kelp *Ecklonia radiata* near the northern limit of its range in Western Australia. *Marine Biology* 95: 63–73.

Hatfield, B. B., D. Marks, M. T. Tinker, and K. Nolan. 1998. Attacks on sea otters by killer whales. *Marine Mammal Science* 14: 888–894.

Hatter, I. W., and D. W. Janz. 1994. Apparent demographic changes in black-tailed deer associated with wolf control on northern Vancouver Island. *Canadian Journal of Zoology* 72: 878–884.

Havlick D. G. 2002. *No place distant: roads and motorized recreation on America's public lands.* Island Press, Covello, California, USA.

Hawkins, J. P., and C. Roberts. 2004. Effects of artisanal fishing on Caribbean coral reefs. *Conservation Biology* 18: 215–226.

Hay, M. E. 1984. Patterns of fish and urchin grazing on Caribbean coral reefs: are previous results typical? *Ecology* 65: 446–454.

Hay, M. E. 1986. Associational plant defenses and maintenance of species diversity: turning competitors into accomplices. *American Naturalist* 128: 617–641.

Hay, M., and P. Taylor. 1985. Competition between herbivorous fishes and urchins on Caribbean reefs. *Oecologia* 65: 591–598.

Hayes, R. D., and J. R. Gunson. 1995. Status and management of wolves in Canada. Pages 21–34 in L. N. Carbyn, S. H. Fritts, and D. R. Seip, editors. *Ecology and conservation of wolves in a changing world.* Canadian Circumpolar Institute, Occasional Publication No. 35. Edmonton, Alberta, Canada.

Hayes, R. D., and A. S. Harestad. 2000a. Demography of a recovering wolf population in the Yukon. *Canadian Journal of Zoology* 78: 36–48.

Hayes, R. D., and A. S. Harestad. 2000b. Wolf functional response and regulation of moose in the Yukon. *Canadian Journal of Zoology* 78: 60–66.

Hayes, R. D., R. Farnell, R. M. P. Ward, J. Carey, M. Dehn, G. W. Kuzyk, A. M. Baer, C. L. Gardner, and M. O'Donoghue. 2003. Experimental reduction of wolves in the Yukon: ungulate responses and management implications. *Wildlife Monographs* 152: 1–35.

Healy, W. M. 1997. Influence of deer on the structure and composition of oak forests in Central Massachusetts. Pages 249–266 in W. J. McShea, H. B. Underwood, and J. H. Rappole, editors. The science of overabundance; deer ecology and management. Smithsonian Institution Press, Washington, District of Columbia, USA.

Healy, W. M., K. W. Gottschalk, R. P. Long, and P. M. Wargo. 1997. Changes in eastern forests: chestnuts are gone. Can oaks be far behind? Transaction of North American Wildlife and Natural Resources Conference 62: 249–263.

Hebblewhite, M., D. H. Pletscher, and P. C. Paquet. 2002. Elk population dynamics in areas with and without predation by recolonizing wolves in Banff National Park, Alberta. Canadian Journal of Zoology 80: 789–799.

Hector, A., J. Joshi, S. P. Lawler, E. M. Spehn, and A. Wilby. 2001. Conservation implications of the link between biodiversity and ecosystem functioning. Oecologia 129: 624–628.

Hedges, L. V., and I. Olkin. 1985. Statistical methods for meta-analysis. Academic Press, Boston, Massachusetts, USA.

Heinrich, B. 1989. Ravens in winter. Random House, New York, New York, USA.

Henke, S. E., and F. C. Bryant. 1999. Effects of coyote removal on the faunal community in western Texas. Journal of Wildlife Management 63: 1066–1081.

Hereu, B. 2004. The role of trophic interactions between fishes, sea urchins and algae in the northwestern Mediterranean rocky infralittoral. Ph.D. dissertation, University of Barcelona, Barcelona, Spain.

Hickerson, H. 1965. The Virginia deer and intertribal buffer zones in the upper Mississippi Valley. Pages 43–65 in A. Leeds and A. Vayda, editors. Man, culture, and animals: the role of animals in human ecological adjustments. American Association for the Advancement of Science, Washington, District of Columbia, USA.

Hill, K., J. Padwe, C. Bejyvagi, A. Bepurangi, F. Jakugi, R. Tykuarangi, T. Tykuarangi. 1997. Impact of hunting on large vertebrates in the Mbaracayu Reserve, Paraguay. Conservation Biology 11: 1339–1353.

Hilty, J., and A. Merenlender. 2000. Faunal indicator taxa selection for monitoring ecosystem health. Biological Conservation 92: 185–197.

Hixon, M. A. 1997. Effects of reef fishes on corals and algae. Pages 230–248 in C. Birkeland, editor. Life and death of coral reefs. Chapman and Hall, New York, New York, USA.

Hixon, M. A., and J. P. Beets. 1993. Predation, prey refuges and the structure of coral-reef fish assemblages. Ecological Monographs 63: 77–101.

Hjorleifsson, E., O. Kassa, and K. Gunnarsson. 1995. Grazing of kelp by green sea urchins in Eyyjafjordu, North Iceland. Pages 593–597 in H. R. Skjoldal, C. Hopkins, K. K. Erikstad, and H. P. Leinass, editors. Ecology of fjords and coastal waters. Elsevier Science, San Diego, USA.

Hobbs, N. T. 1996. Modification of ecosystems by ungulates. Journal of Wildlife Management 60: 695–713.

Hobbs, N. T., D. L. Baker, J. E. Ellis, and D. M. Swift. 1982. Energy- and nitrogen-based estimates of elk winter-range carrying capacity. Journal of Wildlife Management 46: 12–21.

Hobbs, R. J. 2001. Synergisms among habitat fragmentation, livestock, grazing, and biotic invasion in southwestern Australia. *Conservation Biology* 15: 1522–1528.

Hodges, K. E., C. J. Krebs, and A. R. E. Sinclair. 1999. Snowshoe hare demography during a cyclic population low. *Journal of Animal Ecology* 68: 581–594.

Hodges, K. E., C. J. Krebs, D. S. Hik, C. I. Stefan, E. A. Gillis, and C. E. Doyle. 2001. Snowshoe hare demography. Pages 141–178 *in* C. J. Krebs, S. Boutin, and R. Boonstra, editors. *Ecosystem dynamics of the boreal forest: the Kluane project*. Oxford University Press, New York, New York, USA.

Hofer, H., and M. East. 1995. Population dynamics, population size, and the commuting system of Serengeti spotted hyenas. Pages 332–384 *in* A. R. E. Sinclair, and P. Arcese, editors. *Serenget II*. University Press, Chicago, Illinois, USA.

Holling, C. S. 1959. The components of predation as revealed by a study of small mammal predation of the European pine sawfly. *Canadian Entomologist* 91: 293–320.

Holling, C. S. 1965. The functional response of predators to prey density and its role in mimicry and population regulation. *Memoirs of the Entomological Society of Canada* 45: 1–60.

Holling, C. S. 1973. Resilience and stability of ecological systems. *Annual Review of Ecology and Systematics* 4: 1–23.

Holsman, R. H. 2000. Goodwill hunting? Exploring the role of hunters as ecosystem stewards. *Wildlife Society Bulletin* 28: 808–816.

Holt, R. D. 1977. Predation, apparent competition, and the structure of prey communities. *Theoretical Population Biology* 12: 197–229.

Holt, R. D., and J. H. Lawton. 1994. The ecological consequences of shared natural enemies. *Annual Review of Ecology and Systematics* 25: 495–520.

Homolka, M., and Heroldova, M. 2001. Native red deer and introduced chamois: foraging habits and competition in a sub-alpine meadow-spruce forest area. *Folia Zoologica* 50: 89–98.

Hooker, S. K., and L. R. Gerber. 2004. Marine reserves as a tool for ecosystem-based management: the potential importance of megafauna. *BioScience* 54: 27–39.

Hornocker, M. 1970. An analysis of mountain lion predation upon mule deer and elk in the Idaho Primitive Area. *Wildlife Monographs* 21: 1–39.

Horsley, S. B., S. L. Stout, and D. S. deCalesta. 2003. White-tailed deer impact on the vegetation dynamics of a northern hardwood forest. *Ecological Applications* 13: 98–118.

Hough, A. F. 1965. A twenty-year record of understory vegetational change in a virgin Pennsylvania forest. *Ecology* 46: 370–373.

Houston, D. B. 1982. *The northern Yellowstone elk: ecology and management*. Macmillan Publishing, New York, New York, USA.

Howell, J. A., G. C. Brooks, M. Semenoff-Irving, and C. Greene. 2002. Population dynamics of Tule elk at Point Reyes National Seashore, California. *Journal of Wildlife Management* 66: 478–490.

Huber, D., J. Kusak, G. Gužvica, T. Gomercic, and G. Schwaderer. 2002. The effectiveness of green bridge Dedin in Gorski Kotar (Croatia) for brown bears. Abstracts from the 14th International Congress of Bear Research and Management. Nord Trondelag University College, Steinkjer, Norway.

Hughes, T. P. 1994. Catastrophes, phase shifts, and large-scale degradation of a Caribbean coral reef. *Science* 265: 1547–1551.

Hughes, T., A. M. Szmant, R. S. Steneck, R. C. Carpenter, and S. L. Miller. 1999. Algal blooms on coral reefs: what are the causes? *Limnology and Oceanography* 44: 1583–1586.

Hughes, T. P., A. H. Baird, D. R. Bellwood, M. Card, S. R. Connolly, C. Folke, R. Grosberg, et al. 2003. Climate change, human impacts and the resilience of coral reefs. *Science* 301: 929–933.

Hunsberger, A. G. B., R. M. Giblin-Davis, and T. J. Weissling. 2000. Symptoms and population dynamics of *Rhynchophorus cruentatus* (Coleoptera: Curculionidae) in Canary Island date palms. *Florida Entomologist* 83: 290–303.

Hunter, M. D., and P. W. Price. 1992. Playing chutes and ladders: heterogeneity and the relative roles of bottom-up and top-down forces in natural communities. *Ecology* 73: 724–732.

Hunter, N. D. 1993. A message from the director of the Department of Wildlife and National Parks. In *Strategic plan for wildlife research in Botswana*. Department of Wildlife and National Parks, Gaborone, Botswana.

Hunziker, M. 1995. The spontaneous reafforestation in abandoned agricultural lands: perception and aesthetical assessment by locals and tourists. *Landscape and Urban Planning* 31: 399–410.

Hurlbert, S. H. 1971. The nonconcept of species diversity: a critique and alternative parameters. *Ecology* 52: 577–586.

Husseman, J. S., D. L. Murray, G. Power, C. Mack, C. R. Wenger, and H. Quigley. 2003. Assessing differential prey selection patterns between two sympatric large carnivores. *Oikos* 101: 591–601.

Huston, M. A. 1979. A general hypothesis of species diversity. *American Naturalist* 113: 81–101.

Huston, M. A. 1994. *Biological diversity: the coexistence of species on changing landscapes*. Cambridge University Press, New York, New York, USA.

Hutchinson, G. E. 1957. Concluding remarks. Population studies: animal ecology and demography. *Cold Spring Harbor Symposia on Quantitative Biology* 22: 415–427.

Hutchinson, G. E. 1959. Homage to Santa Rosalia: or, why are there so many kinds of animals. *American Naturalist* 93: 145–159.

Hutchinson, G. E. 1980. *An introduction to population ecology*. Yale University Press, New Haven, Connecticut, USA.

Huxham, M., D. Raffaelli, and A. Pike. 1995. Parasites and food web patterns. *Journal of Animal Ecology* 64: 168–175.

Hyrenbach, K. D., K. A. Forney, and P. K. Dayton. 2000. Marine protected areas and ocean basin management. *Aquatic Conservation: Marine and Freshwater Ecosystems* 10: 437–458.

Illueca, J. 2001. The Paseo Pantera agenda for regional conservation. Pages 241–257 *in* A. G. Coates, editor. *Central America: a natural and cultural history.* Yale University Press, New Haven, Connecticut, USA and London, UK.

INDEC. 2002. *Censo Nacional Agropecuario.* Instituto Nacional de Estadística y Censos, Buenos Aires, Argentina.

Iriarte, J. A., W. E. Johnson, and W. L. Franklin. 1991. Feeding ecology of the Patagonia puma in southernmost Chile. *Revista Chilena de Historia Natural* 64: 145–156.

Irons, D. B., R. G. Anthony, and J. A. Estes. 1986. Foraging strategies of glaucous-winged gulls in rocky intertidal communities. *Ecology* 67: 1460–1474.

Irons, J. G. III, J. P. Bryant, and M. W. Oswood. 1991. Effects of moose browsing on decomposition rates of birch leaf litter in a subarctic stream. *Canadian Journal of Fisheries and Aquatic Sciences* 48: 442–444.

Jablonski, D., and D. M. Raup. 1995. Selectivity of end-Cretaceous marine bivalve extinctions. *Science* 268: 389–391.

Jackson, G. A., and C. D. Winant. 1983. Effect of a kelp forest on coastal current. *Continental Shelf Research* 2: 75–80.

Jackson, J. B. C. 1994. Constancy and change of life in the sea. *Philosophical Transactions of the Royal Society Series B.* 344: 55–60.

Jackson, J. B. C. 1997. Reefs since Columbus. *Coral Reefs* 16: S23–S32.

Jackson, J. B. C. 2001. What was natural in the coastal oceans? *Proceedings of the National Academy of Sciences* (USA) 98: 5411–5418.

Jackson, J. B. C., and E. Sala. 2001. Unnatural oceans. *Scientia Marina* 65 (Supplement 2): 275–283.

Jackson, J. B. C., M. X. Kirby, W. H. Berger, K. A. Bjorndal, L. W. Botsford, B. J. Bourque, R. Bradbury, et al. 2001. Historical overfishing and the recent collapse of coastal ecosystems. *Science* 293: 629–638.

Jackson, S. G. 1992. Relationships among birds, willows, and native ungulates in and around northern Yellowstone National Park. M.S. Thesis. Utah State University, Logan, Utah, USA.

Jaksic, F. M., P. Feinsinger, and J. Jiminez. 1993. A long-term study on the dynamics of guild structure among predatory vertebrates at a semi-arid Neotropical site. *Oikos* 67: 87–96.

Jameson, R. J. 1998. Translocated sea otter populations off the Oregon and Washington coasts. Pages 684–686 *in* M. J. Mac, P. A. Opler, C. E. Puckett Haecker, and P. D. Doran, editors. *Status and trends of the nation's biological resources.* Volume 2. U.S. Geological Survey, Washington, District of Columbia, USA.

Jameson, S. C., M. H. Tupper, and J. M. Ridley. 2002. The three screen doors: Can marine "protected" areas be effective? *Marine Pollution Bulletin* 2002: 1177–1183.

Jędrzejewska, B., and W. Jędrzejewski. 1998. *Predation in vertebrate communities: the*

Białowieża Primeval Forest as a case study. Springer Verlag, Berlin, Heidelberg, New York.

Jędrzejewska, B., H. Okarma, W. Jędrzejewski, and L. Miłkowski. 1994. Effects of exploitation and protection on forest structure, ungulate density, and wolf predation in Białowieża Primeval Forest, Poland. *Journal of Applied Ecology* 31: 664–676.

Jędrzejewska, B., W. Jędrzejewski, A. N. Bunevich, L. Miłkowski, and Z. A. Krasiński. 1997. Factors shaping population densities and increase rates of ungulates in Białowieża Primeval Forest (Poland and Belarus) in the 19th and 20th centuries. *Acta Theriologica* 42: 399–451.

Jędrzejewska, B., W. Jędrzejewski, A. N. Bunevich, L. Miłkowski, and H. Okarma. 1996. Population dynamics of wolves *Canis lupus* in Białowieża Primeval Forest (Poland and Belarus) in relation to hunting by humans, 1847–1993. *Mammal* Review 26: 103–126.

Jędrzejewski, W., B. Jędrzejewska, H. Okarma, K. Schmidt, A. N. Bunevich, and L. Miłkowski. 1996. Population dynamics (1869–1994), demography, and home ranges of the lynx in Białowieża Primeval Forest (Poland and Belarus). *Ecography* 19: 122–138.

Jędrzejewski, W., B. Jędrzejewska, H. Okarma, K. Schmidt, K. Zub, and M. Musiani. 2000. Prey selection and predation by wolves in Białowieża Primeval Forest, Poland. *Journal of Mammalogy* 81: 197–212.

Jędrzejewski, W., K. Schmidt, J. Theuerkauf, B. Jędrzejewska, N. Selva, K. Zub, and L. Szymura. 2002. Kill rates and predation by wolves on ungulate populations in Białowieża Primeval Forest (Poland). *Ecology* 83: 1341–1356.

Jennings, S., and N. V. C. Polunin. 1996. Effects of fishing effort and catch rate upon the structure and biomass of Fijian reef fish communities. *Journal of Applied Ecology* 33: 400–412.

Jenssen, S. M. 2003. The eastern coyote and the white-tailed deer in Maine: a history of dollars vs sense. Available from http://www.nosnare.org/articles/sjenssen1.html (accessed October 2004).

Jepson, P., and S. Canney. 2003. Values-led conservation. *Global Ecology and Biogeography* 12: 271–274.

Jhala, Y. V., and R. H. Giles. 1991. The status and conservation of the wolf in Gujarat and Rajasthan India. *Conservation Biology* 5: 476–483.

Johnsingh, A. J. T. 1985. Distribution and status of dhole *Cuon alpinus* Pallas 1811 in South Asia. *Mammalia* 49: 203–208.

Johnson, E. A. 1992. *Fire and vegetation dynamics: studies from the North American boreal forest.* Cambridge University Press, Cambridge, UK.

Johnson, W. E., and W. L. Franklin. 1994a. Role of body size in the diets of sympatric gray and culpeo foxes. *Journal of Mammalogy* 75: 163–174.

Johnson, W. E., and W. L. Franklin. 1994b. Spatial resource partitioning by sympatric grey fox (*Dusicyon griseus*) and culpeo fox (*Dusicyon culpaeus*) in southern Chile. *Canadian Journal of Zoology* 72: 1788–1793.

Johnson, W. E., and W. L. Franklin. 1994c. Conservation implications of South American grey fox (*Dusicyon griseus*) socioecology in the Patagonia of southern Chile. *Vida Silvestre Neotropical* 3: 16–23.

Jones, G. P., M. I. McCormick, M. Srinivasan, and J. V. Eagle. 2004. Coral decline threatens fish biodiversity in marine reserves. *Proceedings of the National Academy of Sciences* (USA) 101: 8251–8253.

Jordan P. A., B. E. McLaren, and S. M. Sell. 2000. A summary of research on moose and related ecological topics at Isle Royale, U.S.A. *Alces* 36: 217–232.

Jorgenson, J. P., and K. H. Redford. 1993. Humans and big cats as predators in the Neotropics. *Symposia of the Zoological Society of London* 65: 367–390.

Jorritsma, I. T. M., A. F. M. van Hees, and G. M. J. Mohren. 1999. Forest development in relation to ungulate grazing: a modeling approach. *Forest Ecology and Management* 120: 23–34.

Kaczensky, P. 1996. *Livestock–carnivore conflicts in Europe.* Munich Wildlife Society, Germany.

Kain, J. M. 1975. Algal recolonization of some cleared subtidal areas. *Journal of Ecology* 63: 739–765.

Kaplanov, L. G. 1948. *Tigers in Sikhote-Alin* (in Russian). Izd. Mosk., Moscow, Russia.

Karanth, K. U. 1991. Ecology and management of the tiger in tropical Asia. Pages 156–159 in N. Maruyama, B. Bobek, Y. Ono, W. Regli, L. Bartos, and R. Ratcliffe, editors. *Wildlife conservation: present trends and perspectives for the 21st century.* Japan Wildlife Research Centre, Tokyo, Japan.

Karanth, K. U. 1993. Predator–prey relationships among large mammals on Nagarhole National Park (India). Ph.D. thesis. Mangalore University, India.

Karanth, K. U., and J. D. Nichols. 1998. Estimation of tiger densities in India using photographic captures and recaptures. *Ecology* 79: 2852–2862.

Karanth, K. U., and J. D. Nichols. 2000. *Ecological status and conservation of tigers in India.* Final Technical Report to the U.S. Fish and Wildlife Service (Division of International Conservation), Washington, D.C., and Wildlife Conservation Society, New York. Centre for Wildlife Studies, Bangalore, India.

Karanth, K. U., and B. M. Stith. 1999. Prey depletion as a critical determinant of tiger population viability. Pages 100–113 in J. Seidensticker, S. Christie, and P. Jackson, editors. *Riding the tiger: tiger conservation in human-dominated landscapes.* Cambridge University Press, Cambridge, UK.

Karanth, K. U., and M. E. Sunquist. 1992. Population-structure, density and biomass of large herbivores in the tropical forests of Nagarahole, India. *Journal of Tropical Ecology* 8: 21–35.

Karanth, K. U., and M. E. Sunquist. 1995. Prey selection by tiger, leopard and dhole in tropical forests. *Journal of Animal Ecology* 64: 439–450.

Karanth, K. U., and M. E. Sunquist. 2000. Behavioural correlates of predation by tiger (*Panthera tigris*), leopard (*Panthera pardus*) and dhole (*Cuon alpinus*) in Nagarahole, India. *Journal of Zoology* 250: 255–265.

Karanth, K. U., J. D. Nichols, N. S. Kumar, W. A. Link, and J. E. Hines. 2004. Tigers and their prey: predicting carnivore densities from prey abundance. *Proceedings of the National Academy of Sciences* (USA) 101: 4854–4858.

Kareiva, P. and S. A. Levin, editors. 2003. *The importance of species*. Princeton University Press, Princeton, New Jersey, USA.

Karns, P. D. 1998. Population distribution, density and trends. Pages 125–139 *in* A. W. Franzmann and C. C. Schwartz, editors. *Ecology and management of the North American moose*. Smithsonian Institution Press. Washington, District of Columbia, USA.

Karr, J. R. 2000. Health, integrity, and biological assessment: the importance of measuring whole things. Pages 209–226 *in* D. Pimentel, L. Westra, and R. F. Noss, editors. *Ecological integrity: integrating environment, conservation and health*. Island Press, Washington, District of Columbia, and Covelo, California, USA.

Kay, C. E. 1985. Aspen reproduction in the Yellowstone Park–Jackson Hole area and its relationship to the natural regulation of ungulates. Pages 131–160 *in* G. W. Workman, editor. *Western elk management: a symposium*. Utah State University, Logan, Utah, USA.

Kay, C. E. 1994. Aboriginal overkill: the role of native Americans in structuring western ecosystems. *Human Nature* 5: 359–398.

Kay, C. E. 1998. Are ecosystems structured from the top-down or bottom-up: a new look at an old debate. *Wildlife Society Bulletin* 26: 484–498.

Keech, M. A., R. T. Bowyer, J. M. Ver Hoef, R. D. Boertje, B. W. Dale, and T. R. Stephenson. 2000. Life-history consequences of maternal condition in Alaskan moose. *Journal of Wildlife Management* 64: 450–462.

Keeling, M. J., H. B. Wilson, and S. W. Pacala. 2000. Reinterpreting space, time lags, and functional responses in ecological models. *Science* 290: 1758–1761.

Keesing, F., and T. Crawford. 2001. Impacts of density and large mammals on space use by the pouched mouse (*Soccostomus mernsi*) in central Kenya. *Journal of Tropical Ecology* 17: 465–472.

Keigley, R. B. 2000. Elk, beaver, and the persistence of willows in national parks: comment on Singer et al. (1998). *Wildlife Society Bulletin* 28: 448–450.

Keith, L. B. 1983a. Population dynamics of wolves. Pages 66–77 *in* L. N. Carbyn, editor. *Wolves in Canada and Alaska: their status, biology, and management*. Canadian Wildlife Service Report Series, Number 45. Canadian Wildlife Service, Ottawa, Ontario, Canada.

Keith, L. B. 1983b. Role of food in hare population cycles. *Oikos* 40: 385–395.

Keith, L. B., and L. A. Windberg. 1978. A demographic analysis of the snowshoe hare cycle. *Wildlife Monographs* 58: 1–70.

Keith, L. B., A. W. Todd, C. J. Brand, R. S. Adamcik, and D. H. Rusch. 1977. An analysis of predation during a cyclic fluctuation of snowshoe hares. *Proceedings of the International Congress of Game Biologists* 13: 151–175.

Keith, L. B., J. R. Cary, O. J. Rongstad, and M. C. Brittingham. 1984. Demography and ecology of a declining snowshoe hare population. *Wildlife Monographs* 90: 1–43.

Kellert, S. R., M. Black, C. R. Rush, and A. J. Bath. 1996. Human culture and large carnivore conservation in North America. *Conservation Biology* 10: 977–990.

Kenyon, K. W. 1969. *The sea otter in the eastern North Pacific Ocean.* U.S. Fish and Wildlife Service, North American Fauna No. 68.

Kerley, L. L., J. M. Goodrich, D. G. Miquelle, E. N. Smirnov, H. B. Quigley, and M. G. Hornocker. 2003. Reproductive parameters of wild female Amur (Siberian) tigers (*Panthera tigris altaica*). *Journal of Mammalogy* 84: 288–298.

Kerr, J. T. 1997. Species richness, endemism, and the choice of areas for conservation. *Conservation Biology* 11: 1094–1100.

Khramtsov, V. S. 1995. The wolves of the Kievka River Valley (Southern Primorie). *Bulletin Moskovskogo Obschestva Ispytateley Prirody Otdyel Biologiskesky* 100: 18–21.

Kie, J. G. 1999. Optimal foraging in a risky environment: life-history strategies for ungulates. *Journal of Mammalogy* 80: 1114–1129.

Kie, J. G., and R. T. Bowyer. 1999. Sexual segregation in white-tailed deer: density dependent changes in use of space, habitat selection, and dietary niche. *Journal of Mammalogy* 80: 1004–1020.

Kie, J .G., R. T. Bowyer, and K. M. Stewart. 2003. Ungulates in western forests: habitat requirements, population dynamics, and ecosystem processes. Pages 296–340 *in* C. J. Zabel and R. G. Anthony, editors. *Mammal community dynamics: management and conservation in the coniferous forests of western North America.* Cambridge University Press, New York, New York, USA.

Kilgo, J. C., R. F. Labisky, and D. E. Fritzen. 1998. Influences of hunting on the behavior of white-tailed deer: implications for conservation of the Florida panther. *Conservation Biology* 12: 1359–1364.

Kinnear, J. E., M. L. Onus, and N. R. Sumner. 1998. Fox control and rock-wallaby population dynamics, II: An update. *Wildlife Research* 25: 81–88.

Kirby, K. J. 2001. The impact of deer on the ground flora of British broadleaved woodland. *Forestry* 74: 219–230.

Kitchen, A. M., E. M. Gese, and E. R. Schauster. 2000. Changes in coyote activity patterns due to reduced exposure to human persecution. *Canadian Journal of Zoology* 78: 853–857.

Kitching, J. A., and F. J. Ebling. 1961. The ecology of Lough Ine, XI: The control of algae by *Paracentrotus lividus* (Echinoidea). *Journal of Animal Ecology* 30: 373–383.

Kleiman, D. G., and J. F. Eisenberg. 1973. Comparisons of canid and felid social systems from an evolutionary perspective. *Animal Behavior* 21: 637–659.

Klein, D. R. 1968. The introduction, increase, and crash of reindeer on St. Matthew Island. *Journal of Wildlife Management* 32: 350–367.

Knight, J. 2003. *Waiting for wolves in Japan: an anthropological study of people–wildlife relations.* Oxford University Press, New York, New York, USA.

Knight, R. L., and P. Landres. 1998. *Stewardship across boundaries.* Island Press, Covello, California, USA.

Knowlton, F. F., E. M. Gese, and M. M. Jaeger. 1999. Coyote depredation control: an interface between biology and management. *Journal of Range Management* 52: 398–412.

Knowlton, N. 1992. Thresholds and multiple stable states in coral reef community dynamics. *American Zoologist* 32: 674–682.

Knox, W. M. 1997. Historical changes in the abundance and distribution of deer in Virginia. Pages 27–36 *in* W. J. McShea, H. B. Underwood, and J. H. Rappole, editors. *The science of overabundance: deer ecology and management.* Smithsonian Institution Press, Washington, District of Columbia, USA.

Körpimaki, E., and C. J. Krebs. 1996. Predation and population cycles of small mammals. *BioScience* 46: 754–765.

Körpimaki, E., K. Norrdahl, T. Klemola, T. Pettersen, and N. C. Stenseth. 2002. Dynamic effects of predators on cyclic voles: field experimentation and model extrapolation. *Proceedings of the Royal Society of London, Series B, Biology* 269: 991–997.

Kortlandt, A. 1984. Vegetation research and the "bulldozer" herbivores of tropical Africa. Pages 205–226 *in* A. C. Chadwick and C. L. Sutton, editors. *Tropical rainforest.* Leeds Philosophical Literature Society, Leeds, UK.

Kozulko, G. A. 1998. Vliyane intensivnogo vedeniya okhotnichego khozaistva na pochvennykh bespozvonochnykh v nacionalnom parke Belovezhskaya Pushcha [Effect of the primacy of game management on soil invertebrates in the National Park, Belovezha Primeval Forest]. *Ekologiya* 1998: 489–492 (in Russian).

Kramer, P. 2003. Synthesis of coral reef health indicators for the western Atlantic: results of the AGRRA program (1997–2000). Pages 1–57 *in* J. C. Lang, editor. *Status of Coral Reefs in the western Atlantic: results of initial surveys, Atlantic and Gulf Rapid Reef Assessment (AGRRA) Program.* Atoll Research Bulletin Number 496, National Museum of Natural History, Smithsonian Institution. Washington, District of Columbia, USA.

Krasiński, Z. 1967. Free living European bisons. *Acta Theriologica* 12: 391–405.

Krebs, C. J. 1989. *Ecological methodology.* Harper and Row, New York, New York, USA.

Krebs, C. J. 2002. Beyond population regulation and limitation. *Wildlife Research* 29: 1–10.

Krebs, C. J., S. Boutin, R. Boonstra, A. R. E. Sinclair, J. N. M. Smith, M. R. T. Dale, K. Martin, and R. Turkington. 1995. Impact of food and predation on the snowshoe hare cycle. *Science* 269: 1112–1115.

Krebs, C. J., R. Boonstra, S. Boutin, and A. R. E. Sinclair. 2001. What drives the 10-year cycle of snowshoe hares? *BioScience* 51: 25–35.

Krebs, C. J., S. Boutin, and R. Boonstra, editors. 2001a. *Ecosystem dynamics of the boreal forest: the Kluane project.* Oxford University Press. New York, New York, USA.

Krebs, C. J., M. R. T. Dale, V. O. Nams, A. R. E. Sinclair, and M. O'Donoghue. 2001b. Shrubs. Pages 92–115 *in* C. J. Krebs, S. Boutin, and R. Boonstra, editors. *Ecosystem dynamics of the boreal forest: the Kluane project.* Oxford University Press, New York, New York, USA.

Krebs, C. J., K. Danell, A. Angerbjörn, J. Agrell, D. Berteaux, K. A. Brathen, O. Danell, et al.

2003. Terrestrial trophic dynamics in the Canadian Arctic. *Canadian Journal of Zoology* 81: 827–843.

Kruuk, H. 1972. *The spotted hyena.* University of Chicago Press, Chicago, Illinois, USA.

Kruuk, H. 2002. *Hunter and hunted: relationships between carnivores and people.* Cambridge University Press, Cambridge, Illinois, UK.

Kucherenko, S. P. 1974. Carnivores (Carnivora) of the Sikhote-Alin. Fauna i ecologiya nazemnykh pozvonochnykh yuga Dalnego Vostoka SSSR. Trudy Biologo-pochvennogo instituta. *Nov. seria. Vladivostok DVNC AN SSSR* 17: 107–119.

Kucherenko, S. P. 2001. Amur tigers at the turn of the century. *Okhota i okhotnichie khozyaistvo* 4: 20–24 (in Russian).

Kunkel, K. E. 2003. Ecology, conservation, and restoration of large carnivores in western North America. Pages 250–295 *in* C. J. Zabel and R. G. Anthony, editors. *Mammal community dynamics: management and conservation in the coniferous forests of western North America.* Cambridge University Press, Cambridge, UK.

Kunkel, K., and D. H. Pletscher. 1999. Species-specific population dynamics of cervids in a multipredator ecosystem. *Journal of Wildlife Management* 63: 1082–1093.

Kunkel, K. E., T. K. Ruth, D. H. Pletscher, and M. G. Hornocker. 1999. Winter prey selection by wolves and cougars in and near Glacier National Park, Montana. *Journal of Wildlife Management* 63: 901–910.

Kurlansky, M. 1997. *Cod: a biography of the fish that changed the world.* Walker and Company. New York, New York, USA.

Kursar, T. A., and P. D. Coley. 1991. Nitrogen content and expansion rate of young leaves of rain forest species: implications for herbivory. *Biotropica* 23: 141–150.

Kurten, B., and E. Anderson. 1980. *Pleistocene mammals of North America.* Columbia University Press, New York, New York, USA.

Labisky, R. F., and M. C. Boulay. 1998. Behaviors of bobcats preying on white-tailed deer in the Everglades. *American Midland Naturalist* 139: 275–281.

Lackey, R. T. 1998. Seven pillars of ecosystem management. *Landscape and Urban Planning* 40: 21–30.

Lambeck, R. J. 1997. Focal species: a multispecies umbrella for nature conservation. *Conservation Biology* 11: 849–856.

Land, E. D. 1991. *Big Cypress deer/panther relationships: deer mortality.* Final Report, Study 7509, E-1IIE-5B. Florida Game and Fresh Water Fish Commission, Tallahassee, Florida, USA.

Land, E. D., D. S. Maehr, J. C. Roof, and J. W. McCown. 1993. Mortality patterns of female white-tailed deer in southwest Florida. *Proceedings of the Annual Conference of Southeastern Fish and Wildlife Agencies* 47: 176–184.

Landa, A., J. D. C. Linnell, J. E. Swenson, E. Røskaft, and Moskness. 2000. Conservation of Scandinavian wolverines in ecological and political landscapes. Pages 1–20 *in* H. I. Griffiths, editor. *Mustelids in a modern world: conservation aspects of small carnivore–human interactions.* Backhuys Publishers, Leiden, Netherlands.

Lange, I. M. 2002. *Ice age mammals of North America*. Mountain Press, Missoula, Montana, USA.

Large Carnivore Initiative for Europe (LCIE). 2004. Large Carnivore Initiative for Europe. Available from http://www.lcie.org (accessed October 2004).

Larsen, E. J. 2002. Aspen age structure and stand conditions on elk winter range in the northern Yellowstone ecosystem. Ph.D. dissertation, Oregon State University, Oregon, USA.

Lauenroth, W. K. 1998. Guanacos, spiny shrubs and the evolutionary history of grazing in the Patagonian steppe. *Ecología Austral* 8: 211–216.

Laundré, J. W., L. Hernandez, and K. B. Altendorf. 2001. Wolves, elk, and bison: reestablishing the "landscape of fear" in Yellowstone National Park, USA. *Canadian Journal of Zoology* 79: 1401–1409.

Laurenson, M. K. 1994. High juvenile mortality in cheetahs (*Acinonyx jubatus*) and its consequences for maternal care. *Journal of Zoology, London* 234: 387–408.

Laurenson, M. K. 1995. Implications for high offspring mortality for cheetah population dynamics. Pages 385–399 *in* A. R. E. Sinclair and P. Arcese, editors. *Serengeti II: dynamics, management, and conservation of an ecosystem*. University of Chicago Press, Chicago, Illniois, USA.

Lavigne, D. M., V. B. Scheffer, and S. R. Kellert. 1999. The evolution of North American attitudes toward marine mammals. Pages 10–47 *in* J. R. Twiss, Jr. and R. R. Reeves, editors. *Conservation and management of marine mammals*. Smithsonian Institution Press, Washington, District of Columbia, USA.

Lawton, J. H. 1991. Are species useful? *Oikos* 62: 3–4.

Laycock, W. A. 1991. Stable states and thresholds of range conditions on North American rangelands: a viewpoint. *Journal of Range Management* 44: 427–433.

Leader-Williams, N., and H. T. Dublin. 2000. Charismatic megafauna as "flagship species." Pages 53–83 *in* A. Entwistle and N. Dunstone, editors. *Priorities for the conservation of mammalian diversity*. Cambridge University Press, Cambridge, UK.

Leader-Williams, N., D. W. H. Walton, and P. A. Prince. 1989. Introduced reindeer on South Georgia: a management dilemma. *Biological Conservation* 47: 1–11.

Leader-Williams, N., J. A. Kayera, and G. L. Overton. 1996. *Tourist hunting in Tanzania*. IUCN/SSC Occasional Papers No 14. Gland, Switzerland.

Lee, R., and I. Devore, editors. 1968. *Man the hunter*. Aldine Press, Chicago, Illinois, USA.

Leigh, E. G., Jr., S. J. Wright, E. A. Herre, and F. E. Putz. 1993. The decline of tree diversity on newly isolated tropical islands: a test of a null hypothesis and the implications. *Evolutionary Ecology* 7: 76–102.

Leland, A. 2002. A new apex predator in the Gulf of Maine? Large, mobile crabs (*Cancer borealis*) control benthic community structure. Master's thesis. University of Maine, School of Marine Sciences, Orono, Maine, U.S.A.

Lemke, T. O., J. A. Mack, and D. B. Houston. 1998. Winter range expansion by the northern Yellowstone elk herd. *Intermountain Journal of Sciences* 4: 1–9.

Leopold, A. 1943. Deer irruptions. *Wisconsin Conservation Bulletin* 8: 3–11.

Leopold, A. 1949. *A Sand County almanac.* Oxford University Press, New York, New York, USA.

Leopold, A. 1993. *Round River.* Oxford University Press, New York, New York, USA.

Lessa, E. P., B. Van Valkenburgh, and R. A. Fariña. 1997. Testing hypotheses of differential mammalian extinctions subsequent to the Great American Biotic Interchange. *Paleogeography, Paleoclimatology, Paleoecology* 135: 157–162.

Lessios, H. A. 1988. Mass mortality of *Diadema antillarum* in the Caribbean: what have we learned? *Annual Review of Ecology and Systematics* 19: 371–393.

Levitan, D. R. 1992. Community structure in times past: influence of human fishing pressure on algal–urchin interactions. *Ecology* 73: 1597–1605.

Lima, S. L. 1998. Nonlethal effects in the ecology of predator–prey interactions. *BioScience* 48: 25–34.

Lindeman, R. L. 1942. The trophic-dynamic aspect of ecology. *Ecology* 23: 399–418.

Linden, H. 1988. Latitudinal gradients in predator–prey interactions, cyclicity and synchronism in voles and small game populations in Finland. *Oikos* 52: 341–349.

Lindenmayer, D. B., and J. F. Franklin. 2002. *Conserving forest biodiversity: a comprehensive multiscale approach.* Island Press, Washington, District of Columbia. USA.

Lindström, E. R., H. Andren, P. Angelstam, G. Cenderlund, B. Hornfeldt, L. Jaderberg, P.-A. Lemnell, B. Martinsson, K. Skold, and J. E. Swenson. 1994. Disease reveals the predator: sarcoptic mange, red fox predation, and prey populations. *Ecology* 75: 1042–1049.

Linnell, J. D. C., and O. Strand. 2000. Interference interactions, coexistence and conservation of mammalian carnivores. *Diversity and Distribution* 6: 169–176.

Linnell, J. D. C., R. Aanes, and R. Anderson. 1995. Who killed Bambi? The role of predation in the neonatal mortality of temperate ungulates. *Wildlife Biology* 14: 209–223.

Linnell, J. D. C., M. E. Smith, J. Odden, P. Kaczensky, and J. E. Swenson. 1996. Strategies for the reduction of carnivore–livestock conflicts: a review. *Norwegian Institute for Nature Research Oppdragsmelding* 443: 1–118.

Linnell, J. D. C., J. Odden, V. Pedersen, and R. Andersen. 1998. Records of intraguild predation by Eurasian lynx, *Lynx lynx. Canadian Field Naturalist* 112: 707–708.

Linnell, J. D. C., J. E. Swenson, and R. Andersen. 2000. Conservation of biodiversity in Scandinavian boreal forests: large carnivores as flagships, umbrellas, indicators, or keystones? *Biodiversity and Conservation* 9: 857–868.

Linnell, J. D. C., R. Andersen, T. Kvam, H. Andrén, O. Liberg, J. Odden, and P. Moa. 2001a. Home range size and choice of management strategy for lynx in Scandinavia. *Environmental Management* 27: 869–879.

Linnell, J. D. C., J. E. Swenson, and R. Andersen. 2001b. Predators and people: conservation of large carnivores is possible at high human densities if management policy is favourable. *Animal Conservation* 4: 345–349.

Linnell, J. D. C., J. Løe, H. Okarma, J. C. Blancos, Z. Andersone, H. Valdmann, L. Balciauskas, et al. 2002. The fear of wolves: a review of wolf attacks on humans. *Norwegian Institute for Nature Research Oppdragsmelding* 731: 1–65.

Linnell, J. D. C., E. B. Nilsen, U. S. Lande, I. Herfindal, J. Odden, K. Skogen, R. Andersen, and U. Breitenmoser. In press. Zoning as a means of mitigating conflicts with large carnivores: principles and reality. In R. Woodroffe, S. Thirgood, and A. Rabinowitz, editors. *People and wildlife: conflict or coexistence?* Cambridge University Press, Cambridge, UK.

Lirman, D. 2001. Competition between macroalgae and corals: effects of herbivore exclusion and increased algal biomass on coral survivorship and growth. *Coral Reefs* 19: 392–399.

Loreau, M., S. Naeem, P. Inchausti, J. Bengtsson, J. P. Grime, A. Hector, D. U. Hooper, et al. 2001. Biodiversity and ecosystem functioning: current knowledge and future challenges. *Science* 294: 804–808.

Lovaas A. L. 1973. A cooperative elk trapping program in Wind Cave National Park. *Wildlife Society Bulletin* 1: 93–100.

Lubchenco, J. 1978. Plant species diversity in a marine intertidal community: importance of herbivore food preference and algal competitive abilities. *American Naturalist* 112: 23–39.

Lucchini, V., E. Fabbri, F. Marucco, S. Ricci, L. Boitani, and E. Randi. 2002. Noninvasive molecular tracking of colonizing wolf (*Canis lupus*) packs in the western Italian Alps. *Molecular Ecology* 11: 857–868.

Ludwig, D. 2001. Can we exploit sustainability? Pages 16–38 in J. D. Reynolds, G. M. Mace, K. H. Redord, and J. G. Robinson, editors. *Conservation of exploited species.* Cambridge University Press, Cambridge, USA.

Ludwig, D., D. D. Jones, and C. S. Holling. 1978. Qualitative analysis of insect outbreak systems in the spruce budworm and forest. *Journal of Animal Ecology* 47: 315–332.

MacArthur, R. H. 1972. Strong, or weak, interactions? *Transactions of the Connecticut Academy of Arts and Sciences* 44: 177–188.

MacArthur, R. H., and J. W. MacArthur. 1961. On bird species diversity. *Ecology* 42: 594–598.

Macdonald, D., J. R. Crabtree, G. Wiesinger, T. Dax, N. Stamou, P. Fleury, J. Gutierrez Lazpita, and A. Gibon. 2000. Agricultural abandonment in mountain areas of Europe: environmental consequences and policy response. *Journal of Environmental Management* 59: 47–69.

Macdonald, D. W. 1983. The ecology of carnivore social behaviour. *Nature* 301: 379–384.

Macdonald, D. W. 2001. Postscript-carnivore conservation: science and compromise and tough choices. Pages 524–538 in J. L. Gittleman, S. M. Funk, D. W. Macdonald, and R. K. Wayne, editors. *Carnivore conservation.* Cambridge University Press, Cambridge, UK.

Macdonald, D. W., G. M. Mace, and G. R. Barretto. 1999. The effects of predators on fragmented prey populations: a case study for the conservation of endangered prey. *Journal of Zoology,* London 247: 487–506.

Mace, G. M., A. Balmford, and J. R. Ginsberg. 1999. *Conservation in a changing world.* Cambridge University Press, Cambridge, UK.

MacFadden, B. J. 2000. Cenozoic mammalian herbivores from the Americas: reconstructing ancient diets and terrestrial communities. *Annual Review of Ecology and Systematics* 31: 33–59.

Madhusudan, M. D., and K. U. Karanth. 2002. Local hunting and the conservation of large mammals in India. *Ambio* 31: 49–54.

Maehr, D. S. 1990. The Florida panther and private lands. *Conservation Biology* 4: 167–170.

Maehr, D. S. 1997a. The comparative ecology of bobcat, black bear, and Florida panther in south Florida. *Bulletin of the Florida Museum of Natural History* 40: 1–176.

Maehr, D. S. 1997b. *The Florida panther: life and death of a vanishing carnivore.* Island Press, Washington, District of Columbia, USA.

Maehr, D. S. 2001. Restoring the large mammal fauna in the east: what follows the elk? *Wild Earth* 11: 50–53.

Maehr, D. S., and J. R. Brady. 1984. Comparison of food habits in two north Florida black bear populations. *Florida Scientist* 47: 171–175.

Maehr, D. S., and J. R. Brady. 1986. Food habits of bobcats in Florida. *Journal of Mammalogy* 67: 133–138.

Maehr, D. S., and J. A. Cox. 1995. Landscape features and panthers in Florida. *Conservation Biology* 9: 1008–1019.

Maehr, D. S., and J. T. DeFazio. 1985. Foods of black bears in Florida. *Florida Field Naturalist* 13: 8–12.

Maehr, D. S., R. C. Belden, E. D. Land, and L. Wilkins. 1990a. Food habits of panthers in southwest Florida. *Journal of Wildlife Management* 54: 420–423.

Maehr, D. S., E. D. Land, J. C. Roof, and J. W. McCown. 1990b. Day beds, natal dens, and activity of Florida panthers. *Proceedings of the Annual Conference of Southeastern Fish and Wildlife Agencies* 44: 310–318.

Maehr, D. S., E. D. Land, and J. C. Roof. 1991. Social ecology of Florida panthers. *National Geographic Research and Exploration* 7: 414–431.

Maehr, D. S., R. T. McBride, and J. J. Mullahey. 1996. Status of coyotes in south Florida. *Florida Field Naturalist* 24:101–107.

Maehr, D. S., T. S. Hoctor, and L. D. Harris. 2001a. Remedies for a denatured biota: restoring landscapes for native carnivores. *International Congress on Wildlife Management* 2: 123–127.

Maehr, D. S., T. S. Hoctor, and L. D. Harris. 2001b. The Florida panther: a flagship for regional restoration. Pages 293–312 *in* D. S. Maehr, R. F. Noss, and J. L. Larkin, editors. *Large mammal restoration: ecological and sociological challenges in the 21st century.* Island Press, Washington, District of Columbia, USA.

Maehr, D. S., J. L. Larkin, and J. J. Cox. 2004. Shopping centers as panther habitat: inferring animal locations from models. *Ecology and Society* 9(2): 9. Available from: http://www.ecologyandsociety.org/vol9/iss2/art9 (accessed December 2004).

Maguire, L. A. 1994. Science, values, and uncertainty: a critique of the wildlands project. Pages 267–272 *in* R. E. Grumbine, editor. *Environmental policy and biodiversity.* Island Press, Washington, Districct of Columbia, and Covelo, California, USA.

Makarov, Y. M., and V. T. Tagirova. 1989. Large carnivores of Bolshekhekhtsirskiy Za-povednik. Pages 134–136 *in Teriologicheskie issledovaniya na yuge Dalnego Vostoka.* Sbornik nauchn. Statey. DVO AN SSSR, Vladivostok, Russia (in Russian).

Makovkin, L. I. 1999. *The sika deer of Lazovsky Reserve and surrounding areas of the Russian Far East.* Almanac "Russki Ostrov," Vladivostok, Russia.

Mann, K. H. 1973. Seaweeds: Their productivity and strategy for growth. *Science* 182: 975–981.

Marcogliese, D. J., and D. K. Cone. 1997. Food webs: a plea for parasites. *Trends in Ecology and Evolution* 12: 320–325.

Margules, C. R., and R. L. Pressey. 2000. Systematic conservation planning. *Nature* 405: 243–253.

Marker, L. L. 2002. Aspects of cheetah (*Acinonyx jubatus*) biology, ecology and conservation on Namibian farmlands. Ph.D. dissertation. University of Oxford, UK.

Marker-Kraus, L., D. Kraus, D. Barnett, and S. Hurlbut. 1996. *Cheetah survival on Namibia farmlands.* Cheetah Conservation Fund, Windhoek, Namibia.

Markgraf, V. 1985. Late Pleistocene faunal extinctions in southern Patagonia. *Science* 228: 1110–1112.

Marshal, J. P., and S. Boutin. 1999. Power analysis of wolf–moose functional response. *Journal of Wildlife Management* 63: 396–402.

Marshall, L. G. 1984. Who killed cock robin? An investigation of the extinction controversy. Pages 785–806 *in* P. S. Martin and R. G. Klein, editors. *Quaternary extinctions: a prehistoric revolution.* University of Arizona Press, Tuscon, Arizona, USA.

Martin, P. S. 1973. The discovery of America. *Science* 179: 969–974.

Martin, P. S., and R. G. Klein, editors. 1984. *Quartenary extinctions: a prehistoric revolution.* University of Arizona Press, Tucson, USA.

Martinez, N. D. 1991. Artifacts or attributes? Effects of resolution on the Little Rock Lake food web. *Ecological Monographs* 61:367–392.

Martinez, N. D. 1993. Effects of resolution on food web structure. *Oikos* 66: 403–412.

Martinez, A., D. G. Hewitt, and M. C. Correa. 1997. Managing overabundant white-tailed deer in northern Mexico. *Wildlife Society Bulletin* 25: 430–432.

Marquard-Petersen, U. 1995. Status of wolves in Greenland. Pages 55–58 *in* L. N. Carbyn, S. H. Fritts, and D. R. Seip, editors. *Ecology and conservation of wolves in a changing world.* Canadian Circumpolar Press, Edmonton, Alberta, Canada.

Marquist, R. J., and C. J. Whelan. 1994. Insectivorous birds increase growth of white oak through consumption of leaf-chewing insects. *Ecology* 75: 2007–2014.

Matkin, C. O. 1994. *The killer whales of Prince William Sound.* Prince William Sound Books, Valdez, Alaska, USA.

Matson, P., K. A. Lohse, and S. J. Hall. 2002. The globalization of nitrogen deposition: consequences for terrestrial ecosystems. *Ambio* 31: 113–119.

Mattson, D. J. 1997. Use of ungulates by Yellowstone grizzly bears. *Biological Conservation* 81: 161–177.

Mattson, D. J., and J. J. Craighead. 1994. The Yellowstone grizzly bear recovery program:

uncertain information, uncertain policy. Pages 101–129 *in* T. W. Clark, R. P. Reading, and A. L. Clarke, editors. *Endangered species recovery: finding the lessons, improving the process.* Island Press, Washington, District of Columbia, USA.

Mattson, D.J., B. M. Blanchard, and R. R. Knight. 1992. Yellowstone grizzly bear mortality, human habituation, and white bark pine seed crops. *Journal of Wildlife Management* 56: 432–442.

Mattson, D.J., R. R. Knight, and B. M. Blanchard. 1994. The effects of developments and primary roads on grizzly bear habitat use in Yellowstone National Park, Wyoming. International Conference on Bear Research and Management 7: 259–273.

Mattson, W. J., and N. D. Addy. 1975. Phytophagous insects as regulators of forest primary production. *Science* 190: 515–522.

Matyushkin, E. N., A. A. Astafiev, V. A. Zaitsev, V. E. Kostoglod, V. A. Palkin, E. N. Smirnov, and R. G. Yudt. 1981. The history, current state, and prospects for tiger conservation in the Sikhote-Alinsky Zapovednik. In *Khischnye mlekopitaiuschie. Sbornik nauchn. Trudov* (in Russian). CNIL Glavokhoty RF, Moscow, Russia.

Matyushkin, E. N., D. G. Pikunov, Y. M. Dunishenko, D. G. Miquelle, I. G. Nikolaev, E. N. Smirnov, G. P. Salkina, et al. 1999. Distribution and numbers of Amur tigers in the Russian Far East in the mid-1990s. Pages 242–271 *in* A. A. Aristova, editor. *Rare mammal species of Russia and neighbouring territories.* Russian Academy of Sciences Therological Society, Moscow, Russia (in Russian).

May, R. M. 1974. Models for two interacting populations. Pages 78–104 *in* R. M. May, editor. *Theoretical ecology: principles and applications.* Sinauer Associates, Sunderland, Massachusetts, USA.

May, R. M. 1973. *Stability and complexity in model ecosystems.* Monographs in Population Biology. Princeton University Press, Princeton, New Jersey, USA.

Mazák, V. 1981. *Panthera tigris. Mammalian Species Accounts* 152: 1–8.

Mbano, B. N. N., R. C. Malpas, M. K. S. Maige, P. A. K. Symonds, and D. M. Thompson. 1995. The Serengeti regional conservation strategy. Pages 605–616 *in* A. R. E. Sinclair and P. Arcese, editors. *Serengeti II: dynamics, management, and conservation of an ecosystem.* University of Chicago Press, Chicago, Illinois, USA.

McCabe, T. R., and R. E. McCabe. 1997. Recounting whitetails past. Pages 11–26 *in* W. J. McShea, H. B. Underwood, and J. H. Rappole, editors. *The science of overabundance; deer ecology and management.* Smithsonian Institution Press, Washington, District of Columbia, USA.

McClanahan, T. R. 1988. Coexistence in a sea urchin guild and its implications to coral reef diversity and degradation. *Oecologia* 77: 210–218.

McClanahan, T. R. 1994. Kenyan coral reef lagoon fish: effects of fishing, substrate complexity, and sea urchins. *Coral Reefs* 13: 231–241.

McClanahan, T. R. 1997. Primary succession of coral-reef algae: differing patterns on fished versus unfished reefs. *Journal of Experimental Marine Biology and Ecology* 218: 77–102.

McClanahan, T. R. 1998. Predation and the distribution and abundance of tropical sea urchin populations. *Journal of Experimental Marine Biology and Ecology* 221: 231–255.

McClanahan, T. R. 1999. Predation and the control of the sea urchin *Echinometra viridis* and fleshy algae in the path reefs of Glovers Reef, Belize. *Ecosystems* 2: 511–523.

McClanahan, T. R. 2000. Recovery of the coral reef keystone predator, *Balistapus undulatus*, in East African marine parks. *Biological Conservation* 94: 191–198.

McClanahan, T. R. 2002. A comparison of the ecology of shallow subtidal gastropods between western Indian Ocean and Caribbean coral reefs. *Coral Reefs* 21: 399–406.

McClanahan, T. R., and B. Kaunda-Arara. 1996. Fishery recovery in a coral-reef marine park and its effect on the adjacent fishery. *Conservation Biology* 10: 1187–1199.

McClanahan, T. R., and J. Maina. 2003. Response of coral assemblages to the interaction between natural temperature variation and rare warm-water events. *Ecosystems* 6: 551–563.

McClanahan, T. R., and N. A. Muthiga. 1989. Patterns of predation on a sea urchin, *Echinometra mathaei* (de Blainville), on Kenyan coral reefs. *Journal of Experimental Marine Biology and Ecology* 126: 77–94.

McClanahan, T. R., and N. A. Muthiga. 1998. An ecological shift in a remote coral atoll of Belize over 25 years. *Environmental Conservation* 25: 122–130.

McClanahan, T. R., and S. H. Shafir. 1990. Causes and consequences of sea urchin abundance and diversity in Kenyan coral reef lagoons. *Oecologia* 83: 362–370.

McClanahan, T. R., V. Hendrick, M. J. Rodrigues, and N. V. C. Polunin. 1999a. Varying responses of herbivorous and invertebrate-feeding fishes to macroalgal reduction on a coral reef. *Coral Reefs* 18: 195–203.

McClanahan, T. R., N. A. Muthiga, A. T. Kamukuru, H. Machano, and R. Kiambo. 1999b. The effects of marine parks and fishing on the coral reefs of northern Tanzania. *Biological Conservation* 89: 161–182.

McClanahan, T. R., N. A. Muthiga, and S. Mangi. 2001a. Coral and algal response to the 1998 coral bleaching and mortality: interaction with reef management and herbivores on Kenyan reefs. *Coral Reefs* 19: 380–391.

McClanahan, T. R., M. McField, M. Huitric, K. Bergman, E. Sala, M. Nystrom, I. Nordemer, T. Elfwing, and N. A. Muthiga. 2001b. Responses of algae, corals and fish to the reduction of macro algae in fished and unfished patch reefs of Glovers Reef Atoll , Belize. *Coral Reefs* 19: 367–379.

McClanahan, T. R., B. A. Cokos, and E. Sala. 2002. Algal growth and species composition under experimental control of herbivory, phosphorus and coral abundance in Glovers Reef, Belize. *Marine Pollution Bulletin* 44: 441–451.

McClennen, N., R. R. Wigglesworth, S. H. Anderson, and D. G. Wachob. 2001. The effect of suburban and agricultural development on the activity patterns of coyotes (*Canis latrans*). *American Midland Naturalist* 146: 27–36.

McCown, J. W., D. S. Maehr, and J. Roboski. 1990. A portable cushion as a wildlife capture aid. *Wildlife Society Bulletin* 18: 34–36.

McCullough, D. R. 1979. *The George Reserve deer herd: population ecology of a K-selected species.* University of Michigan Press, Ann Arbor, Michigan, USA.

McCullough, D. R. 1992. Concepts of large herbivore population dynamics. Pages 967–984 *in* D. R. McCullough and R. H. Barrett, editors. *Wildlife 2001: populations.* Elsevier Science Publishers Ltd., London, UK.

McCullough, D. R. 1997. Irruptive populations of ungulates. Pages 69–98 *in* W. J. McShea, H. B. Underwood, and J. H. Rappole, editors. *The science of overabundance; deer ecology and management.* Smithsonian Institution Press, Washington, District of Columbia, USA.

McCullough, D. R. 1999. Life-history strategies of ungulates. *Journal of Mammalogy* 79: 1130–1146.

McCullough, D. R. 2001. Population manipulations of North American deer *Odocoileus* spp.; balancing high yield with sustainability. *Wildlife Biology* 7:161–170.

McCullough, D. R., K. W. Jennings, N. B. Gates, B. G. Elliott, and J. E. DiDonato. 1997. Overabundant deer populations in California. *Wildlife Society Bulletin* 25: 478–483.

McDonald, J. N. 1984. The reordered North American selection regime and late Quarternary megafaunal extinctions. Pages 404–439 *in* P. Martin and R. G. Klein, editors. *Quarternary extinctions: a prehistoric revolution.* University of Arizona Press, Tucson, USA.

McDonnell, M. J., and T. A. Pickett. 1993. *Humans as components of ecosystems: the ecology of subtle human effects and populated areas.* Spring-Verlag, New York, New York, USA.

McInnes, P. F., R. J. Naiman, J. Pastor, and Y. Cohen. 1992. Effects of moose browsing on vegetation and litter of the boreal forest, Isle Royale, Michigan, USA. *Ecology* 73: 2059–2075.

McLaren, B. E., and R. O. Peterson. 1994. Wolves, moose and tree rings on Isle Royale. *Science* 266: 1555–1558.

McNaught, D. C. 1999. The indirect effects of macroalgae and micropredation on the postsettlement success of the green sea urchin in Maine. Ph.D. dissertation. University of Maine, Orono, Maine, USA.

McNaughton, S. J. 1984. Grazing lawns: animals in herds, plant form, and coevolution. *American Naturalist* 124: 863–886.

McNaughton, S. J., R. W. Ruess, and S. W. Seagle. 1988. Large mammals and process dynamics in African ecosystems. *BioScience* 38: 794–800.

McNaughton, S. J., M. Oesterheld, D. A. Frank, and K. J. Williams. 1989. Ecosystem-level patterns of primary productivity and herbivory in terrestrial habitats. *Nature* 341: 101–105.

McNeely, J. A. 2000. Practical approaches for including mammals in biodiversity conservation. Pages 355–367 *in* A. Entwistle and N. Dunstone, editors. *Priorities for the conservation of mammalian diversity. Has the panda had its day?* Cambridge University Press, Cambridge, UK.

McPhaden, M. J. 1999. Genesis and evolution of the 1997–98 El Niño. *Science* 283: 950–954.

McShea, W. J. 2000. The influence of acorn crops on annual variation in rodent and bird populations. *Ecology* 81:228–238.

McShea, W. J., and W. M. Healy. 2002. *Oak forest ecosystems: ecology and management for wildlife.* Johns Hopkins University Press, Baltimore, Maryland, USA.

McShea, W. J., and J. H. Rappole. 2000. Managing the abundance and diversity of breeding bird populations through manipulation of deer populations. *Conservation Biology* 14: 1161–1170.

McShea, W. J., M. V. McDonald, E. S. Morton, R. Meier, and J. H. Rappole. 1995. Long-term trends in habitat selection by Kentucky Warblers. *Auk* 112: 375–381.

McShea, W. J., H. B. Underwood, and J. H. Rappole, editors. 1997. *The science of overabundance: deer ecology and population management.* Smithsonian Institution Press, Washington, District of Columbia, USA.

Mduma, S. A. R., A. R. E. Sinclair, and R. Hilborn. 1999. Food regulates the Serengeti wildebeest: a 40-year record. *Journal of Animal Ecology* 68: 1101–1122.

Meagher, M. 1973. *The bison of Yellowstone National Park.* National Park Service Scientific Monograph Series 1, Washington, District of Columbia, USA.

Meagher, M. 1989. Range expansion by bison of Yellowstone National Park. *Journal of Mammalogy* 70: 670–675.

Mech, L. D. 1974. *Canis lupus. Mammalian Species Accounts* 37: 1–6.

Mech, L. D. 1977. Wolf-pack buffer zones as prey reservoirs. *Science* 198: 320–321.

Mech, L. D. 1981. *The wolf: the ecology and behavior of an endangered species.* University of Minnesota Press, Minneapolis, Minnesota, USA.

Mech, L. D. 1984. Predators and predation. Pages 189–200 *in* L. K. Halls, editor. *White-tailed deer ecology and management.* Stackpole Books, Harrisburg, Pennsylvania, USA.

Mech, L. D. 1996. A new era for carnivore conservation. *Wildlife Society Bulletin* 24: 397–401.

Mech, L. D., and M. E. Nelson. 2000. Do wolves affect white-tailed deer buck harvest in northeastern Minnesota? *Journal of Wildlife Management* 64: 129–136.

Mech, L. D., and R. O. Peterson. 2003. Wolf–prey relations. Pages 131–160 *in* L. D. Mech and L. Boitani, editors. *Wolves: behavior, ecology, and conservation.* University of Chicago Press, Chicago, USA.

Mech, L. D., L. G. Adams, T. J. Meier, J. W. Burch, and B. W. Dale. 1998. *The wolves of Denali.* University of Minnesota Press, Minneapolis, Minnesota, USA.

Mech, L. D., D. W. Smith, K. M. Murphy, and D. R. MacNulty. 2001. Winter severity and wolf predation on a formerly wolf-free elk herd. *Journal of Wildlife Management* 65: 998–1003.

Meffe, G. K., and C. R. Carroll. 1997. *Principles of conservation biology.* 2nd edition. Sinauer Associates, Inc., Sunderland, Massachusetts, USA.

Melgaard, M. 1988. The Greenland caribou: zoogeography, taxonomy, and population dynamics. *Journal of Archaeological Science* 15: 348–349.

Menge, B. A. 1995. Indirect effects in marine rocky intertidal interaction webs: patterns and importance. *Ecological Monographs* 65: 21–74.

Menge, B. A. 2003. The overriding importance of environmental context in determining the outcome of species-deletion experiments. Pages 16–43 *in* P. Kareiva and S. A. Levin, editors. *The importance of species.* Princeton University Press, Princeton, New Jersey, USA.

Menge, B. A., and J. P. Sutherland. 1976. Species diversity gradients: synthesis of the roles of predation, competition, and temporal heterogeneity. *American Naturalist* 110: 351–369.

Menge, B. A., and J. Sutherland. 1987. Community regulation: variation in disturbance, competition, and predation in relation to environmental stress and recruitment. *American Naturalist* 130: 730–757.

Menge, B., E. Berlow, C. Blanchette, S. Navarrete, and S.Yamada. 1994. The keystone species concept: variation in interaction strength in a rocky intertidal habitat. *Ecological Monographs* 64: 249–286.

Mercer, W. E., and B. E. McLaren. 2002. Evidence of carrying capacity effects in Newfoundland moose. *Alces* 38: 123–124.

Meserve, P. L., D.A. Kelt, B. Milstead, and J. R. Guitterez. 2003. Thirteen years of shifting top-down and bottom-up control. *BioScience* 53: 633–646.

Messier, F. 1991. The significance of limiting and regulating factors on the demography of moose and white-tailed deer. *Journal of Animal Ecology* 60: 377–393.

Messier, F. 1994. Ungulate population models with predation: a case study with the North American moose. *Ecology* 75: 478–488.

Messier, F. 1995. On the functional and numerical responses of wolves to changing prey density. Pages 187–198 *in* L. N. Carbyn, S. H. Fritts, and D. R. Seip, editors. *Ecology and conservation of wolves in a changing world.* Canadian Circumpolar Institute, Occasional Publication Series, Number 35. Canadian Circumpolar Institute, Edmonton, Alberta, Canada.

Messier, F., J. Huot, D. LeHenaff, and S. Lettich. 1988. Demography of the George River caribou herd : evidence of population regulation by forage exploitation and range expansion. *Arctic* 41: 279–287.

Messmer, T. A., D. Reiter, and B. C. West. 2001. Enhancing wildlife sciences' linkage to public policy: lessons from the predator-control pendulum. *Wildlife Society Bulletin* 29: 1253–1259.

Micheli, F., G. A. Polis, P. D. Boersma, M. A. Hixon, E. A. Norse, P. V. R. Snelgrove, and M. E. Soulé. 2000. Human alteration of food webs. Pages 31–57 *in* M. E. Soulé and G. H. Orians, editors. *Conservation biology: research priorities for the next decade.* Island Press, Washington, District of Columbia, USA.

Miller, M. W., M. E. Hay, S. L. Miller, D. Malone, E. E. Sotka, and A. M. Szmant. 1999. Effects of nutrients versus herbivores on reef algae: a new method for manipulating nutrients on coral reefs. *Limnology and Oceanography* 44: 1847–1861.

Miller, S. G., S. P. Bratton, and J. Hadidian. 1992. Impacts of white-tailed deer on endangered and threatened vascular plants. *Natural Areas Journal* 12: 67–74.

Mills, M. G. L. 1984. Prey selection and feeding habits of the large carnivores in the southern Kalahari. *Koedoe* 27 (Supplement): 281–294.

Mills, M. G. L. 1989. Comparative behavioral ecology of hyenas: the importance of diet and food dispersion. Pages 125–142 in J. L. Gittleman, editor. *Carnivore behavior, ecology, and evolution.* Volume 1. Cornell University Press, Ithaca, New York, USA.

Mills, M. G. L. 1990. *Kalahari hyaenas.* Unwin Hyman, London, UK.

Mills, M. G. L., and H. C. Biggs. 1993. Prey apportionment and related ecological relationships between large carnivores in Kruger National Park. *Symposia of the Zoological Society of London* 65: 253–268.

Mills, M. G. L., and P. J. Funston. 2003. Large carnivore ecology and savanna heterogeneity. Pages 370–388 in J. Du Toit, K. H. Rogers, and H. C. Biggs, editors. *The Kruger experience.* Island Press, Washington, District of Columbia, USA.

Mills, M. G. L., and M. L. Gorman. 1997. Factors affecting the density and distribution of wild dogs in the Kruger National Park. *Conservation Biology* 11: 1397–1406.

Mills, M. G. L., and P. F. Retief. 1984. The response of ungulates to rainfall along the riverbeds of the southern Kalahari, 1972–1982. *Koedoe* 27 (Supplement): 129–142.

Mills, M. G. L., and T. M. Shenk. 1992. Predator–prey relationships: the impact of lion predation on wildebeest and zebra populations. *Journal of Animal Ecology* 61: 693–702.

Mills, M. G. L., H. C. Biggs, and I. J. Whyte. 1995. The relationship between rainfall, lion predation and population trends in African herbivores. *Wildlife Research* 22: 75–88.

Mills, M. G. L., S. Ellis, R. Woodroffe, D. Macdonald, P. Fletcher, M. Bruford, D. Wildt, and U. S. Seal, editors. 1998. *African wild dog (Lycaon pictus) population and habitat viability assessment.* IUCN/SSC Conservation Breeding Specialist Group, Apple Valley, California, USA.

Mills, M. G. L., S. Freitag, and A. S. Van Jaarsveld 2001. Geographic priorities for carnivore conservation in Africa. Pages 467–483 in J. L. Gittleman, S. M. Funk, D. W. Macdonald, and R. K. Wayne, editors. *Carnivore conservation.* Cambridge University Press, Cambridge, UK.

Mills, M. G. L., J. Lubchenco, W. Robertson IV, H. C. Biggs, and D. Mabunda. 2003. Reflections on the Kruger experience and reaching forward. Pages 488–501 in J. Du Toit, K. H. Rogers, and H. C. Biggs, editors. *The Kruger experience.* Island Press, Washington, District of Columbia, USA.

Mitchell, F. J. G., and E. Cole. 1998. Reconstruction of long-term successional dynamics of temperate woodland in Białowieża Forest, Poland. *Journal of Ecology* 86: 1042–1059.

Mittermeier, R. A., N. Myers, J. B. Thomsen, G. A. DaFonseca, and S. Oliver. 1998. Biodiversity hotspots and major tropical wilderness areas: approaches to setting conservation priorities. *Conservation Biology* 12: 516–520.

Miquelle, D. 1998. Tigers overestimated. *Oryx* 32: 15.

Miquelle, D. G., E. N. Smirnov, H. B. Quigley, M. G. Hornocker, I. G. Nikolaev, and E. N. Matyushkin. 1996. Food habits of Amur tigers in Sikhote-Alin Zapovednik and the Russian Far East, and implications for conservation. *Journal of Wildlife Research* 1: 138–147.

Miquelle, D. G., E. N. Smirnov, R. W. Merrill, A. E. Myslenkov, H .B. Quigley, M. G. Hornocker, and B. Schleyer. 1999. Hierarchical spatial analysis of Amur tiger relationships to habitat and prey. Pages 71–99 in J. Seidensticker, S. Christie, and P. Jackson, editors. *Riding the tiger: tiger conservation in human-dominated landscapes.* Cambridge University Press, Cambridge, UK.

Miquelle, D. G., I. G. Nikolaev, J. M. Goodrich, V. P. Litvinov, E. N. Smirnov, and E. Suvorov. In press. *Searching for the coexistence recipe: a case study of conflicts between people and tigers in the Russian Far East.* Cambridge University Press, Cambridge, UK.

Mladenoff, D. J., and F. Stearns. 1993. Eastern hemlock regeneration and deer browsing in the Northern Great Lakes Region: a reexamination and model stimulation. *Conservation Biology* 7: 889–900.

Mladenoff, D. J., R. G. Haight, T. A. Sickley, and A. P. Wydeven. 1997. Causes and implications of species restoration in altered ecosystems. *BioScience* 47: 21–31.

Mladenoff, D. J., T. A. Sickley, and A. P. Wydeven. 1999. Predicting grey wolf landscape recolonization: logistic regression models vs. new field data. *Ecological Applications* 9: 37–44.

Moen, A., L. S. Nilsen, D. I. øien, and T. Arnesen. 1999. Outlying haymaking lands at Sølendet, central Norway: effects of scything and grazing. *Norsk Geografisk Tidsskrift* 53: 93–102.

Molvar, E. M., and R. T. Bowyer. 1994. Costs and benefits of group living in a recently social ungulate: the Alaskan moose. *Journal of Mammalogy* 75: 621–630.

Molvar, E. M., R. T. Bowyer, and V. Van Ballenberghe. 1993. Moose herbivory, browse quality, and nutrient cycling in an Alaskan treeline community. *Oecologia* 94: 472–479.

Morecroft, M. D., M. E. Taylor, S. A. Ellwood, and S. A. Quinn. 2001. Impacts of deer herbivory on ground vegetation at Wytham Woods, central England. *Forestry* 74: 251–257.

Morin, P. J., and S. P. Lawler. 1995. Food web architecture and population dynamics: theory and empirical evidence. *Annual Review of Ecology and Systematics* 26: 505–529.

Mouritsen, K. N., and R. Poulin. 2002. Parasitism, community structure and biodiversity in intertidal systems. *Parisitology* 124: S101–S117.

Munro, J. L. 1996. The scope of tropical reef fisheries and their management. Pages 1–14 in N. V. C. Polunin and C. M. Roberts, editors. *Reef fisheries.* Chapman and Hall, London, UK.

Murdoch, W. W. 1966. Community structure, population control, and competition: a critique. *American Naturalist* 100: 219–226.

Murphy, K. M. 1998. The ecology of the cougar (*Puma concolor*) in the northern Yellowstone ecosystem: interactions with prey, bears, and humans. Ph.D. dissertation. University of Idaho, Idaho, USA, and Moscow, Russia.

Musiani, M., and P. C. Paquet. 2004. The practices of wolf persecution, protection, and restoration in Canada and the United States. *BioScience* 54: 50–60.

Musters, G. C. 1964. *Vida entre los patagones*. Editorial Solar/Hachette, Buenos Aires, Argentina.

Myers, N., and A. H. Knoll. 2001. The biotic crisis and the future of evolution. *Proceedings of the National Academy of Sciences* 98: 5389–5392.

Myers, N., R. A. Mittermeier, C. G. Mittermeier, G. A. B. da Fonesca, and J. Kent. 2000. Biodiversity hotspots for conservation priorities. *Nature* 403: 853–858.

Myers, R. A., and B. Worm. 2003. Rapid worldwide depletion of predatory fish communities. *Nature* 423: 280–283.

Nagy, K. A., I. A. Girard, and T. K. Brown. 1999. Energetics of free-ranging mammals, reptiles, and birds. *Annual Review of Nutrition* 19: 247–277.

Naiman, R. J., G. Pinay, C. A. Johnston, and J. Pastor. 1994. Beaver influences on the long-term biogeochemical characteristics of boreal forest drainage networks. *Ecology* 75: 905–921.

National Research Council (NRC). 1996. *The Bering Sea ecosystem*. National Academy Press, Washington, District of Columbia, USA.

National Research Council (NRC). 1997. *Wolves, bears, and their prey in Alaska: biological and social challenges in wildlife management*. National Academy Press, Washington, District of Columbia, USA.

National Research Council (NRC). 2003. *The decline of the Steller sea lion in Alaskan waters: untangling food webs and fishing nets*. National Academy Press, Washington, District of Columbia, USA.

Nellemann, C., L. Kullerud, I. Vistnes, B. C. Forbes, T. Foresman, E. Husby, G. P. Kofinas, et al. 2001. *GLOBIO. Global methodology for mapping human impacts and the biosphere*. United Nations Environment Program, Environmental Information and Assessment Technical Report UNEP/DEWA/TR.01–3.

Nelson, G. 1994. *The trees of Florida*. Pineapple Press, Sarasota, Florida, USA.

Nelson, M. E., and L. D. Mech. 1981. Deer social organization and wolf predation in northeastern Minnesota. *Wildlife Monographs* 77: 1–53.

Nemethy, A. 2002. Cry wolf. *Yankee* (October): 57–63.

Nietvelt, C. G. 2001. Herbivory interactions between beaver (*Castor canadensis*) and elk (*Cervus elaphus*) on willow (*Salix* spp.) in Banff National Park, Alberta. Master's thesis, University of Alberta, Edmonton, Alberta, Canada.

Norberg, J. 2000. Resource-niche complementarity and autotrophic compensation determines ecosystem-level responses to increased cladoceran species richness. *Oecologia* 122: 264–272.

Norse, E., editor. 1993. *Global marine biological diversity: a strategy for building conservation into decision making*. Island Press, Washington, District of Columbia, USA.

Northridge, S. P., and R. J. Hofman. 1999. Marine mammal interactions with fisheries. Pages 99–119 in J. R. Twiss, Jr. and R. R. Reeves, editors. *Conservation and management of marine mammals.* Smithsonian Institution Press, Washington, District of Columbia, USA.

NoSnare. 2003. Available at http://www.nosnare.org (accessed October 2004).

Noss, A. J. 1998. The impacts of cable snare hunting on wildlife populations in the forests of the CAR. *Conservation Biology* 12: 390–98.

Noss, R. F. 1990. Indicators for monitoring biodiversity: a hierarchial approach. *Conservation Biology* 4: 355–364.

Noss, R. F. 1996. Ecosystems as conservation targets. *Trends in Ecology and Evolution* 11: 351.

Noss, R. F. 2000. Maintaining the ecological integrity of landscapes and ecoregions. Pages 191–208 in D. Pimentel, L. Westra, and R. F. Noss, editors. *Ecological integrity: integrating environment, conservation and health.* Island Press, Washington, District of Columbia, USA.

Noss, R. F. 2001. Why restore large mammals? Pages 1–21 in D. S. Maehr, R. F. Noss, and J. L. Larkin, editors. *Large mammal restoration: ecological and sociological challenges in the 21st century.* Island Press, Washington, District of Columbia, USA.

Noss, R. 2003. A checklist for Wildlands Network Designs. *Conservation Biology* 17: 1–7.

Noss, R. F., H. B. Quigley, M. G. Hornocker, T. Merrill, and P. C. Paquet. 1996. Conservation biology and carnivore conservation in the Rocky Mountains. *Conservation Biology* 10: 949–963.

Noss, R. F., J. R. Strittholt, K. Vance-Borland, C. Carroll, and P. Frost. 1999. A conservation plan for the Klamath-Siskiyou ecoregion. *Natural Areas Journal* 19: 392–411.

Novaro, A. J. 1997. Source-sink dynamics induced by hunting: case study of culpeo foxes on rangelands in Patagonia, Argentina. Ph.D. dissertation, University of Florida, Gainesville, Florida, U.S.A.

Novaro, A. J., M. C. Funes, M. B. Bongiorno, O. B. Monsalvo, E. Donadío, R. S. Walker, G. Sanchez, and O. Pailacura. 1999. *Proyecto integrado de investigación sobre especies predadoras y perjudiciales de la producción ganadera en la provincia del Neuquén.* I Report to Secretaría de Agricultura y Ganadería de la Nación. Centro de Ecología Aplicada del Neuquén.

Novaro, A. J., M. C. Funes, and R. S. Walker. 2000. Ecological extinction of native prey of a carnivore assemblage in Argentine Patagonia. *Biological Conservation* 92: 25–33.

Novaro, A. J., M. C. Funes, E. Donadío, O. Pailacura, G. Sanchez, O. B. Monsalvo, M. Monteverde, M. V. Pía, and R. S. Walker. 2003. *Dinámica poblacional de la liebre europea y el zorro colorado en el noroeste de Patagonia: efectos de la depredación y la cacería.* Report to the Secretaría de Estado de Coordinación y Producción, Ministerio de Jefatura de Gabinete, Provincia del Neuquén.

Novaro A. J., M. C. Funes, and Jiménez, J. 2004. Selection for introduced prey and conservation of culpeo and chilla foxes in Patagonia. Pages 243–254 in D. W. MacDonald and

C. Sillero, editors. *The biology and conservation of wild canids*. Oxford University Press, Oxford, UK.

Nowak, R. M. 2002. The original status of wolves in Eastern North America. *Southeastern Naturalist* 1: 95–130.

Nowell, K., and P. Jackson. 1996. *Wild cats: status survey and conservation action plan*. IUCN, Gland, Switzerland.

O'Donoghue, M., S. Boutin, C. J. Krebs, G. Zuleta, D. L. Murray, and E. J. Hofer. 1998. Functional responses of coyotes and lynx to the snowshoe hare cycle. *Ecology* 79: 1193–1208.

O'Donoghue, M., S. Boutin, D. L. Murray, C. J. Krebs, E. J. Hofer, U. Breitenmoser, C. Breitenmoser-Wuersten, G. Zuleta, C. Doyle, and V. O. Nams. 2001a. Coyotes and lynx. Pages 275–323 *in* C. J. Krebs, S. Boutin, and R. Boonstra, editors. *Ecosystem dynamics of the boreal forest: the Kluane project*. Oxford University Press, New York, New York, USA.

O'Donoghue, M., S. Boutin, E. J. Hofer, and R. Boonstra. 2001b. Other mammalian carnivores. Pages 324–336 *in* C. J. Krebs, S. Boutin, and R. Boonstra, editors. *Ecosystem dynamics of the boreal forest: the Kluane project*. Oxford University Press, New York, New York, USA.

Odum, E. P. 1953. *Fundamentals of ecology*. W. B. Saunders, Philadelphia, Pennsylvania, USA.

Ojeda, F. P., and J. H. Dearborn. 1991. Feeding ecology of benthic mobile predators: experimental analyses of their influence in rocky subtidal communities of the Gulf of Maine. *Journal of Experimental Marine Biology and Ecology* 149: 13–44.

Ojeda, F., and B. Santelices. 1984. Ecological dominance of *Lessonia nigrescens* (Phaeophyta) in central Chile. *Marine Ecology Progress Series* 19: 83–91.

Ojeda, R. A., and M. A. Mares. 1982. Conservation of South American mammals: Argentina as a paradigm. Pages 505–521 *in* M. A. Mares and H. H. Genoways, editors. *Mammalian biology in South America*. Pymatuning Symposia in Ecology, volume 6. Special Publication Series. University of Pittsburgh, Pittsburgh, Pennsylvania, USA.

Okarma, H., B. Jędrzejewska, W. Jędrzejewski, Z. A. Krasiński, and L. Miłkowski. 1995. The roles of predation, snow cover, acorn crop, and man-related factors on ungulate mortality in Białowieża Primeval Forest, Poland. *Acta Theriologica* 40: 197–217.

Okarma, H., W. Jędrzejewski, K. Schmidt, R. Kowalczyk, and B. Jędrzejewska. 1997. Predation of Eurasian lynx on roe deer and red deer in Białowieża Primeval Forest, Poland. *Acta Theriologica* 42: 203–224.

Okarma, H., W. Jędrzejewski, K. Schmidt, S. Śnieżko, A. N. Bunevich, and B. Jędrzejewska. 1998. Home ranges of wolves in Białowieża Primeval Forest, Poland, compared with other Eurasian populations. *Journal of Mammalogy* 79: 842–852.

Okarma, H., Y. Dovhanych, S. Findo, O. Ionescu, P. Kaubek, and L. Szemethy. 2000. Status of carnivores in the Carpathian ecoregion. Report of the Carpathian Ecoregion Initiative: 1–37.

Oksanen, L. 1990. Predation, herbivory, and plant strategies along gradients of primary productivity. Pages 445–474 in J. B. Grace and D. Tilman, editors. *Perspective on plant competition.* Academic Press, New York, New York, USA.

Oksanen, L., and T. Oksanen. 2000. The logic and realism of the hypothesis of exploitation. *American Naturalist* 155: 703–723.

Oksanen, L, S. D. Fretwell, J. Arruda, and P. Niemelä. 1981. Exploitation of ecosystems in gradients of primary productivity. *American Naturalist* 118: 240–261.

Oksanen, M. 1997. The moral value of biodiversity. *Ambio* 26: 541–545.

Oliveria, J. F. S., S. J. Passos de Carvalho, R. M. Giblin-Davis, and F. W. Howard. 1989. Vulnerability of stressed palms to attack by *Rhynchophorus cruentatus* (Coleoptera: Curculionidae) and insecticidal control of the pest. *Journal of Economic Entomology* 82: 1185–1190.

Olson, D. M., and E. Dinerstein. 2002. The Global 200: priority ecoregions for global conservation. *Annals of the Missouri Botanical Gardens* 89: 199–224.

Opitz, S. 1996. *Trophic interactions in Caribbean coral reefs.* International Center for Living Aquatic Resources Management, Makati, City Philippines.

Orians, G .H., P. A. Cochran, J. W. Duffield, T. K. Fuller, R. J. Gutierrez, W. M. Hanemann, F. C. James, et al. 1997. Wolves, bears, and their prey in Alaska: biological and social challenges in wildlife management. National Academy Press, Washington, District of Columbia, USA.

Ostfeld, R. S. 1997. The ecology of Lyme-disease risk. *American Scientist* 85:338–346.

Ostfeld, R. S. 2002. Ecological webs involving acorns and mice: basic research and its management implications. Pages 196–214 in W. J. McShea and W. M. Healy, editors. *Oak forest ecosystem: ecology and management for wildlife.* Johns Hopkins University Press, Baltimore, Maryland, USA.

Ostfeld, R. S., and R. D. Holt. 2004. Are predators good for your health? Evaluating evidence for top-down regulation of zoonotic disease reservoirs. *Frontiers in Ecology and Environment* 2: 13–20.

Ostfeld, R. S., and F. Keesing. 2000. Biodiversity and disease risk: the case of Lyme disease. *Conservation Biology* 14: 722–728.

Ostrander, G. K., K. M. Armstrong, E. T. Knobbe, D. Gerace, and E. P. Scully. 2000. Rapid transition in the structure of a coral reef community: the effects of coral bleaching and physical disturbance. *Ecology* 97: 5297–5302.

Otten, C.F., editor. 1986. *A lycanthropy reader. werewolves in Western culture.* Syracuse University Press, Syracuse, New York, USA.

Owen-Smith, R. N. 1988. *Megaherbivores: the influence of very large body size on ecology.* Cambridge University Press, Cambridge, UK.

Pace, M. L., J. J. Cole, S. R. Carpenter, and J. F. Kitchell. 1999. Trophic cascades revealed in diverse ecosystems. *Trends in Ecology and Evolution* 14: 483–488.

Packer, C. 1983. Sexual dimorphism: the horns of African antelope. *Science* 221: 1191–1193.

Packer, C., A. E. Pusey, H. Rowley, D. A. Gilbert, J. Martenson, and S. J. O'Brien. 1991. Case study of a population bottleneck: lions of the Ngorongoro Crater. *Conservation Biology* 5: 219–230.

Paine, R. T. 1966. Food web complexity and species diversity. *American Naturalist* 100: 65–75.

Paine, R. T. 1980. Food webs, linkage interaction strength, and community infrastructure. *Journal of Animal Ecology* 49: 667–685.

Paine, R. 1992. Food-web analysis through field measurement of per capita interaction strength. *Nature* 355: 73–75.

Paine, R. T., and R. L. Vadas. 1969. The effect of grazing by sea urchins *Strongylocentrotus* spp. on benthic algal populations. *Limnology and Oceanography* 14: 710–719.

Palma, A. T., R. S. Steneck, and C. Wilson. 1999 Settlement-driven, multiscale demographic patterns of large benthic decapods in the Gulf of Maine. *Journal of Experimental Marine Biology and Ecology* 241: 107–136.

Palomares, F., and T. M. Caro. 1999. Interspecific killing among mammalian carnivores. *American Naturalist* 153: 492–508.

Pandolfi, J. M., R. H. Bradbury, E. Sala, T. P. Hughes, K. A. Bjorndal, R. G. Cooke, D. McArdle, et al. 2003. Global trajectories of the long-term decline of coral reef ecosystems. *Science* 301: 955–958.

Paquet, P.C., and A. Hackman. 1995. *Large carnivore conservation in the Rocky Mountains.* World Wildlife Fund–Canada and World Wildlife Fund–U.S., Toronto, Ontario, Canada and Washington, District of Columbia, USA.

Paradiso, J. L., and R. M. Nowak. 1972. *Canis rufus. Mammalian Species* 22: 1–4.

Parks Canada 1997. State of the parks 1997 report. Available from http://www.parkscanada .gc.ca/docs/pc/rpts/etat-state/state-etat1_e.asp (accessed October 2004).

Parrish, F. A., and R. C. Boland. 2004. Habitat and reef-fish assemblages of banks in the Northwestern Hawaiian Islands. *Marine Biology* 144: 1065–1073.

Pascual, M., and S. A. Levin. 1999. From individuals to population densities: searching for the intermediate scale of nontrivial determinism. *Ecology* 80: 2225–2236.

Pastor, J., and Y. Cohen. 1997. Herbivores, the functional diversity of plant species and the cycling of nutrients in ecosystems. *Theoretical Population Biology* 51: 165–179.

Pastor, J., and R. J. Naiman. 1992. Selective foraging and ecosystem processes in boreal forests. *American Naturalist* 139: 690–705.

Pastor, J., B. Dewey, R. Moen, D. J. Mladenoff, M. White, and Y. Cohen. 1998. Spatial patterns in the moose–forest–soil ecosystem on Isle Royal, Michigan, USA. *Ecological Applications* 8: 411–424.

Pastor, J., B. Dewey, R. J. Naiman, P. F. McInnes, and Y. Cohen. 1993. Moose browsing and soil fertility of Isle Royale National Park. *Ecology* 74: 467–480.

Paulay, G. 1997. Diversity and distribution of reef organisms. Pages 298–353 in C. E. Birkeland, editor. *Life and death of coral reefs.* Chapman and Hall, New York, New York, USA.

Pauly, D., and J. Maclean. 2003. *In a perfect ocean: the state of fisheries and ecosystems in the North Atlantic ocean*. Island Press, Washington, District of Columbia, USA.

Pauly, D., V. Christensen, J. Dalsgaard, R. Froese, and F. Torres, Jr. 1998. Fishing down marine food webs. *Science* 279: 860–863.

Pauly, D., M. L. Palomares, R. Froese, P. Sa-a, M. Vakily, D. Preikshot, and S. Wallace. 2001. Fishing down Canadian aquatic food webs. *Canadian Journal of Fisheries and Aquatic Sciences* 58: 51–62.

Pauly, D., V. Christensen, S. Guénette, T. J. Pitcher, U. R. Sumaila, C. J. Walters, R. Watson, and D. Zeller. 2002. Towards sustainability in world fisheries. *Nature* 418: 689–695.

Pech, R. P., A. R. E. Sinclair, A. E. Newsome, and P. C. Catling. 1992. Limits to predator regulation of rabbits in Australia: evidence from predator-removal experiments. *Oecologia* 89: 102–112.

Pelliza-Sbriller, A. P. Willems, V. Nakamatsu, and A. Manero. 1997. *Atlas dietario de herbívoros patagónicos*. San Carlos de Bariloche, Argentina.

Pennings, S. C. 1997. Indirect interactions on coral reefs. Pages 249–272 *in* C. Birkeland, editor. *Life and death of coral reefs*. Chapman and Hall, New York, New York, USA.

Perkins, J. S., and A. Carr. 1985. The Belize barrier reef: status and prospects for conservation management *Biological Conservation* 31: 291–301.

Perrins, C. M., and R. Overall. 2001. Effects of increasing numbers of deer on bird populations in Wytham woods, central England. *Forestry* 74: 299–309.

Person, D. K., R. T. Bowyer, and V. Van Ballenberghe. 2001. Density dependence of ungulates and functional responses of wolves: effects on predator–prey ratios. *Alces* 37: 253–273.

Persson, L. 1999. Trophic cascades: abiding heterogeneity and the trophic level concept at the end of the road. *Oikos* 85: 385–397.

Peterken, G. F., and C. R. Tubbs. 1965. Woodland regeneration in the New Forest, Hampshire, since 1650. *Journal of Applied Ecology* 2: 159–170.

Peters, R. H. 1983. *The ecological implications of body size*. Cambridge University Press. Cambridge, UK.

Petersen, D. 2002. Rewilding the hunt: a conversation with Michael Soulé. *Bugle* (September/October): 60–71.

Peterson, R. O. 1995. Wolves as interspecific competitors in canid ecology. Pages 315–323 *in* L. N. Carbyn, S. H. Fritts, and D. R. Seip, editors. *Ecology and conservation of wolves in a changing world*. Canadian Circumpolar Institute, University of Alberta, Edmonton, Alberta, Canada.

Peterson, R. O. 1999. Wolf–moose interaction on Isle Royale: the end of natural regulation? *Ecological Applications* 9: 10–16.

Peterson, R. O. 2001. Wolves as top carnivores: new faces in new places. Pages 151–160 *in* Sharpe, V. A., B. G. Norton, and S. Donnelley, editors. *Wolves and human communities: biology, politics, and ethics*. Island Press, Washington, District of Columbia, USA.

Peterson, R. O., and P. Ciucci. 2003. The wolf as a carnivore. Pages 104–130 *in* L. D. Mech and L. Boitani, editors. *Wolves: behavior, ecology, and conservation*. University of Chicago Press, Chicago, Illinois, USA.

Peterson, R. O., and R. E. Page. 1988. The rise and fall of Isle Royale wolves, 1975–1986. *Journal of Mammalogy* 69: 89–99.

Peterson, R. O., N. J. Thomas, J. M. Thurber, J. A. Vucetich, and T. A. Waite. 1998. Population limitation and the wolves of Isle Royale. *Journal of Mammalogy* 79: 828–841.

Pfister, C., and M. Hay. 1988. Associational plant refuges: convergent patterns in marine and terrestrial communities result from differing mechanisms. *Oceologia* 77: 118–129.

Phillips, M. K., and D. W. Smith. 1996. *The wolves of Yellowstone*. Voyageur Press, Stillwater, Minnesota, USA.

Pierce, B. M., V. C. Bleich, J. D. Wehausen, and R. T. Bowyer. 1999. Migratory patterns of mountain lions: implications for social regulation and conservation. *Journal of Mammalogy* 80: 986–992.

Pierce, B. M., V. C. Bleich, and R. T. Bowyer. 2000a. Social organization of mountain lions: does a land-tenure system regulate population size? *Ecology* 81: 1533–1543.

Pierce, B. M., V. C. Bleich, and R. T. Bowyer. 2000b. Selection of mule deer by mountain lions and coyotes: effects of hunting style, body size, and reproductive status. *Journal of Mammalogy* 81: 462–472.

Pinnegar, J. K., N. V. C. Polunin, P. Francour, F. Badalamenti, R. Chemello, M. L. Harmelin-vivien, B. Hereu, et al. 2000. Trophic cascades in benthic marine ecosystems: lessons for fisheries and protected-area management. *Environmental Conservation* 27: 179–200.

Pole, A. 2000. The behaviour and ecology of African wild dogs (*Lycaon pictus*) in an environment with reduced competitor density. Ph.D. thesis. University of Aberdeen, Aberdeen, UK.

Polis, G. A. 1999. Why are parts of the world green? Multiple factors control productivity and the distribution of biomass. *Oikos* 86: 3–15.

Polis, G. A., and R. D. Holt. 1992. Intraguild predation: the dynamics of complex trophic interactions. *Trends in Ecology and Evolution* 7: 151–154.

Polis, G. A., and D. R. Strong. 1996. Food web complexity and community dynamics. *American Naturalist* 147: 813–846.

Polis, G. A., C. A. Myers, and R. D. Holt. 1989. The ecology and evolution of intraguild predation: potential competitors that eat each other. *Annual Review of Ecology and Systematics* 20: 297–330.

Polis, G. A., A. L. W. Sears, G. R. Huxel, D. R. Strong, and J. Maron. 2000. When is a trophic cascade a trophic cascade? *Trends in Ecology and Evolution* 15: 473–475.

Poole, K. G. 1994. Characteristics of an unharvested lynx population during a snowshoe hare decline. *Journal of Wildlife Management* 58: 608–618.

Post, E., N. C. Stenseth, R. O. Peterson, J. A. Vucetich, and A. M. Ellis. 2002. Phase dependence and population cycles in a large-mammal predator–prey system. *Ecology* 83: 2997–3002.

Powell, R. A. 2001. Who limits whom? Predators or prey? *Endangered Species Update* 18: 98–102.

Power, M. E. 1990. Effects of fish in river food webs. *Science* 250: 811–814.

Power M. E. 1992. Top-down and bottom-up forces in food webs: do plants have primacy? *Ecology* 73: 733–746.

Power, M., D. Tilman, J. Estes, B. Menge, W. Bond, L. Mills, G. Daily, J. Castilla, J. Lubchenco and R. Paine. 1996. Challenges in the quest for keystones. *BioScience* 46: 609–620.

Prakash, S., and A. M. de Roos. 2002. Habitat destruction in a simple predator–prey patch model: how predators enhance prey persistence and abundance. *Theoretical Population Biology* 62: 231–249.

Price, A. R. G. 2002. Simultaneous "hotspots" and "coldspots" of marine biodiversity and implications for global conservation. *Marine Ecology Progress Series* 241: 23–27.

Pringle, J. D., G. J. Sharp, and J. F. Caddy, editors. 1980. *Proceedings of the Workshop on the Relationship between sea urchin grazing and commercial plant / animal harvesting.* Canadian Technical Report of Fisheries and Aquatic Sciences, Number 954. Fisheries and Oceans Canada. Halifax, Nova Scotia.

Prins, H. H. T. 1996. *Ecology and behaviour of the African buffalo: social inequality and decision making.* Chapman and Hall, New York, New York, USA.

Pritchard, J. A. 1999. *Preserving Yellowstone's natural conditions: science and the perception of nature.* University of Nebraska Press, Lincoln, Nebraska, USA.

Pucek, Z. 1991. Conservation strategy for European bison. *Proceedings of the International Symposium* (Touluse, France, 2–6 September 1991), *Ongoles / Ungulates* 91: 589–594.

Pucek, Z., W. Jędrzejewski, B. Jędrzejewska, and M. Pucek. 1993. Rodent population dynamics in a primeval deciduous forest (Białowieża National Park) in relation to weather, seed crop, and predation. *Acta Theriologica* 38: 199–232.

Putman, R. J., and N. P. Moore. 1998. Impact of deer in lowland Britain on agriculture, forestry and conservation habitats. *Mammalian Review* 28: 141–164.

Putman, R. J., P. J. Edwards, J. C. E. Mann, R. C. How, and S. D. Hill. 1989. Vegetational and faunal changes in an area of heavily grazed woodland following relief from grazing. *Biological Conservation* 47: 13–32.

Pyare, S., and J. Berger. 2003. Beyond demography and delisting: ecological recovery for Yellowstone's grizzly bears and wolves. *Biological Conservation* 113: 63–73.

Pyare, S., S. Cain, D, Moody, C. Schwartz, and J. Berger. 2004. Distributional changes and re-colonization by grizzly bears: 100 years of change. *Animal Conservation* 7: 1–7.

Quammen, D. 2003. *Monster of God: the man-eating predator in the jungles of history and the mind.* W.W. Norton, New York, New York, USA.

Raedeke, K. J. 1979. Population dynamics and socioecology of the guanaco (*Lama guanicoe*) of Magallanes, Chile. Ph.D. dissertation, University of Washington, Seattle, Washington, USA.

Ramakrishnan, U., R. G. Coss, and N. W. Pelkey. 1999. Tiger decline caused by the reduction of large ungulate prey: evidence from a study of leopard diets in southern India. *Biological Conservation* 89: 113–120.

Randall, J. E. 1967. Food habits of reef fishes of the West Indies. *Studies in Tropical Oceanography* 5: 665–847.

Rao, M. 2000. Variation in leaf-cutter ant (*Atta* sp.) densities in forest isolates: the potential role of predation. *Journal of Tropical Ecology* 16: 209–225.

Rapport, D. J., and W. G. Whitford. 1999. How ecosystems respond to stress: common properties of arid and aquatic systems. *BioScience* 49: 193–203.

Rasa, O. 1983. Dwarf mongoose and hornbill mutualism in the Tara desert, Kenya. *Behavioral Ecology and Sociobiology* 12: 181–190.

Rausch, R. L. 1953. Studies of the helminth fauna of Alaska, XIII: Disease in the sea otter with special reference to helminth parasites. *Ecology* 34: 584–604.

Ray, J. C. 2001. Carnivore biogeography and conservation in the African rainforest: a community perspective. Pages 214–232 *in* W. Weber, L. J. T. White, A. Vedder, and L. Naughton-Treves, editors. *African rainforest ecology and conservation.* Yale University Press. New Haven, Connecticut, USA.

Redford, K. H. 1992. The Empty Forest. *BioScience* 42: 412–422.

Redford, K. H. 2000. Natural areas, hunting, and nature conservation in the neotropics. *Wild Earth* 10: 41–48.

Redford, K., and J. F. Eisenberg 1992. *Mammals of the Neotropics: the Southern Cone.* Volume 2. University of Chicago Press, Chicago, Illinois, USA.

Redford, K. H., and P. Feinsinger. 2001. The half-empty forest: sustainable use and the ecology of interactions. Pages 370–399 *in* J. D. Reynolds, G. M. Mace, K. H. Redford, and J. G. Robinson, editors. *Conservation of exploited species.* Cambridge University Press, Cambridge, London, UK.

Redford, K. H., and B. D. Richter. 1999. Conservation of biodiversity in a world of use. *Conservation Biology* 13: 1246–1256.

Redford, K. H., and J. G. Robinson. 2002. Introduction. Pages 21–24 *in* R. A. Medellin, C. Equihua, C. L. B. Chetkiewicz, P. G. Crawshaw, A. Rabinowitz, K. H. Redford, J. G. Robinson, E. W. Sanderson, and A. B. Taber, editors. *El Jaguar en el Nuevo Milenio.* Fondo de Cultura Economica, Mexico D.F.

Redford, K. H., P. B. Coppolillo, E. W. Sanderson, G. A. B. da Fonesca, E. Dinerstein, C. Groves, G. M. Mace, et al. 2003. Mapping the conservation landscape. *Conservation Biology* 17: 116–131.

Reed, D. C., and M. S. Foster. 1984. The effects of canopy shading on algal recruitment and growth in a giant kelp forest. *Ecology* 65: 937–948.

Reisewitz, S. 2002. Changes in Aleutian Island fish communities associated with recent shifts in nearshore and offshore ecosystems. Master's thesis. Ocean Sciences, University of California, Santa Cruz, California, USA.

Reneker, L. A., and C. C. Schwartz. 1998. Food habitat and feeding behavior. Pages 403–439 *in* A. W. Franzmann and C. C. Schwartz , editors. *Ecology and management of the North American moose.* Smithsonian Institution Press. Washington, District of Columbia, USA.

Rice, W. R. 1989. Analyzing tables of statistical test. *Evolution* 43: 223–225.

Riley, G. A., and R. T. McBride. 1972. A survey of the red wolf (*Canis rufus*). Special Scientific Report: Wildlife, no. 162. U. S. Department of the Interior, Bureau of Sport Fisheries and Wildlife. Washington, District of Columbia, USA.

Riney, T. 1982. *Study and management of large mammals.* John Wiley and Sons, New York, New York, USA.

Ripple, W. J., and R. L. Beschta. 2003. Wolf reintroduction, predation risk, and cottonwood recovery in Yellowstone National Park. *Forest Ecology and Management* 184: 299–313.

Ripple, W. J., and E. J. Larsen. 2000. Historic aspen recruitment, elk, and wolves in northern Yellowstone National Park, USA. *Biological Conservation* 95: 361–370.

Ripple, W. J., E. J. Larsen, R. A. Renkin, and D. W. Smith. 2001. Trophic cascades among wolves, elk, and aspen on Yellowstone National Park's northern range. *Biological Conservation* 102: 227–234.

Roberge, J.-M., and P. Angelstam. 2004. Usefulness of the umbrella species concept as a conservation tool. *Conservation Biology* 18: 76–85.

Roberts, G. R., T. F. Flannery, L. K. Ayliffe, H. Yoshida, J. M. Olley, G. J. Prideaux, G. M. Laslett, et al. 2001. New ages for the last Australian megafauna: continent-wide extinction about 46,000 years ago. *Science* 292: 1888–1892.

Robinson, J. G. 2001. Using sustainable use approaches to conserve exploited populations. Pages 485–498 *in* J. D. Reynolds, G. M. Mace, K. H. Redford, and J. G. Robinson, editors. *Conservation of exploited species.* Cambridge University Press, Cambridge, UK.

Robinson, J. G., and E. I. Bennett, editors. 2000. *Hunting for sustainability in tropical forests.* Columbia University Press, New York, New York, USA.

Robinson, J. G., and K. H. Redford. 1992. *Neotropical wildlife use and conservation.* University of Chicago Press, Chicago, Illinois, USA.

Roemer, G. W., C. J. Donlan, and F. Courchamp. 2002. Golden eagles, feral pigs, and insular carnivores: how exotic species turn native predators into prey. *Proceedings of the National Academy of Sciences* 99: 791–796.

Rohner, C., and C. J. Krebs. 1996. Owl predation on snowshoe hares: consequences of antipredator behaviour. *Oecologia* 108: 303–310.

Rohner, C., F. I. Doyle, and J. N. M. Smith. 2001. Great horned owls. Pages 339–376 *in* C. J. Krebs, S. Boutin, and R. Boonstra, editors. *Ecosystem dynamics of the boreal forest: the Kluane project.* Oxford University Press, New York, New York, USA.

Rolley, R. E., and L. B. Keith. 1980. Moose population dynamics and winter habitat use at Rochester, Alberta, 1965–1979. *Canadian Field Naturalist* 94: 9–18.

Romme, W. H., and D. G. Despain. 1989. Historical perspective on the Yellowstone fires of 1988. *BioScience* 39: 695–699.

Rooney, T. P. 2001. Deer impacts on forest ecosystems: a North American perspective. *Forestry* 74: 201–208.

Rooney, T. P., and D. M. Waller. 2003. Direct and indirect effects of white-tailed deer in forest ecosystems. *Forest Ecology and Management* 181: 165–176.

Rosenfeld, J. S. 2002. Functional redundancy in ecology and conservation. *Oikos* 98: 156–162.

Rosenzweig, M. L. 2003. *Win–win ecology.* Oxford University Press, Oxford, UK.

Ross, P. I., and N. G. Jalkotzy. 1996. Cougar predation on moose in southwest Alberta. *Alces* 32: 1–8.

Rowe, J. S. 1972. *Forest regions of Canada.* Canadian Forest Service Publication no. 1300, Ottawa, Ontario, Canada.

Rowland, M. M., M. J. Wisdom, B. K. Johnson, and J. G. Kie. 2000. Elk distribution and modeling in relation to roads. *Journal of Wildlife Management* 64: 672–684.

Ruesink, J. L., and K. E. Hodges. 2001. Trophic mass flow models of the Kluane Boreal Forest Ecosystem. Pages 463–490 *in* C. J. Krebs, S. Boutin, and R. Boonstra, editors. *Ecosystem dynamics of the boreal forest: the Kluane project.* Oxford University Press, New York, New York, USA.

Ruess, R. W., and S. J. McNaughton. 1987. Grazing and the dynamics of nutrient and energy regulated microbial processes in the Serengeti grasslands. *Oikos* 49: 101–110.

Ruggiero, L. F., K. B. Aubry, S. W. Buskirk, G. M. Koehler, C. J. Krebs, K. S. McKelvey, and J. R. Squires. 2000. *Ecology and conversation of lynx in the United States.* University Press of Colorado, and the USDA, Rocky Mountain Research Station.

Runkle, J. R. 1990. Gap dynamics in an Ohio *Acer-Fagus* forest and speculations on the geography of disturbance. *Canadian Journal of Forest Resources* 20: 632–641.

Runyoro, V. A., H. Hofer, E. B. Chausi, and P. D. Moehlman. 1995. Long-term trends in the herbivore populations of the Ngorongoro Crater, Tanzania. Pages 146–168 *in* A. R. E. Sinclair and P. Arcese, editors. *Serengeti II: Dynamics, Management, and Conservation of an Ecosystem.* University of Chicago Press, Chicago, Illinois, USA.

Russ, G. R., and A. C. Alcala. 2003. Marine reserves: rates and patterns of recovery and decline of predatory fish, 1983–2000. *Ecological Applications* 13: 1553–1565.

Ruth, T. K., D. W. Smith, M. A. Haroldson, P. C. Buotte, C. C. Schwartz, H. B. Quigley, S. Cherry, K. M. Murphy, D. Tyers, and K. Frey. 2003. Large carnivore response to recreational big game hunting along the Yellowstone National Park and Absaroka–Beartooth Wilderness boundary. *Wildlife Society Bulletin* 31: 1150–1161.

Sæther, B.-E. 1997. Environmental stochasticity and population dynamics of large herbivores: a search for mechanisms. *Trends in Ecology and Evolution* 12: 143–149.

Sala, E. 1997. Fish predators and scavengers of the sea urchin *Paracentrotus lividus* in protected areas of the north-west Mediterranean Sea. *Marine Biology* 129: 531–539.

Sala, E., and M. H. Graham. 2002. Community-wide distribution of predator–prey interaction strength in kelp forests. *Proceedings of the National Academy of Sciences* 99: 3678–3683.

Sala, E., and G. Sugihara. 2005. Food web theory provides guidelines for marine conservation. Pages 170–183 *in* A. Belgrano, U. Scharler, J. Dunne, and B. Ulanowicz, editors. *Aquatic Food Webs: An Ecosystem Approach.* Oxford University Press, Oxford, UK.

Sala, E., and M. Zabala. 1996. Fish predation and the structure of the sea urchin *Paracentrotus lividus* (Lamarck) populations in the NW Mediterranean. *Marine Ecology Progress Series* 140: 71–81.

Sala, E., C. F. Boudouresque, and M. Harmelin-Vivien. 1998. Fishing, trophic cascades, and the structure of algal assemblages: evaluation of an old but untested paradigm. *Oikos* 82: 425–439.

Sala, E., E. Ballesteros, and R. M. Starr. 2001. Rapid decline of Nassau grouper spawning aggregations in Belize: fishery management and conservation needs. *Fisheries* 26: 23–30.

Salafsky, N., R. Margoluis, and K. Redford. 2001. *Adaptive management: a tool for conservation practitioners.* Biodiversity Support Program, Washington District of Columbia, USA.

Salafsky, N., R. Margoluis, K. H. Redford, and J. G. Robinson. 2002. Improving the practice of conservation: a conceptual framework and research agenda for conservation science. *Conservation Biology* 16: 1469–1479.

Sall, J., A. Lehmaan, and L. Creighton. 2001. *JMP Start Statistics.* 2nd edition. Brooks Cole, Duxbury/Thomson Learning, Florence, Kentucky, USA.

Sánchez Roig, M., and F. Gómez de la Maza. 1952. *La pesca en Cuba.* Ministerio de Agricultura, República de Cuba, La Habana, Cuba.

Sanderson, E. W., M. Jaiteh, M. A. Levy, K. H. Redford, A. V. Wannebo, and G. Woolmer. 2002a. The human footprint and the last of the wild. *BioScience* 52: 891–904.

Sanderson, E. W., K. H. Redford, C. L. B. Chetkiewicz, R. A. Medellin, A. R. Rabinowitz, J. G. Robinson, and A. B. Taber. 2002b. Planning to save a species: the jaguar as a model. *Conservation Biology* 16: 58–72.

Sanderson, E. W., K. H. Redford, A. Vedder, P. B. Coppolillo, and S. E. Ward. 2002c. A conceptual model for conservation planning based on landscape species requirements. *Landscape and Urban Planning* 58: 41–56.

Santelices, B., and F. P. Ojeda. 1984. Population dynamics of coastal forests of Macrocystis pyrifera in Puerto Toro, Isla Navarino, Southern Chile. *Marine Ecology Progress Series* 14: 175–183.

Saunders, N. J. 1990. Tezcatlipoca: jaguar metaphors and the Aztec mirror of nature. Pages 159–177 *in* R. Willis, editor. *Signifying Animals: Human Meaning in the Natural World.* One World Archaeology vol. 16. Unwin Hyman Ltd., London, UK.

Schaller, G. B. 1967. *The deer and the tiger: a study of wildlife in India.* University of Chicago Press, Chicago, Illinois, USA.

Schaller, G. B. 1972. *The Serengeti lion: a study of predator–prey relations*. University of Chicago Press, Chicago, Illinois, USA.

Scheffer, M., and S. R. Carpenter. 2003. Catastrophic regime shifts in ecosystems: linking theory to observation. *Trends in Ecology and Evolution* 18: 648–656.

Scheffer, M., S. Carpenter, J. A. Foley, C. Folke, B. Walker. 2001. Catastrophic shifts in ecosystems. *Nature* 413: 591–596.

Schiel, D. R. 1990. Macroalgal assemblages in New Zealand: structure, interactions and demography. *Hydrobiologia* 192: 59–76.

Schiel, D. R., N. L. Andrew, and M. S. Foster. 1995. The structure of subtidal algal and invertebrate assemblages at the Chatham Islands, New Zealand. *Marine Biology* 123: 355–367.

Schindler, D. E., T. E. Essington, J. F. Kitchell, C. Boggs, and R. Hilborn. 2002. Sharks and tunas: fisheries impacts on predators with contrasting life histories. *Ecological Applications* 12: 735–748.

Schmiegelow, F. K. A., and M. Mönkkönen. 2002. Habitat loss and fragmentation in dynamic landscapes: avian perspectives from the boreal forest. *Ecological Applications* 12:375–389.

Schmidt, K., W. Jędrzejewski, and H. Okarma. 1997. Spatial organization and social relations in the Eurasian lynx population in Białowieża Primeval Forest, Poland. *Acta Theriologica* 42: 289–312.

Schmitz O. J., A. P. Beckerman, and K. M. O'Brien. 1997. Behaviorally mediated trophic cascades: effects of predation risk on food web interactions. *Ecology* 78: 1388–1399.

Schmitz, O. J., and A. R. E. Sinclair. 1997. Rethinking the role of deer in forest ecosystem dynamics. Pages 201–223 *in* W. J. McShea, H. B. Underwood, and J. H. Rappole, editors. *The science of overabundance: deer ecology and population management*. Smithsonian Institution Press. Washington, District of Columbia, USA.

Schmitz, O. J., P. A. Hambäck, and A. P. Beckerman. 2000. Trophic cascades in terrestrial systems: a review of the effects of carnivore removal on plants. *American Naturalist* 155: 141–153.

Schneider, M. F. 2001. Habitat loss, fragmentation and predator impact: spatial implications for prey conservation. *Journal of Applied Ecology* 38: 720–735.

Schneider, R. R., J. B. Stelfox, S. Boutin, and S. Wasel. 2003. Managing the cumulative impacts of land uses in the Western Canadian sedimentary basin: a modeling approach. *Conservation Ecology* 7: 8.

Schoener, T. W., and D. A. Spiller. 2003. Effects of removing a vertebrate versus an invertebrate predator on a food web, and what is their relative importance? Pages 69–84 *in* P. Kareiva and S. A. Levin, editors. *The importance of species*. Princeton University Press, Princeton, New Jersey, USA.

Schreiner, E. G., K. A. Kruger, P. J. Happe, and D. B. Houston. 1996. Understory patch dynamics and ungulate herbivory in old-growth forests of Olympic National Park, Washington. *Canadian Journal of Forestry Research* 26:255–265.

Schullery, P. 1997. *Searching for Yellowstone: ecology and wonder in the last wilderness.* Houghton Mifflin Company, Boston, Massachusetts, USA.

Schwartz, M. W. 1999. Choosing the appropriate scale of reserves for conservation. *Annual Review of Ecology and Systematics* 30: 83–108.

Schwartz, M. W., C. A. Brigham, J. D. Hoeksema, K. G. Lyons, M. H. Mills, and P. J. van Mantgem. 2000. Linking biodiversity to ecosystem function: implications for conservation ecology. *Oecologia* 122: 297–305.

Scognamillo, D., I. E. Maxit, M. Sunquist, and J. Polisar. 2003. Coexistence of jaguar (*Panthera onca*) and puma (*Puma concolor*) in a mosaic landscape in the Venezuelan llanos. *Journal of Zoology* 259: 269–279.

Seagle, S. W., and S. Y. Liang. 1997. Bottomland forest composition and seedling diversity; simulating succession and browsing by overabundant deer. Pages 346–365 *in* W. J. McShea, H. B. Underwood, and J. H. Rappole, editors. *The science of overabundance: deer ecology and population management.* Smithsonian Institution Press, Washington, District of Columbia, USA.

Seidensticker, J. 1976. On the ecological separation between tigers and leopards. *Biotropica* 8: 225–234.

Seidensticker, J. 1986. Large carnivores and the consequences of habitat insuralisation: Ecology and conservation of tigers in Indonesia and Bangladesh. Pages 1–41 *in* D. D. Everett, editor. *Cats of the world: biology, conservation and management.* National Wildlife Federation, Washington, District of Columbia, USA.

Seidensticker, J., M. G. Hornocker, W. V. Wiles, and J. P. Messick. 1973. Mountain lion social organization in the Idaho Primitive Area. *Wildlife Monographs* 35:1–60.

Seidensticker, J., M. E. Sunquist, and C. McDougal. 1990. Leopards living at the edge of the Royal Chitwan National Park Nepal. Pages 415–423 *in* J. C. Daniel and J. S. Serrao, editors. *Conservation in developing countries: problems and prospects.* Bombay Natural History Society, Bombay.

Seidensticker, J., S. Christie, and P. Jackson. 1999. *Riding the tiger: tiger conservation in human-dominated landscapes.* Cambridge University Press, Cambridge, UK.

Seidman, V. M., and C. J. Zabel. 2001. Bat activity along intermittent streams in Northwestern California. *Journal of Mammology* 82: 738–747.

Selva, N., B. Jędrzejewska, W. Jędrzejewski, and A. Wajrak. 2003. Scavenging on European bison carcasses in Białowieża Primeval Forest (eastern Poland). *Ecoscience* 10: 303–311.

Serret, A. 1995. *Estado de conservación del huemul en el canal Moyano, Parque Nacional Los Glaciares.* Boletín Técnico 25. Fundación Vida Silvestre. Buenos Aires, Argentina.

Shannon, C. E., and W. Weaver. 1949. *The mathematical theory of communication.* University of Illinois Press, Urbana, Illinois, USA.

Shaw, J. H. 1975. Ecology, behavior, and systematics of the red wolf (*Canis rufus*). Ph.D. dissertation, Yale University, New Haven, Connecticut, USA.

Shears, N. T., and R. C. Babcock. 2003. Continuing trophic cascade effects after 25 years of no-take marine reserve protection. *Marine Ecology Progress Series* 246: 1–16.

Shiomoto, A., K. Tadokoro, K. Nagasawa, and Y. Ishida. 1997. Trophic relations in the subarctic North Pacific ecosystem: possible feeding effects from pink salmon. *Marine Ecology Progress Series* 150: 75–85.

Shulman, M. J., and D. R. Robertson. 1996. Changes in the coral reef of San Blas, Caribbean Panama: 1983 to 1990. *Coral Reefs* 15: 231–236.

Shurin, J. B., E. T. Borer, E. W. Seabloom, K. Anderson, C. A. Blanchette, B. Broitman, S. D. Cooper, and B. S. Halpern. 2002. A cross-ecosystem comparison of the strength of trophic cascades. *Ecology Letters* 5: 785–791.

Silliman, B. R., and M. D. Bertness. 2002. A trophic cascade regulates salt marsh primary production. *Proceedings of the National Academy of Sciences* 99: 10500–10505.

Simberloff, D. 1998. Flagships, umbrellas, and keystones: is single species management passé in the landscape era. *Biological Conservation* 83: 247–257.

Simberloff, D. 2003. Community and ecosystem impacts of single-species extinctions. Pages 221–233 *in* P. Kareiva and S. A. Levin, editors. *The importance of species.* Princeton University Press, Princeton, New Jersey, USA.

Simenstad, C. A., J. A. Estes, and K. W. Kenyon. 1978. Aleuts, sea otters and alternate stable-state communities. *Science* 200: 403–411.

Simpson, C. T. 1920. *In lower Florida wilds.* G. P. Putnam's Sons, New York, New York, USA.

Simpson, E. H. 1949. Measurement of diversity. *Nature* 163: 688.

Sinclair, A. R. E. 1977a. *The African buffalo.* University of Chicago Press, Chicago, Illinois, USA.

Sinclair, A. R. E. 1977b. Lunar cycle and timing of mating season in Serengeti wildebeest. *Nature* 267: 832–833.

Sinclair, A. R. E. 1979a. The Serengeti environment. Pages 31–45 *in* A. R. E. Sinclair and M. Norton-Griffiths, editors. *Serengeti: dynamics of an ecosystem.* University of Chicago Press, Chicago, Illinois, USA.

Sinclair, A. R. E. 1979b. The eruption of the ruminants. Pages 82–103 *in* A. R. E. Sinclair and M. Norton-Griffiths, editors. *Serengeti: dynamics of an ecosystem.* University of Chicago Press, Chicago, Illinois, USA.

Sinclair, A. R. E. 1989. Population regulation in animals. Pages 197–241 *in* J. M. Cherrett, editor. *Ecological concepts.* Symposium of the British Ecological Society, Blackwell Scientific Publishers, Oxford, England.

Sinclair, A. R. E. 1997. Carrying capacity and the overabundance of deer: a framework for management. Pages 380–394 *in* W. J. McShea, H. B. Underwood, and J. H. Rappole, editors. *The science of overabundance: deer ecology and population management.* Smithsonian Institution Press, Washington, District of Columbia, USA.

Sinclair, A. R. E., and P. Arcese, editors. 1995a. *Serengeti II: dynamics, management, and conservation of an ecosystem.* University Press, Chicago, Illinois, USA.

Sinclair, A. R. E., and P. Arcese. 1995b. Serengeti in the context of worldwide conservation efforts. Pages 31–46 in A. R. E. Sinclair and P. Arcese, editors. *Serengeti II: dynamics, management, and conservation of an ecosystem*. University of Chicago Press, Chicago, Illinois, USA.

Sinclair, A. R. E., and C. J. Krebs. 2001. Trophic interactions, community organization, and the Kluane system. Pages 25–48 in C. J. Krebs, S. Boutin, and R. Boonstra, editors. *Ecosystem dynamics of the boreal forest: the Kluane project*. Oxford University Press, New York, New York, USA.

Sinclair, A. R. E., and M. Norton-Griffiths, editors. 1979. *Serengeti: dynamics of an ecosystem*. Chicago University Press, Chicago, Illinois, USA.

Sinclair, A. R. E., J. M. Gosline, G. Holdsworth, C. J. Krebs, S. Boutin, J. N. M. Smith, R. Boonstra, and M. R. T. Dale. 1993. Can the solar cycle and climate synchronize the snowshoe hare cycle in Canada? Evidence from tree rings and ice cores. *American Naturalist* 141: 173–198.

Sinclair, A. R. E., R. P. Pech, C. R. Dickman, D. Hik, P. Mahon, and A. E. Newsome. 1998. Predicting effects of predation on conservation of endangered prey. *Conservation Biology* 12: 564–575.

Sinclair, A. R. E., C. J. Krebs, J. M. Fryxell, R. Turkington, S. Boutin, R. Boonstra, P. Seccombe-Hett, P. Lundberg, and L. Oksanen. 2000. Testing hypotheses of trophic level interactions: a boreal forest ecosystem. *Oikos* 89: 313–328.

Sinclair, A. R. E., S. Mduma, and J. S. Brashares. 2003. Patterns of predation in a diverse predator–prey system. *Nature* 425: 288–290.

Singer, F. J., and J. Mack. 1999. Predicting the effects of wildfire and carnivore predation on ungulates. Pages 189–238 in T. W. Clark, A. P. Curlee, S. C. Minta, and P. M. Kareiva, editors. *Carnivores in ecosytems: the Yellowstone experience*. Yale University Press, New Haven, Conneticut, USA.

Singer, F. J., J. Mark, and R. G. Cates. 1994. Ungulate herbivory of willows on Yellowstone's northern winter range. *Journal of Range Management* 47:435–443.

Singer, F. J., L. C. Zeigenfuss, and D. T. Barnett. 2000. Elk, beaver, and the persistence of willows in national parks: response to Keigley. *Wildlife Society Bulletin* 28: 451–453.

Singer, F. J., L. C. Zeigenfuss, B. Lubow, and M. J. Rock. 2002. Ecological evaluation of potential abundance of ungulates in U. S. National Parks: a case study. Pages 205–250 in F. J. Singer, and L. C. Zeigenfuss, editors. *Ecological evaluation of the abundance and effects of elk herbivory in Rocky Mountain National Park, Colorado, 1994–1999*. USGS-BRD Open File Rep, 02-208, Fort Collins, Colorado, USA.

Singer, F., G. Wang, and N. Hobbs. 2003. The role of ungulates and large predators on plants, community structure, and ecosystem processes in national parks. Pages 444–486 in C. J. Zabel and R. G. Anthony, editors. *Mammal community dynamics*. Cambridge University Press, Cambridge, UK.

Singleton, P. H., W. L. Gaines, and J. F. Lehmkuhl. 2002. *Landscape permeability for large car-*

nivores in Washington: a geographic information system weighted-distance and least-cost corridor assessment. Research Paper PNW-RP-549. U.S. Department of Agriculture, Forest Service, Pacific Northwest Research Station, Portland, Oregon, USA.

Sivertsen, K. 1997. Geographic and environmental factors affecting the distribution of kelp beds and barren grounds and changes in biota associated with kelp reduction at sites along the Norwegian coast. Canadian Journal of Fisheries and Aquatic Sciences 54: 2872–2887.

Skogen, K. 2001. Who's afraid of the big, bad wolf? Young peoples responses to the conflicts over large carnivores in eastern Norway. Rural Sociology 66: 203–226.

Skogen, K., H. Haaland, S. Brainerd, and H. Hustad. 2003. Lokale syn på rovvilt og rovviltforvaltning. En undersøkelse i fire kommuner: Aurskog-Høland, Lesja, Lierne og Porsanger. NINA fagrapport 070.

Skogland, T. 1991. What are the effects of predators on large ungulate populations? Oikos 61: 401–411.

Slobodkin, L. B., F. E. Smith, and N. G. Hariston. 1967. Regulation in terrestrial ecosystems, and the implied balance of nature. American Naturalist 101: 109–124.

Slough, B. G., and G. Mowat. 1996. Lynx population dynamics in an untrapped refugium. Journal of Wildlife Management 60: 946–961.

Smallidge, P. J., and D. J. Leopold. 1997. Vegetation management for the maintenance and conservation of butterfly habitats in temperate human-dominated landscapes. Landscape and Urban Planning 38: 259–280.

Smirnov, E. N., and D. G. Miquelle. 1999. Population dynamics of the Amur tiger in Sikhote-Alin Zapovednik, Russia. Pages 61–70 in J. Seidensticker, S. Christie, and P. Jackson, editors. Riding the tiger: tiger conservation in human-dominated landscapes. Cambridge University Press, Cambridge, UK.

Smit, R., J. Bokdam, J. den Ouden, H. Olff, H. Schot-Opschoor, and M. Schrijvers. 2001. Effects of introduction and exclusion of large herbivores on small rodent communities. Plant Ecology 155: 119–127.

Smith, B. L. 2001. Winter feeding of elk in western North America. Journal of Wildlife Management 65:173–190.

Smith, B. L., and S. H. Anderson. 1998. Juvenile survival and population regulation of the Jackson Hole elk herd. Journal of Wildlife Management 62: 1036–1045.

Smith, C. 1992. Late stone age hunters of the British Isles. Routledge, London, UK.

Smith, D. W., W. G. Brewster, and E. E. Bangs. 1999. Wolves in the Greater Yellowstone Ecosystem: Restoration of a top carnivore in a complex management environment. Pages 103–125 in T. W. Clark, A. P. Curlee, S. C. Minta, and P. M. Kareiva, editors. Carnivores in ecosystems: the Yellowstone experience. Yale University Press, New Haven, Connecticut, USA, and London, UK.

Smith, D. W., L. D. Mech, M. Meagher, W. E. Clark, R. Jaffe, M. K. Phillips, and J. A. Mack. 2000.

Wolf–bison interactions in Yellowstone National Park. *Journal of Mammalogy* 81: 1128–1135.

Smith, D. W., R. O. Peterson, and D. B. Houston. 2003. Yellowstone after wolves. *BioScience* 53: 330–340.

Smith, D. W., T. D. Drummer, K. M. Murphy, D. S. Guernsey, and S. B. Evans. 2004. Winter prey selection and estimation of wolf kill rates in Yellowstone National Park, 1995–2000. *Journal of Wildlife Management* 68: 153–166.

Smith, J. M. N., and N. F. G. Folkard. 2001. Other herbivores and small predators: arthropods, birds and mammals. Pages 261–274 *in* C. J. Krebs, S. Boutin, and R. Boonstra, editors. *Ecosystem dynamics of the boreal forest: the Kluane project.* Oxford University Press, New York, New York, USA.

Smith, K. 1998. Return of the deer menace. *Forest and Bird* 289: 36–39.

Smith, T. B., M. W. Bruford, and R. K. Wayne. 1993. The preservation of process: the missing element of conservation programs. *Biodiversity Letters* 1993: 164–167.

Smith, T. R., C. G. Hunter, J. F. Eisenberg, and M. E. Sunquist. 1996. Ecology of white-tailed deer in eastern Everglades National Park: an overview. *Bulletin of the Florida Museum of Natural History* 39: 141–172.

Smith-Flueck, J. A. M., and W. T. Flueck. 2001. Natural mortality patterns in a population of southern Argentina huemal (*Hippocamelus bisulcus*), an endangered Andean cervid. *Zeitschrift für Jagdwissenschaft* 47: 187–188.

Smuts, G. L. 1978. Interrelations between predators, prey and their environment. *BioScience* 28: 316–320.

Sokal, R. R., and F. J. Rohlf. 1981. *Biometry.* 2nd edition. Freeman, New York, New York, USA.

Solberg, E. J., P. Jordhoy, O. Strand, A. R. Loison, B.-E. Sæther, and J. D. C. Linnell. 2001. Effects of density-dependence and climate on the dynamics of a Svalbard reindeer population. *Ecography* 24: 441–451.

Sörlin, S. 1999. The articulation of territory: landscape and the constitution of regional and national identity. *Norsk Geografisk Tidsskrift* 53: 103–111.

Soulé, M. E. 1985. What is conservation biology? *BioScience* 35: 727–734.

Soulé, M., and R. Noss. 1998. Rewilding and biodiversity: complementary goals for continental conservation. *Wild Earth* 8: 19–27.

Soulé, M., and J. Terborgh. 1999. *Continental conservation: scientific foundations of regional reserve networks.* Island Press, Washington, District of Columbia, USA.

Soulé, M. E., J. A. Estes, J. Berger, and C. Martinez del Rio. 2003. Ecological effectiveness: conservation goals for interactive species. *Conservation Biology* 17: 1238–1250.

Sousa, W. P. 1991. Can models of soft-sediment community structure be complete without parasites? *American Zoologist* 31:821–830.

Springer, A. M., J. A. Estes, G. B. van Vliet, T. M. Williams, D. F. Doak, E. M. Danner, K. A.

Forney, and B. Pfister. 2003. Sequential megafaunal collapse in the North Pacific Ocean: an ongoing legacy of industrial whaling? *Proceedings of the National Academy of Sciences* 100: 12223–12228.

Stacey, P. B. 1995. Biodiversity of rangeland bird populations. Pages 33–41 *in* N. West, editor. *Biodiversity of rangelands.* Utah State University Press, Logan, Utah, USA.

Stahler, D. R., B. Heinrich, and D. Smith. 2002. Common ravens *Corvus corax* preferentially associate with gray wolves *Canis lupus* as a foraging strategy. *Animal Behavior* 64: 283–290.

Stander, P. E. 1990. Notes on foraging habits of cheetah. *South African Journal of Wildlife Research* 20: 130–132.

Stander, P. E. 1991. Aspects of the ecology and scientific management of large carnivores in sub-Saharan Africa. M. Phil. thesis. University of Cambridge, Cambridge, UK.

Stander, P. E. 1992. Foraging dynamics of lions in a semiarid environment. *Canadian Journal of Zoology* 70: 8–21.

Stanley, S. M. 1973. An explanation for Cope's Rule. *Evolution* 27: 1–26.

Starfield, A., M. Qadling, and J. Venter. 1992. *A management orientated buffalo population dynamics model for the Kruger National Park.* Report to National Parks Board, Skukuza, South Africa.

Steneck, R. S. 1994. Is herbivore loss more damaging to reefs than hurricanes? Case studies from two Caribbean reef systems (1978–1988). Pages 220–226 *in* R. N. Ginsburg, editor. *Proceedings Colloquium on Global Aspects of Coral Reefs: Health, Hazards, and History, 1993.* Rosenstiel School of Marine and Atmospheric Science, University of Miami, Miami, Florida, USA.

Steneck, R. S. 1997. Fisheries-induced biological changes to the structure and function of the Gulf of Maine ecosystem. Pages 151–165 *in Proceedings of the Gulf of Maine Ecosystem Dynamics Scientific Symposium and Workshop.* Regional Association for Research in the Gulf of Maine (RARGOM) Report 91–1. Hanover, New Hampshire, USA.

Steneck, R. S. 1998. Human influences on coastal ecosystems: does overfishing create trophic cascades? *Trends in Ecology and Evolution* 13: 429–430.

Steneck, R. S., and J. T. Carlton. 2001. Human alterations of marine communities: students beware! Pages 445–468 *in* M. Bertness, S. Gaines, and M. Hay, editors. *Marine community ecology.* Sinauer Press, Sunderland, Massachusetts, USA.

Steneck, R. S., and M. N. Dethier. 1994. A functional group approach to the structure of algal-dominated communities. *Oikos* 69: 476–498.

Steneck, R. S., and M. N. Dethier. 1995. Are functional classifications different for marine vs terrestrial plants? A reply to Grime. *Oikos* 73: 122–124.

Steneck, R. S., and C. J. Wilson. 2001. Long-term and large scale spatial and temporal patterns in demography and landings of the American lobster, *Homarus americanus,* in Maine. *Journal of Marine and Freshwater Research* 52: 1302–1319.

Steneck, R. S., M. H. Graham, B. J. Bourque, D. Corbett, J. M. Erlandson, J. A. Estes, and M. J. Tegner. 2002. Kelp forest ecosystem: biodiversity, stability, resilience and future. *Environmental Conservation* 29: 436–459.

Steneck, R. S., J. Vavrinec, and A. V. Leland. 2004. Accelerating trophic level dysfunction in kelp forest ecosystems of the western North Atlantic. *Ecosystems* 7: 323–331.

Stephens, P. A., O. J. Zaumyslova, A. E. Myslenkov, G. D. Hayward, and D. G. Miquelle. In press. Population dynamics of prey species in Sikhote-Alin Zapovednik: 1962–2003. *In* D. G. Miquelle, E. N. Smirnov, and J. M. Goodrich, editors. *Tigers of Sikhote-Alin Zapovednik: ecology and conservation*. (in Russian).

Stevens, J. D., R. Bonfil, N. K. Dulvy, and P. A. Walker. 2000. The effects of fishing on sharks, rays, and chimaeras (chondrichthyans), and the implications for marine ecosystems. *ICES Journal of Marine Science* 57: 476–494.

Stevenson, G. B. 1996. *Palms of south Florida*. University Press of Florida, Gainesville, Florida, USA.

Stewart, A. J. A. 2001. The impact of deer on lowland woodland invertebrates: a review of the evidence and priorities for future research. *Forestry* 74: 259–270.

Stewart, K. M., T. E. Fulbright, and D. L. Drawe. 2000. White-tailed deer use of clearings relative to forage availability. *Journal of Wildlife Management* 64: 733–741.

Stewart, K. M., R. T. Bowyer, J. G. Kie, N. J. Cimon, and B. K. Johnson. 2002. Temporospatial distributions of elk, mule deer, and cattle: resource partitioning and competitive displacement. *Journal of Mammalogy* 83: 229–244.

Støen, O. G., and P. Wegge. 1996. Prey selection and prey removal by tiger (*Panthera tigris*) during the dry season in lowland Nepal. *Mammalia* 60: 363–373.

Strauss, S. Y. 1991. Indirect effects in community ecology: their definition, study, and importance. *Trends in Ecology and Evolution* 6: 206–210.

Stromayer, K. A. K., and R. J. Warren. 1997. Are overabundant deer herds in the eastern United States creating alternative stable states in forest plant communities? *Wildlife Society Bulletin* 25: 227–234.

Stromayer, K. A. K., R. J. Warren, A. S. Johnson, P. E. Hale, C. L. Rogers, and C. L. Tucker. 1998. Chinese privet and the feeding ecology of white-tailed deer: the role of an exotic plant. *Journal of Wildlife Management* 62: 1321–1329.

Strong D. R. 1992. Are trophic cascades all wet? Differentiation and donor-control in speciose ecosystems. *Ecology* 73: 747–754.

Struhsaker, T. T., J. S. Lwanga, and J. M Kasenene. 1996. Elephants, selective logging and forest regeneration in the Kabale Forest, Uganda. *Journal of Tropical Ecology* 12: 45–64.

Sudekum, A. E., J. D. Parrish, R. L. Radtke, and S. Ralston. 1991. Life-history and ecology of large jacks in undisturbed, shallow, oceanic communities. *Fishery Bulletin* 89: 493–513.

Sukumar, R. 2003. *The living elephants*. Oxford University Press, Oxford, UK.

Sunquist, M. E. 1981. The social organization of tigers (*Panthera tigris*) in Royal Chitawan National Park, Nepal. *Smithsonian Contributions to Zoology* 336: 1–98.

Sunquist, M., and F. Sunquist. 2002. *Wild cats of the world*. University of Chicago Press, Chicago, Illinois, USA.

Sunquist, M., K. U. Karanth, and F. Sunquist. 1999. Ecology, behaviour and resilience of the tiger and its conservation needs. Pages 5–18 in J. Seidensticker, S. Christie, and P. Jackson, editors. *Riding the tiger: tiger conservation in human-dominated landscapes*. Cambridge University Press, Cambridge, UK.

Suominen, O., K. Danell, and R. Bergstrom. 1999a. Moose, trees, and ground-living invertebrates: indirect interactions in Swedish pine forests. *Oikos* 84: 215–226.

Suominen, O., K. Danell, and J. P. Bryant. 1999b. Indirect effects of mammalian browsers on vegetation and ground-dwelling insects in an Alaskan floodplain. *Ecoscience* 6: 505–510.

Sutherland, W. J. 1996. *From individual behaviour to population ecology*. Oxford University Press, Oxford, UK.

Sutherland, W. J. 2002. Restoring a sustainable countryside. *Trends in Ecology and Evolution* 17: 148–150.

Sutherland, W. J. 2003. Parallel extinction risk and global distribution of languages and species. *Nature* 423: 276–279.

Sweitzer, R. A., S. H. Jenkins, and J. Berger. 1997. Near-extinction of porcupines by mountain lions and consequences of ecosystem change in the Great Basin Desert. *Conservation Biology* 11: 1407–1417.

Swenson, J. E., F. Sandegren, M. Heim, S. Brunberg, O. J. Sørensen, A. Söderberg, A. Bjärvall, et al. 1996. Er den skandinavisk bjørnen farlig? *NINA Oppdragsmelding* 404: 1–26.

Swenson, J. E., F. Sandegen, and A. Soderberg. 1998. Geographic expansion of an increasing bear population: evidence for presaturation dispersal. *Journal of Animal Ecology* 66: 819–826.

Swenson, J. E., N. Gerstl, B. Dahle, and A. Zedrosser. 2000. Action plan for the conservation of the brown bear (*Ursus arctos*) in Europe. Council of Europe, *Nature and Environment* 114: 1–69.

Swihart, R. K., Z. L. Feng, N. A. Slade, D. M. Mason, and T. M. Gehring. 2001. Effects of habitat destruction and resource supplementation in a predator–prey metapopulation model. *Journal of Theoretical Biology* 210: 287–303.

Tabachnick, B. G., and L. S. Fidell. 1983. *Using multivariate statistics*. Harper and Row, New York, USA.

Takatsuki, S., K. Suzuki, and I. Suzuki. 1994. A mass mortality of Sika deer on Kinkazan Island, northern Japan. *Ecological Research* 9: 215–223.

Tamang, K. M. 1982. The status of the tiger (*Panthera tigris*) and its impact on principal prey populations in Royal Chitwan National Park, Nepal. East Lansing, Michigan State University, Ph.D. Dissertation, Michigan, USA.

Tannerfeldt, M., B. Elmhagen, and A. Angerbjorn. 2002. Exclusion by interference competition? The relationship between red and arctic foxes. *Oecologia* 132: 213–220.

Taylor, R. J. 1984. *Predation*. Chapman and Hall, New York, New York, USA.

Taylor, T. N., and E. L. Taylor. 1993. *The biology and evolution of fossil plants*. Prentice-Hall, Inc., Englewood Cliffs, New Jersey, USA.

Terborgh, J. 1988. The big things that run the world. *Conservation Biology* 2: 402–403.

Terborgh, J. 1990. The role of felid predators in neotropical forests. *Vida Silvestre Neotropical* 2: 3–5.

Terborgh, J., and M. E. Soulé. 1999. Why we need mega-reserves—and how to design them. Pages 199–209 *in* M. E. Soulé and J. Terborgh, editors. *Continental conservation: scientific foundations of regional reserve networks*. Island Press, Washington, District of Columbia, USA.

Terborgh, J., and S. J. Wright. 1994. Effects of mammalian herbivores on plant recruitment in two Neotropical forests. *Ecology* 75: 1829–1833.

Terborgh, J., L. Lopez, and J. Tello. 1997a. Bird communities in transition: the Lago Guri islands. *Ecology* 78: 1494–1501.

Terborgh, J., L. Lopez, J. Tello, D. Yu, and A. R. Bruni. 1997b. Transitory states in relaxing land bridge islands. Pages 256–274 *in* W. F. Laurance and R. O. Bierregaard Jr., editors. *Tropical forest remnants: ecology, management, and conservation of fragmented communities*. University of Chicago Press, Chicago, USA.

Terborgh, J., J. A Estes, P. Paquet, K. Ralls, D. Boyd-Heger, B. J. Miller, and R. F. Noss. 1999. The role of top carnivores in regulating terrestrial ecosystems. Pages 39–64 *in* M. E. Soulé and J. Terborgh, editors. *Continental conservation*. Island Press, Washington, District of Columbia, USA.

Terborgh, J., L. Lopez, P. Nuñez V., M. Rao, G. Shahabuddin, G. Orijuela, M. Riveros, et al. 2001. Ecological meltdown in predator-free forest fragments. *Science* 294: 1923–1926.

Thapar, V. 1986. *Tiger: portrait of a predator*. Collins, London, UK.

Theberge, J. B. 1990. Potentials for misinterpreting impacts of wolf predation through prey:predator ratios. *Wildlife Society Bulletin* 18: 188–192.

Theberge, J. B. 2000. An ecologist's perspective on wolf recovery in the northeastern United States. Pages 22–63 *in* J. Elder, editor. *The return of the wolf: reflections on the future of wolves in the northeast*. Middlebury College Press, Hanover, New Hampshire, USA.

Thomas, C. D. 2000. Dispersal and extinction in fragmented landscapes. *Proceedings of the Royal Society of London, Series B* 267: 139–145.

Thurber, J. M., R. O. Peterson, J. D. Woolington, and J. A. Vucetich. 1992. Coyote coexistence with wolves on the Kenai Peninsula, Alaska. *Canadian Journal of Zoology* 70: 2494–2498.

Tilghman, N. G. 1989. Impacts of white-tailed deer on forest regeneration in northwestern Pennsylvania. *Journal of Wildlife Management* 53: 524–532.

Tilman, D. 1982. *Resource competition and community structure.* Monographs in Population Biology. Princeton University Press. Princeton, New Jersey, USA.

Todd, A. W., L. B. Keith, and C. A. Fischer. 1981. Population ecology of coyotes during a fluctuation of snowshoe hares. *Journal of Wildlife Management* 45: 629–640.

Torres, H. 1985. *Guanaco: distribución y conservación del guanaco.* Informe Especial No. 2. IUCN, Cambridge, UK.

Treves, A., R. Jurewicz, L. Naughton-Treves, R. Rose, R. Willging, and A. Wydeven. 2002. Wolf depredation on domestic animals in Wisconsin, 1976–2000. *Wildlife Society Bulletin* 30: 231–241.

Treves, A., and K. U. Karanth. 2003. Human–carnivore conflict and perspectives on carnivore management worldwide. *Conservation Biology* 17: 1491–1499.

Trombulak, S. C., and C. A. Frissell. 2000. Review of ecological effects of roads on terrestrial and aquatic communities. *Conservation Biology* 14:18–30.

Turchi, G. M., P. L. Kennedy, D. Urban, and D. Hein. 1995. Bird species richness in relation to isolation of aspen habitats. *Wilson Bulletin* 107: 463–474.

Tyers, D. B. 2003. Winter ecology of moose on the northern Yellowstone winter range. Ph.D. dissertation, Montana State University, Bozeman, Montana, USA.

Tyson, P. D., and T. G. J. Dyer. 1975. Mean annual fluctuations of annual precipitation in the summer rainfall regions of South Africa. *South African Geographic Journal* 57: 104–110.

Uhl, N. W., and J. Dransfield. 1987. *Genera Palmarum.* International Palm Society, Allen Press, Lawrence, Kansas, USA.

Underwood, H. B., and W. F. Porter. 1997. Reconsidering paradigms of overpopulation in ungulates. Pages 185–198 *in* W. J. McShea, H. B. Underwood, and J. H. Rappole, editors. *The science of overabundance; deer ecology and management.* Smithsonian Institution Press, Washington, District of Columbia, USA.

Urban, D. L., and T. M. Smith. 1989. Microhabitat pattern and the structure of forest bird communities. *American Naturalist* 133: 811–829.

Vadas, R.L., and R. S. Steneck. 1988. Zonation of deep water benthic algae in the Gulf of Maine. *Journal of Phycology* 24: 338–346.

Vadas, R. L., and R. S. Steneck. 1995. Overfishing and inferences in kelp–sea urchin interactions. Pages 509–524 *in* H. R. Skjoldal, C. Hopkins, and K.E. Erikstad, editors. *Ecology of fjords and coastal waters.* Elsevier Science, Amsterdam, Netherlands.

Valone, T. J., M. Meyer, J. H. Brown, and R. M. Chew. 2002. Timescale of perennial grass recovery in desertified arid grasslands following livestock removal. *Conservation Biology* 16: 995–1002.

Van Ballenberghe, V. 1987. Effects of predation on moose numbers: a review of recent North American studies. *Swedish Wildlife Research* 1 (Supplement): 431–460.

Van Ballenberghe, V., and W. B. Ballard. 1994. Limitation and regulation of moose populations: the role of predation. *Canadian Journal of Zoology* 72: 2071–2077.

Van der Walt, P. T., P. F. Retief, E. A. N. le Riche, M. G. L. Mills, and G. de Graaff. 1984. Features of habitat selection by larger herbiverous mammals and the ostrich in the southern Kalahari conservation areas. *Koedoe* 27 (Supplement): 119–128.

Van Dreische, J., and R. Van Dreische. 2000. *Nature out of place: biological invasions in the global age.* Island Press, Washington, District of Columbia, USA.

Van Orsdol, K. G., J. P. Bygott, and J. D. Bygott. 1986. Ecological correlates of lion social organization (*Panthera leo*). *Journal of Zoology, London* 206: 97–112.

Van Valkenburgh, B. 1999. Major patterns in the history of carnivorous mammals. *Annual Review of Earth and Planetary Sciences* 27: 463–493.

Van Valkenburgh, B. 2001. The dog-eat-dog world of carnivores: a review of past and present carnivore community dynamics. Pages 101–121 *in* C. Stanford and H. T. Bunn, editors. *Meat-eating and human evolution.* Oxford University Press, Oxford, UK.

Van Valkenburgh, B., and F. Hertel. 1993. Tough times at La Brea: tooth breakage in large carnivores of the late Pleistocene. *Science* 261: 456–459.

Vanderbilt, C. F., R. M. Giblin-Davis, and T. J. Weissling. 1998. Mating behavior and sexual response to aggregation pheromone of *Rhynchophorus cruentatus* (Coleoptera: Curculionidae). *Florida Entomologist* 81: 351–360.

Vanni, M. J., C. D. Layne, and S. E. Arnott. 1997. "Top-down" trophic interactions in lakes. *Ecology* 78: 1–20.

Vartan, S. 2002. Overpopulation and inbreeding in small game reserves: the lion *Panthera leo* as a case study. Unpublished master's thesis. University of Cape Town, Cape Town, South Africa.

Vasquez, J. A. 1993. Abundance, distributional patterns and diets of main herbivorous and carnivorous species associated with *Lessonia trabeculata* kelp beds in northern Chile. *Serie Ocasional Universidad Catolica del Norte* 2: 213–229.

Vasquez, J. A., and A. H. Buschmann. 1997. Herbivore–kelp interactions in Chilean subtidal communities: a review. *Revista Chilena de Historia Natural* 70: 41–52.

Vasquez, J. A., J. C. Castilla, and B. Santelices. 1984. Distributional patterns and diets of four species of sea urchins in giant kelp forest (*Macrocystis pyrifera*) of Puerto Toro, Navarino Island, Chile. *Marine Ecology Progress Series* 19: 55–63.

Vavrinec, J. 2003. Resilience of green sea urchin (*Strongylocentrotus droebachiensis*) populations following fishing mortality: marine protected areas, larval ecology and post-settlement survival. Ph.D. dissertation. University of Maine, School of Marine Sciences, Maine.

Veblen, T. T., M. Mermoz, C. Martin, and T. Kitzberger. 1992. Ecological impacts of introduced animals in Nahuel Huapi National Park, Argentina. *Conservation Biology* 6: 71–83.

Vellend, M. 2002. A pest and an invader: white-tailed deer (*Odocoileus virginianus* Zimm.) as a seed dispersal agent for honeysuckle shrubs (*Lonicera* L.). *Natural Areas Journal* 22: 230–234.

Venkataraman, A. B. 1995. Do dholes (*Cuon alpinus*) live in packs in response to competition with or predation by large cats? *Current Science* 69: 934–936.

Venkataraman, A. B., R. Arumugam, and R. Sukumar. 1995. The foraging ecology of dhole (*Cuon alpinus*) in Mudumalai Sanctuary, Southern India. *Journal of Zoology* 237: 543–561.

Verheyen, K., B. Bossuyt, M. Hermy, and G. Tack. 1999. The land use history (1278–1990) of a mixed hardwood forest in western Belgium and its relationship with chemical soil characteristics. *Journal of Biogeography* 26: 1115–1128.

Verity, P. G., and V. Smetacek. 1996. Organism life cycles, predation, and the structure of marine pelagic ecosystems. *Marine Ecology Progress Series* 130: 277–293.

Vilà, C., V. Urios, and J. Castroviejo. 1995. Observations on the daily activity patterns in the Iberian wolf. Pages 335–340 *in* L. N. Carbyn, S. H. Fritts, and D. R. Seip, editor. *Ecology and conservation of wolves in a changing world*. Canadian Circumpolar Institute, Edmonton, Alberta, Canada.

Viljoen, P. C. 1993. The effects of changes in prey availability on lion predation in a large natural ecosystem in northern Botswana. *Symposia of the Zoological Society of London* 65: 193–213.

Viranta, S. 2003. Geographic and temporal ranges of Middle and Late Miocene carnivores. *Journal of Mammalogy* 84: 1267–1278.

Vos, J. 2000. Food habits and livestock depredation of two Iberian wolf packs (*Canis lupus signatus*) in the north of Portugal. *Journal of Zoology, London* 251: 457–462.

Vucetich, J. A., and S. Creel. 1999. Ecological interactions, social organisation, and extinction risk in African wild dogs. *Conservation Biology* 13: 1172–1182.

Vucetich, J. A., R. O. Peterson, and C. L. Schaefer. 2002. The effect of prey and predator densities on wolf predation. *Ecology* 83: 3003–3013.

Vucetich, J. A., R. O. Peterson, and T. A. Waite. 2004. Raven scavenging favours group foraging in wolves. *Animal Behaviour* 67: 1117–1126.

Wabakken, P., H. Sand, O. Liberg, and A. Bjärvall. 2001. The recovery, distribution, and population dynamics of wolves on the Scandinavian peninsula, 1978–1998. *Canadian Journal of Zoology* 79: 710–725.

Wahle, R. A., and R. S. Steneck. 1991. Recruitment habitats and nursery grounds of the American lobster (*Homarus americanus* Milne Edwards): a demographic bottleneck? *Marine Ecology Progress Series* 69: 231–243.

Wahle, R. A., and R. S. Steneck. 1992. Habitat restrictions in early benthic life: experiments on habitat selection and in situ predation with the American lobster. *Journal of Experimental Marine Biology and Ecology* 157: 91–114.

Walker, B. H. 1992. Biodiversity and ecological redundancy. *Conservation Biology* 6: 18–23.

Walker, B. H. and I. Noy-Meir. 1982. Aspects of the stability and resilience of savanna ecosystems. Pages 556–590 *in* B. J. Huntly and B. H. Walker, editors. *Ecology of tropical savannas*. Springer-Verlag, Berlin, Germany.

Walker, B. H., R. H. Emslie, R. N. Owen-Smith, and R. J. Scholes. 1987. To cull or not to cull: lessons from a southern African drought. *Journal of Applied Ecology* 24: 381–401.

Walker, R. S., G. Ackermann, J. Schachter-Broide, V. Pancotto, and A. J. Novaro. 2000. Habitat use by mountain vizcachas (*Lagidium viscacia* Molina, 1782) in the Patagonian steppe. *Zeitschrift für Säugetierkunde* 65: 293–300.

Wallace, A. R. 1858. On the tendency of varieties to depart indefinitely from the original type. *Journal of the Proceedings of the Linnean Society (Zoology)* 3: 53–62.

Wallace, R. J. 1975. A reconnaissance of the sedimentology and ecology of Glovers Reef Atoll, Belize (British Honduras). Ph.D. dissertation, Princeton University, Princeton, New Jersey, USA.

Waller, D. M., and W. S. Alverson. 1997. The white-tailed deer: a keystone herbivore. *Wildlife Society Bulletin* 25: 217–226.

Wallis de Vries, M. F. 1995. Large herbivores and the design of large-scale nature reserves in western Europe. *Conservation Biology* 9: 25–33.

Walters, C. J. 1986. *Adaptive management of renewable resources*. Macmillan, New York, New York, USA.

Walters, C. J., R. Hilborn, and R. Peterman. 1975. Computer simulation of barren-ground caribou dynamics. *Ecological Modeling* 1: 303–315.

Walters, C. J., and C. S. Holling. 1990. Large-scale management experiments and learning by doing. *Ecology* 71: 2060–2068.

Warren, J. T. 1998. Conservation biology and agroecology in Europe. *Conservation Biology* 12: 499–500.

Warren, R. J. 1997. An emerging management tool: large mammal predator reintroductions. Pages 393–395 *in* G. K. Meffe and C. R. Carroll, editors. *Principles of Conservation Biology*. 2nd edition. Sinauer Associates, Sunderland, Massachusetts, USA.

Warren, R. J., C. J. Conroy, W. E. James, L. A. Baker, and D. R. Diefenbach. 1990. Reintroduction of bobcats on Cumberland Island, Georgia: a biopolitical lesson. *Transactions of the 55th North American Wildlife and Natural Resources Conference* 55: 580–589.

Watkinson, A. R., A. E. Riding, and N. R. Cowie. 2001. A community and population perspective of the possible role of grazing in determining the ground flora of ancient woodlands. *Forestry* 74: 231–239.

Watson, J. C. 1993. Effects of sea otter foraging on rocky sublittoral communities off northwestern Vancouver Island, British Columbia. Ph.D. dissertation. University of California, Santa Cruz, California, USA.

Watson, R., and D. Pauly. 2001. Systematic distortions in world fisheries catch trends. *Nature* 414: 534–536.

Watts, S. 2001. The end of the line? Global threats to sharks. WildAid, San Francisco, California, USA. Available from http://www.wildaid.org/PDF/reports/TheEndoftheLine(1).pdf (accessed December 2004).

Weaver, J. 1978. *The wolves of Yellowstone*. U.S. Department of the Interior, National Park Service, Natural Resources Report Number 14.

Weber, W., and A. Rabinowitz. 1996. A global perspective on large carnivore conservation. *Conservation Biology* 10: 1046–1055.

Weber, W. In press. Culturally determined wildlife populations: the problem with the designer ark. In *State of the Wild*. Island Press. Washington, District of Columbia, USA.

Webster, C. R., and G. R. Parker. 2000. Evaluation of *Osmorhiza claytonii* (Michx.) C. B. Clarke, *Arisaema triphyllum* (L.) Schott, and *Actaea pachypoda* Ell. As potential indicators of white-tailed deer overabundance. *Natural Areas Journal* 20: 176–188.

Weckerly, F. W. 1998. Sexual size dimorphism: influence of mass and mating systems in the most dimorphic mammals. *Journal of Mammalogy* 79: 33–52.

Wehausen, J. D. 1996. Effects of mountain lion predation on bighorn sheep in the Sierra Nevada and Granite Mountains of California. *Wildlife Society Bulletin* 24: 471–479.

Weissling, T. J., and R. M. Giblin-Davis. 1994. Fecundity and fertility of *Rhynchophorus cruentatus* (Coleoptera: Curculionidae). *Florida Entomologist* 77: 373–376.

Wemmer, C., editor. 1998. *Deer: status survey and conservation action plan*. IUCN/SSC Deer Specialist Group. IUCN Gland, Switzerland, and Cambridge, UK.

Werner, E. E., and J. F. Gilliam. 1984. The ontogenetic niche and species interactions in size-structured populations. *Annual Review of Ecology and Systematics* 15: 393–425.

Wessing, R. 1986. *The soul of ambiguity: the tiger in Southeast Asia*. Center for Southeast Asian Studies. Monograph Series on Southeast Asia. Special Report 24. Northern Illinois University, Dekalb, Illinois, USA.

West, D. 1997. Hunting strategies in central Europe during the last glacial maximum. *British Archaeological Reports International Series* 672: 1–153.

Western, D. 2001. Human-modified ecosystems and future evolution. *Proceedings of the National Academy of Sciences* 98: 5458–5465.

White, C. A., C. E. Olmsted, and C. E. Kay. 1998. Aspen, elk, and fire in the Rocky Mountain national park of North America. *Wildlife Society Bulletin* 26: 449–462.

White, P. A., and D. K. Boyd. 1989. A cougar *Felis concolor* kitten killed and eaten by gray wolves *Canis lupus* in Glacier National Park Montana USA. *Canadian Field Naturalist* 103: 408–409.

White, P. J., and R. A. Garrott. 1997. Factors regulating kit fox populations. *Canadian Journal of Zoology* 75: 1982–1988.

White, P. J., and R. A. Garrott. 1999. Population dynamics of kit foxes. *Canadian Journal of Zoology* 77: 486–493.

Whittaker, R. H. 1975. *Communities and ecosystems*. 2nd edition. Macmillan, New York, New York, USA.

Whyte, I. L. 1985. The present ecological status of the blue wildebeest (*Connochaetes taurinus taurinus* Burchell, 1823) in the central district of the Kruger National Park. Unpublished master's thesis. University of Natal, Pietermaritzburg, South Africa.

Wiebe, W., H. Chansang, C. Birkeland, C. R. Wilkinson, G. J. Vermeij, P. W. Sammarco, J. C. Ogden, J. D. Parrish, R. E. Thresher, R. H. Richmond. 1987. Comparison between

Atlantic and Pacific tropical marine coastal ecosystems: community structure, ecological processes, and productivity. *Unesco Reports in Marine Science* 46: 1–262.

Wikramanayake, E. D., E. Dinerstein, J. G. Robinson, U. Karanth, A. Rabinowitz, D. Olson, T. Mathew, et al. 1998. An ecology-based method for defining priorities for large mammal conservation: the tiger as case study. *Conservation Biology* 12: 865–878.

Wikramanayake, E., E. Dinerstein, C. J. Loucks, D. M. Olson, J. Morrison, J. Lamoreux, M. McKnight, and P. Hedao. 2002. *Terrestrial ecoregions of the Indo-Pacific: a conservation assessment.* Island Press, Washington, District of Columbia and Covelo, California, USA.

Wilcove, D. S., D. Rothstein, J. Dubow, A. Phillips, and E. Losos. 1998. Quantifying threats to imperiled species in the United States. *BioScience* 48: 607–615.

Williams, C. K., G. Ericsson, and T. A. Heberlein. 2002. A quantitative summary of attitudes toward wolves and their reintroduction (1972–2000). *Wildlife Society Bulletin* 30: 575–584.

Williams, I. D., and N. V. C. Polunin. 2000. Differences between protected and unprotected reefs of the western Caribbean in attributes preferred by dive tourists. *Environmental Conservation* 27: 382–391.

Williams, T. M., J. A. Estes, D. F. Doak, and A. M. Springer. 2004. Killer appetites: assessing the role of predators in ecological communities. *Ecology* 85: 3373–3384.

Willis, K. J., and R. J. Whittaker. 2002. Species diversity-scale matters. *Science* 295: 1245–1246.

Willis, T. J., R. B. Millar, and R. C. Babcock. 2003. Protection of exploited fish in temperate regions: high density and biomass of snapper *Pagrus auratus* (Sparidae) in northern New Zealand marine reserves. *Journal of Applied Ecology* 40: 214–227.

Wilmers, C. C., R. L. Crabtree, D. W. Smith, K. M. Murphy, and W. M. Getz. 2003a. Trophic facilitiation by introduced top predators: grey wolf subsidies to scavengers in Yellowstone National Park. *Journal of Animal Ecology* 72: 909–916.

Wilmers, C. C., D. R. Stahler, R. L. Crabtree, D. W. Smith, and W. M. Getz. 2003b. Resource dispersion and consumer dominance: scavenging at wolf- and hunter-killed carcasses in Greater Yellowstone, USA. *Ecology Letters* 6: 996–1003.

Wilson, E. O., and F. M. Peter, editors. 1988. *Biodiversity.* National Academy Press, Washington, District of Columbia, USA.

Wilson, P. 1984. Puma predation on guanacos in Torres del Paine National Park, Chile. *Mammalia* 48: 515–522.

Wing, S. R., and E. S. Wing. 2001. Prehistoric fisheries in the Caribbean. *Coral Reefs* 20: 1–8.

Witman, J. D., and K. P. Sebens. 1992. Regional variation in fish predation intensity: a historical perspective in the Gulf of Maine. *Oecologia* 90: 305–315.

Wolff, J. O. 1980. The role of habitat patchiness in the population dynamics of snowshoe hares. *Ecological Monographs* 50:111–130.

Wood, D., and J. M. Lenne. 1997. The conservation of agrobiodiversity on-farm: questioning the emerging paradigm. *Biodiversity and Conservation* 6: 109–129.

Woodley, S. 1993. Monitoring and measuring ecosystem integrity in Canadian national parks. Pages 155–176 in S. Woodley, J. Kay and G. Francis, editors. *Ecological integrity and the management of ecosystems.* St. Lucie Press, Delray Beach, Florida, USA.

Woodley, S. 1997. Science and protected area management: an ecosystem-based perspective. *NATO ASI Series G* 40: 11–21.

Woodroffe, R. 2000. Predators and people: using human densities to interpret declines of large carnivores. *Animal Conservation* 3: 165–173.

Woodroffe, R. 2001. Strategies for carnivore conservation: lessons for contemporary extinctions. Pages 61–92 in J. L. Gittleman, S. M. Funk, D. W. Macdonald, and R. K. Wayne, editors. *Carnivore conservation.* Cambridge University Press, Cambridge, UK.

Woodroffe, R. 2003. Conserving a cooperative species. *Trends in Ecology and Evolution* 18: 109–110.

Woodroffe, R., and J. R. Ginsberg. 1998. Edge effects and the extinction of populations inside protected areas. *Science* 280: 2126–2128.

Woodroffe, R., and J. R. Ginsberg. 1999. Conserving the African wild dog, *Lycaon pictus,* I: Diagnosing and treating causes of decline. *Oryx* 33: 132–142.

Woodroffe, R., and J. R. Ginsberg. 2000. Ranging behaviour and extinction in carnivores: how behaviour affects species vulnerability. Pages 125–140 in W. J. Sutherland, editor. *Behaviour and conservation.* Cambridge University Press, Cambridge, UK.

Woodroffe, R., J. W. McNutt, and M. G. L. Mills. 2004. African wild dog. Pages 174–183. In C. Sillero-Zubiri and D. W. Macdonald, editors. *Foxes, wolves, jackals and dogs: status survey and conservation action plan.* 2nd edition. IUCN, Gland, Switzerland.

Wootton, J. T. 1994. The nature and consequences of indirect effects in ecological communities. *Annual Review of Ecology and Systematics* 25: 433–466.

Wootton, J. T. 1995. Effects of birds on sea urchins and algae: a lower-intertidal trophic cascade. *Ecoscience* 2: 321–328.

World Wildlife Fund (WWF). 2003. Bismarck Solomon Seas. Available from http://www.panda.org/downloads/marine/wwfbsse2.pdf (accessed October 2004).

World Wildlife Fund (WWF). 2004. Threatened Species. Available at http://www.panda.org/about_wwf/what_we_do/species/our_solutions/endangered_species/index.cfm (accessed October 2004).

Worm, B., and J. E. Duffy. 2003. Biodiversity, productivity and stability in real food webs. *Trends in Ecology and Evolution* 18: 628–632.

Worm, B., H. K. Lotze, H. Hillebrand, and U. Sommer. 2002. Consumer versus resource control of species diversity and ecosystem functioning. *Nature* 417: 848–851.

Wright, S. J. 2003. The myriad consequences of hunting for vertebrates and plants in tropical forests. *Perspectives in Plant Ecology, Evolution and Systematics* 61: 73–86.

Wright, S. J., and H. C. Duber. 2001. Poachers and forest fragmentation alter seed dispersal, seed survival, and seedling recruitment in the palm *Attalea butraceae* with implications for tropical tree diversity. *Biotropica* 33: 583–595.

Wright, S. J., M. E. Gompper, and B. DeLeon. 1994. Are large predators keystone species in Neotropical forests? The evidence from Barro Colorado Island. *Oikos* 71: 279–294.

Yochim, M. J. 2001. Aboriginal overkill overstated. *Human Nature* 12: 141–167.

Yodzis, P. 2001. Must top predators be culled for the sake of fisheries? *Trends in Ecology and Evolution* 16: 78–84.

Young, S. P. 1946. *The wolf in North American history.* Caxton Printers Ltd., Caldwell, Idaho, USA.

Young, S. P., and E. A. Goldman. 1944. *The wolves of North America.* American Wildlife Institute, Washington, District of Columbia, USA.

Yudakov, A. G., and I. G. Nikolaev. 1992. *Ecology of the Amur tiger: winter long-term observations in the western part of the middle Sikhote-Alin in 1970–1973* (in Russian). Nauka, Moscow, Russia.

Yudin, V. G. 1992. *The wolf of the Russian Far East* (in Russian). Russian Academy of Sciences Far Eastern Branch, Blogaveshchensk, Russia.

Zangerl, A. R., and C. E. Rutledge. 1996. The probability of attack and patterns of constitutive and induced defense: a test of optimal defense theory. *American Naturalist* 147: 599–608.

List of Contributors

Reidar Andersen, Norwegian Institute for Nature Research, Tungasletta-2, N-7485 Trondheim, Norway

Joel Berger, Teton Field Office, Wildlife Conservation Society, 96 Canyon Crest, Victor, Idaho 83455, USA

Luigi Boitani, Department of Animal and Human Biology, University of Rome, Viale Univesita'32, I-00185 Rome, Italy

Stan Boutin, University of Alberta, Department of Biological Sciences, Edmonton, Alberta T6G 2E9, Canada

R. Terry Bowyer, Department of Biological Sciences, Idaho State University, Pocatello, Idaho 83209-8007, USA

Urs Breitenmoser, Swiss Rabies Center, Institute of Veterinary Virology, University of Bern, Laengass-Sts. 122, CH-3012 Bern, Switzerland

John J. Cox, University of Kentucky Department of Forestry, 205 T. P. Cooper Building, Lexington, Kentucky 40546-0073, USA

James A. Estes, U.S. Geological Survey, University of California, 100 Shaffer Road, Santa Cruz, California 95060, USA

Joshua R. Ginsberg, Wildlife Conservation Society, Asia Program, 2300 Southern Boulevard, Bronx, New York 10460, USA

John M. Goodrich, Wildlife Conservation Society, Russian Far East Program, Sikhote-Alin, Biosphere Reserve, Terney, Primorye Kria 692150, Russia

Bogumiła Jędrzejewska, Mammal Research Institute, Polish Academy of Sciences, 17-230 Białowieża, Poland

Włodzimierz Jędrzejewski, Mammal Research Institute, Polish Academy of Sciences, 17-230 Białowieża, Poland

John D. C. Linnell, Norwegian Institute for Nature Research, Tungasletta 2, N-7485 Trondheim, Norway

David S. Maehr, University of Kentucky, Department of Forestry, 205 T. P. Cooper Building, Lexington, Kentucky 40546-0073, USA

Tim R. McClanahan, Wildlife Conservation Society, Coral Reef Conservation Project, P.O. Box 99470, Mombasa, 80107, Kenya

William J. McShea, National Zoological Park and Wildlife Conservation Society Joint Appalachian Forest Program, Conservation and Research Center, 1500 Remount Road, Front Royal, Virginia 22630, USA

M. G. L. Mills, SAN Parks, Endangered Wildlife Trust and University of Pretoria, P. Bag X402, Skukuza, 1350, South Africa

Dale G. Miquelle, Wildlife Conservation Society, Russian Far East Program, Sikhote-Alin, Biosphere Reserve, Terney, Primorye Kria 692150, Russia

Alexander E. Myslenkov, Sikhote-Alin State Biosphere Zapovednik, Terney, Terneiski Raion, Primorski Krai, Russia

Andrés J. Novaro, Wildlife Conservation Society, Southern Cone Program, Curruhue s/n, Junin de los Andes, 8371 Provincia del Neuquen, Argentina

Michael A. Orlando, University of Kentucky Department of Forestry, 205 T. P. Cooper Building, Lexington, Kentucky 40546-0073, USA

David K. Person, Alaska Department of Fish and Game, 2030 Sea Level Drive, Ketchikan, Alaska 99901, USA

Becky M. Pierce, Sierra Nevada Bighorn Sheep Recovery Program, California Department of Fish and Game, 407 West Line Street, Bishop, California 93514, USA

Christoph Promberger, Carpathian Large Carnivore Project, Str. Dr. Ioan Senchea 162, RO-2223 Zarnesti, Romania

Justina C. Ray, Wildlife Conservation Society Canada, 720 Spadina Avenue, Suite 600, Toronto ON M5S 2T9, Canada

Kent H. Redford, Wildlife Conservation Society, WCS Institute, 2300 Southern Boulevard, Bronx, New York 10460, USA

Enric Sala, Center for Marine Biodiversity and Conservation, Scripps Institution of Oceanography, La Jolla, California 92093-0202, USA

Evgeny N. Smirnov, Sikhote-Alin State Biosphere Zapovednik, Terney, Terneiski Raion, Primorski Krai, Russia

Douglas W. Smith, Yellowstone Wolf Project, Yellowstone Center for Resources, P.O. Box 168, Yellowstone National Park, Wyoming 82910, USA

Robert S. Steneck, University of Maine, School of Marine Sciences, Darling Marine Center, Walpole, Maine 04573, USA

Philip A. Stephens, Department of Mathematics, University of Bristol, University Walk, Bristol BS8 1TW, United Kingdom

Jon E. Swenson, Norwegian Institute for Nature Research, Tungasletta-2, N-7485 Trondheim, Norway

John Terborgh, Center for Tropical Conservation, P. O. Box 90381, Nicholas School of the Environment and Earth Sciences, Duke University, Durham, North Carolina 27708, USA

R. Susan Walker, Wildlife Conservation Society, Southern Cone Program, Curruhue s/n, Junin de los Andes, 8371 Provincia del Neuquen, Argentina

Rosie Woodroffe, University of California, Davis, Wildlife Fish and Conservation Biology, One Shields Avenue, Davis, California 95616, USA

Olga J. Zaumyslova, Sikhote-Alin State Biosphere Zapovednik, Terney, Terneiski Raion, Primorski Krai, Russia

Index

Abramov, K.G., 187
Acropora coral, 250
African savanna ecosystems: biodiversity implications, 223–224; conservation, 225–228; hunting in, 224; intraguild relationships, 223; Kalahari predator-prey relationships, 220–221; overview, 208–209, 228–229; predator-prey relationships, 212–222, *215, 217, 221*; study areas, 209–212, *210*; without carnivores, 227–228
Agaricia coral, 250
Alder, black (*Alnus glutinosa*), 231
Alewife (*Alosa pseudoharengus*), 15
Alligator (*Alligator mississippiensis*), 305
"Alpha" (within-habitat) diversity, 12
Andrew, N.L., 134
Andrewartha, 11
Angelfish, 252
Angelstam, P., 373
Antelope, pronghorn (*Antilocapra americana*), 331; at YNP, 103
Antelope, roan (*Hippotraggus equinus*): in carnivore absence, 227; decline of, 222
Ants, carpenter (*Campanotus floridanus*), 299
Ants, leaf cutter: *Acromyrmex* sp., 90, 91, 93; *Atta* spp., 90, 91, 93
Apex predators: description/examples of, 16, 18, 19; loss of, 30, 31; size and, 21; as targets of fisheries, 30
Apparent competition, 276–277, 283
Armadillo: hairy (*Chaetophractus villosus*), 271; nine-banded (*Dasypus novemcinctus*), 328
Ash (*Fraxinus excelsior*), 231
Aspen: *Populus* sp., 145; *Populus tremuloides*, 105–106, 372
ATV effects, 332, *335*
Auroch (*Bos primigenius*), 383

Ballard, W.B., 241, 374
Barnacle (*Chthamalus stellatus*), 11
Barracuda (*Sphyraena* spp.), 252

Bartram, W., 306
Bass, freshwater, 15–16
Bear Island Unit, BCNP, 295, 297
Bears (*Ursus* spp.) myths, 5
Bears, black (*Ursus americanus*): boreal forest study, 364, 370–371; diet of, 294, 298; giant palm weevil and, 297–300, *298*, 309, 311–312; as habitat generalist, 294–295; as indicator species, 44–45; ungulate predation by, 148
Bears, grizzly/brown (*Ursus arctos*): Białowieża Primeval Forest/study, 232; boreal Forest/study, 364, 370–371; as conservation symbol, 40–41; effects on moose, 319, 322; in European forests, 383, 387–388; Greater Yellowstone Ecosystem and, 308; as indicator species, 44–45; loss of, 76, 243; in Russian Far East, 181; scavenging and, 330; YNP wolves and, 104
Bears, Himalayan black (*Ursus thibetanus*), 181
Beavers (*Castor canadensis*), 102, 104, 106
Berger, J., 58, 150, 290, 372, 418, 420–421
Bern Convention (Council of Europe/1979), 389
Bertram, B.C.R., 304
Beshta, R.L., 243
Bessie bugs (*Odontotaenius disjunctus*), 299
"Beta" (between-habitat) diversity, 12
Between-habitat ("beta") diversity, 12
Białowieża National Park, 231
Białowieża Primeval Forest/study: conservation implications, 244–245; datasets of, 232–233, 236–240, *237–239*; hunting in, 233, 234, 236; overview, 230–231, 245–246; study area, 231–232; temperature and, *237, 238*, 239–240, *239*; top-down/bottom-up forces in, 240–242; trophic cascades, 242–244; wolf/lynx predation on ungulates, 233–236, *234, 235*
Big Cypress National Preserve, 295
Big Fish Eat Little Fish (Brugel), 126

Biodiversity: African savanna ecosystems implications, 223–224; as conservation target, 420; defining, 12–13, 363, 382, 401–402; habitat loss and, 378–379; information indices and, 12; levels of, 208, 362, 401–402; scope of, 223–224; status monitoring, 44–45; ungulate populations without carnivores and, 142–147, 150–152

Biodiversity/large carnivore conservation: compatibility of, 411–415; confounding factors and, 407–408; conservation recommendations, 419–425; human attitudes/carnivore presence and, 415; Intermediate Disturbance Hypothesis, 11, 13, 14; knowledge needs on, 412–413; link between, 401–402; link recognition, 406–409; low link situations, 405–406; main issues about, 401; marine/terrestrial differences in, 413–414; needs overlap with, 413–415; overview, 400, 425–427; researcher bias and, 408–409; research measuring the right things, 406–407; strong evidence for link, 403–404; unknown/unknowable factors, 409–411; weaker evidence for link, 405–409. See also Large carnivores as conservation tools

Birch: Betula glandulosa, 373; Betula pendula, 231; Betula pubesces, 231

Birch (ecologist), 11

Bird biodiversity, 145–146, 243–244, 363

Bismarck-Solomon Sea, 43

Bison (Bison bison): behavior with predation threat, 328; bottom-up regulation with, 95; reduction of, 407; wounding losses of, 324; at YNP, 102, 103

Bison, European (Bison bonasus): Białowieża Primeval Forest/study, 232, 233, 236, 237, 239; in European forests, 383

Bluefish (Pomatomus saltator), 134

Boars, wild (Sus scrofa): Białowieża Primeval Forest/study, 232, 233, 234, 234, 236, 237, 239; in European forests, 383; high density populations and, 138–139; hunting of, 297; in Patagonia steppe, 271, 283; in Russian Far East, 181, 182, 184, 190, 191–193, 192, 194, 195, 200–201; in South Florida, 305

Bobcat (Lynx rufus): before humans, 27; habitat in South Florida, 301–306, 303; reintro-

duction of, 40, 54, 424; space/prey availability, 294; ungulate predation by, 148, 301–306, 303, 307, 308, 309

Bonito (Sarda sarda), 134

Boomgaard, P., 6

Boreal forest carnivores/biodiversity conservation: biodiversity conservation approach, 378–379; biodiversity effects, 374–376; biomass of predator/prey, 366–370, 367, 368; boreal forest context, 364–366; carnivore effects, 366–376, 367–368; carnivore loss/trophic cascade, 371–374; carnivore prey limitation, 370–371; carnivores as umbrella species, 377–378; forest description, 362–363; loss/changes to, 362–363; overview, 362–364, 379–380; population cycles in, 363

Bottom-up regulation: description, 14, 62–63, 62; as ecologists view, 77–78; with herd-forming migratory ungulates, 95; as layperson's view, 61–62; loss of marine predators and, 116, 134; 135–136. See also Top-down/bottom-up regulation

Boutin, S., 290, 352–353

Bowyer, R.T., 150, 290

BPF. See Białowieża Primeval Forest

Braided cascades, 248, 262

Bromley, G.F., 183–184, 187, 188

Brooks, J.L., 15

Brugel, Pieter the Elder, 126

Buffalo (Syncerus caffer), 210, 214–216, 215

Bugle, 293–294

Bushwillow (Combretum), 210

Butterflyfish, 252

Camels (Camelus spp.), 97

Canada Large Carnivore Conservation Strategy, 38

Canine distemper, 411

Canine parvovirus, 411

Capuchin (Cebus apella), 328

Capybaras (Hydrochaeris hydrochaeris), 92–93

Carbone, C., 201

Caribbean coral reef Ecopath model, 253

Caribou (Rangifer tarandus): boreal forest study, 364; high-density populations of, 139; in predator-free areas, 324; wolves and, 241

Carnivore exclusion experiments, 88–94, 364, 371
Carnivores. *See* Large Carnivores
Carpenter, R.C., 126, 134
Carroll, C., 377
Carrying capacity: populations in agricultural/suburban areas, 140–141, 150; populations near/exceeding, 140–141; predator-prey dynamics, 346–349, *348*, 352, 353–354, *353*, 355–358, *356*, 360; ungulate populations without carnivores, 140–142, 152–153
Catbird, gray (*Dumetella carolinensis*), 243–244
Caterpillars, tent (*Malacosoma disstria*), 369–370
Cats, Geoffroy's (*Oncifelis geoffroyi*), 270
Cauquens (*Chloephaga* spp.), 271, 280
Cave drawings, 28
Cedar, white (*Thuja occidentalis*), 152
Chamois (*Rupicapra rupicapra*), 139, 383
Cheetahs (*Acinonyx jubatus*): African savanna ecosystem study, 220, *221*; behavior with predator threat, 329; guilds and, 157, 164, 173; intraguild relationships, 159–160, 223, 226; restoration of, 97; scavenging and, 163
Chesapeake Bay, 116
Chesson's index of selectivity, 184
Chestnut, American (*Castanea dentata*), 407
Chinese privet (*Ligustrum sinense*), 143
Choique (*Pterocnemia pennata*), 270–271, 272–273, 275, 280, 281, 285
Cod, Atlantic (*Gadus morhua*), 118–119
Colinvaux, P., 26
Compensatory community changes, 20
Competition: apparent competition, 276–277, 283; as density-dependent process, 10–11; exploitation competition, 331; interference competition, 307, 330–331; predation and, 11, 13, 14, 22, *23*
Connell, J.H., 11
Conover, M.R., 148
Conservation: adaptive management for, 423–424; course filter-fine filter approach, 35; ecosystem function and, 30–32; ecotourism and, 226; in fragmented landscapes, 169–172; humans as predators and, 420–421; interventions in protected areas, 225; natural fluctuations and, 225; political

will and, 31–32; predation as threat, 282–284; predator importance, 421–422; public communication and, 424–425; quality vs. quantity, 227; restoring carnivore function with, 422–423; of single species vs. ecosystems, 34–35; site-based conservation planning, 43–44; spatial scale and, 225–226; target specification with, 420; thresholds of change and, 423; values and, 424–425. *See also* Large carnivores as conservation tools
Conservation biology, 34
Conservation implications/recommendations: African savanna ecosystems, 225–228; Białowieża Primeval Forest/study, 244–245; biodiversity/large carnivore conservation, 419–425; coral reef marine parks study, 264–266; Lago Guri land-bridge islands study, 97; large carnivore guilds, 172–174, 175; marine carnivores, 134–136, 419; Patagonia steppe study, 284–287; ungulate populations without carnivores study, 150–152
Coral reef marine parks/study: complex trophic cascades of, 247–248, 261–264, 266–267; conservation recommendations, 264–266; data analysis, 252–254; ENSO event (1998) and, 251; field sampling, 251–252; fished vs. unfished areas, 261, *262*; fishing pressure and, 264–265, 266; food web diagram, *250*; frondose algae as producer, 250; overview, 266–267; physiochemical conditions and, 265, 266; problem statement, 249; research findings, 254–261, *256–261*; study sites/history, 249–251. *See also* Glovers Reef Atoll; Mombasa Marine National Park
Coral reefs: apex predators and, 130, 131–132; in Caribbean, 127–131, *129*; decline in, 127–130, *129*, 132; fishing down food webs in, 128; herbivory loss in, 127–130, *129*; in Indo-Pacific, 131–132; macroalgal abundance and, 127–130, *129*, 132; marine carnivores and, 127–132; species diversity in, 131; trophic cascades in, 130
Coral trout (*Plectropomus leopardus*), 131
Cottonwood (*Populus* spp.), 243
Cougars. *See* Pumas

Council of Europe, 389

Cowen, R.K., 407

Coyotes (Canis latrans): before humans, 27; behavior with predation threat, 329; biomass of, 366, 367–368; conflicting messages on, 54; effects on prey, 241; intraguild predation and, 159, 331; limitation of prey by, 371; local extinction/recolonization of, 77; lynx and, 376; population cycles and, 363; predator exclusion experiments, 88–89; ungulate predation by, 148, 307, 308, 308, 310, 331; wolves and, 104–105, 374, 417

Crabs: modifying herbivores foraging behavior by, 121, 126; nursery habitat controls on, 120; predator effects on, 119, 120, 126–127

Crabs, Jonah (Cancer borealis), 120

Crabtree, R.L., 374

Crawford, 5–6

Créte, M., 241, 373

Crocodiles, 112–113

Culpeo, fox (Pseudalopex culpaeus), 270, 273, 274, 275, 276, 276–277, 280–282, 282–284, 285

Damselfish, 130, 131, 252, 263

Danilkin, A.A., 183–184

DeCalesta, D.S., 143, 145

Deer, black-tailed (Odocoileus hemionus columbianus), 241

Deer, brocket (Mazama spp.), 328

Deer, Chinese water (Hydropotes inermis), 139

Deer, fallow (Dama dama), 139

Deer, mule (Odocoileus hemionus): high-density populations of, 139, 146; puma predation on, 278, 350, 351, 359; road kills of, 148–149; wolf predation on, 241; at YNP, 103

Deer, red (Cervus elaphus): Białowieża Primeval Forest/study, 232, 233, 234, 235–236, 235, 236, 237, 237, 240; density variability of, 325, 326; in European forests, 383; in Patagonia steppe, 271, 272, 276, 277, 284, 285; in Russian Far East, 181, 184, 185, 190, 191–195, 192, 194–195, 200–201; wounding losses of, 324. See also Elk (Cervus elaphus)

Deer, roe (Capreolus capreolus): Białowieża Primeval Forest/study, 232, 233–234, 234,

235, 236, 237, 237, 240; in European forests, 383; high-density populations of, 139; lynx predation on, 365; in Russian Far East, 181, 182, 190, 191–193, 192

Deer, sika (Cervus nippon): high-density populations of, 139; in Russian Far East, 181

Deer, white-tailed (Odocoileus virginianus): bird biodiversity and, 145–146; as coyote prey, 307, 308, 308, 309; as felid prey in South Florida, 301–306, 303, 307, 308, 309; habitat in South Florida, 301–306, 303; high-density populations of, 139, 141, 142–143, 144; increase in, 96; as red wolf prey, 306–308, 308; tick diseases and, 147; vegetation effects, 96, 142–143, 144, 146–147; wolves and, 40, 52, 96, 103, 147, 241, 329, 349; at YNP, 102, 103

Deforestation, 23–24, 25, 32

Density-dependent processes, 10–11

Density-independent processes, 10

Desertification, 271–272

Dhole (Cuon alpinus): area requirements of, 171; guild redundancy/compensation and, 173; intraguild predation and, 159, 160, 162; tiger killings of, 189

Diploria coral, 250

Directive on the Conservation of Natural Habitats and of Wild Flora and Fauna (EU/1992), 389

Dodson, S.I., 15

Dolphins (Coryphaena hippurus), 134

"Dominant" species, 16

Duever, M.J., 296

Duggins, D.O., 118

Eagles, golden (Aquila chrysaetos), 104–105, 286

Eagles, Bald (Haliaeetus leucocephalus), 104–105

Eagles, harpy (Harpia harpyja), 77

Eberhardt, L.L., 196, 201, 343

Ebling, E.J., 126, 134

Ecosystem conservation, 2, 38–39

Ecosystem restoration, 39–40

Ecotourism, 226

Edibility and trophic cascades, 19, 25

Einstein, Albert, 10

Elephants (Loxodonta africana): African savanna ecosystem study, 210; effects on biodiversity, 138; effects on vegetation, 86–87, 95

Elk (*Cervus elaphus*): behavior with predation threat, 328, 329; density variability of, 325, 326; high-density populations of, 139, 143–144; RMEF and, 293–294; trophic cascades and, 25; vegetation effects, 143, 144; at YNP, 101–102, 103–104, 105. *See also* Deer, red (*Cervus elaphus*)
Elton, C., 14
"Eltonian food pyramid," 14, 26
Energy balance model, 185, 196
Environmental stress and predation, 22, 23
Equitability (*E*) index, 183, 190
Errington, P.L., 311
Estes, J.A., 18, 57, 118, 362
European forests: biodiversity threats, 394–395; carnivore conservation goals, 389–394, 390–391; carnivore conservation in, 389, 394; carnivore habitat, 395; carnivore persecution, 395; carnivore populations, 386–389, 386, 387, 389; carnivore prey needs, 395–396; carnivore requirements, 395–396; carnivores as umbrella species, 396; carnivores-biodiversity link, 394–396; conservation ambition levels, 390–391, 392; conservation recommendations, 397–398; European approach transfer value, 396–397; forestry and, 384; grazing/hay production and, 384–385; human-carnivore conflicts, 388, 392; human influences on, 382–385, 393–394; overview, 381–382, 398–399; political landscape and, 388–389; predominant species, 383; protected areas, 388, 389, 394; threats to carnivores, 394–395
European Landscape Convention, 385
European temperate forest ecosystem. *See* Białowieża Primeval Forest
European Union, 388, 389
Everglades: deer/felid predators, 304; saving of, 1
Exotic/introduced species: effects of, 139, 268; island vegetation and, 84–85; Patagonia steppe, 269, 271, 277, 280, 281, 282
Exploitation competition, 331
Exploitation ecosystems, 230–231

Farrow, E.P., 63
Favia coral, 251
Favites coral, 251

Feral goats, 139
Feral pigs. *See* Boars, wild
Feral sheep, 139
Fern, hay-scented (*Dennstaetia punctilobula*), 143
Fever tree (*Acacia xanthophloea*), 211
Fir, balsam (*Abies balsamea*): browsing of, 372, 373; on Lake Superior island, 16, 25
Fisher (*Martes pennanti*), 365
Fisheries Department, Belize, 249
"Fishing down food webs," 113–114, 128
Fishing pressure: body size and, 112, 128, 131–132, 132–134, 134–135; carnivore loss with, 29–30, 112–116, 113, 114, 115, 128, 131–137; herbivory and, 31
Florida. *See* South Florida study
Florida Fish and Wildlife Conservation Commission (FWC), 297
Food webs: direct vs. indirect pathways, 63, 64; issue of spatial/temporal scales, 63. *See also* Sea otter-kelp forest ecosystem
"Foundation" species, 16
Fox, chilla (*Pseudalopex chilla*), 270, 273, 282, 284
Fox, red (*Vulpes vulpes*): intraguild predation and, 159; population cycles and, 363; predation by, 148, 375
Fraenkel, G., 82
Franklin, J.F., 379
Fretwell, S.D., 63, 74, 75
Fretwell model, 63, 74, 75, 79
Frontier Forest Initiative, 43
Fryxell, J.M., 214
Fuller, T.K., 186, 197, 368
Functionality: of carnivores as conservation tools, 45, 46, 47–48; complexity/unpredictability and, 52–53; spatial/temporal variation in, 51–52; testing assumption, 50–53
Functional redundancy: definition, 315–316. *See also* Hunting (carnivore/human functional redundancy)
Functional viability, 50–51

Galaxea coral, 251
Game management: game species as conservation target, 420; top predators and, 403. *See also* Wildlife management
"Gamma" diversity, 12

Gazelles: Grant's (*Gazella granti*), 211; Thomson's (*Gazella thomsonii*), 211
Gemsbok (*Oryx gazella*): African savanna ecosystem, 212, 221; horns of, 221
Geoffroy's cats (*Oncifelis geoffroyi*), 270
Gerber, L.R., 43
Ghoral (*Nemorhaedus caudatus*), 181
Ginsberg, J.R., 59, 171, 177, 200, 223, 422
Gittleman, J.L., 201
Global Species Programme, 41
Glovers Reef Atoll: bleaching event, 251, 252; description/history, 249–250; research findings, 255, *256–257*; trophic cascades of, 262–263. *See also* Coral reef marine parks/study
Glyptodonts (*Glyptotherium* spp.), 299, 310
Goats, feral, 139
Goats, mountain (*Oreamnos americanus*), 139
Grazing/overgrazing: biodiversity loss with, 139; carrying capacity and, 141; in Patagonia steppe, 271–272, 273, 274, 277, 285; Plant Self-Defense Hypothesis and, 86
Great Barrier Reef, 131
Greater Yellowstone Ecosystem/area: consumption of mammalian biomass in, *155*; moose densities in, 372; roads/human hunting season effects, 332, *333*; wolves and, 51–52
Great Lakes region wolves, 417–418
Green World Hypothesis: criticism of, 343; description, 10, 30; Lago Guri land-bridge islands study and, 94; Plant Self-Defense Hypothesis vs., 82–84, 94–98. *See also* Top-down regulation
Grison (*Galictis cuja*), 270
Gromov, E.I., 183, 188, 190
Grouper, goliath (*Epinephelus itajara*), 131
Groupers, 252
Guanacos (*Lama guanicoe*), 270–271, 272, 274–275, 277, 285
Guilds. *See* Large carnivore guilds
Gulls, glaucous-winged (*Larus glaucescens*), 67–68, 68

Habitat loss/modification: in Patagonia steppe, 271–272, 273, 274, 277, 282; sea otter study and, 418–419; Yellowstone and, 418–419
Hairston, N.G., 10, 11, 14, 32, 63, 74, 82. *See also* HSS paper

Hares: European (*Lepus europaeus*), 271, 272, 276, 281, 282, 283, 284, 285; snowshoe (*Lepus americanus*), 88–89, 363, 365, 367–368, 373–374, 375. *See also* Rabbits
Hayes, R.D., 241
Hemlock (*Tsuga canadensis*), 144, 152
Herbivores: biomass/rainfall relationship, 94; denuding landscapes by, 31; epizootic diseases and, 31; megaherbivores, 19. *See also specific herbivores*
Herd-forming migratory ungulates: bottom-up regulation with, 95; HSS and, 95; human effects on, 96
Hereu, B., 126
High density populations: introduced species and, 139; overview, 138–139. *See also* Ungulate populations (without carnivores)
Hobblebush (*Viburnum alnifolium*), 143
Hogfish (*Lachnolaimus maximus*), 263
Holocentridae, 252
Honeysuckle (*Lonicera* sp.), 143
Hooker, S.K., 43
Hornbeam (*Carpinus betulus*), 231
Horses: effects of, 332, 334, *335*; *Equus* spp., 299; wild (*Equus ferus*), 383
Hairston, Smith & Slobotkin (HSS), 63, 74, 75, 79
HSS paper, 10–11, 17–18, 27, 28, 63, 82, 95
Huemuls (*Hippocamelus bisulcus*), 283, 285
Humans: attitudes towards carnivores, 38–39, 293–294, 414, 415; biodiversity loss and, 225; carnivore conflicts with, 388, 392; carnivore loss with arrival of, 28–29; current carnivore loss with, 1, 9, 26, 84, 96–97, 269, 377; effects on migratory ungulates, 96; European forests influences by, 382–385, 393–394; herbivory regulation forces and, 96–97; large carnivore guilds loss and, 161–162, 163, 170; Patagonia carnivores impacts by, 273, 287; Patagonia herbivores impacts by, 273, 287
Hunting (carnivore/human functional redundancy): biodiversity effects (tier 3), 334–337; competition for prey, 316; concordance in effects prediction, 319–337, *320–321, 323, 326, 333, 335*; conservation planning and, 316; conservation recommendations, 337–340; current overlap,

Hunting (continued)
317–318; differences summary, 421; exploitation competition, 331; functional redundancy prerequisites, 317, 318; importance of understanding, 338–340; indirect vs. "subtle" effects, 319, 332, 333, 334, 335; interference competition, 330–331; intraguild predation/mesocarnivore release, 330–331; making human/carnivore hunting equivalent, 338; overview, 315–317, 340–341; predation timing/meat distribution, 322, 323, 324; predator-free areas, 322, 323, 324; predator-prey interactions (tier 1), 322–327, 323, 326; prey activity patterns, 329; prey behavior, 327–329; prey densities through time, 325–327; prey ecological dynamics (tier 2), 327–334, 333; prey foraging rates/predator detection, 328; prey habitat shifts/refuge use, 329; scavenging, 330; ungulate densities and herbivory, 327; ungulate populations study, 148–150, 153; wounding losses, 324–325
Hunting: culture and, 316–317; human changes in, 315; human hunting methods, 316–317; human hunting reasons, 316; in African savanna ecosystems, 224; in Białowieża Primeval Forest, 233, 234, 236; in Patagonia steppe, 274, 275, 277, 278, 282; of herbivores/effects, 287; Russian Far East, 203–204; traditional vs. modern, 224; trophy animals and, 224
Huroncito (Lyncodon patagonicus), 270
Hutchinson, G.E, 11, 12, 13, 14
Hyenas, brown (Hyaena brunnea): intraguild relationships, 223; prey density and, 171–172
Hyenas, spotted (Crocuta crocuta): African savanna ecosystem, 209, 214, 216–217, 217, 218, 219, 221, 241; canine distemper effects, 411; as dominant predator, 157; guild redundancy/compensation and, 163–164, 164–165; loss of, 168; poaching/poisoning of, 163; prey density and, 171–172
Hypercarnivores, 27

Ibex, Alpine (Capra ibex), 383
Iguanas (Iguana iguana), 90, 91
Impala (Aepyceros melampus), 210

Interaction strength, 111
Interference competition, 307, 330–331
Intermediate browsing hypothesis, 151–152
Intermediate Disturbance Hypothesis, 11, 13, 14
Intraguild interactions: in African savanna ecosystems, 223; overview, 158–160, 161–162, 173–174, 180; Patagonia steppe, 284
Introduced species. See Exotic/introduced species
Isard (Rupicapra pyrenaica), 383

Jackals (Canis spp.), 163
Jacks: Carnax ignobilis, 131–132; coral reef marine parks, 252; in Indo-Pacific reefs, 131–132
Jaguars (Panthera onca): conservation priority areas identification and, 42; intraguild predation and, 159; myths on, 5, 6
Jameson, S.C., 408
Japan's reintroduction of wolves, 40
Jays, gray (Perisoreus canadensis), 374–375
Jędrzejewska, B., 178, 183–184, 190, 192, 200
Jędrzejewska, W., 178, 183–184, 190, 192, 200
Jorritsma, I.T.M., 242

Kalahari predator-prey relationships, 220–221
Karanth, K.U., 184
Keith, L.B., 371
Kelp forests: in Alaska, 117–118; global patterns/processes, 121–127, 122–125; influences of, 65; marine carnivores and, 117–127, 122–125; photo of, 66; Western North Atlantic, 118–121. See also Sea otter-kelp forest ecosystem
Kenya Wildlife Service, 249, 251
Keystone species: as top-down force, 15–16, 17; variability of effects, 32
Kgalagadi Transfrontier Park: description, 209, 210, 211–212. See also African savanna ecosystems
Kie, J.G., 343
Killer whales (Orcinus orca): industrial whaling effects on, 73, 74; sea otter-kelp forest ecosystem, 18–19, 65, 69–74
Kitching, J.A., 126, 134
Kleptoparasitism, 307
Knobthorn (Acacia nigrescens), 210

Krebs, C., 88–89
Kruger National Park: consumption of mammalian biomass in, *155*; description, 209–211, *210*; predator-prey relationships/rainfall, 212–216, *215*; roan antelope decline, 222. *See also* African savanna ecosystems
Kruuk, H., 216, 219
Kucherenko, S.P., 183–184
Kunkel, K.E., 184

Lago Guri land-bridge islands/study: conservation implications, 97; Green World Hypothesis and, 94; herbivores on, 90–91, 92–94, 98; island description, 89–90; overview, 89–94, 98; vegetation change/causes, 91–94; vegetation projection, 92, *93*
Land, E.D., 301
Land-bridge islands, 89–94
Landscape ecology, 34
Landscapes, multiple-use, 420
Large carnivore guilds: collapsing effects, 160–162; conservation in fragmented landscapes, 169–172; conservation needs, 168–169; conservation recommendations, 172–174, 175; consumption of mammalian biomass by, *155*, 157; dominant species effects, 170, 172–173, 174; dominant species loss effects, 160, 168; extinction-prone species in, 168–169, 170, 171; feeding competition, 157–158, 180; intraguild predation, 158–160, 161–162, 173–174, 180; loss causes, 160–161; loss of, 156, 157, 160–162, 169–170; overview, 154–157, *155*, 174–175; prey density and, 171–172, 174; prey partitioning by, 158; redundancy/compensation, 162–169, *166*, *167*, 173, 179–180; redundancy/compensation effects on activity patterns, 167; redundancy/compensation effects on diet, 165, *166–167*; redundancy/compensation effects on population density, 163–165; restoration and, 172; structuring forces, 157–160; top-down regulation and, 158, 172–173; umbrella species for, 170–171, 174; wide-ranging behavior/extinction link, 161–162, 171
Large Carnivore Initiative for Europe, 38, 389, 391

Large carnivore loss: arrival of humans and, 28–29; current human interactions and, 1, 9, 26, 84, 96–97, 269, 377; effects overview, 289; extinction rates and, 26–27; with fishing pressure, 29–30, 112–116, *113*, *114*, *115*, 128, 131–137; forestry and, 377, 378; habitat loss/disruption and, 413–414; marine systems, 112–116, *113*, *114*, *115*, 128, 131–137, 419; overview, 33, 83; prior to humans, 27–28; in the sea, 29–30; trophic cascade assessment and, 76
Large carnivore predation: competition and, 11, 13, 14, 22, *23*; as density-independent process, 10; environmental stress and, 22, *23*
Large carnivores: as conservation target, 420; densities of, 269; direct vs. indirect effects, 111; as ecologically pivotal species, 44; ecological role of (overview), 1, 2–3; as epiphenomenon, 82; how to value, 1–6; humans attitudes towards, 38–39, 293–294, 414, 415; as indicator species, 44–45; needs of, 35; prey densities and, 268; prey impact factors, 184–185; prey size and, 241, 242; resource-limited competition among, 28; size effects, 21–22, 25; as umbrella species, 61, 377–378, 396. *See also* Marine carnivores; *specific carnivores*
Large carnivores as conservation tools: assumptions underlying, 45–48, *46*; assumption testing, 48–53; biodiversity status monitoring, 44–45; carnivore restoration and, 97; conservation implications, 53–55, 78–80; ecosystem conservation, 38–39; ecosystem restoration, 39–40; efficiency of, 45, *46*, 47; functionality of, 45, *46*, 47–48; lessons from sea otter-kelp forest ecosystems, 79–80; meeting efficiency assumptions, 49–50; meeting functionality assumptions, 50–53; meeting value assumptions, 48–49; overview, 55–56, 415–419; potential of, 415–416; priority areas identification, 42–43; project scale with, 416; restoration/healing broken systems, 416–419; site-based conservation planning, 43–44; as symbols, 40–41; tools summary table, *37*; as untested assumption, 35; value of, 45, *46*, 47
Laurenson, M.K., 226

Leopards (*Panthera pardus*): in African sa-
vanna ecosystem study, 209, 220, *221*; diet
of, 165, *167*; guild redundancy / compensa-
tion and, 163, 165, *167*; intraguild predation and, 159; myths on, 5; prey of, 158
Leopards, Far Eastern (*Panthera pardus orien-
talis*), 181, 182
Leopold, A., 294
Levitan, D.R., 129
Liang, S.Y., 144
Lime (*Tilia cordata*), 231
Lindenmayer, D.B., 379
Linnell, J., 290–291
Lions (*Panthera leo*): African savanna ecosystem study, 209, 212–216, *215*, 216–218, *217*,
219, 220, 222, 241; canine distemper ef-
fects, 411; diet of, 165, *166*; as dominant
African predator, 157; guilds and, 157, 159,
162, 164–165, *166*; intraguild predation
and, 159, 162; loss of, 168; myths on, 5;
poaching / poisoning of, 163; prey density
and, 172
Lions, marsupial (*Thylacoleo carnifex*), 157
Lions *Panthera atrox*, 27
Llamas (*Hemiauchenia* spp.), 299
Lobsters: nursery habitat controls on, 120;
predator effects on, 119, 120, 121, 126–127
Lotka-Volterra equations, 343, 358–359
Low-density equilibrium model, 344–345,
345, 346–347
Lyme disease, 147
Lynx (*Felis lynx*), 181, 182
Lynx (*Lynx canadensis*): biomass of, 366–368,
368–369; coyotes and, 376; limitation of
prey by, 371; population cycles and, *363*;
predator exclusion experiments, 88–89;
ungulate predation by, 148, 242
Lynx, European (*Lynx lynx*): Białowieża
Primeval Forest / study, 232, 233–234, *234*,
236, 237, 239, 240; prey of, 365
Lynx, Iberian (*Lynx pardinus*), 383, 387–388

MacArthur, R.H., 11
Mackerel (*Scomber scombrus*), 134
Macroalgae: in Mediterranean, 133, *133*, 134;
in tropical coral reefs, 127–130, *129*, 132
Maehr, D., 289–290, 294, 301, 310
Magpies (*Pica pica*), 104–105, 330
Maina, J., 264

Man and Biosphere Reserve, 231
Manu National Park, Peru, *155*
Maple (*Acer planatoides*), 231
Maras (*Dolichotis patagonum*), 271, 285
Marine carnivores: in Black Sea, 134; conservation implications, 134–136, 419; ecological importance of, 134–137; effects
overview, 110–111; fisheries-induced declines, 112–116, *113, 114, 115*, 128, 131–137,
419; "fishing down food webs," 113–114,
128; fishing pressure / body size, 112, 128,
131–132, 132–134, 134–135; kelp forests
and, 117–127, *122–125*; in Mediterranean,
132–134, *133*; methods of controlling herbivory, 121, 126; modifying herbivores foraging behavior by, 121, 126; New Zealand
marine reserves and, 121; overview, 110–
111, 136–137; predation theory / evidence
of effects, 111–112; top-down effects evidence, 110–111, 116–132, *122–125, 129*;
"tragedy of the commons" in, 136; tropi-
cal coral reefs and, 127–132; zooplankton
abundance and, 134. *See also* Coral reefs
Marine protected areas (MPAs), 131
Marker-Kraus, L., 226
Marshal, J.P., 352–353
Marten (*Martes americana*), 365
Martin, P.S., 28
Marula (*Sclerocarya birrea*), 210
Matyushkin, E.N., 183, 188, 190
May, R.M., 343
McClanahan, T.R., 130, 178, 251, 264
McCown, J.W., 301
McCullough, D.R., 347
McInnes, P.F., 372
McLaren, B.E., 372
McNaughton, S.J., 230
McShea, W., 58–59
Mech, L.D., 200, 306
Megaherbivores: definition of, 86; humans'
effects on, 88, 96–97; loss of, 88, 96–97,
287, 309; predation and, 87
Menge, B.A., 21
Mesoamerican Biological Corridor (Paseo
Panthera), 41
Mesopredator description, 111
Mesopredator release, 126, 154, 311, 330–331,
376
Millepora coral, 250

Miller, S.G., 142
Mills, M.G.L., 177–178, 212
Miquelle, D., 163, 177
Mladenoff, D.J., 144
Mombasa Marine National Park: description/history, 249, 251; research findings, 255, 258–261, 261; trophic cascades of, 264. *See also* Coral reef marine parks/study
Mongooses (*Helogale* spp.), 327–328
Monkeys, howler (*Alouatta seniculus*), 90, 91
Montastrea coral, 250
Moose (*Alces alces*): behavior with predation threat, 327–328, 329; Białowieża Primeval Forest/study, 232, 234, 237, 239; biomass of, 367, 369; boreal forest and, 364, 367, 369; in European forests, 383; grizzly bear effects on, 319, 322; high-density populations of, 139, 141–142, 143–144, 145, 372; human hunting and, 322; on Lake Superior island, 16, 25, 147, 411; in Scandinavia, 336; trophic cascades and, 16, 25; vegetation effects, 143–144, 145; wolves and, 16, 25, 241, 319, 322, 370, 371; YNP and, 102, 103, 106, 309
Moose, Manchurian (*Alces alces cameloidus*), 181
Mopane (*Colophospermum mopane*), 210
Mountain lions. *See* Pumas
Mouse, white-footed (*Peromyscus leucopus*), 147
Mouse, wood (*Apodemus sylvaticus*), 243
Multiple-equilibrium model, 344–345, 345, 346–347
Multiple-use landscapes, 420
Muntjacs (*Muntiacus reevesi*), 139–140
Musk deer (*Moschus moschiferus*), 181
Muskrats (*Ondatra zibethicus*), 375
Mussels (*Mytilus californianus*), 15, 22
Mustard, garlic (*Allaria officinalis*), 143
Myrtle, wax (*Myrica cerifera*), 295

Nagarhole National Park, India, 155
"Natural" systems: defining, 100; rarity of, 381
Neotropical migrant birds, 363
New Zealand marine reserves, 121
Ngorongoro Crater: description, 209, 210, 211; predator-prey relationships/habitat, 216–219, 217. *See also* African savanna ecosystems
Nightingale (*Luscinia megarhynchos*), 146

Nikolaev, I.G., 202
Nongovernmental organizations (NGOs), 389
Nonlethal attributes of predation, 404
Noss, R.F., 228
Novaro, Andrés, 178
Nutrient cycling, 145, 359, 375

Oak (*Quercus* sp.) forests, 146–147
Oaks: live (*Quercus virginianus*), 295; Mongolian (*Quercus mongolica*), 180; *Quercus robur*, 231; red (*Quercus rubra*), 144
Odum, E.P., 13–14
Opitz, S., 253
Overall, R., 146
Owls, great horned (*Bubo virginianus*), 369, 371, 376
Oystercatchers, 75–76

Paine, R.T., 11, 13, 15, 16–17, 21, 63, 64, 230
Palatability and trophic cascades, 19
Palm, sabal (*Sabal pametto*), 295, 297, 298, 299, 309–310
Palmetto, saw (*Serenoa repens*), 295, 298, 299, 309–310
Pampas (*Lynchailurus colocolo*), 270
Panther, Florida (*Puma concolor coryi*): as deer predator, 301–306, 303, 307, 308, 309; ecological function of, 311, 312; as flagship species, 310; habitat in South Florida, 301–306, 303, 307; palm use by, 309; space/prey availability, 294
Parasites predator role, 410–411
Parrotfish, 128, 131, 252
Paseo Panthera (Mesoamerican Biological Corridor), 41
Patagonian Shelf Large Marine Ecosystem, 44
Patagonia steppe/study: apparent competition, 276–277, 283; carnivore/prey communities, 269–271; conservation recommendations, 284–287; culpeo predation effects, 280–282; desertification in, 271–272; exotic/introduced species, 269, 271, 277, 280, 281, 282; fragmentation of prey populations, 283; generalist predators in, 283; guanaco populations/pumas, 277–280, 279, 280; habitat degradation in, 271–272, 273, 274, 277, 282; human impact on carnivores, 273, 287; human impact on

Patagonia steppe / study (continued)
native herbivores, 271–273, 287; hunting
effects, 271, 272, 280, 281, 282, 285, 287;
hunting reduction effects, 274, 275, 277,
278, 282; intraguild interactions, 284; map,
270; other ecosystems and, 287; overgrazing effects, 271–272, 273, 274, 277, 285;
overview, 269, 287–288; plant defenses,
282–283; predation as conservation threat,
282–284; predator-prey population densities, 269; predator reproductive potential
and, 283; prey switching, 276–277, 283;
sheep introduction / effects, 271, 272, 277;
sheep reduction effects, 273–275, 276,
278–279, 281–282; top-down processes,
276–282, 279, 280, 287
Pathogens predator role, 410–411
Pauly, D., 113–114
Pavona coral, 251
Peccaries: Mylohyus spp., 299; Platygonus
spp., 299
Perrins, C.M., 146
Person, D.K., 343, 352–353, 359
Peterken, G.F., 141
Peterson, R.O., 201, 372
Pichi (Zaedyus pichiy), 271
Pierce, B.M., 148–149, 359
Pigeons, passenger (Ectopistes migratorius),
407
Pine: Korean (Pinus koraiensis), 180; Pinus
silvestris, 231; slash (Pinus elliottii var.
densa), 295; white (Pinus strobus), 144
Pinnegar, J.K., 126, 410
Plant secondary compounds, 82
Plant Self-Defense Hypothesis: crops and, 84;
current vertebrate herbivores, 87; Green
World Hypothesis vs., 82–84, 94–98; hy-
perabundant herbivores and, 90–94, 95–96,
98; insect outbreaks and, 85; island vegeta-
tion / introduced herbivores, 84–85; mega-
herbivore effects, 86–88; overgrazing by
cattle and, 86; Pleistocene vertebrates,
87–88; tests of, 84–88
Pleistocene megafauna, 28–29, 87–88, 157,
287, 309
Pocillopora coral, 251
Polis, G.A., 372, 374, 405
Population interaction strength, 111
Porcupines: human killing of, 334; Lago Guri

land-bridge islands study, 92; puma predation on, 278
Porgy, jolthead (Calamus bajonado), 263
Porites coral, 250, 251
"Predator chain," 13–14
Predator-prey (ungulates) dynamics: conceptual models of, 344–346, 345; ecosystem
processes-biodiversity links, 359–360; future predator-prey modeling, 358–359; K
effects, 346–349, 348, 352, 353–354, 353,
355–358, 356, 360; kill rates, 347, 351–354,
353; overview, 342–344, 360–361; prey-
based approach / top-down and bottom-up
processes, 354–358, 356, 357; prey to predator ratio studies, 349–351, 351; ungulate
life-history traits, 355–358, 357
Predators. See Large carnivores
Prey switching, 276–277, 283
Project Tiger, 171
Pufferfish, 252
Pumas (Puma concolor): as conservation symbol, 41; deer and, 96, 148–149, 278, 350,
351, 359; guild redundancy / compensation
and, 164; as indicator species, 44–45;
myths on, 5; in Patagonia steppe, 269–270,
273, 274, 274, 276, 276–280, 279–280,
282–284, 285; prey of, 158, 278; prior to
humans, 27; restoration of, 97; scavenging
and, 330; YNP and, 102, 104, 330. See also
Panther, Florida (Puma concolor coryi)

Quammen, D., 5, 6

Rabbits (Oryctolagus cuniculus), 271. See also
Hares
Randall, J.E., 254
Raptors predator role, 410
Rat, cotton (Sigmodon hispidus), 304–305
Ravens (Corvus corax), 104–105, 330,
374–375
Ray, J., 7–8
Recurrent fluctuation model, 344–345, 345,
346–347
Redundancy / compensation, 162–169, 166,
167, 173, 179–180
Reindeer: Rangifer tarandus, 383; Rangifer
tarandus platyrhynchus, 324
Reintroduction reasons, 39–40
Reptiles predator role, 410

Restoration: carnivores as conservation tools, 97; large carnivore guilds and, 172. *See also specific examples*

Rhea, Pampas (*Rhea americana*), 281

Rinderpest: eradication effects, 172, 218; outbreak effects, 211

Ripple, W.J., 105–106, 243

Rocky Mountain Elk Foundation (RMEF), 293–294

Roemer, G.W., 286

Russian Far East/study (tigers/wolves): biomass production, 196, *197*; BPF data for, 183, 184; competitive exclusion/functional redundancy, 200–203; conservation implications, 203–205; correlation between tiger/wolf numbers, 187–188, *189*, 200; data analysis/modeling methods, 182–186; diet breadth/prey selection, 190–195, *191–195*; energy balance model, 185, 196; equilibrium model, 201–202; hunters and, 203–204; life-history parameters used, *207*; other carnivores at site, 181–182; overview, 205–206; predation on prey populations, 195–200, *197–199*; predator-prey simulation model, 185–186, 196–200, *197–199*; predator territory size-prey densities, 186, 196–200, *198–199*, 203; prey impacts, 189; prey impacts data, 183–186; research findings, 186–200, *191–195*, *197–199*; study area, 180–182, *181*; tiger-wolf relationship, 187–189, *189*; ungulate species, 181; wolf-tiger relationship information, 182–183

Sala, E., 58, 126, 132–134

Salamanders, redback (*Plethodon cinereus*), 146

Salt marsh system, 77–78

Sambar (*Cervus unicolor*), 139–140

San Bushmen, 224

Sawflies, leaf-galling (*Phyllocolpa bozemanii*), 337

Scavenging, 104–105, 244, 330, 374–375

Schaller, G.B., 216, 218, 311

Schmitz, O.J., 24–25, 372, 374

Sea cows: *Dugong dugon*, 113; Steller's (*Hydrodamalis gigas*), 30

Seagle, S.W., 144

Sea lions, Stellar (*Eumetopias jubatus*), 72–73, *73*, 117

Seals: decline in, 112–113; harbor (*Phoca vitulina*), 72–73, *73*, 117

Sea otter-kelp forest ecosystem: Alaskan overview, 117–118; biodiversity and, 417; conservation implications, 78–80, 81; direct effects, 65, *66*; effects on fishes, 67; effects on gulls, 67–68, *68*; feedback on otters, 69; fish parasites and, 69; indirect effects, 65, 67–69, *68*; killer whales and, 18–19, 65, 69–74, *73*, 117; odd-numbered/even-numbered trophic levels and, 74, *75*; otter-dominated/otter-free systems production, 67; overview, 63–65, 371–372; pinniped population decline and, 72–74, *73*; processes overview, 80–81; sea urchins and, 15, 18–19, 24, 29, 64, 65, *66*, 68–69, *68*; top-down forces with, 15, 17, 18, 24, 29, 64, 83

Sea otters (*Enhydra lutris*): behavior changes with predation threat, 328; as keystone species, 118; population trends, 69–70, *70*, 77; reintroduction of, 40; as top-down force, 15, *17*, 18, 24, 29, 64

Sea stars (*Pisaster ochraceus*), 15, *17*, 22, 75–76

Sea urchins: *Diadema*, 128–130, *129*, 263; *Echinometra viridis*, 255, 263; harvesting effects on, 119–120; in Mediterranean system, 133–134, *133*; mouthpart changes, 129; nursery habitat controls on, 120; otter decreases and, 29; otters/kelp forests and, 15, 18–19, 24, 118; "plague" of, 119; predator effects on, 119, 121, 126, 248; purple (*Paracentrotus lividus*), 121, 126; red (*Strongylocentrotus franciscanus*), 407; in sea otter-kelp forest ecosystem, 15, 18–19, 24, 29, 64, 65, *66*, 68–69, *68*; tropical coral reefs and, 127–130, *129*

Selva, N., 244

Serengeti Ecosystem: description, 209, *210*, 211; predator-prey relationships/habitat, 216–219, *217*. *See also* African savanna ecosystems

Shannon's diversity (*H*) index, 183, 190

Sharks: arresting decline in, 40; *Carcharhinus* spp., 131–132; as conservation symbol, 40–41; great white (*Carcharodon carcharias*), 113; in Indo-Pacific reefs, 131–132; loss of, 114–115; reduction of human interactions with, 39; tiger, 115–116, 419

Sheep: bighorn (*Ovis canadensis*), 103, 278; Dall (*Ovis dalli*), 241; effects in Patagonia steppe, 271, 272, 273–275, 276, 277, 278–279, 281–282; feral, 139
Sheephead (*Semicossyphus pulcher*), 29, 407
Sheldon, J.W., 374
Shenk, T.M., 212
Sikhote-Alin Mountains, 180, *181*
Sikhote-Alin Zapovednik (SAZ), 182. *See also* Russian Far East/study (tigers/wolves)
Sinclair, A.R.E., 165, 219, 241–242
Site-based conservation planning, 43–44
Skunk, hog-nosed: *Conepatus chinga*, 270, 273; *Conepatus humboldti*, 270, 273
Slobodkin, L.B., 10, 32, 63, 82. *See also* HSS paper
Sloths, ground (*Megalonyx* spp.), 299
Smith, D., 58
Smith, F.E., 10, 15, 32, 63, 82. *See also* HSS paper
Smithsonian Tropical Research Institute, Panama, 91
Smuts, G.L., 214
Snails (*Concholepas concholepas*), 15–16, *17*
Soulé, M., 293, 402
Sutherland, W.J., 347
South Florida/study: black bear/weevil, 297–300, *298*; felid predators/deer, 301–306, *303*; habitat loss/fragmentation in, 311; management challenges in, 309–312; map, *296*; overview, 295, 313–314; palms significance, 309–310; predator "rebound effect," 307; prescribed burns timing, 311; study area, 295–297, *296*; wolves/deer, 306–309
Sparids, 135
Sparrow, Lincoln's (*Melospiza lincolnii*), 106
"Species diversity," 12
Species richness, 12
Sphinx, 5
Springbuck (*Antidorcas marsupialis*): African savanna ecosystem study, 212, 220–221, *221*; age class of prey, *221*; sex ratio of prey, 220–221
Spruce: *Picea abies*, 231; white (*Picea glauca*), 365
Squirrels: gray (*Sciurus carolinensis*), 300; red (*Tamiasciurus hudsonicus*), 367, 369, 375
Stable-limit cycles model, 344–345, *345*, 346–347

Stearns, F., 144
Stellar, G., 328
Steneck, R., 7, 58
Stomoxys biting flies, 219
Succession processes, 144, 145, 359–360
Sunquist, M.E., 184
Surgeonfish (*Acanthurus* spp.), 128, 252
Sutherland, J.P., 21
Sweitzer, R.A., 418

Tahr (*Hemitragus jemlahicus*), 139
Tapir (*Tapirus terrestris*), 328
Tar pits of La Brea, California, 27
Temperate forest ecosystems: trophic cascades and, 242–244. *See also* Białowieża Primeval Forest
Terborgh, J., 58, 372, 422–423
Theberge, J.B., 350
Thurber, J.M., 159
Tick diseases, 147
Ticks (*Ixodes scapularis*), 147
Tiger, saber-toothed (*Smilodon fatalis*), 27, 157
Tigers (*Panthera tigris*): conservation priority areas identification and, 42; diet of, *167*; as dominant Asian predator, 157; forest-tiger relationship story, 5–6; guild redundancy/compensation and, 163, *167*; intraguild predation and, 159, *162*; loss of, 168; myths on, 5–6
Tigers, Amur (*Panthera tigris altaica*): as ambush predator, 201, *202*; distribution of, *181*; dog killings by, 188–189; overview, 180; population trends of, 187–189, *189*; sex-age classes of prey, 192, 194, *194–195*; wolf killings by, 188–189. *See also* Russian Far East/study
Top-down/bottom-up regulation: in Białowieża Primeval Forest, 240–242; prey-based approach to studying, 354–358, *356, 357*
Top-down regulation: in aquatic systems, 75–76; carnivore essentiality and, 61; description, 14–15, *62*; evidence for, *63*; generalities and, 76–78; keystone species and, 15–16, *17*; large carnivore guilds and, 158, 172–173; marine carnivores, 110–111, 116–132, *122–125*, *129*; Patagonia steppe, 276–282, *279, 280*, 287; with sea otter-kelp forest ecosystem, 15, *17*, 18, 24, 29, 64, 83;

trophic cascades and, 16–17, *17*. *See also*
specific examples; Trophic cascades
"Tragedy of the commons," 136
Triggerfish, 135, 252, 263; *Balistes vetula*, 263;
Canthidermis sufflamen, 263
Trophic cascades: aquatic systems overview,
247–248; Białowieża Primeval Forest
study, 242–244; braided cascades, 248, 262;
description, 111; elk and, 25; factors affecting, 20–23, *23*; marine systems vs. terrestrial systems, 24–26, *24*, 32, 372, 374, 404,
408; positive/negative effects with, 247;
research/conservation and, 76; top-down
forces and, 16–17, *17*; in tropical coral
reefs, 130; variability of, 17–19, *18*. *See also*
Coral reef marine parks; Top-down regulation
"Trophic dynamics," 14
Trophic level dysfunction, 31
Trophic levels: in European temperate forest
ecosystems, 230–231; odd-
numbered/even-numbered effects, 17–18,
18, 74, *75*, 118
Trophic structures: classic three-level structure, 17–18, 31, 118; as poorly defined,
20–21
Trophy animals, 224
Tubbs, C.R., 141
Tuco-tucos (*Ctenomys* spp.), 271
Turkeys (*Melagris gallopavo*), 297
Turtles: decline in, 112–113; green (*Chelonia
mydas*), 113

Umbrella species: carnivores as, 61, 377–378,
396; large carnivore guilds and, 170–171,
174
Ungulate populations (without carnivores):
biodiversity effects of, 142–147, 150–152;
browsing benefits, 244; carrying capacity
and, 140–142, 152–153; conservation recommendations, 150–152; deer overview,
139; exotic/invasive plant species and, 143,
151; extent of problem, 138–140; forest
birds and, 145–146, 243–244; forest succession and, 144, 145; hunters vs. large carnivores, 148–150, 153; intermediate browsing
hypothesis, 151–152; irruption/crash type
of, 140; management and population densities, 152; numerical/functional responses

of, 149–150, 152, 153; nutrient flow and,
145; overview, 138, 152–153; plant communities and, 142–145; reduction of popu-
lations/biodiversity, 147–148; road kills of,
148–149; rodent population and, 146–147
Ungulates: life-history traits, 355–358, *357*. *See
also* Predator-prey (ungulates) dynamics
United States Fish and Wildlife Service,
69–70

Vadas, R.L., 64
Value, intrinsic: of large carnivores, 1, 55
Van Valkenburgh, B., 27
Vireo, warbling (*Vireo gilvus*), 106
Vizcachas, mountain (*Lagidium* spp.), 271,
272, 275–276, 280, 281, 285
Voles: bank (*Clethrionomys glareolus*), 146, 243;
field (*Microtus agrestis*), 243; population
cycles of, *363*
Vucetich, J.A., 352

Walker, Susan, 178
Walking sticks, giant (*Anisomorpha
buprestoides*), 299
Wallace, A.R., 14
Warblers: Kentucky (*Oporornis formusus*),
145–146; MacGillivray's (*Oporornis tolmiei*),
243–244; Wilson's (*Wilsonia pusilla*), 106;
yellow (*Dendroica petechia*), 106
Warren, R.J., 54, 424
Weasels (*Mustela* spp.), 363, 365
Weevil, giant palm (*Rhynchophorus cruentatus*): black bears and, 297–300, *298*, 309; life
cycle, *298*
Wehausen, J.D., 418
Whales: North Atlantic gray (*Eschrichtius
robustus*), 30; northern right (*Eubalaena
glacialis*), 30
Whaling industry indirect effects, 73–74, *73*
Whittaker, R.H., 12
Why Big Fierce Animals Are Rare (Colinvaux),
26
Wild dogs, African (*Lycaon pictus*): African
savanna ecosystem study, 209; diet of, 165,
166; energy demands of, 202; guild
redundancy/compensation and, 164, 165,
166, 173; intraguild relationships and,
159–160, 162, 223; local extinction of, 162,
163; prey density and, 172

Wildebeest (*Connochaetes* spp.), 325
Wildebeest, blue (*Connochaetes taurinus*):
    African savanna ecosystem study, 210, 211,
    212–213, *215*, 216, 217–218, *217*, 219, 220,
    222, 241; bottom-up regulation with, 95;
    food supplies and, 241
Wild hog. *See* Boars, wild
Wildlife Conservation Society: coral reef
    marine parks, 249; description / mission
    of, 2
Wildlife corridors, 41
Wildlife management: "real world" focus of,
    294. *See also* Game management
Wild swine. *See* Boars, wild
Willows: *Salix caroliniana*, 295; *Salix* spp.,
    106, 107, 243, 372
Wilmers, C.C., 244
Within-habitat ("alpha") diversity, 12
Wolverines (*Gulo gulo*): as apex carnivore,
    364; as conservation symbol, 49; in
    European forests, 383, 387; as indicator
    species, 44–45
Wolves (*Canis lupus*): boreal forest study and,
    364, 366; buffer zones between packs, 306,
    329; canine parvovirus effects, 411;
    conflicts with people / livestock, 107–108;
    as conservation symbol, 40–41; coyotes
    and, 104–105, 374, 417; deer populations
    and, 40, 52, 96, 103, 147, 241, 329, 349; as
    dominant predator, 157; in European
    forests, 383, 387–388; in Great Lakes
    region, 417–418; intraguild predation and,
    159, 331; on Lake Superior island, 16, 25,
    411; limiting of prey, 370, 371; loss of, 76,
    243; moose and, 16, 25, 241, 319, 322, 370,
    371; prey size of, 158; restoration of, 40,
    51–52, 77, 97, 148, 150, 168, 417–418;
    scavenging and, 104–105, 244, 330,
    374–375; study on Vancouver Island, 349;
    temporal / spatial variation in functionality,
    51–52; as top-down force, 16, 25; ungulate
    densities and, 240–241
Wolves, dire (*Canis dirus*), 27, 157
Wolves, red (*Canis rufus*): as deer predator,

295, 306–308, *308*; Florida restoration of,
    310; habitat of, 307
Wolves / Białowieża Primeval Forest study:
    densities of, 236, *237*, 239; predation by,
    232, 233–234, *234*, 235–236, *235*, 236,
    240, 242
Wolves in Russian Far East study: as cursorial
    predator, 202; energy needs of, 201–202;
    overview, 180; population trends of,
    187–189, *189*
Wolves / Yellowstone National Park: birds
    and, 106, 148, 150, 372; ecology
    with / without wolves, 101–107, 107–108,
    243; guild redundancy / compensation and,
    164; history of wolves in, 51, 100, 148, 150,
    308–309; landscape and, 105–107;
    overview, 100–101, 108–109, 418
Woodroffe, R., 59, 162, 164, 177, 200, 223,
    422
Woolly mammoths, 87–88
World Heritage Site, 231
World Resources Institute, 43
World Wildlife Fund: Canada Large
    Carnivore Conservation Strategy, 38;
    ecoregional priority-setting initiative,
    42–43; Global Species Programme, 41
Wrasses: in coral reef marine park study, 252;
    as mesopredator, 131

Yellow jackets (*Vespula* spp.), 299
Yellowstone National Park: native ungulates
    list, 103; scavenging in, 104–105, 330;
    trophic level interactions summary, *103*.
    See also Greater Yellowstone Ecosystem;
    Wolves / Yellowstone National Park
Yellowthroat, common (*Geothlypis trichas*),
    106
Yew, Canada (*Taxus canadensis*), 143
Yudakov, A.G., 202
Yudin, V.G., 187

Zebra, plains (*Equus burchelii*), 210, 211,
    212–214, *215*, 216, *217*, 218, 222